Medicinal Plants in the Asia Pacific for Zoonotic Pandemics

Medicinal Plants in the Asia Pacific for Zoonotic Pandemics

Series Editor:
Christophe Wiart

This series provides an unprecedented comprehensive overview of the pharmacological activity of more than 100 Asian medicinal plants. It discusses their actions against viruses and bacteria representing a threat of epidemic and pandemic diseases, with an emphasis on the molecular basis and cellular pathways. Scientific names, botanical classifications and descriptions, medicinal uses and chemical constituents are presented, as well as chemical structures, and a vast selection of bibliographical references. This series is a critical reference for anyone involved in the discovery of leads for the treatment of pandemic viral bacterial and fungal infections.

Medicinal Plants in the Asia Pacific for Zoonotic Pandemics,
Volume 1: Family Amborellaceae to Vitaceae, Christophe Wiart

Medicinal Plants in the Asia Pacific for Zoonotic Pandemics,
Volume 2: Family Zygophyllaceae to Salvadoraceae, Christophe Wiart

Medicinal Plants in the Asia Pacific for Zoonotic Pandemics,
Volume 3: Family Bixaceae to Portulacaceae, Christophe Wiart

Medicinal Plants in the Asia Pacific for Zoonotic Pandemics

Family Bixaceae to Portulacaceae

Volume 3

Christophe Wiart

CRC Press
Taylor & Francis Group
Boca Raton London New York

CRC Press is an imprint of the
Taylor & Francis Group, an **informa** business

First edition published 2022
by CRC Press
6000 Broken Sound Parkway NW, Suite 300, Boca Raton, FL 33487-2742

and by CRC Press
2 Park Square, Milton Park, Abingdon, Oxon, OX14 4RN

CRC Press is an imprint of Taylor & Francis Group, LLC

ISBN: 978-1-032-01858-4 (hbk)
ISBN: 978-1-032-01033-5 (pbk)
ISBN: 978-1-003-18037-1 (ebk)

DOI: 10.1201/9781003180371

Typeset in Times
by codeMantra

A ma grand mère, Renée Monllor

A ma mère, Flora Monllor

A mon épouse, Shirley Pita

A mon fils, Christophe Pita-Wiart

Contents

Foreword

We live in a time where the prevalence of zoonotic pandemics has a significant impact on various aspects of our livelihoods. Zoonoses in this sense are diseases transmissible between animals and humans, where the prominent risk factors include animal slaughter, handling and preparing food of animal origin, and particularly the consumption of such food when raw or undercooked. Not just in terms of health and well-being, the "side effects" of these detrimental periods zoonotic pandemics have a bearing on socioeconomic aspects too. For instance, as a result of the current COVID-19 outbreak, schools and other academic institutes have resorted to online teaching and learning; e-commerce has reigned supreme during periods of lockdown. According to the joint statement by ILO, FAO, IFAD, and WHO, the pandemic has been affecting the entire food system and has laid bare its fragility. Border closures, trade restrictions and confinement measures have been preventing farmers from accessing markets, including for buying inputs and selling their produce, and agricultural workers from harvesting crops, thus disrupting domestic and international food supply chains and reducing access to healthy, safe and diverse diets. The pandemic has decimated jobs and placed millions of livelihoods at risk. As breadwinners lose jobs, fall ill and die, the food security and nutrition of millions of women and men are under threat, with those in low-income countries, particularly the most marginalized populations, which include small-scale farmers and indigenous peoples, being hardest hit.

Given these dire straits, there is a whiff of haste and acceleration in the air to find a cure and go back to how things were before the pandemic, rather than settling down under what is referred to as "new normal". The question always plays at the back of any mind investigating the "new normal" whether the periods where there were no pandemics were minor phases of abnormalities. For instance, a worldwide analysis of pathogenic diseases done from 1940 to 2004 shows that 60.3% of diseases were zoonotic and have spread vigorously. How we have dealt with the recent outbreaks of zoonotic diseases of viral origin [i.e. severe acute respiratory syndrome (SARS), Middle East respiratory syndrome (MERS), yellow fever, West Nile, dengue, Zika, Chikungunya, and Ebola)] demonstrates a tell-tale sign of human interaction (or rather, destruction) with wildlife and natural habitats, preserving terrestrial ecosystems and biodiversity, and overhauling food production.

It would only seem logical to seek a remedy for zoonotic pandemics from nature itself – to be specific, from medicinal plants. Ethnopharmacological evidence suggests that human beings have dealt with zoonotic diseases by relying on herbs and their potency to combat viral and bacterial infections in particular, of animal origin. Over the past decade or so nevertheless, the health of humans and animals has been threatened by increasing resistance to antibiotics, environmental pollution, and a higher likelihood of developing chronic diseases. Together with this trend, the interest in phytotherapy has increased because of the observed decline in the field of antibiotic research and increased concerns about the spread of antibiotic resistance. It is without a doubt that plants are rich in bioactive substances that have displayed and proven protective effects against harmful microorganisms. This is the very reason as to why they are being extensively researched as promising materials that can be used in the development of antibiotics and alternative medicines for zoonotic diseases. On the other hand, almost all of the traditional medicinal systems are based on holistic principles, which treat the human body as a whole rather than in fragmented mechanisms. Therefore, traditional herbal treatments target toward boosting immunity rather than extermination of zoonotic pathogens.

The ancient traditional herbal system of medicines is maintained in many countries due to the easy availability of medicinal plants. The use of traditional medicinal plants is widely spread in China, India, Japan, Pakistan, Sri Lanka, and Thailand, and this is very much the case when it comes to zoonotic diseases. China alone accounts for approximately 40% of total herbal medicinal consumption for zoonotic diseases. Treatment with herbal drugs is easily available and cost-effective for many developing countries. It is worthwhile to mention that herbal drugs are the best and most affordable way to overcome infectious diseases for these countries. From the perspective of side effects associated with synthetic drugs, herbal drugs represent a better placement of human diseases.

It is heartening to usher in this four-part volume by Prof. Christophe Wiart, introducing in-depth content on zoonotic diseases and the herbal remedies on offer from the Asia Pacific region. It is a timely initiative and it is without a doubt that volume 3 would garner a readership of not just scientists, researchers, and academics but also a lay audience who would be curious as a whole about alternative remedies for zoonotic diseases. It is hoped that the book would not just serve as an introduction to the plethora of medicinal plants of the Asia Pacific region, their specific mechanisms of action, bioactive compounds but also an update to any existing knowledge on the herbs covered in this volume. There is a place and time for everything, and there would be no objections in considering the current times as most suitable of all to launch an almanac containing valuable information, which would be of value to combat the health issues and challenges of modern times.

Dr. Viduranga Y. Waisundara

Preface

The recent emergence of life-threatening waves of viral zoonosis of increasing severity, culminating today with the COVID-19 pandemic, marks the beginning of an era that might be the last before the ultimate disappearance of human beings. Throughout history, plagues threatened human survival. The Roman Empire was struck from about 166 to 189 A.D. by a smallpox pandemic known as "the Antonine plague" claiming the lives of half of the population of the Empire including the emperor Marcus Aurelius and contributing to the socioeconomic decline of the Western Roman Empire until its fall in 476 A.D. The zoonotic bacterium *Yersinia pestis* accounted for two major plagues in the Middle Ages: "the Justinian plague" (6th century) and the "Black Death" (14th century) claiming the lives of millions. The Spanish flu that followed World War I due to a zoonotic Influenza A (H1N1) virus made more than 20 million victims. For the last about 50 years, novel life-threatening zoonotic viruses have emerged and spread globally including the Human immunodeficiency virus, the Ebola virus, the Zika virus, the Middle East respiratory syndrome coronavirus, and today the ignominious severe acute respiratory syndrome coronavirus 2 (SARS-CoV-2) which according to the WHO has infected so far 204 million people, caused 4.31 million deaths, and paralyzes the whole world as never seen before. Among the reasons for the emergence of zoonotic pandemics is the increasing contact of humans with animal reservoirs due to deforestation and industrial animal farming. According to Wolfe et al. (2005), the richness of microbes in a region, environmental changes, and increased human and animal contact with wildlife result in the emergence of new zoonotic pathogenic microorganisms. Thus, one could envisage the increasing waves of zoonosis to be biological "negative feedback loops" meant to compel humans to put an end to primary rainforest burning, carbon dioxide emission, global warming, as well as intensive poultry, swine, and other animal farming. The pharmaceutical industry has managed to produce vaccines offering some levels of protection against SARS-CoV-2. However, if pressures on wildlife and nature continue, other new zoonotic pandemics will come with the possibility of the emergence of a completely untreatable fast killing virus, bacteria, or even fungi. Furthermore, between the time of appearance of a new virus and the time of production of a vaccine, people are left vulnerable. The intention behind the writing of this book is to offer researchers, academics, and students a comprehensive and interrelated set of botanical, ethnopharmacological, and pharmacological evidence to facilitate the discovery of natural products or even herbal remedies for the treatment or prevention of viral, bacterial, or fungal zoonotic pandemics. It is also to offer an intellectual rationale or tool to understand and thus cogently approach the fascinating realm of drug discovery from medicinal plants. There is a law for the rational selection and use of plants for the treatment of diseases. This third volume covers the medicinal plants traditionally used for the treatment of microbial infection in Asia and the Pacific classified in the Clade Malvids started in Volume 2. Within this Clade, plants are presented according to their respective orders and families. Scientific and common names are given as well as synonyms, habitat, distribution, botanical observation, pharmacology, commentaries, personally handmade botanical plates, as well as carefully selected references. We are now in lack of drugs for the treatment of COVID-19 and of lead molecules to face the upcoming pandemics. Our peer the Greek physician Hippocrates (460-370 BC) said "Foolish the doctor who despises the knowledge acquired by the ancients" and it is now time for us in the face of these pandemics to join efforts to discover new antimicrobial agents from the medicinal plants in the Asia Pacific for zoonotic pandemics.

Christophe Wiart
Kuala Lumpur
August 11, 2021

REFERENCES

Claas, E.C., 2000. Pandemic influenza is a zoonosis, as it requires introduction of avian-like gene segments in the human population. *Veterinary Microbiology, 74*(1–2), pp. 133–139.

Littman, R.J. and Littman, M.L., 1973. Galen and the Antonine plague. *The American Journal of Philology, 94*(3), pp. 243–255.

Morse, S.S., 2001. Factors in the emergence of infectious diseases. In Price-Smith, A.T. (ed) *Plagues and Politics* (pp. 8–26). Palgrave Macmillan, London.

Sabbatani, S. and Fiorino, S., 2009. The antonine plague and the decline of the Roman Empire. *Le infezioni in medicina: rivista periodica di eziologia, epidemiologia, diagnostica, clinica e terapia delle patologie infettive, 17*(4), pp. 261–275.

Slingenbergh, J., Gilbert, M., Balogh, K.D. and Wint, W., 2004. Ecological sources of zoonotic diseases. *Revue scientifique et technique-Office international des épizooties, 23*(2), pp. 467–484.

Taylor, L.H., Latham, S.M. and Woolhouse, M.E., 2001. Risk factors for human disease emergence. *Philosophical Transactions of the Royal Society of London. Series B: Biological Sciences, 356*(1411), pp. 983–989.

Wagner, D.M., Klunk, J., Harbeck, M., Devault, A., Waglechner, N., Sahl, J.W., Enk, J., Birdsell, D.N., Kuch, M., Lumibao, C. and Poinar, D., 2014. Yersinia pestis and the Plague of Justinian 541–543 AD: a genomic analysis. *The Lancet Infectious Diseases, 14*(4), pp. 319–326.

Wolfe, N.D., Daszak, P., Kilpatrick, A.M. and Burke, D.S., 2005. Bushmeat hunting, deforestation, and prediction of zoonotic disease. *Emerging Infectious Diseases, 11*(12), p. 1822.

Author

Christophe Wiart, PharmD, PhD, is currently an associate professor of the School of Pharmacy of the University of Nottingham, Malaysia Campus. His fields of expertise are Asian ethnopharmacology, chemotaxonomy, and ethnobotany. He has collected, identified, classified, and examined the therapeutic potentials of several hundred species of medicinal plants from Asia and the Pacific.

Dr. Wiart appeared on HBO's Vice (TV series) in season 3, episode 6 (episode 28 of the series), titled "The Post-Antibiotic World and Indonesia's Palm Bomb," April 17, 2015. It highlighted the need to find new treatments for infections that were previously treatable with antibiotics but are now resistant to multiple drugs. "The last hope for the human race's survival, I believe, is in the rainforests of tropical Asia," said ethnopharmacologist Christophe Wiart. "The pharmaceutical wealth of this land is immense."

8.4 Order Malvales Juss. ex Bercht. & J. Presl (1820)

8.4.1 FAMILY BIXACEAE KUNTH (1822)

The family Bixaceae consists of the single genus *Bixa* L.

8.4.1.1 *Bixa orellana* L.

Synonyms: *Bixa acuminata* Bojer; *Bixa americana* Poir.; *Bixa odorata* Ruiz & Pav. ex G. Don; *Bixa platycarpa* Ruiz & Pav. ex G. Don; *Bixa tinctoria* Salisb.; *Bixa upatensis* Ram. Goyena; *Bixa urucurana* Willd.; *Orellana americana* Kuntze; *Orellana orellana* (L.) Kuntze

Common names: Annatto, lipstick tree; chompuh chralok (Cambodia); cham pu, xiem phung (China); avam sinduri, kurangumanjal, sinduri, virpushpa (India); sati (Laos); thi dew, thideng, thidin (Myanmar); nauksinu, pop, tar (Papua New Guinea); achiote, anate, sijotes (the Philippines); kam set (Thailand); dieu nhuom (Vietnam)

Habitat: Cultivated

Distribution: Tropical Asia and Pacific

Botanical observation: This shrub or small tree native to northern South America grows up to about 6 m tall. The leaves are simple, spiral, and stipulate. The petiole is 5–7.5 cm long. The blade is 10–20×6.3–12.5 cm, dark green, ovate, acute or acuminate, truncate or subcordate at the base and glabrous. The flowers are 5 cm in diameter, showy, and arranged in terminal panicles. The calyx consists of 2 small concave sepals and 3 large, up to 1 cm long, oblong sepals. The 5 petals are pinkish, red or white, obovate, membranaceous, and up to 3 cm long. The androecium comprises numerous stamens. Capsules are dehiscent, somewhat ovoid, prickly, reddish, with somewhat a negative aura, and containing about 15–20 trigonous seeds embedded in a bright red pulp.

Medicinal uses: Cough, dysentery (India; the Philippines); fever (India; Myanmar); ear, eyes, skin infection, jaundice (India)

Moderate antibacterial (Gram-positive) essential oil: Essential oil inhibited the growth of *Streptococcus mutans* with the MIC of 310 μg/mL (Miller et al., 2015).

Strong broad-spectrum antibacterial polar extracts: Methanol extract of seeds inhibited the growth of *Streptococcus mutans* (ATCC 25175) and *Streptococcus sanguinis* (ATCC 10556) with the MIC values of 31.2 and 125 μg/mL, respectively (Medina-Flores et al., 2016). Methanol extract of leaves inhibited the growth of *Staphylococcus aureus* (MTCC 3160), *Enterococcus faecalis* (MTCC 3159), *Brucella* sp. (MTCC 685), *Vibrio cholera* (MTCC 3906), *Klebsiella pneumoniae* (MTCC 3384), *Pseudomonas aeruginosa* (MTCC 2582), *Moraxella catarrhalis* (MTCC 445), *Acinetobacter* sp. (MTCC 1611), *Aeromonas hydrophila* (MTCC 1739), and *Salmonella typhi* (MTCC 3216) with the MIC values of 15.6, 31.2, 65.5, 31.2, 31.2, 31.3, 250, 31.2, 62.5, and 31.2 μg/mL, respectively (Tamil Selvi et al., 2011).

Antibacterial halo developed by apocarotenoid: Bixin (750 μg/disc) inhibited the growth of *Staphylococcus aureus* and *Escherichia coli* with the inhibition zone diameters of 8.4 and 8 mm, respectively (Kurniawati et al., 2007).

Anticandidal polar extract: Ethanolic extract of seeds (10 mg/mL/well) inhibited the growth of *Candida albicans* with the inhibition zone diameter of 22 mm (Fleischer et al., 2003).

Commentary: The plant produces the flavonol hypolaetin (also known as 8-hydroxyluteolin) (Harborne, 1975).

DOI: 10.1201/9781003180371-1

REFERENCES

Fleischer, T.C., Ameade, E.P.K., Mensah, M.L.K. and Sawer, I.K., 2003. Antimicrobial activity of the leaves and seeds of *Bixa orellana*. *Fitoterapia*, *74*(1–2), pp. 136–138.

Harborne, J.B., 1975. Flavonoid bisulphates and their co-occurrences with ellagic acid in the bixaceae, frankeniaceae and related families. *Phytochemistry*, *14*(5–6), pp. 1331–1337.

Kurniawati, P.T., Soetjipto, H. and Limantara, L., 2007. Antioxidant and antibacterial activities of Bixin pigment from Annatto (*Bixa orellana* L.) seeds. *Indonesian Journal of Chemistry*, *7*(1), pp. 88–92.

Medina-Flores, D., Ulloa-Urizar, G., Camere-Colarossi, R., Caballero-GarcÝa, S., Mayta-Tovalino, F. and del Valle-Mendoza, J., 2016. Antibacterial activity of *Bixa orellana* L. (achiote) against *Streptococcus mutans* and *Streptococcus sanguinis*. *Asian Pacific Journal of Tropical Biomedicine*, *6*(5), pp. 400–403.

Miller, A.B., Cates, R.G., Lawrence, M., Soria, J.A.F., Espinoza, L.V., Martinez, J.V. and Arbizú, D.A., 2015. The antibacterial and antifungal activity of essential oils extracted from Guatemalan medicinal plants. *Pharmaceutical Biology*, *53*(4), pp. 548–554.

Tamil Selvi, A., Dinesh, M.G., Satyan, R.S., Chandrasekaran, B. and Rose, C., 2011. Leaf and seed extracts of *Bixa orellana* L. exert anti-microbial activity against bacterial pathogens. *Journal of Applied Pharmaceutical Science*, *1*(9), pp. 116–120.

8.4.2 FAMILY BOMBACACEAE KUNTH (1822)

The family Bombaceae consists of about 30 genera and 250 species of mostly trees, mostly growing in savannahs. The bole is massive and sometimes spiny or buttressed. The leaves are simple, alternate, and stipulate. The stipules are minute. The petiole is often straight. The blade is entire or palmately compound. The flowers are often showy, axillary, and solitary. The calyx is tubular and 3–5-lobed or comprises 5 sepals. The 5 petals are fused at base. The androecium includes numerous stamens merged into a column at base. The ovary is 5-loculed, each locule sheltering mostly a pair of ovules. The capsules are dehiscent or not and shelter numerous seeds. Bombacaceae, Malvaceae, Sterculiaceae, and Tiliaceae have been recently merged into the Malvaceae but since these clearly differ morphologically and chemically they are treated separately in this volume.

8.4.2.1 *Adansonia digitata* L.

Synonyms: *Adansonia bahobab* L.; *Adansonia scutula* Steud.; *Adansonia integrifolia* Raf.; *Adansonia situla* (Lour.) Spreng.; *Adansonia sphaerocarpa* A. Chev.; *Adansonia sulcata* A. Chev.; *Baobabus digitata* (L.) Kuntze; *Ophelus sitularius* Lour.

Common names: Baobab tree; monkey bread-tree, upside-down tree; kattio daghor (Bangladesh); hou mian bao shu (China); gorakh imli, papparappuli (India)

Habitat: Savannah, cultivated

Distribution: Pakistan, India, Bangladesh, the Philippines, Australia

Botanical observation. This very large, strange, somewhat eerie, and massive tree native to Africa can grow to 20m tall. The bole can reach 10m across and look like a huge smooth cylinder and the crown looks like if it was roots. The leaves are compound and spiral. The petiole is slender and about 10cm long. The blade is palmately compound, edible, 5–7 foliate, the folioles elliptic to oblanceolate, dull green, somewhat fleshy, 5–15×3–7cm, and hairy below when young. The flowers are massive and pendulous. The calyx is cup-shaped and develops 5 lobes. The corolla is membranaceous and develops 5 lobes which are orbiculate, about 7cm long, and with a mushroom-like white coloration. The numerous stamens are partially united in a showy, white, and about 10cm long column. The capsules are ovoid, up to 25×11cm, hairy, edible, and contain numerous seeds which are pure white and polyhedral.

Medicinal uses: Dysentery (India, Bangladesh); fever, diarrhea (India)

Moderate broad-spectrum antibacterial polar extract: Methanol extract of leaves inhibited the growth of *Escherichia coli* (ATCC 10536), *Enterobacter aerogenes* (EA 298), *Klebsiella pneumoniae* (K2), and *Pseudomonas stuartii* (PS 299645) with the MIC values of 512, 256, 1024, and

256 μg/mL, respectively (Seukep et al., 2013). Methanol extract of stem bark inhibited the growth of *Staphylococcus aureus* (ATCC 6538), *Enterococcus faecalis* (ATCC 29212), methicillin-resistant *Staphylococcus aureus*, and *Staphylococcus epidermidis* with the MIC values of 1250, 1250, 78, and 1250 μg/mL, respectively (Lagnika et al., 2012).

Moderate antifungal (filamentous) essential oil: Essential oil of seeds inhibited the growth of *Fusarium proliferatum* with the MFC value of 480 μg/mL (Samie & Nefefe, 2012).

Strong antiviral (enveloped, linear, single-stranded (−) RNA) polar extract: Methanol extract of leaves inhibited the replication of the Influenza virus and the replication of the Respiratory syncytial virus (Selvarani & Hudson, 2009).

Strong antiviral (enveloped, linear, double-stranded DNA) polar extract: Methanol extract of leaves inhibited the replication of the Herpes simplex virus the MIC value of 11.7 μg/mL (Selvarani & Hudson James, 2009). Ethanol extract of fruit at the concentration of 125 μg/mL inhibited the replication of the Herpes simplex virus type-1 (clinical strain) by 56% (Silva et al., 1997). Ethanol extract of fruits at the concentration of 125 μg/mL inhibited the replication of the African swine fever virus (strain Lisbon 60) by 80% (Silva et al., 1997).

Commentaries

i. The plant assembles the triterpenes α-amyrin, β-amyrin palmitate, ursolic acid, the steroid campestrol, the alkaloid adansonin (Refaat et al., 2013) as well as series of flavonoids of which (−)-epicatechin, flavanone, flavone glycosides, and procyanidins (Kamatou et al., 2011) of which procyanidin B1, B2, and C1 (Ismail et al., 2020). Procyanidins are often antiviral and have some levels of oral absorption (Zhang et al., 2016) and as such are interesting antiviral candidates. For instance, procyanidin A2 and procyanidin B1 isolated from the bark of a member of the genus *Cinnamomum* Schaeff. (family Lauraceae) inhibited the replication of Severe acute respiratory syndrome coronavirus with the IC_{50} values of 29.9 and 41.3 μM and selectivity indexes of 37.3 and 15.6, respectively (Zhuang et al., 2009).

ii. Arab legend regarding this tree is that the devil plucked up the tree and placed the branches into the soil and left the roots in the air (Wickens, 2008). Legends about this tree are also plentiful in India.

REFERENCES

Ismail, B.B., Yusuf, H.L., Pu, Y., Zhao, H., Guo, M. and Liu, D., 2020. Ultrasound-assisted adsorption/desorption for the enrichment and purification of flavonoids from baobab (*Adansonia digitata*) fruit pulp. *Ultrasonics Sonochemistry*, *65*, p. 104980.

Kamatou, G.P.P., Vermaak, I. and Viljoen, A.M., 2011. An updated review of *Adansonia digitata*: a commercially important African tree. *South African Journal of Botany*, *77*(4), pp. 908–919.

Lagnika, L., Amoussa, M., Adjovi, Y. and Sanni, A., 2012. Antifungal, antibacterial and antioxidant properties of *Adansonia digitata* and *Vitex doniana* from *Bénin pharmacopeia*. *Journal of Pharmacognosy and Phytotherapy*, *4*(4), pp. 44–52.

Refaat, J., Desoky, S.Y., Ramadan, M.A. and Kamel, M.S., 2013. Bombacaceae: a phytochemical review. *Pharmaceutical Biology*, *51*(1), pp. 100–130.

Samie, A. and Nefefe, T., 2012. Antifungal activities of essential oils from Southern African medicinal plants against five Fusarium species. *Journal of Medicinal Plant Research*, *6*(3), pp. 465–478.

Selvarani, V. and Hudson James, B., 2009. Multiple inflammatory and antiviral activities in *Adansonia digitata* (Baobab) leaves, fruits and seeds. *Journal of Medicinal Plants Research*, *3*(8), pp. 576–582.

Seukep, J.A., Fankam, A.G., Djeussi, D.E., Voukeng, I.K., Tankeo, S.B., Noumdem, J.A., Kuete, A.H. and Kuete, V., 2013. Antibacterial activities of the methanol extracts of seven Cameroonian dietary plants against bacteria expressing MDR phenotypes. *Springerplus*, *2*(1), p. 363.

Silva, O., Barbosa, S., Diniz, A., Valdeira, M.L. and Gomes, E., 1997. Plant extracts antiviral activity against Herpes simplex virus type 1 and African swine fever virus. *International Journal of Pharmacognosy*, *35*(1), pp. 12–16.

Wickens, G.E., 2008. *The baobabs: pachycauls of Africa, Madagascar and Australia*. Springer Science & Business Media, Kew.

Zhang, L., Wang, Y., Li, D., Ho, C.T., Li, J. and Wan, X., 2016. The absorption, distribution, metabolism and excretion of procyanidins. *Food & Function*, *7*(3), pp. 1273–1281.

Zhuang, M., Jiang, H., Suzuki, Y., Li, X., Xiao, P., Tanaka, T., Ling, H., Yang, B., Saitoh, H., Zhang, L. and Qin, C., 2009. Procyanidins and butanol extract of Cinnamomi Cortex inhibit SARS-CoV infection. *Antiviral Research*, *82*(1), pp. 73–81.

8.4.2.2 *Bombax ceiba* L.

Synonyms: *Bombax heptaphyllum* L.; *Bombax malabaricum* DC.; *Gossampinus malabarica* (DC.) Merr.; *Salmalia malabarica* (DC.) Schott & Endl.

Common names: Cottonwood, Indian bombax, red cotton tree, silk cotton tree, simul (Bangladesh; India); simaful (Bangladesh); ken hao (China); shaalmali, semala, semal ka phool (India); kapuk hutan (Indonesia); ngiou ban (Laos); kapok (Malaysia); kadung (Myanmar); simal, simaltun (Nepal); malabulak (the Philippines); ngui (Thailand); moc nien (Vietnam)

Habitat: Cultivated, roadsides

Distribution: Pakistan, India, Sri Lanka, Nepal, Bhutan, Bangladesh, Myanmar, Laos, Malaysia, Indonesia, the Philippines, Papua New Guinea

Botanical observation: This magnificent tree grows up to 25 m tall. The bole is buttressed and spiny. The leaves are spiral, 5–7 palmate, and stipulate. The stipules are minute. The petiole is slender and up to 20 cm long. The folioles are oblong, 10–15 × 3.5–5.5 cm, with 15–17 pairs of secondary nerves, and acuminate at apex. The flowers are solitary and terminal. The calyx is cup-shaped, fleshy, glossy, coriaceous, 2–3 cm long and develops with 5 lobes which are triangular. The 5 petals are oblong, up to 10 cm long, red to burgundy, recurved, and glossy. The stamens are numerous and showy. The capsules are oblong, 10–15 × 4.5–5 cm, and open to release a cottony mass (used by birds to make their nests) and numerous seeds.

Medicinal uses: Boils (Bangladesh; India); gonorrhea (Bangladesh); dysentery, fever (India; Nepal); diarrhea, fever (Bangladesh; Nepal); tuberculosis (Pakistan, India); cuts, tuberculosis, wounds (Nepal); acne, mouth ulcers (India); sores (China); chicken pox, cough, cystitis, smallpox (India); diarrhea, dysentery (India; Nepal); wounds (China).

Broad-spectrum antibacterial halo developed by polar extract: Methanol extract of stem (150 µg/well) inhibited the growth of *Bacillus subtilis*, *Staphylococcus aureus* (ATCC 6538), *Escherichia coli* (clinical strain), *Escherichia coli* (clinical strain), *Pseudomonas aeruginosa* (ATCC 9087), and *Salmonella typhi* (clinical strain) (Masood-ur-Rehman & Mustafa, 2017).

Broad-spectrum antibacterial halo developed by mid-polar extract: Aqueous (10% chloroform) extract of bark (25 µg/mL/6 mm well) inhibited the growth of *Bacillus subtilis*, *Bacillus aureus*, *Staphylococcus aureus*, *Escherichia coli*, *Klebsiella pneumonia*, and *Pseudomonas aeruginosa* with the inhibition zones of 10, 11, 15, 14, 14, and 12 mm, respectively (Digge et al., 2015).

Antibacterial xanthone glycoside: Mangiferin (Faizi & Ali, 1999; Shahat et al., 2003).

Strong antiviral (enveloped monopartite linear double-stranded DNA) polar extract: Methanol extract of flowers inhibited the replication of Herpes simplex type-1 extract with the IC_{50} value of 62 µg/mL (Rajbhandari, 2001).

Moderate antiviral (enveloped, monopartite, linear, single-stranded (−) RNA) flavonols: Quercetin and kaempferol-3-*O*-(6″-*O*-E-*p*-coumaroyl)-β-D-glucopyranoside from the flowers inhibited the replication of the Respiratory syncytial virus (Long strain) in Hep-2 cells with IC_{50} values of 20 and 6.3 µM, respectively, and the selectivity indexes of 12.9 and >79.3, respectively (ribavirin IC_{50} of 10 µM and selectivity index of 25.6) (Zhang et al., 2015).

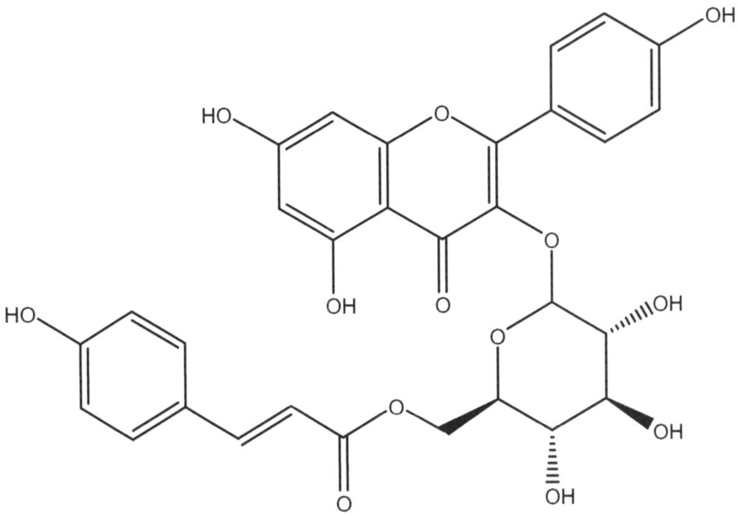

Kaempferol-3-*O*-(6”-*O*-E-*p*-coumaroyl)-β-D-glucopyranoside

Weak antiviral (enveloped monopartite linear single-stranded (+) RNA) flavonol: Quercetin inhibited the Severe acute respiratory syndrome-associated coronavirus in Vero cells with EC_{50} value of 83.4 μM (Yi et al., 2004).

Weak antiviral (enveloped, monopartite, linear, single-stranded (−) RNA) xanthone glycoside: Mangiferin isolated from the flowers inhibited the Respiratory syncytial virus (Long strain) in Hep-2 cells plaque formation with IC_{50} value of 40 μM, and a selectivity index above12.5 (Zhang et al., 2015).

Moderate antiviral (monopartite, linear, dimeric, single stranded (+) RNA) xanthone glycoside: Mangiferin inhibited the replication of the Human immunodeficiency virus type-1 (strain IIIB) with the IC_{50} value of at 16.9 μM and a selectivity index above 140, via, at least in part, inhibition of protease (Wang et al., 2011).

Weak antiviral (enveloped monopartite circular double-stranded DNA) lignans: Bombasinol A, 5,6-dihydroxymatairesinol, (+) pinoresinol, and maitaresinol isolated from the roots inhibited the secretion of Hepatitis B virus surface antigen with IC_{50} values of 118.3, 123.7, 118.7, and 218.2 μM, respectively (Wang et al., 2013).

Commentary: Phenolics stored in this plant may act as a mild astringent in dysentery.

REFERENCES

Digge, V.G., Kuthar Sonali, S., Hogade Maheshwar, G., Poul, B.N. and Jadge Dhanraj, R., 2015. Screening of antibacterial activity of aqueous bark extract of *Bombax ceiba* against some gram positive and gram negative bacteria. *Screening*, *5*, p. 8.

Faizi, S. and Ali, M., 1999. Shamimin: a new flavonol C-glycoside from leaves of *Bombax ceiba*. *Planta Medica*, *65*(4), pp. 383–385.

Masood-ur-Rehman, N.A. and Mustafa, R., 2017. Antibacterial and antioxidant potential of stem bark extract of *Bombax ceiba* collected locally from south Punjab Area of Pakistan. *African Journal of Traditional, Complementary, and Alternative Medicines*, *14*(2), p. 9.

Shahat, A.A., Hassan, R.A., Nazif, N.M., Van Miert, S., Pieters, L., Hammuda, F.M. and Vlietinck, A.J., 2003. Isolation of mangiferin from *Bombax malabaricum* and structure revision of shamimin. *Planta Medica*, *69*(11), pp. 1068–1070.

Rajbhandari, M., Wegner, U., Jülich, M., Schoepke, T. and Mentel, R., 2001. Screening of Nepalese medicinal plants for antiviral activity. *Journal of Ethnopharmacology*, *74*(3), pp. 251–255.

Wang, G.K., Lin, B.B., Rao, R., Zhu, K., Qin, X.Y., Xie, G.Y. and Qin, M.J., 2013. A new lignan with anti-HBV activity from the roots of *Bombax ceiba*. *Natural Product Research*, *27*(15), pp. 1348–1352.
Wang, R.R., Gao, Y.D., Ma, C.H., Zhang, X.J., Huang, C.G., Huang, J.F. and Zheng, Y.T., 2011. Mangiferin, an anti-HIV-1 agent targeting protease and effective against resistant strains. *Molecules*, *16*(5), pp. 4264–4277.
Yi, L., Li, Z., Yuan, K., Qu, X., Chen, J., Wang, G., Zhang, H., Luo, H., Zhu, L., Jiang, P. and Chen, L., 2004. Small molecules blocking the entry of severe acute respiratory syndrome coronavirus into host cells. *Journal of Virology*, *78*(20), pp. 11334–11339.
Zhang, Y.B., Wu, P., Zhang, X.L., Xia, C., Li, G.Q., Ye, W.C., Wang, G.C. and Li, Y.L., 2015. Phenolic compounds from the flowers of *Bombax malabaricum* and their antioxidant and antiviral activities. *Molecules*, *20*(11), pp. 19947–19957.

8.4.2.3 *Ceiba pentandra* (L.) Gaertn.

Synonyms: *Bombax pentandrum* L.; *Ceiba casearia* Medik.; *Eriodendron anfractuosum* DC.; *Eriodendron caribaeum* G. Don; *Eriodendron occidentale* (Spreng.) G. Don; *Gossampinus rumphii* Schott; *Xylon pentandrum* (L.) Kuntze

Common names: Kapok, silk-cotton tree; koo (Cambodia); kapu (Indonesia); ngiu (Laos); pohon kapok (Malaysia); balios (the Philippines); nun (Thailand); gon (Vietnam)

Habitat: Roadsides, villages

Distribution: Tropical Asia

Botanical observation: This massive tree, native to Central America, grows to a height of 40 m and a girth of 3 m. The wood is whitish and soft. The bark is smooth and greenish and produces a few thorns which are conical. The leaves are palmate, up to 40 cm diameter, and consist of 7–8 folioles which are 5 × 1.5 cm. The flowers are tubular, whitish to pink in axillary fascicles. The capsules are green, fleshy, fusiform, 8–14 × 4.5–7 cm, and containing numerous seeds. The seeds are up to 8 mm long and comose. The seeds are used by birds to make nests and by humans to stuff cushions.

Medicinal uses: Fever, gonorrhea, chicken pox (Cambodia, Laos, Vietnam); urinary tract infection (Indonesia); colds, syphilis (Malaysia); gonorrhea (Myanmar); dysentery (the Philippines).

Moderate antibacterial (Gram-positive) polar extract: Ethanol extract of stem bark inhibited the growth of *Micrococcus luteus* (MTCC (106), *Bacillus subtilis* (MTCC (121), *Bacillus cereus* (MTCC (430)), and *Streptococcus pneumoniae* (MTCC (2672)) with the MIC values of 300, 1200, 1200, and 600 µg/mL, respectively (Singh et al., 2010).

Weak antibacterial (Gram-positive) sesquiterpene lactones: 2,7-Dimethary-5-isopropyl -3-methyl-8,1-naphthalene carbolactone and 2-hydroxy-5-isopropyl-7-methoxy-3-methyl-8,1-naphthalene carbolactone isolated from the root bark inhibited the growth of *Bacillus subtilis* (ATCC 6633) with the MIC values of 250 and 350 µg/mL and *Staphylococcus aureus* (ATCC 6538) with the MIC values of 275 and 300 µg/mL (Rao et al., 1993.

Weak broad-spectrum antifungal sesquiterpene lactones: 2,7-Dimethary-5-isopropyl -3-methyl-8,1-naphthalene carbolactone and 2-hydroxy-5-isopropyl-7-methoxy-3-methyl-8,1-naphthalene carbolactone isolated from the root bark inhibited the growth of *Aspergillus niger* (NRKL 530) with the MIC values of 200 and 250 µg/mL and *Candida albicans* (ATCC 10231) with the MIC values of 175 and 225 µg/mL (Rao et al., 1993).

Strong antiviral (enveloped, monopartite, linear, single-stranded (+) RNA) extract: Extract of leaves inhibited the replication of the Dengue virus serotype 2 (strain NGC) in Huh7it-1 cells with the IC_{50} value of 19.8 µg/mL and a selectivity index of 6.7 (Dewi et al., 2019).

Commentary: One could examine the antiviral (and antibacterial) properties of the sesquiterpenes and sesquiterpene lactones in this plant because they have planar naphthalene frameworks substituted with hydroxyl and ketone groups and as such most probably interfere with viral RNA or DNA machinery, in particular, 8-formyl-7-hydroxy-5-isopropyl-2-methoxy-3-methyl-1,4-naphthaquinone.

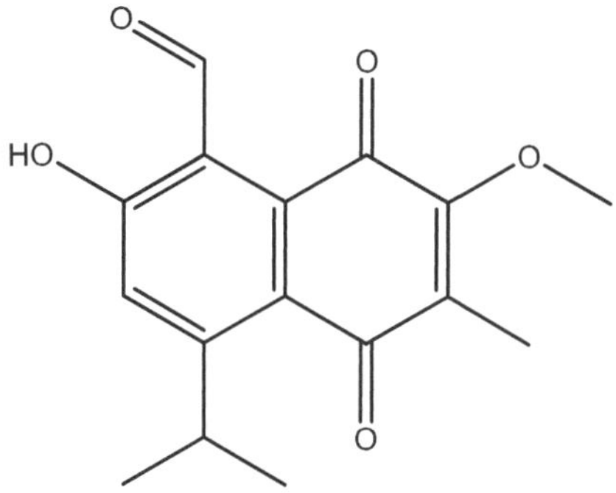

8-Formyl-7-hydroxy-5-isopropyl-2-methoxy-3-methyl-1,4-naphthaquinone

REFERENCES

Dewi, B.E., Angelina, M., Ardiantara, S., Prakoso, A.R., Desti, H. and Sudiro, T.M., 2019, December. Antiviral activity of *Ceiba pentandra* and *Eugenia uniflora* leaf extracts to dengue virus serotype-2 in Huh 7it-1 cell line. In *AIP Conference Proceedings* (Vol. 2193, No. 1, p. 030003). AIP Publishing LLC.

Rao, K.V., Sreeramulu, K., Gunasekar, D. and Ramesh, D., 1993. Two new sesquiterpene lactones from *Ceiba pentandra*. *Journal of Natural Products*, *56*(12), pp. 2041–2045.

Singh, M., Khatoon, S., Singh, S., Kumar, V., Rawat, A.K.S. and Mehrotra, S., 2010. Antimicrobial screening of ethnobotanically important stem bark of medicinal plants. *Pharmacognosy Research*, 2(4), p. 254.

8.4.3 FAMILY DIPTEROCARPACEAE BLUME (1825)

The family Dipterocarpaceae consists of about 20 genera and 550 species of mostly magnificent and giant trees, mostly growing in primary forests (soon to vanish under caterpillars of the palm oil and logging industries). The wood is often very valuable (looted by British colonialists in India and Malaysia with the help of locals), and exudes aromatic resins or basalms. The bole is straight and sometimes buttressed. The leaves are simple, alternate, coriaceous, and stipulate. The inflorescences are terminal or axillary racemes or panicles. The calyx is tubular and 5-lobed or comprises 5 sepals. The 5 petals are contorted. The androecium includes about 10–15 stamens. The ovary is 3-loculed, each locule sheltering mostly a pair of ovules. The nuts often develop wings (the trees are so tall that without wings which rotate during fall to slow down the fall the fruits would break). Members in this family generate a fascinating array of antimicrobial oligostilbenes as well as monoterpenes.

8.4.3.1 *Dipterocarpus turbinatus* Gaertn.

Synonyms: *Dipterocarpus jourdainii* Pierre; *Dipterocarpus laevis* Buch.-Ham.

Common names: Gurjun tree, wood oil tree; teli garjan (Bangladesh); jie bu luo xiang (China); asvakarna (India); kauyeen (Myanmar)

Habitat: Forests

Distribution: India, Bangladesh, Myanmar, Cambodia, Laos, Vietnam, Thailand

Botanical observation: This majestic timber tree grows to 30 m tall. The bark is brown, fissured, flaky, and yield an aromatic exudate (known as Gurjun basalm obtained upon cutting and burning). The leaves are simple, alternate, and stipulate. The stipules are 2–6 cm long and puberulous. The petiole is 2–3 cm long. The blade is ovate-oblong, 20–30 × 8–13 cm, coriaceous, marked with 15–20 pairs of secondary nerves, rounded at base, and acuminate or acute at apex. The racemes are axillary and 3–6-flowered. The calyx comprises 5 sepals. The 5 lobes of the corolla are linear, contortate, and pinkish-white. The androecium includes 30 stamens. The ovary is pubescent and develops a terete style. The nuts are oblong and present a pair of wings, which are purplish-brown, 12–15 × 3 cm, and reflexed.

Medicinal uses: Leprosy, skin affections, ulcers (India); ringworm (Bangladesh; India)

Broad-spectrum antibacterial halo developed by polar extract: Ethanol extract (30 mg/mL/6 mm well) inhibited the growth of *Pseudomonas aeruginosa*, *Staphylococcus aureus*, *Escherichia coli*, and *Enterococcus faecalis* with the zones of inhibition of 22.8, 21.3, 22.3, and 24 mm, respectively (Biswas et al., 2020).

Broad-spectrum antibacterial halo developed by monoterpene: Borneol (300 µg/5 mm disc) inhibited the growth of *Clavibacter michiganense*, *Agrobacterium tumefaciens*, *Xanthomonas campestris*, and *Enterobacter cloacae* (Kotan et al., 2007).

Broad-spectrum antifungal halo developed by polar extract: Ethanol extract (30 mg/mL/6 mm well) inhibited the growth of various fungi (Biswas et al., 2020).

Commentaries

i. The plant produces borneol (Chen et al., 2011), which being neuroactive (Quintans-Junior et al., 2010) might be able to block bacterial and or fungal efflux pumps.
ii. The plant was incorporated in the 1868 edition of the Pharmacopoeia of India during the time of British occupation.

REFERENCES

Biswas, D., Mahalakshmi, B., Samelu, B., Devadas, M. and Suresehkumar, S.V., 2020. Antimicrobial activity of ethanol extracts and isolated fractions Shorea robusta Gaertn f and *Dipterocarpus turbinatus* Gaertn f. *Pharmacognosy Communications*, *10*(2), pp. 102–107.

Chen, L., Su, J., Li, L., Li, B. and Li, W., 2011. A new source of natural D-borneol and its characteristic. *Journal of Medicinal Plants Research*, *5*(15), pp. 3440–3447.

Kotan, R., Kordali, S. and Cakir, A., 2007. Screening of antibacterial activities of twenty-one oxygenated monoterpenes. *Zeitschrift für Naturforschung C*, *62*(7–8), pp. 507–513.

Quintans-Júnior, L.J., Guimarães, A.G., Araújo, B.E., Oliveira, G.F., Santana, M.T., Moreira, F.V., Santos, M.R., Cavalcanti, S.C., Júnior, W.L., Botelho, M.A. and Ribeiro, L.A., 2010. Carvacrol, (–) borneol and citral reduce convulsant activity in rodents. *African Journal of Biotechnology*, *9*(39), pp. 6566–6572.

8.4.3.2 *Dryobalanops aromatica* C.F. Gaertn.

Synonyms: *Dryobalanops camphora* Colebr.; *Dryobalanops sumatrensis* (J.F. Gmel.) Kosterm.

Common names: Borneo camphorwood; bhimseni, kapoor (India); pokok kapur barus (Indonesia); kapur (Malaysia).

Habitat: Forests, cultivated

Distribution: Malaysia, Indonesia

Botanical observation: This magnificent timber tree grows up to a height of 45 m. The bole is straight, covered with a flaky bark, and exudes some sort of glassy aromatic masses upon incision. The stipules are about 1 cm long, linear, and ephemeral. The leaves are simple, alternate, and stipulate. The blade is 4–6 × 2–4 cm, broadly lanceolate, coriaceous, somewhat dull dark green, smooth, cuneate to round at base, and acuminate at apex. The panicle is about 10 cm long, terminal or axillary, and supports numerous beautiful flowers. The calyx is 5-lobed. The corolla consists of 5 pure white and oblong petals. The androecium is made of about 30 bright yellow stamens. The ovary is minute and somewhat ovoid. The nut is about 1 cm long and develops 5 light green to brownish-pink spathulate wings.

Medicinal use: Apparently none for infectious diseases

Antibacterial (Gram-positive) halo developed by oiligostilbenoids: ε-Viniferin (200 µL/well from 1 mg/mL solution) inhibited the growth of *Escherichia coli* and *Staphylococcus aureus* with inhibition zone diameter of 11 and 7 mm, respectively (Sahidin et al., 2017). α-Viniferin inhibited

(200 μL/well from 1 mg/mL solution) the growth of *Escherichia coli* and *Staphylococcus aureus* with inhibition zone diameter of 8 and 8 mm, respectively (Sahidin et al 2017).

Very weak anticandidal monoterpene: Borneol inhibited the growth of *Candida albicans* with the MIC/MBC values of 320/640 μg/mL (Liu et al., 2017).

Strong antiviral (enveloped monopartite single-stranded (+) RNA) oligostilbenoid: ε-Viniferin inhibited the replication of the Hepatitis C virus in huh7.5 cells with the EC_{50} value of 0.1 μM and a CC_{50} value above 10 μM (Lee et al., 2019).

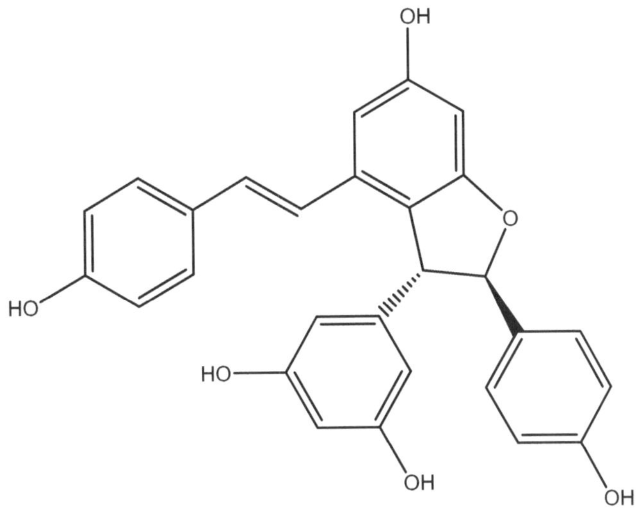

ε-Viniferin

Viral enzyme inhibition: ε-Viniferin inhibited the Hepatitis C virus NS3 helicase with the IC_{50} value of 58.7 nM (Lee et al., 2019).

Commentaries

i. The plant has not been much apparently been much examined for antimicrobial properties. Borneo camphor was used as medicine during plagues including the Black Plague and the Plague of Marseille both due to the zoonotic bacterium *Yersina pestis* from Asian gerbils and rats transmitted by fleas (Eisen & Gage, 2012). Camphor was a fumigant during the 14th century Black plague (Donkin, 1999) brought to Italy ports by merchant ships from the Black Sea (Langer, 194). The Black Plague from the year 1348 to 1350 incurred to about a fourth of the European population high fever, agonizing pain, prostration, and death within 6 days (Langer, 1964). Mixed with vinegar, Borneo camphor was a medicine of repute in the case of the Plague of Marseille of 1720 brought by a merchant ship from presumably Turkey (Huxham, 1764).
ii. Borneo camphor consists mainly of α-pinene, γ-terpinene, terpinen-4-ol, α and β-caryophyllene, borneol, and camphor (Le al., 2016).
iii. Other interesting natural products from *Dryobalanops aromatica* C.F Gaertn are oligostilbenoids of which laevifonol and ampelopsin E (Wibowo et al., 2011).
iv. *Dryobalanops lanceolata* Burck generates a series of antibacterial oligostilbenoids: upunaphenol D inhibited the growth of *Staphylococcus epidermidis*, *Staphylococcus aureus* and *Staphylococcus xylosus* with the MIC values of 25, 50, and 25 μM, respectively (Wibowo et al., 2013). Flexuosol A inhibited the growth of *Staphylococcus epidermidis*, *Staphylococcus aureus* and *Staphylococcus xylosus* with the MIC values of 75, 100, and

75 μM, respectively (Wibowo et al., 2013). Balanocarpol (200 μL per well from 1 mg/mL solution) inhibited the growth of *Escherichia coli* and *Staphylococcus aureus* with inhibition zone diameter of 9 and 13 mm respectively (Sahidin et al., 2017). Vaticanol B isolated from the stem bark of *Dryobalanops lanceolata* (200 μL per well from 1 mg/mL solution) inhibited the growth of *Escherichia coli* and *Staphylococcus aureus* with inhibition zone diameter of 5 and 4 mm, respectively (Sahidin et al., 2017).

v. One could examine the anti-COVID-19 property of ε-viniferin because it inhibits the replication of the Hepatitis C virus, which is, like the Severe acute respiratory syndrome-associated coronavirus 2 virus, an enveloped, monopartite, single-stranded (+) RNA virus. There is the gut-feeling that oligostilbenoids may have potentials as antiviral agents.

REFERENCES

Eisen, R.J. and Gage, K.L., 2012. Transmission of flea-borne zoonotic agents. *Annual Review of Entomology*, *57*, pp. 61–82.

Huxham, J., 1764. *An essay on fevers: to which is now added, a dissertation on the malignant, ulcerous sore-throat* (J. Hinton).

Langer, W.L., 1964. The black death. *Scientific American*, *210*(2), pp. 114–121.

Le, T.-X., Ho, A.S.H., Mah, S.H., Wong, T.W., Ong, H.C., Loh, P.H.M. and Lim, Y.M., 2016. Determination of borneol and other chemical compounds of essential oil of *Dryobalanops aromatica* exudate from Malaysia. *Tropical Journal of Pharmaceutical Research*, *15*(6), pp. 1293–1297.

Lee, S., Mailar, K., Kim, M.I., Park, M., Kim, J., Min, D.H., Heo, T.H., Bae, S.K., Choi, W. and Lee, C., 2019. Plant-derived purification, chemical synthesis, and *in vitro/in vivo* evaluation of a resveratrol dimer, viniferin, as an HCV replication inhibitor. *Viruses*, *11*(10), p. 890.

Liu, Y., Zhao, Y., Guo, D.L., Liu, W.W. and Liu, Y.X., 2017. Synergistic antimicrobial activity of berberine hydrochloride, baicalein and borneol against *Candida albicans*. *Chinese Herbal Medicines*, *9*(4), pp. 353–357.

Sahidin, I., Wahyuni, W., Malaka, M.H., Imran, I., 2017. Antibacterial and cytotoxic potencies of stilbene oligomers from stem barks of baoti (*Dryobalanops lanceolata*) growing in Kendari, Indonesia. *Asian Journal of Pharmaceutical and Clinical Research*, *10*(8), pp. 139–143.

Wibowo, A., Ahmat, N., Hamzah, A.S., 2011. Oligostilbenoids from the stem bark of *Dryobalanops aromatica*. *Planta Medica*, *77*(12), p. PG9.

Wibowo, A., Ahmat, N., Hamzah, A.S. and Low, A.L.M., 2013. Antibacterial activity of oligostilbenoid derivatives from the stem bark of *Dryobalanops lanceolata*. *The Open Conference Proceedings Journal*, *4*, p. 116.

8.4.3.3 *Shorea robusta* Gaertn.

Common names: Bengal dammer, sal tree, red balau; sajamdare, shal (Bangladesh); suo luo chuang (China); roli shaala, uswukunida (India); agrath, sal, jhesin, agras, phoskin (Nepal)

Habitat: Forests

Distribution: Pakistan, India, Bhutan, Nepal, Bangladesh, China

Botanical observation: This majestic timber tree grows to 40 m tall. The bark is greyish brown and fissured. The trunk exudes a resin. The wood is very hard and dense. The leaves are simple, alternate, and stipulate. The stipules are small and lanceolate. The petiole is 2–2.5 cm long. The blade is 8–20×5–15 cm, ovate to oblong, marked with about 12 pairs of secondary nerves, glossy, base rounded to cordate, and acuminate at apex. The inflorescences are axillary racemes which grow up to 20 cm long. The calyx includes 5 sepals which are ovate, subequal, and pubescent. The corolla produces 5 lobes which are contorted, creamy-yellow to pinkish, 1–1.5 cm×5 mm, and linear. The androecium includes about 15 anthers. The ovary is ovoid and pubescent. The nut is ovoid, 5×12 mm with 5 wings of unequal size which grow up to 3.5–8 cm long.

Medicinal uses: Boils, fever, wounds (India); ulcers, small pox, tuberculosis, syphilis (Nepal); diarrhea, (India; Nepal); dysentery (Bangladesh; India)

Broad-spectrum antibacterial polar extract: Methanol extract of resin (30 µg/6 mm disc) inhibited the growth of *Staphylococcus aureus*, *Bacillus subtilis*, *Pseudomonas aeruginosa*, and *Escherichia coli* with the inhibition zones diameters of 12, 10, 10, and 12 mm, respectively (Vashisht et al., 2016).

Moderate antibacterial (Gram-negative) polar extract: Aqueous extract of leaves inhibited the growth of *Salmonella enterica* subsp. *enterica* serovar *typhi* (clinical isolates), *Salmonella typhi* (MTCC 734), and *Salmonella typhimurium* (NCTC 74) with the MIC/MBC values of 256/768, 700/2800, and 256/1028 µg/mL, respectively (Chattopadhyay et al., 2018).

Antifungal (filamentous) halo developed by polar extract: Ethanol extract of flowers (250 mg/mL/8 mm well) inhibited the growth of *Fusarium oxysporum* with the inhibition zone diameter of 21 mm (Prasad & Khalkho).

Antifungal (filamentous) resin: The resin inhibited the growth of *Rhizoctonia solani*, and *Sclerotium rolfsii* (Gaurea et al., 2017).

Strong antiviral (enveloped, monopartite, linear, double-stranded DNA) extract: Extract inhibited the replication of the Herpes simplex type-1 (7 ACV-R strain) and Herpes simplex type-2 (11 ACV-R strain) with the IC_{50} values of 10 and 5 µg/mL, respectively (Thompson et al., 2005).

Commentary: The plant brings to being oligostilbenoids of which shoreaphenol (Patra et al., 1992) which could be examined for its antimicrobial effects. In fact, oligostilbenoids could be responsible for the anti-Herpes property. Distichol and ε-viniferin isolated from the bark of a member of the genus *Shorea* Roxb. ex C.F. Gaertn. inhibited the growth of *Escherichia coli* and *Staphylococcus* (Sultanbawa et al., 1987).

REFERENCES

Chattopadhyay, D., Ojha, D., Mukherjee, H., Bag, P., Vaidya, S.P. and Dutta, S., 2018. Validation of a traditional preparation against multi-drug resistant *Salmonella Typhi* and its protective efficacy in *Salmonella typhi* murium infected mice. *Biomedicine & Pharmacotherapy*, *99*, pp. 286–289.

Gaurea, S.H., Smrity, P. and Bapat, U.C., 2017. Antifungal activity of gum-resin extracts of Boswellia serrata, *Commiphora mukul*, *Gardenia resinifera* and *Shorea robusta* against some plant pathogenic fungi. *Indian Phytopathology*, *70*(2), pp. 224–227.

Patra, A., Dey, A.K., Kundu, A.B., Saraswathy, A. and Purushothaman, K.K., 1992. Shoreaphenol, a polyphenol from *Shorea robusta*. *Phytochemistry*, *31*(7), pp. 2561–2562.

Prasad, B. and Khalkho, A.S., Study of antifungal activities of *Shorea robusta* Gaertn against *Fusarium oxyporum*. *International Journal of Scientific Development and Research*, *5*(4), pp. 387–389.

Sultanbawa, M.U.S., Surendrakumar, S. and Bladon, P., 1987. Distichol an antibacterial polyphenol from *Shorea disticha*. *Phytochemistry*, *26*(3), pp. 799–801.

Thompson, K.D., Ather, A. and Khan, M.T.H., 2005. Antiviral activities of three Bangladeshi medicinal plant extracts against herpes simplex viruses. *Minerva Biotecnologica*, *17*(3), p. 193.

Vashisht, S., Singh, M.P. and Chawla, V., 2016. *In-vitro* antioxidant and antibacterial activity of methanolic extract of *Shorea robusta* Gaertn. F. resin. *International Journal of Pharmaceutical and Phytopharmacological Research*, *6*(4), pp. 68–71.

8.4.3.4 *Vatica lanceifolia* (J. Roxb.) Blume

Common names: Chooa oil tree; xi zang qing mei (China); morsal (India)

Habitat: Forests

Distribution: India, Nepal, Bangladesh, Myanmar, China

Botanical observation: This timber tree grows up to 10 m tall. The bark is pale greyish green and exudes a resin upon incision. The leaves are simple, alternate or spiral, and stipulate. The stipules are ephemeral and minute. The petiole grows up to about 2 cm long. The blade is glaucous below, elliptic to lanceolate, 10–23 × 3–10 cm, acute to rounded at base, somewhat coriaceous, marked with 10–15 pairs of secondary nerves, and acuminate at apex. The racemes are axillary and terminal, lax, and up to about 15 cm long. The 5 sepals are triangular, hairy, and 3 mm long. The 5 petals are whitish, narrowly oblong, and about 2.5 cm long. The androecium is made of 15 stamens which are about 3 mm long. The ovary is ovoid and develops a stout style and a clavate stigma. The nut is ovoid to globose, up to about 3 cm long, and seated on 5 accrescent sepals which reach about 1 cm long.

Medicinal use: Apparently none for the treatment of microbial infections

Commentaries

i. Members of the genus *Vatica* L. are not uncommonly able to build natural products inhibiting the replication of the Human immunodeficiency virus *in vitro:* ethanol extract of stems of *Vatica astrotricha* Hance inhibited the replication of the Human immunodeficiency virus type-1 at the concentration of 10 µg/mL via a mechanism involving, at least in part, inhibition of virus entry in host cell (Park et al., 2009). The lanostane-type triterpene (24*E*)-3-oxo-lanostane-8,24-dien-26-oic acid inhibited the replication of the Human immunodeficiency virus type-1 with the IC_{50} value of 3.9 µg/mL and a selectivity index of 1.7, the dammarane-type triterpenes dammara-20,25-dien-3β, 24-diol and (23*E*)-dammara-20,23-dien-3β, 25-diol inhibited the replication of the Human immunodeficiency virus type-1 with the IC_{50} values of 2.7 and 10 µg/mL and selectivity indexes of 2.3 and 1.3, respectively, and the lupane-type triterpenes betulinic acid and betulin inhibited the replication of the Human immunodeficiency virus type-1 with the IC_{50} values of 16.4 and 8.6 µg/mL and selectivity indexes of 1.1 and 1.4, respectively (Zhang et al., 2003).

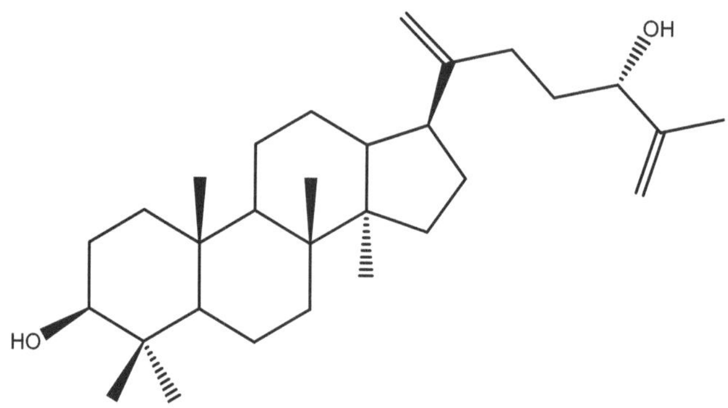

Dammara-20,25-dien-3β,24-diol

ii. Oligostibelnoids in this genus are antibacterial and antimycobacterial such as hopeaphenol A and vaticaphenol A (Zgoda-Pols et al., 2002).
iii. The primary forests of Southeast Asia will be (if nothing is done) within less than 100 years eradicated for palm oil and logging industries (Nellemann, 2007). This will have global hellish climatological, geological, and human impacts and generate the emergence of further pandemic viruses, bacteria, or fungi. It also means the destruction of trees, climbers, herbs, mosses, lichens, microorganisms, amphibians, and molluscs producing molecules of clinical therapeutic usefulness. One must grasp the point that zoonotic pandemics are consequences of deforestation.

REFERENCES

Nellemann, C. ed., 2007. *The Last Stand of the Orangutan: State of Emergency: Illegal Logging, Fire and Palm Oil in Indonesia's National Parks.* UNEP/Earthprint, London.

Park, I.W., Han, C., Song, X., Green, L.A., Wang, T., Liu, Y., Cen, C., Song, X., Yang, B., Chen, G. and He, J.J., 2009. Inhibition of HIV-1 entry by extracts derived from traditional Chinese medicinal herbal plants. *BMC Complementary and Alternative Medicine*, *9*(1), p. 29.

Zgoda-Pols, J.R., Freyer, A.J., Killmer, L.B. and Porter, J.R., 2002. Antimicrobial resveratrol tetramers from the stem bark of *Vatica oblongifolia* ssp. oblongifolia. *Journal of Natural Products*, *65*(11), pp. 1554–1559.
Zhang, H.J., Tan, G.T., Hoang, V.D., Hung, N.V., Cuong, N.M., Soejarto, D.D., Pezzuto, J.M. and Fong, H.H., 2003. Natural anti-HIV agents. Part IV. Anti-HIV constituents from vatica cinerea. *Journal of Natural Products*, *66*(2), pp. 263–268.

8.4.4 FAMILY MALVACEAE A.L. DE JUSSIEU (1789)

The family Malvaceae consists of about 100 genera and 1000 species of herbs, shrubs, or trees. The stems are terete, often hairy, and fibrous. The leaves are simple, alternate, or spiral, and stipulate. The petiole is often slender and long. The blades in this family often present characteristic secondary nerve patterns such as the emergence of secondary nerves from the base and are serrate. The inflorescences are solitary, or cymes, racemes or panicles. The flowers are often showy. An epicalyx is often present. The calyx presents 5 sepals. The corolla comprises 5 petals, often obovate and often pink. The androecium consists of numerous anthers characteristically partially fused in a column. The gynoecium consists of 2–5 fused carpels forming a plurilocular ovary, with 1 to many ovules per locules, and develops numerous pistils. The fruits are loculicidal capsules or schizocarps containing numerous reniform seeds. Plants in this taxon produce a bewildering array of antimicrobial flavonoids and sesquiterpenes.

8.4.4.1 *Abelmoschus manihot* (L.) Medik.

Synonyms: *Hibiscus manihot* L.; *Hibiscus papyriferus* Salisb.

Common names: Aibika, sunset Hibiscus; be nin doi; huang shu ku (China); momoke (Fiji); jangli bhindi, dehradun, hriilomodzii (India); degi (Indonesia); saka (Papua New Guinea); lagikuway (the Philippines); neka (Solomon Islands); po fai (Thailand); kwere (Vuanatu)

Habitat: Fields, riverbanks, wastelands, cultivated

Distribution: Tropical Asia and Pacific

Botanical observation: This herb grows up to 2 m tall. The stems are terete and hairy. The leaves are simple, spiral, and stipulate. The stipules are linear and up to 1.5 cm long. The petiole is up to about 15 cm long. The blade is up to about 30 cm across, edible, hairy, producing 5–9 lobes which are oblong lanceolate and serrate. The flowers are axillary, solitary, and on a 4 cm long peduncle. The calyx is tubular and 5-lobed. The corolla comprises 5 membranaceous petals which are yellow, about 6 cm long, and purple at base. The androecium includes numerous stamens joined in a staminal column which is up to 2.5 cm long. The capsules are ellipsoid, coriaceous, hairy, 5 cm long, dehiscent, and contain numerous reniform seeds.

Medicinal uses: Abscesses, blennorrhea, dysentery (Cambodia, Laos, Vietnam); boils (Cambodia, Laos, Vietnam, China, Indonesia); bronchitis, sores (China); pneumonia, shampoo (India); dysentery (Fiji); abscesses, gonorrhoea (Korea); colds (Papua New Guinea); cough, pharyngitis, fever (Bangladesh).

Broad-spectrum antibacterial halo developed by flavonol glycoside: Hyperoside (50 μg/disc) inhibited the growth of *Bacillus subtilis*, *Haemophilus influenza*, and *Moraxella catarrhalis* with inhibition zone diameter of 8, 8, and 11 mm, respectively (Pretorius et al., 2003).ç

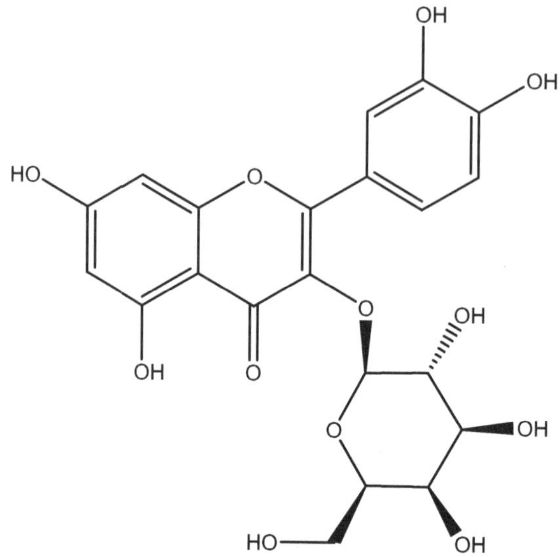

Hyperoside

Moderate antifungal flavonol glycoside: Hyperoside inhibited the growth of *Pestalotia guepinii* and *Drechslera* sp. with MIC values ranging from 100 to 150 µg/mL (Li et al., 2005).

Strong antiviral (enveloped circular double stranded DNA) flavonol glycoside: Hyperoside (also known as hyperin or quercetin-3-*O*-galactoside; Log D = −1.1 at pH 7.4; molecular mass = 464.3 g/mol) from the flowers inhibited the secretion of the Hepatitis B virus e-antigen and surface antigen by HepG2.2.2.15 cells with IC_{50} values of 0.01 and 0.015 g/L (Wu et al., 2007).

In vivo antiviral (enveloped circular double stranded DNA) flavonol glycoside: At 0.1 g/Kg twice a day orally, hyperoside decreased Hepatitis B virus DNA by about 70% after 7 days in ducks (Wu, 2007). This treatment improved hepatic cytoarchitecture with reduction of necrosis and vacuolization (Wu, 2007).

Commentary: One could examine the antiviral activity of hyperoside against coronaviruses.

REFERENCES

Li, S., Zhang, Z., Cain, A., Wang, B., Long, M. and Taylor, J., 2005. Antifungal activity of camptothecin, trifolin, and hyperoside isolated from *Camptotheca acuminata*. *Journal of Agricultural and Food Chemistry*, *53*(1), pp. 32–37.

Pretorius, J.C., Magama, S., Zietsman, P.C. and Jäger, A.K., 2003. Purification and identification of Santibacterial compounds from *Euclea crispa* subsp. crispa (Ebenaceae) leaves. *South African Journal of Botany*, *69*(4), pp. 579–586.

Wu, L.L., Yang, X.B., Huang, Z.M., Liu, H.Z. and Wu, G.X., 2007. *In vivo* and *in vitro* antiviral activity of hyperoside extracted from *Abelmoschus manihot* (L.) medik. *Acta Pharmacologica Sinica*, *28*(3), pp. 404–409.

8.4.4.2 *Abelmoschus moschatus* Medik.

Synonyms: *Abelmoschus abelmoschus* (L.) H. Karst.; *Abelmoschus officinalis* Endl.; *Hibiscus abelmoschus* (L.) H. Karst.; *Hibiscus abelmoschus* L.; *Hibiscus chinensis* Roxb.; *Hibiscus collinsianus* Nutt. ex Torr. & A. Grey; *Hibiscus moschatus* (Medik.) Salisb.

Common names: Musk mallow; koni gaas, mushak-dana, paranga ludi (Bangladesh); gumbo (Haiti); kasturi dana, katuka (India); kasturi (Indonesia); kipas hantu (Malaysia); tawa (Myanmar); kastuli (the Philippines); fai fee (Thailand); bup van (Vietnam)

Habitat: Fields, riverbanks, wastelands, cultivated

Distribution: Tropical Asia and Pacific

Botanical observation: This herb grows to 2 m tall. The stems are terete and hairy. The leaves are simple, spiral, and stipulate. The stipules are linear and 6–12 mm long. The petiole is 2–30 cm long and hairy. The blade is up to about 20 cm long and produces 3–7 lobes which are deltoid and serrate. The flowers are axillary, solitary, and on 2–8 cm long peduncles. The calyx is tubular and 2–3.5 cm long. The 5 petals are membranaceous, yellow, 7–9 cm long, and purple at base. The androecium includes numerous stamens joined in a staminal column which is 1.5-2 cm long. The capsule is bullet-shaped, coriaceous, 5–8 cm long, dehiscent, and contains numerous globose and aromatic seeds.

Medicinal uses: Boils (China; Myanmar); cuts, wounds (Bangladesh); urinary tract infection (India); fever (Malaysia); gonorrhoea (the Philippines); cough (Thailand)

Broad-spectrum antibacterial halo developed by polar extract: Methanol extract of leaves of at the concentration of 1% inhibited the growth of *Staphylococcus aureus*, *Proteus vulgaris*, and *Corynebacterium dyphteriae* (Maheshwari & Kumar, 2009).

Moderate broad-spectrum antibacterial polar extract: Aqueous extract of seeds inhibited the growth of *Bacillus subtilis* (ATCC 5740), *Staphylococcus aureus* (ATCC 25923), *Escherichia coli* (ATCC 25922), *Pseudomonas aeruginosa* (ATCC-27853), *Proteus vulgaris* (ATCC 6380), and *Salmonella enterica paratyphi* (ATCC 9150) with the MIC values of 101.6, 109.6, 406.4, 109.6, 401.7, and 296.7 µg/mL, respectively (Gul et al., 2011).

Weak broad-spectrum antibacterial flavonol: Myricetin inhibited the growth of *Escherichia coli*, *Staphylococcus aureus*, and *Bacillus subtilis* with the MIC values of 400, 200, and 200 µg/mL (Xu et al., 2015).

Strong antiviral (enveloped, segmented, linear, single stranded (−) RNA) hydrophilic flavonol: Myricetin (Log D = 0.2 at pH 7.4; molecular mass = 318.2 g/mol) inhibited the replication of the Influenza virus A/chicken/Rostock/34 (H7N1) in CEF cells with the EC_{50} value of 18.8 µg/mL and the selectivity index of 5.3 (Liu et al., 2005; Pantev et al., 2006).

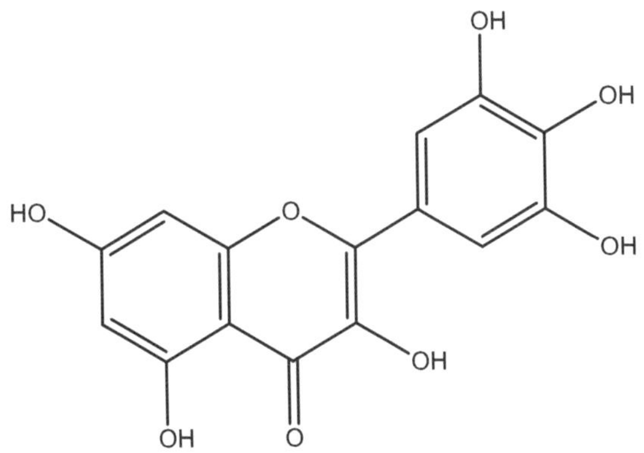

Myricetin

Antifungal (filamentous) essential oil: Essential oil from the seeds was fungicidal for *Aspergillus flavus* at the concentration of 1750 ppm (Dwivedy et al., 2017).

Viral enzyme inhibition by flavonol: Myricetin inhibited the Severe acute respiratory syndrome-associated coronavirus ATPase activity of nsP13helicase by more than 90% at the concentration of 10 µM (Yu et al., 2012).

Commentaries

i. The potentials of myricetin to inhibit the replication of the Severe acute respiratory syndrome-associated coronavirus 2 have been discussed earlier.
ii. *Abelmoschus ficulneus* (L.) Wight & Arn. (local name = *Polud polud*) is used to treat diarrhea in Indonesia. In India, *Abelmoschus esculentus* (L.) Moench. is used to treat bronchitis, cough, and dysentery (local name = *Bhaandi*) and in Pakistan (local name = *Bhindi*) the plant is used to treat cough and diarrhea.

REFERENCES

Dwivedy, A.K., Kumar, M., Kedia, A., Mishra, P.K., Chanotiya, C.S. and Dubey, N.K., 2017. Assessment of chemically characterised *Abelmoschus moschatus* Medik. seed essential oil as shelf life enhancer of herbal raw materials based on antifungal, antiaflatoxigenic and antioxidant efficacy. *Plant Biosystems – An International Journal Dealing with all Aspects of Plant Biology, 151*(1), pp. 139–147.

Gul, M.Z., Bhakshu, L.M., Ahmad, F., Kondapi, A.K., Qureshi, I.A. and Ghazi, I.A., 2011. Evaluation of *Abelmoschus moschatus* extracts for antioxidant, free radical scavenging, antimicrobial and antiproliferative activities using *in vitro* assays. *BMC Complementary and Alternative Medicine, 11*(1), p. 64.

Liu, I.M., Liou, S.S., Lan, T.W., Hsu, F.L. and Cheng, J.T., 2005. Myricetin as the active principle of *Abelmoschus moschatus* to lower plasma glucose in streptozotocin-induced diabetic rats. *Planta Medica, 71*(7), pp. 617–621.

Maheshwari, P. and Kumar, A., 2009. Antimicrobial activity of *Abelmoschus moschatus* leaf extracts. *Current Trends in Biotechnology and Pharmacy, 3*, pp. 260–266.

Pantev, A., Ivancheva, S., Staneva, L. and Serkedjieva, J., 2006. Biologically active constituents of a polyphenol extract from *Geranium sanguineum* L. with anti-influenza activity. *Zeitschrift für Naturforschung C, 61*(7–8), pp. 508–516.

Xu, T., Wang, Z., Lei, T., Lv, C., Wang, J. and Lu, J., 2015. New flavonoid glycosides from *Sedum aizoon* L. *Fitoterapia, 101*, pp. 125–132.

Yu, M.S., Lee, J., Lee, J.M., Kim, Y., Chin, Y.W., Jee, J.G., Keum, Y.S. and Jeong, Y.J., 2012. Identification of myricetin and scutellarein as novel chemical inhibitors of the SARS coronavirus helicase, nsP13. *Bioorganic & Medicinal Chemistry Letters, 22*(12), pp. 4049–4054.

8.4.4.3 *Abutilon indicum* (L.) Sweet

Synonyms: *Abutilon asiaticum* (L.) Sweet; *Abutilon badium* S.A. Husain & Baquar; *Abutilon cavaleriei* H. Lév.; *Abutilon cysticarpum* Hance ex Walp.; *Abutilon populifolium* (Lam.) G. Don; *Abutilon populifolium* (Lam.) Sweet; *Sida asiatica* L.; *Sida indica* L.; *Sida populifolia* Lam.

Common names: Country mallow, Indian Abutilon; jhumka, bon kapas, bon tepari, petari (Bangladesh); dok toc lai (Cambodia); fa nang; mo pan cao (China); kankhani, kansaki, pusi kata daru, thutthi (India); belangan sumpa (Indonesia); huk fao ton (Laos); bunga kisar (Malaysia); bauk hkway (Myanmar); dalupang (the Philippines); khrop fan si (Thailand); dằng xay (Vietnam)

Habitat: Abandonned fields, roadsides

Distribution: Tropical Asia

Botanical observation: This herb, with an aura of dryness, grows to about 1 m tall. The leaves are simple, spiral, and stipulate. The stipules are minute. The petiole is 2–4 cm long and hairy. The blade is ovate-orbicular, 3–9 × 2.5–7 cm, hairy below, cordate at base, serrate, and acute or acuminate at apex. The flowers are solitary, axillary on a 4 cm long and hairy pedicel. The calyx is 5-lobed, about 1 cm in diameter, and hairy. The 5 petals are yellow, about 1 cm long, and somewhat obovate. The androecium includes numerous stamens joined in a staminal column. The ovary includes numerous carpels. The schizocarp is discoid, light brown, somewhat looking like a liliputian cake or button, includes 15–20 mericarps.

Medicinal uses: Abscesses, leucorrhea, cuts, diabetes, toothache, boils, fever, urethiritis (Bangladesh); colds, diarrhea, dog bite, fever, gun shot, mump (China); bronchitis, diarrhea, gonorrhea, mouthwash, tuberculosis, venereal diseases (India); fever (Malaysia); wounds (the Philippines); ulcers (Sri Lanka).

Broad-spectrum antibacterial halo developed by polar extract: Methanol fraction of roots inhibited the growth of *Bacillus licheniformis* (PCSIR-B-252), *Bacillus subtilis* (PCSIR-B-248), *Micrococcus luteus* (NREL-B-287), *Nocardia asteroides* (NRRL-178), and *Escherichia coli* (PCSIR-B-67), with the MIC values of 225, 254, 274, 379, and 239 µg/mL (Yasmin et al., 2008).

Broad-spectrum antifungal halo developed by non-polar extract: Carbon tetrachloride extract of leaves (400 µg/ 6 mm disc) inhibited the growth of *Candida albicans*, *Aspergillus niger*, and *Saccharomyces cerevisiae* with the inhibition zone diameters of 6.1, 8.7, and 7.6 mm, respectively (Abdul et al., 2010).

Weak antifungal (filamentous) polar extract: Ethanol extract of leaves inhibited the growth of *Aspergillus niger*, *Aspergillus flavus*, and *Aspergillus fumigatus* with the MIC value of 2500 µg/mL (Vairavasundaram & Senthil, 2009).

Antifungal (filamentous) sterol: 20,23-Dimethylcholestan-6.22-diene-3β-ol inhibited the growth of *Aspergillus parasiticus* var. globosus and *Asperillus ferreus* var. *aureus* by 100% at the concentration of 5000 ppm (Prabhuji et al., 2010).

Strong broad-spectrum antifungal eudesmane-type amphiphilic sesquiterpene lactone: Isoalantolactone (LogD = 3.7 at pH 7.4; molecular mass = 232.3 g/mol) inhibited the growth of *Aspergillus flavus*, *Aspergillus niger*, *Geotrichum candidum*, *Candida tropicalis*, and *Candida albicans* with the MIC values of 50, 50, 25, 25, and 25 µg/mL, respectively (Picman & Schneider, 1993).

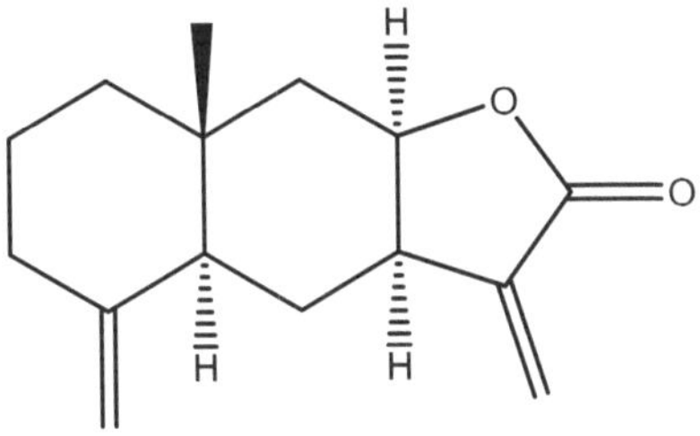

Isoalantolactone

Strong Antiviral (enveloped, monopartite, linear, double-stranded DNA) polar extract: Methanol extract of aerial parts inhibited the growth of the Herpes simplex virus type-1 with the MIC value of about 0.5 µg/mL (Vimalanathan et al., 2009).

Strong Antiviral (enveloped, monopartite, linear, single-stranded (+) RNA) polar extract: Methanol extract of aerial parts inhibited the growth of the Mouse coronavirus with the MIC values of about 2 µg/mL (Vimalanathan et al., 2009).

Commentaries

i. The plant assembles series of flavonols of which luteolin, quercetin, and chrysoeriol and corresponding glycosides (Matlawska & Sikorska, 2002; Hussain et al., 2012) as well as gallic acid (Hussain et al., 2012), the sesquiterpene lactones alantolactone and isoalantolactone (Sharma & Ahmad, 1989). Isoalantolactone is probably antibacterial and virucidal because the α-methylene lactone pharmacophore reacts with secondary amines in proteins and peptides to generate a β-amino lactone (Lawrence et al., 2001).
ii. Luteolin inhibited the replication of the Severe acute respiratory syndrome-associated coronavirus in Vero cells with the EC_{50} values of 10.6 µM (Yi et al., 2004).
iii. In Thailand, *Abutilon hirtum* (Lam) Sweet (local name = *Krob chakkawan*) is used to treat fever and chicken pox.

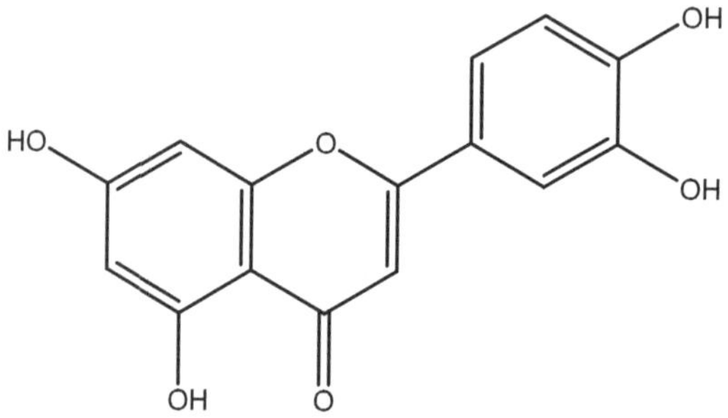

Luteolin

REFERENCES

Abdul, M.M., Sarker, A.A., Saiful, I.M. and Muniruddin, A., 2010. Cytotoxic and antimicrobial activity of the crude extract of *Abutilon indicum*. *International Journal of Pharmacognosy and Phytochemical Research*, *2*(1), pp. 1–4.

Hussain, M.S., Fareed, S., Ali, M. and Rahman, M.A., 2012. Validation of the method for the simultaneous estimation of bioactive marker gallic acid and quercetin in Abutilon indicum by HPTLC. *Asian Pacific Journal of Tropical Disease*, *2*, pp. S76–S83.

Lawrence, N.J., McGown, A.T., Nduka, J., Hadfield, J.A. and Pritchard, R.G., 2001. Cytotoxic michael-type amine adducts of α-methylene lactones alantolactone and isoalantolactone. *Bioorganic & Medicinal Chemistry Letters*, *11*(3), pp. 429–431.

Matlawska, I.R.E.N.A. and Sikorska, M.A.R.I.A., 2002. Flavonoid compounds in the flowers of *Abutilon indicum* (L.) Sweet (Malvaceae). *Acta Poloniae Pharmaceutica*, *59*(3), pp. 227–229.

Picman, A.K. and Schneider, E.F., 1993. Inhibition of fungal growth by selected sesquiterpene lactones. *Biochemical Systematics and Ecology*, *21*(3), pp. 307–314.

Prabhuji, S.K., Singh, D.K., Srivastava, A.K. and Sinha, R., 2010. Antifungal activity of a new steroid isolated from *Abutilon indicum* (L.) Sw. *Medicinal Plants-International Journal of Phytomedicines and Related Industries*, *2*(3), pp. 215–218.

Sharma, P.V. and Ahmad, Z.A., 1989. Two sesquiterpene lactones from *Abutilon indicum*. *Phytochemistry*, *28*(12), p. 3525.

Vairavasundaram, R.P. and Senthil, K., 2009. Antimycotic activity of the components of *Abutilon indicum* (Malvaceae). *Drug Invent Today*, *1*(2), pp. 37–139.

Vimalanathan, S., Ignacimuthu, S. and Hudson, J.B., 2009. Medicinal plants of Tamil Nadu (Southern India) are a rich source of antiviral activities. *Pharmaceutical Biology*, *47*(5), pp. 422–429.

Yasmin, S., Kashmiri, M.A., Ahmad, I., Adnan, A. and Ahmad, M., 2008. Biological activity of extracts in relationship to structure of pure isolates of *Abutilon indicum*. *Pharmaceutical Biology*, *46*(10–11), pp. 673–676.

Yi, L., Li, Z., Yuan, K., et al., 2004. Small molecules blocking the entry of severe acute respiratory syndrome coronavirus into host cells. *Journal of Virology*, *78*(20), pp. 11334–11339.

8.4.4.4 *Abutilon theophrasti* Medik.

Synonyms: *Abutilon abutilon* (L.) Rusby, *Abutilon avicennae* Gaertn.; *Abutilon behrianum* F. Muell.; *Abutilon californicum* Benth.; *Abutilon pubescens* Moench; *Abutilon tiliifolium* (Fisch.) Sweet; *Sida abutilon* L.; *Sida tiliifolia* Fisch.

Common names: Abutilon-hemp, China jute, butterweed; jaya (Bangladesh); qing ma (China); ichibi (Japan); imamkavuğu (Turkey)

Habitat: Abandonned lands, roadsides

Distribution: Turkey, Iran, Afghanistan, Kazakhstan, Kyrgyzstan, Tajikistan, Turkmenistan Uzbekistan, Pakistan, India, Sri Lanka, Bhutan, Bangladesh, China, Korea, Southeast Asia, Taiwan, Japan.

Botanical observation: This herb grows to about 2 m tall, with somewhat an aura of dryness. The barks are used as source of fibers. The leaves are simple, spiral, and stipulate. The stipules are minute. The petiole is up to 12 cm long and hairy. The blade is broadly lanceolate, up to 10 cm across, hairy below, cordate at base, serrate, and acuminate at apex. The flowers are solitary, axillary on a 3 cm long and hairy pedicel. The calyx is 5-lobed, the lobes about 5 mm long, and hairy. The 5 petals are yellow, 1 cm long and obovate. The androecium includes numerous stamens joined in a staminal column. The ovary includes numerous carpels. The schizocarp is cylindrical, light brown, 2 cm across, somewhat looking like a small cake, and includes 15–20 awned mericarps and numerous minute seeds.

Medicinal uses: Dysentery (China); urethritis (Bangladesh)

Antibacterial (Gram-negative) naphthalenol: Hibisculide C inhibited the growth of *Pseudomonas aeruginosa* via DNA fragmentation and increased the sensitivity of multidrug resistant *Pseudomonas aeruginosa* to fluoroquinolones and rifampicin (Lee et al., 2016).

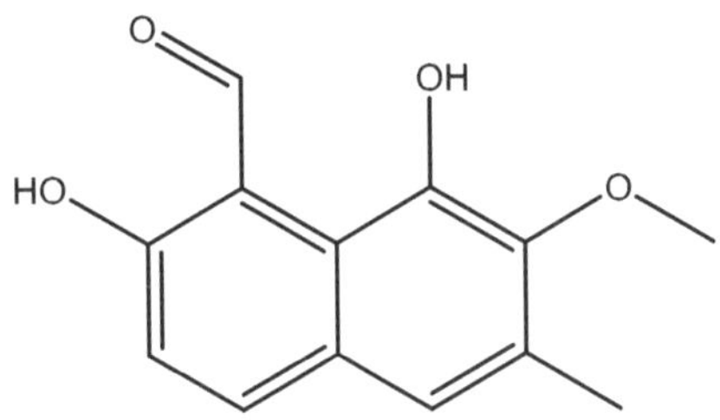

Hibisculide C

Strong broad-spectrum antifungal naphthalenol: Hibisculide C isolated from the aerial parts inhibited the growth of *Candida albicans* (ATCC 90028), *Candida parapsilosis* (ATCC 22019), *Trichosporon beigelii* (KCTC 7707), and *Malassezia furfur* (KCTC 7744) with the MIC values of 5, 5, 10, and 5 µg/mL, respectively (Hwang et al., 2013). This sesquiterpene evoked membrane insults, inhibition on membrane potential, and shrinkage in *Candida albicans* (Hwang et al., 2013).

Viral enzyme inhibition by polar extract: Methanol extract of fruits at the concentration of 100 µg/mL inhibited RNA-dependent DNA polymerase activity of the Human immunodeficiency virus type-1 reverse transcriptase by 35.7% (Min et al., 2001).

Commentary: Once could examine the activities of hibisculide C against coronaviruses.

REFERENCES

Hwang, J.H., Jin, Q., Woo, E.R. and Lee, D.G., 2013. Antifungal property of hibicuslide C and its membrane-active mechanism in *Candida albicans*. *Biochimie*, *95*(10), pp. 1917–1922.

Lee, H., Choi, H., Lee, J.C., Lee, Y.C., Woo, E.R. and Lee, D.G., 2016. Antibacterial activity of Hibicuslide C on multidrug-resistant *Pseudomonas aeruginosa* isolates. *Current Microbiology*, *73*(4), pp. 519–526.

Min, B.S., Kim, Y.H., Tomiyama, M., Nakamura, N., Miyashiro, H., Otake, T. and Hattori, M., 2001. Inhibitory effects of Korean plants on HIV-1 activities. *Phytotherapy Research*, *15*(6), pp. 481–486.

8.4.4.5 *Althaea officinalis* L.

Synonyms: *Althaea kragujevacensis* Pančić ex Diklić & Stevan.; *Althaea micrantha* Wiesb. ex Borbás; *Althaea sublobata* Stokes; *Althaea taurinensis* DC.; *Althaea vulgaris* Bubani; *Malva althaea* E.H.L. Krause; *Malva maritima* Salisb.; *Malva officinalis* (L.) Schimp. & Spenn. ex Schimp. & Spenn.

Common names: Marsh-mallow; khatmi (Afghanistan); shemai-tutti (India); gol-e-khatmi, (Iran); hitmi (Turkey).

Habitat: Fields, riverbanks, roadsides

Distribution: Turkey, Iran, Afghanistan, Pakistan

Botanical observation: This beautiful herb emits a fresh aura of life and grows up to a height of 1 m. The stems are hairy. The leaves are simple, spiral, and stipulate. The stipules are linear and up to 5 mm long. The petiole is up to 5 cm long and hairy. The blade is 3–5 ×2–4 cm, ovate to cordate or 3-lobed, hairy, serrate, and acute to acuminate at apex. The inflorescences are axillary fascicles. The epicalyx presents 8–12 lobes, which are up to 5 mm long. The calyx is tubular, up to 1 cm long, 5-lobed, the lobes lanceolate. The 5 petals are obovate, light pink, up to 2 cm long. The numerous stamens are partially fused into a staminal column. The schizocarp is globular, comprises up to 25 mericarps, and conceals numerous minute kidney shaped seeds.

Medicinal uses: Bronchitis, cough (India); diarrhea, sore throat (Japan); pneumonia (Pakistan, India); bronchitis, cystitis, pharyngitis (Iran); pneumonia, cough, typhoid (Afghanistan).

Broad-spectrum antibacterial halo developed by polar extract: Methanol extract (50 mL of 1 mg/5 mL solution/well) inhibited the growth of *Escherichia coli*, *Klebsiella pneumoniae*, *Pseudomonas aeruginosa*, *Proteus mirabilis*, *Salmonella typhi*, *Salmonella paratyphi* A, *Salmonella paratyphi* B, *Salmonella typhimurium*, *Shigella boydii*, *Shigella flexneri*, *Shigella sonnei*, *Staphylococcus aureus*, and *Streptococcus faecalis* with inhibition zones ranging from 12 to 19.9 mm (Limsuwan et al., 2009).

Broad-spectrum antibacterial halo developed by non-polar extract: Hexane extract of roots of inhibited the growth of *Bacillus subtilis* (ATCC 12228), *Staphylococcus epidermidis*, *Enterococcus faecalis* (ATCC 15753), *Staphylococcus aureus* (ATCC 25923), *Pseudomonas aeruginosa* (ATCC 27852), and *Escherichia coli* (ATCC 25922) with the inhibition zone diameters of 18.9, 17.3, 11.1, 16.8, 14.5, and 16 mm, respectively (Valiei et al., 2011).

Moderate antibacterial (Gram-positive) polar extract: Methanol extract of aerial parts of inhibited the growth of *Bacillus subtilis* and *Micrococcus luteus* with the MIC of 125 and 250 μg/mL, respectively (Ozturk & Ercisli, 2007).

Weak antibacterial (Gram-negative) polar extract: Methanol extract of seeds inhibited the growth of *Haemophilus influenzae* (MTCC 3826) with the MIC value of 3100 μg/mL (Gautam et al., 2015).

Antiviral (enveloped, segmented, linear, single-stranded (−) RNA) polar extract: Aqueous extract (2 g/100 mL) evoked after 10 mins of pretreatment a reduction in Log10 titer of the Influenza H5N1 virus (Chicken/yamaguchi/7/04) by ≥4.6 (Baatartsogt et al., 2016). This extract evoked after 10 mins of treatment a reduction in Log10 titer of Avian/Japan/9UO0036/09, Avian/Japan/11OG1083/11, chicken/Ibaraki/8/05, Whistlingswan/Shimane/499/83 and Duck/Hong Kong/820/80 by ≥4, ≥4, ≥4.7, ≥4, and ≥3.3 in Madin-Darby canine kidney cells (Baatartsogt et al., 2016).

Commentaries

i. The plant builds flavonoids, phenolic acids, and coumarins (Gudej, 1991).
ii. *Althaea rosea* (L.) Cav. is used in China to treat fever. Note that BALB/c mice orally pretreated with an extract of *Althaea rosea* (L.) Cav. (200 μL, 0.1 mg/mL) were protected against influenza A virus (Kim et al., 2018). One could examine the potentials of *Althaea officinalis* L. for the treatment of coronavirus infections. What is (are) the antiviral principles in the activities reported by Baatartsogt et al. (2016)?
iii. Members of the genus *Althaea* L. are used in Iran (local name: *Gul-i-khatmi*) to treat cough.

REFERENCES

Baatartsogt, T., Bui, V.N., Trinh, D.Q., Yamaguchi, E., Gronsang, D., Thampaisarn, R., Ogawa, H. and Imai, K., 2016. High antiviral effects of hibiscus tea extract on the H5 subtypes of low and highly pathogenic avian influenza viruses. *Journal of Veterinary Medical Science*, *78*(9), pp. 1405–1411.

Gautam, S.S., Kumar, S. and Chauhan, R., 2015. Antimicrobial efficacy of *Althaea officinalis* Linn. seed extracts and essential oil against respiratory tract pathogens. *Journal of Applied Pharmaceutical Science*, *5*(9), pp. 115–119.

Gudej, J., 1991. Flavonoids, phenolic acids and coumarins from the roots of *Althaea officinalis*. *Planta Medica*, *57*(3), pp. 284–285.
Kim, M.S., Chathuranga, K., Kim, H., Lee, J.S. and Kim, C.J., 2018. Anti-influenza properties of herbal extract of Althaea rosea in mice. *Korean Journal of Veterinary Research*, *58*(3), p. 153.
Limsuwan, S., Subhadhirasakul, S. and Voravuthikunchai, S.P., 2009. Medicinal plants with significant activity against important pathogenic bacteria. *Pharmaceutical Biology*, *47*(8), pp. 683–689.
Ozturk, S. and Ercisli, S., 2007. Antibacterial activity of aqueous and methanol extracts of *Althaea officinalis* and *Althaea cannabina* from Turkey. *Pharmaceutical Biology*, *45*(3), pp. 235–240.
Valiei, M., Shafaghat, A. and Salimi, F., 2011. Chemical composition and antimicrobial activity of the flower and root hexane extracts of *Althaea officinalis* in Northwest Iran. *Journal of Medicinal Plants Research*, *5*(32), pp. 6972–6976.

8.4.4.6 *Gossypium arboreum* L.

Common names: Bengal cotton, Ceylon cotton, tree cotton; krabas tes (Cambodia); shu mian (China); kaarpaasi (India); kapas merah (Indonesia); kapas (Malaysia, the Philippines; fai daeng (Thailand)

Habitat: Cultivated

Distribution: Tropical and Subtropical Asia

Botanical observation: This handsome shrub grows to about 2 m tall. The stems are hairy. The leaves are simple, spiral, and stipulate. The stipules are linear and caducous. The petiole is 1.5–10 cm long. The blade produces 3–5 lobes which are oblong-lanceolate, hairy below, and about 4–10 cm long. The flowers are solitary, axillary, and on a 1.5–2.5 cm long and hairy pedicel. The calyx is tubular, about 5 mm long, and 5-lobed. The corolla is beautiful and comprises 5 yellowish petals which are 3–5 cm long, purplish at base, and membranaceous. The androecium includes numerous stamens joined in a staminal column which is 1.5-2 cm long. The ovary develops a slender style. The capsules are ovoid, about 3 cm long, dehiscent, and containing numerous seeds in a white wool.

Medicinal uses: Leucorrhoea (Banglasdesh); cystitis, fever, gonorrhea (India)

Broad-spectrum antibacterial (Gram-positive) polar extract: Aqueous extract inhibited the growth of *Staphylococcus aureus* (NCTC 4163), *Bacillus subtilis* (NCTC 10073), *Micrococcus flavus* (NCTC 7743), *Escherichia coli* (NCTC 9002), *Pseudomonas aeruginosa* (NCTC 10662), *Staphylococcus aureus* (SA 1199B: Nor A), and *Staphylococcus aureus* (XU 212) (TetK, resistant to the tetracyclines and overexpresses the Tet K efflux proteins), and *Staphylococcus aureus* (RN 4220 (methicillin-resistant *Staphylococcus aureus*) with the MIC/MBC values of 128/256, 256/256, 128/256, 128/256, 256/512, 128 > 1000, 128 > 1000, and 256/1000 µg/mL, respectively (Annan & Houghton, 2008).

Broad-spectrum antibacterial halo developed by flavonol glycosides: Quercetin 7-*O*-glucoside and gossypetin 8-*O*-rhamnoside (disc impregnated with 20 µL of a 1 mg/20 µL solution) isolated from the flowers inhibited the growth of *Stenotrophomonas maltophilia* and *Enterobacter cloacae* with the inhibition zone diameters of 7 mm (Waage & Hedin, 1984).

Strong antibacterial (Gram-positive) amphiphilic bis-naphthalenol: Gossypol (Log D = 5.1 at pH 7.4; molecular mass = 518.5 g/mol) inhibited the growth of *Bacillus cereus* and *Staphylococcus aureus* (TISTR517) (Boonsri et al., 2008). Gossypol inhibited *Staphylococcus aureus* (NCTC 4163), *Staphylococcus aureus* (ATCC 25923), *Staphylococcus aureus* (ATCC 6538), *Staphylococcus aureus* (ATCC 2921), *Staphylococcus epidermidis* (ATCC 12228), *Bacillus subtilis* (ATCC 6633), *Bacillus cereus* (ATCC 11778), *Enterococcus hirae* (ATCC 10541), *Micrococcus luteus* (ATCC 9341), and *Micrococcus luteus* (ATCC 10240), with the MIC values of 50, 25, 25, 50, 25, 12.5, 12.5, 100, 100, and 25 µg/mL, respectively (Przybylski et al., 2009). Gossypol inhibited the growth of 10 clinical methicillin-sensitive *Staphylococcus aureus* isolates with the MIC values ranging from 12.5 to 50 µg/mL and inhibited the growth of 10 methicillin-resistant *Staphylococcus aureus* clinical isolates with the MIC values of 50 µg/mL (Przybylski et al., 2009). Gossypol inhibited the growth of vancomycin-resistant *Enterococcus faecium* (VRE) with the MIC/MBC value of 4.8/19.3 µM, and *Staphylococcus aureus* and methicillin-resistant *Staphylococcus aureus* with the

IC_{50} values of 9.1 and 8.9 μM, respectively (Masila et al., 2015a). Gossypol inhibited vancomycin-resistant *Enterococcus faecium* (VRE) with the MIC/MBC values of 2.5/10 μg/mL, and methicillin-resistant *Staphylococcus aureus* with that IC_{50} value of 4.6 μg/mL (Masila et al., 2015b).

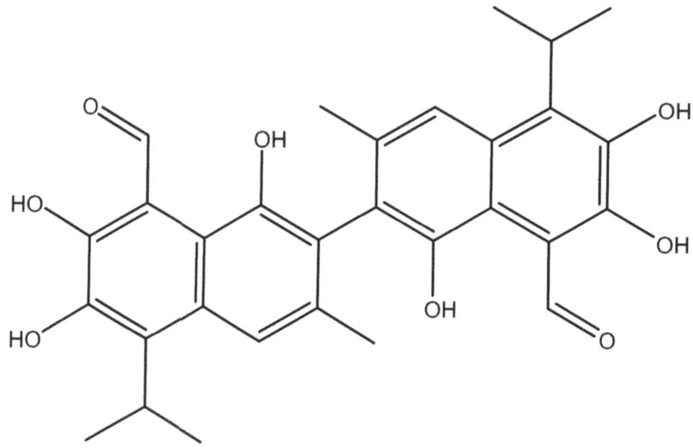

Gossypol

Strong anticandidal bis-naphthalenol: Gossypol inhibited the growth of *Candida glabrata* with the IC_{50} values of 0.7 μM (Masila et al., 2015a).

Strong antiviral (enveloped monopartite linear double stranded DNA) bis-naphthalenol: Gossypol at the concentration of 2 μM inhibited the replication of Herpes simplex virus type-1 and plaque formation by about 99% (Radloff et al., 1986).

Strong antiviral (monopartite linear dimeric single stranded (+) RNA) amphiphilic bis-naphthalenol: Gossypol inhibited the replication of the Human immunodeficiency virus type-1 (LAV-A isolate) and Human immunodeficiency virus type-1 (IIIB strain) in H9 cells with the IC_{50} values of 1 and 0.6 μM, respectively (Royer et al., 1991).

Antiviral (segmented, enveloped, linear, single-stranded (−) RNA) bis-naphthalenol: Gossypol is active against Influenza virus (Dorsett et al., 1975).

Commentaries

i. Gossypol inhibits DNA polymerase (Rosenberg et al., 1986).
ii. *Gossypium barbadense* L. is used to treat fever in Indonesia and essential oil (containing mainly tricyclene, α-pinene, α-terpinene, and bornyl acetate) of leaves inhibited the growth of *Staphylococcus aureus* (NCTC 6571), *Gardnerella* spp., *Escherichia coli* (UCH 2306), and *Escherichia coli* (NCTC 9001). (Essien et al., 2011).
iii. *Gossypium herbaceum* L. (Bengal cotton) is used in Cambodia, Laos, and Vietnam to treat fever and dysentery. Aqueous extract of *Gossypium arboreum* L. inhibited the growth of *Staphylococcus aureus* (NCTC 4163), *Bacillus subtilis* (NCTC 10073), *Micrococcus flavus* (NCTC 7743), *Escherichia coli* (NCTC 9002), *Pseudomonas aeruginosa* (NCTC 10662), *Staphylococcus aureus* (SA1199B: Nor A), *Staphylococcus aureus* (XU212 (TetK, resistant to the tetracyclines and overexpresses the Tet K efflux proteins), and *Staphylococcus aureus* (RN4220 (methicillin-resistant *Staphylococcus aureus*)) with the MIC/MBC values of 128/256, 256/256, 128/256, 128/256, 256/512, 128 > 1000, 128 > 1000, and 256/1000 μg/mL, respectively (Annan & Houghton, 2008). *Gossypium barbadense* L. is used in Indonesia and in India (local name = *Kaarpaasa*) to treat fever, and essential oil from the leaves of this plant (containing mainly tricyclene, α-pinene;

α-terpinene and bornyl acetate) inhibited the growth of *Staphylococcus aureus* (NCTC 6571), *Gardinella* spp., *Escherichia coli* (UCH 2306), and *Escherichia coli* (NCTC 9001) (Essien et al., 2011).

iv. Imai and co-workers (2002) made the interesting observation that estrone and 17β-estradiol increased the cellular accumulation of topotecan in K562/BCRP cells via efflux pump inhibition. Thus, one could wonder if estrogenic or antiestrogenic natural products (including gossypol (Baker, 1984) and isoflavonoids) could be of value in combating antibiotic resistance by inhibition of bacterial efflux pumps. This needs to be examined.

REFERENCES

Annan, K. and Houghton, P.J., 2008. Antibacterial, antioxidant and fibroblast growth stimulation of aqueous extracts of *Ficus asperifolia* Miq. and *Gossypium arboreum* L., wound-healing plants of Ghana. *Journal of Ethnopharmacology*, *119*(1), pp. 141–144.

Baker, M.E., 1984. Gossypol inhibits estrogen binding to rat/-fetoprotein. *FEBS Letters*, *175*(1), pp. 41–44.

Boonsri, S., Karalai, C., Ponglimanont, C., Chantrapromma, S. and Kanjana-opas, A., 2008. Cytotoxic and antibacterial sesquiterpenes from *Thespesia populnea*. *Journal of Natural Products*, *71*(7), pp. 1173–1177.

Dorsett, P.H., Kerstine, E.E. and Powers, L.J., 1975. Antiviral activity of gossypol and apogossypol. *Journal of Pharmaceutical Sciences*, *64*(6), pp. 1073–1075.

Essien, E.E., Aboaba, S.O. and Ogunw, I.A., 2011. Constituents and antimicrobial properties of the leaf essential oil of *Gossypium barbadense* (Linn.). *Journal of Medicinal Plants Research*, *5*(5), pp. 702–705.

Imai, Y., Tsukahara, S., Ishikawa, E., Tsuruo, T. and Sugimoto, Y., 2002. Estrone and 17β-estradiol reverse breast cancer resistance protein-mediated multidrug resistance. *Japanese Journal of Cancer Research*, *93*(3), pp. 231–235.

Masila, V.M., Midiwo, J.O., Zhang, J., Gisacho, B.M., Munayi, R., Omosa, L.K., Wiggers, F.T., Jacob, M.R., Walker, L.A. and Muhammad, I., 2015a. Anti-vancomycin-resistant *Enterococcus faecium* and *Enterococcus faecalis* activities of (–) gossypol and derivatives from *Thespesia garckeana*. *Natural Product Communications*, *10*(4), p. 1934578X1501000419.

Masila, V.M., Midiwo, J.O., Zhang, J., Gisacho, B.M., Munayi, R., Omosa, L.K., Wiggers, F.T., Jacob, M.R., Ibrahim, M.A., Walker, L.A. and Muhammad, I., 2015b. Anti-vancomycin-resistant Enterococcus faecium and *Enterococcus faecalis* activities of (–) gossypol and derivatives from *Thespesia garckeana*. *Planta Medica*, *81*(5), p. PC17.

Przybylski, P., Pyta, K., Stefańska, J., Ratajczak-Sitarz, M., Katrusiak, A., Huczyński, A. and Brzezinski, B., 2009. Synthesis, crystal structures and antibacterial activity studies of aza-derivatives of phytoalexin from cotton plant–gossypol. *European Journal of Medicinal Chemistry*, *44*(11), pp. 4393–4403.

Radloff, R.J., Deck, L.M., Royer, R.E. and Vander Jagt, D.L., 1986. Antiviral activities of gossypol and its derivatives against herpes simplex virus type II. *Pharmacological Research Communications*, *18*(11), pp. 1063–1073.

Rosenberg, L.J., Adlakha, R.C., Desai, D.M. and Rao, P.N., 1986. Inhibition of DNA polymerase α by gossypol. *Biochimica et Biophysica Acta (BBA)-Gene Structure and Expression*, *866*(4), pp. 258–267.

Royer, R.E., Mills, R.G., Deck, L.M., Mertz, G.J. and Vander Jagt, D.L., 1991. Inhibition of human immunodeficiency virus type I replication by derivatives of gossypol. *Pharmacological Research*, *24*(4), pp. 407–412.

Waage, S.K. and Hedin, P.A., 1984. Biologically-active flavonoids from *Gossypium arboreum*. *Phytochemistry*, *23*(11), pp. 2509–2511.

8.4.4.7 *Hibiscus rosa-sinensis* L.

Common names: China rose; joba (Bangladesh); dok mai (Cambodia); zhu jin (China); kembang sepatu, waren gegeh (Indonesia); kokdok may chiey (Laos); bunga raya (Malaysia); wavu wavu (Papua New Guinea); gumamela (the Philippines); ba (Thailand); hồng cận (Vietnam)

Habitat: Cultivated

Distribution: Tropical Asia and Pacific

Botanical observation: This shrub grows to about 3 m tall. It is commonly grown as ornamental. The stems are terete and hairy. The leaves are simple, spiral, and stipulate. The stipule is linear and up to about 1.2 cm long and hairy. The petiole is about 2 cm long and hairy. The blade is ovate, 4–10×2–5 cm, rounded at base, serrate, and acuminate at apex. The flowers are showy and solitary. The epicalyx develops about 6 linear lobes, which are about 1.5 cm long and deep red. The calyx is tubular, about 2 cm long, and develops 5 lobes. The corolla presents 5 reddish and pinkish petals which are obovate, membranaceous, and about 5 cm long. The androecium presents a column which is about 7 cm long. The style develops a 5-lobed stigma. The capsules are ovoid and about 2.5 cm long.

Medicinal uses: Dysentery (Bangladesh); abscesses (China); fever (Malaysia); boils (the Philippines)

Broad-spectrum antibacterial halo developed by polar extract: Methanol extract of flowers (saturated disc with solution at 40 mg/mL) inhibited the growth of *Staphylococcus aureus*, *Streptococcus* sp., *Bacillus subtilis*, and *Escherichia coli* (Ruban & Gajalakshmi, 2012).

Commentaries

i. The plant produces quercetin, kaempferol, scutellarein, and gossypetin glycosides, as well as the triterpene taraxeryl acetate (Ragasa & Rufino, 2011).
ii. In China, *Hibiscus mutabilis* L. (local name: *Mu fu rong*) is used to treat cough, abscesses, wounds, and an extract of leaves inhibited the replication of the Respiratory syncytial virus (Zhang & Zhou, 2013). In China, *Hibiscus syriacus* L. (local name: *Mu jin*) is used for colds. *Hibiscus tiliaceus* L. (Malay name: *Baru*) is used for fever in Malaysia and Indonesia and dysentery in the Philippines (local name: *Malabugo*). The hybrid *Hibiscus×archeri* is used for fever in Indonesia.

REFERENCES

Ragasa, C.Y. and Rufino, L.A.A., 2011. Antimicrobial flavonoid from *Hibiscus rosa-sinensis* Linn. *The Manila Journal of Science*, *7*(1), p. 1.

Ruban, P. and Gajalakshmi, K., 2012. *In vitro* antibacterial activity of *Hibiscus rosa-sinensis* flower extract against human pathogens. *Asian Pacific Journal of Tropical Biomedicine*, *2*(5), pp. 399–403.

Zhang, L. and Zhou, C.Z., 2013. Antiviral actions of the extracts of *Hibiscus mutabilis* L. *Journal of Changchun University of Traditional Chinese Medicine*, *1*, p. 13.

8.4.4.8 *Hibiscus sabdariffa* L.

Synonyms: *Hibiscus cannabinus* L.; *Sabdariffa rubra* Kostel.

Common names: India sorrel, roselle; ranga-amla (Bangladesh); slok chu (Cambodia); mei gui qie (China); mesta (India); rosela (Indonesia); soms fux di (Laos); asam susur (Malaysia); roselle (the Philippines); krachiap-prieo (Thailand); cây dấm (Vietnam)

Habitat: Cultivated

Distribution: Tropical Asia and Pacific

Botanical observation: This shrub grows to about 2 m tall. It is commonly grown as ornamental and medicine in Asia. The stems are terete and purplish. The leaves are simple, spiral, and stipulate. The stipule is linear and about 1 cm long. The petiole is about 8 cm long. The blade is ovate or trilobate, 2–8×0.5–1.5 cm, rounded at base, serrate, and acuminate at apex. The flowers are showy, axillary, and solitary. The epicalyx develops about 12 lobes, which are about 1.5 cm long and deep red. The calyx is tubular, edible, purplish, and develops 5 lobes which are triangular and up to about 2 cm long. The 5 petals are yellow, red at base, obovate, membranaceous, and about 3.5 cm long. The androecium presents a column. The capsules are ovoid and about 1.5 cm across.

Medicinal use: Jaundice (Bangladesh)

Broad-spectrum antibacterial halo developed by polar extract: Ethanol extract of flowers inhibited the growth of *Bacillus subtilis* (ATCC 6633), *Staphylococcus aureus* (ATCC 6538), and *Escherichia coli* (ATCC 8739) (Jung et al., 2013).

Moderate broad-spectrum antibacterial polar extract: Aqueous extract of calyx inhibited the growth of *Staphylococcus aureus*, *Streptococcus faecalis*, *Escherichia coli* and *Klebsiella pneumoniae* and with the MIC values of 500, 500, 1000, and 1000 μg/mL, respectively (Navarro García et al., 2006).

Very weak broad-spectrum antibacterial hydroxycinnamic acid: Ferulic acid inhibited *Staphylococcus aureus*, *Bacillus cereus*, *Staphylococcus epidermidis*, *Escherichia coli*, *Salmonella typhimurium*, *Shigella flexneri*, and *Pseudomonas aeruginosa* (Fu et al., 2016).

Antiviral (non-enveloped monopartite linear single-stranded (+) RNA) polar extract: Aqueous extract of calyces at 40 mg/mL caused complete reduction of Feline calicivirus (FCV-F9) after 15 minutes (Joshi et al., 2015). Murine norovirus (MNV-1) was reduced by 1.7 PFU/mL after 6 hours by the extract at 40 mg/mL (Joshi et al., 2015). Hepatitis A virus (strain HM175) was reduced to undectable levels after 24 hours by the extract at 40 mg/mL (Joshi et al., 2015).

Strong antiviral enveloped linear double stranded DNA simple hydrophilic phenol: Protocatechuic acid (Log D = –1.8 at pH 7.4; molecular mass = 154.1 g/mol) isolated from the calices inhibited the replication of the Herpes simplex virus type-2 replication in Vero cells with the EC_{50} of 0.9 μg/mL (selectivity index > 200) (Hassan et al., 2017).

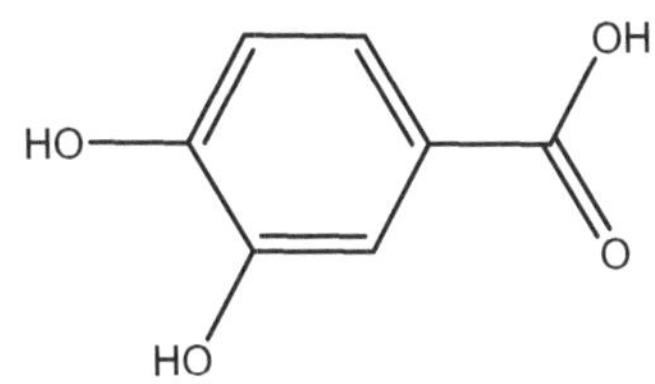

Protocatechuic acid

Commentaries

i. The plant produces the anthocyanins delphinidin-3-*O*-sambubioside and cyaniding-3-*O*-sambubioside which inhibited angiotensin-converting enzyme 2 (Ojeda et al., 2010) which is a receptor for the Severe acute respiratory syndrome-associated coronavirus spike protein (Li et al., 2003). It also produces ferulic acid the antiviral property of which has been described earlier.
ii. Hydro-alcoholic extract of *Hibiscus micranthus* L.f. inhibited the growth of *Staphylococcus aureus* (ATCC 2923), *Streptococcus pneumonia* (ATCC 137348), *Pseudomonas aeruginosa* (ATCC 27853), and *Klebsiella pneumoniae* (ATCC 70060) with the MIC values of 2.5, 0.6, 1.2, and 5 mg/mL, respectively. MBC values for this extract against *Staphylococcus aureus* (ATCC 2923), *Streptococcus pneumonia* (ATCC 137348), *Pseudomonas aeruginosa* (ATCC 27853), and *Klebsiella pneumoniae* (ATCC 70060) were 5, 1.2, 5, and 5 mg/mL, respectively (Begashaw et al., 2017).

REFERENCES

Begashaw, B., Mishra, B., Tsegaw, A. and Shewamene, Z., 2017. Methanol leaves extract *Hibiscus micranthus* Linn exhibited antibacterial and wound healing activities. *BMC Complementary and Alternative Medicine*, *17*(1), p. 337.

Fu, L., Lu, W. and Zhou, X., 2016. Phenolic compounds and *in vitro* antibacterial and antioxidant activities of three tropic fruits: persimmon, guava, and sweetsop. *BioMed Research International*, *2016*. doi:10.1155/2016/4287461.

Hassan, S.T., Švajdlenka, E. and Berchová-Bímová, K., 2017. *Hibiscus sabdariffa* L. and its bioactive constituents exhibit antiviral activity against HSV-2 and anti-enzymatic properties against urease by an ESI-MS based assay. *Molecules*, *22*(5), p. 722.

Jung, E., Kim, Y. and Joo, N., 2013. Physicochemical properties and antimicrobial activity of Roselle (*Hibiscus sabdariffa* L.). *Journal of the Science of Food and Agriculture*, *93*(15), pp. 3769–3776.

Li, W., Moore, M.J., Vasilieva, N., Sui, J., Wong, S.K., Berne, M.A., Somasundaran, M., Sullivan, J.L., Luzuriaga, K., Greenough, T.C. and Choe, H., 2003. Angiotensin-converting enzyme 2 is a functional receptor for the SARS coronavirus. *Nature*, *426*(6965), pp. 450–454.

Navarro García, V.M., Rojas, G., Gerardo Zepeda, L., Aviles, M., Fuentes, M., Herrera, A. and Jiménez, E., 2006. Antifungal and antibacterial activity of four selected Mexican medicinal plants. *Pharmaceutical Biology*, *44*(4), pp. 297–300.

Ojeda, D., Jiménez-Ferrer, E., Zamilpa, A., Herrera-Arellano, A., Tortoriello, J. and Alvarez, L., 2010. Inhibition of angiotensin convertin enzyme (ACE) activity by the anthocyanins delphinidin-and cyanidin-3-*O*-sambubiosides from Hibiscus sabdariffa. *Journal of Ethnopharmacology*, *127*(1), pp. 7–10.

8.4.4.9 *Malva verticillata* L.

Synonyms: *Malva crispa* (L.) L.; *Malva meluca* Graebn. ex P. Medw.

Common name: Whorled mallow; ye kui (China); lofa (Bangladesh; India)

Habitat: Fields, wastelands, roadsides, cultivated

Distribution: Pakistan, India, Bhutan, Myanmar, Korea, China

Botanical observation: This herb grows up to a height of 2.5 m. The stems are hairy. The leaves are simple, spiral, and stipulate. The stipules are lanceolate and 5 mm long. The petiole is up to about 20 cm long and hairy. The blade is 3–20×4–25 cm, edible, cordate at base, dull green, and 5-lobed. The inflorescences are axillary fascicles. The epicalyx presents 3 linear lobes which are up to 6 mm long. The calyx includes 5 sepals which are about 5 mm long and hairy. The 5 petals are up to about 1 cm long, purplish, obovate, bifid, and membranaceous. The androecium comprises about 15 stamens fused in a 5 mm long column. The schizocarp is about 7 mm across and comprises about 12 mericaps and dark brown seeds.

Medicinal uses: Cough, wounds (China)

Commentaries

i. The antimicrobial properties of this plant are apparently unknown. The plant produces benzyl-α-D-galactopyranoside, the lignan (−)-secoisolariciresinol-9′-*O*-β-D-glucopyranoside, and ferulic acid (Ko et al., 2018).

ii. *Malva sylvestris* L. is used for bronchitis and sore throat in China. Methanol fraction of leaves of *Malva sylvestris* L. inhibited *Escherichia coli* (PTCC 1047), *Staphylococcus aureus* (PTCC 1112), *Enterococcus faecalis* (PTCC 1190), and *Streptococcus agalactiae* (PTCC 1321) (Razavi et al., 2011). In Iran *Malva parviflora* L. is used to heal wounds. Methanol roots of leaves of *Malva parviflora* L. (25 mg/mL, wells of 11 mm diameter, 100 μL/well) inhibited the growth of *Staphylococcus aureus* (ATCC 6538) with the inhibition zone of 15 mm (Tadeg et al., 2005).

REFERENCES

Ko, J.H., Nuankaew, W., Joo, S.W., Kim, H.G., Thi, N.N., Lee, Y.G., Kang, T.H. and Baek, N.I., 2018. Phenolic compounds from the aerial parts of *Malva verticillata* and their anti-diabetic effect. *Natural Product Communications*, *13*(7), p. 1934578X1801300719.

Razavi, S.M., Zarrini, G., Molavi, G. and Ghasemi, G., 2011. Bioactivity of *Malva sylvestris* L., a medicinal plant from Iran. *Iranian Journal of Basic Medical Sciences*, *14*(6), p. 574.

Tadeg, H., Mohammed, E., Asres, K. and Gebre-Mariam, T., 2005. Antimicrobial activities of some selected traditional Ethiopian medicinal plants used in the treatment of skin disorders. *Journal of Ethnopharmacology*, *100*(1–2), pp. 168–175.

8.4.4.10 *Malvastrum coromandelianum* (L.) Garcke

Synonyms: *Malva coromandeliana* L.; *Malva domingensis* Spreng. ex DC.; *Malva havanensis* Sessé & Moc.; *Malva lindheimeriana* Scheele; *Malva subhastata* Cav.; *Malva tricuspidata* R.

Br.; *Malvastrum carpinifolium* (Medik.) A. Grey; *Malvastrum lindheimerianum* (Scheele) Walp.; *Malvastrum ruderale* Hance ex Walp.; *Malvastrum tricuspidatum* (R. Br.) A. Grey

Common names: False Mallow, prickly malvastrum, three lobe false mallow; sai kui (China); baddi barear, sonchal buti (India); taw pilaw (Myanmar); kinay-lumpay (the Philippines)

Habitat: Roadsides, wastelands

Distribution: Tropical Asia and Pacific

Botanical observation: This herb grows to about 1 m tall. The stems are hairy and somewhat stiff. The leaves are simple, spiral, and stipulate. The stipules are lanceolate and up to 7 mm long. The petiole is up to 3 cm long, somewhat reddish, and hairy. The blade is broadly lanceolate, 3–7 × 1–4 cm, round to cuneate at base, serrate, dark green and glossy, and acute to obtuse apex. The flowers are axillary and solitary. The calyx is 5-lobed, the lobes ovate, hairy, and about 1 cm long. The 5 petals are yellowish, veined, about 1cm long, obovate, somewhat asymmetrical, and irregular. The androecium includes numerous stamens joined partially in a tube which is about 6 mm long. The schizocarp is discoid, about 6 mm across, seated in persistent calyx, and contain numerous tiny seeds.

Medicinal uses: Dysentery (India; Pakistan); carbuncles (the Philippines)

Weak antibacterial (Gram-positive) polar extract: Aqueous extract of aerial parts inhibited the growth of methicillin-resistant *Staphylococcus aureus* (clinical strain) and *Staphylococcus aureus* (ATCC 25923) with the MIC value of 2.5 and 5 mg/mL (Sittiwet et al., 2008).

Strong antibacterial (Gram-negative) essential oil: Essential oil of aerial parts inhibited the growth of *Staphylococcus aureus*, *Bacillus subtilis*, and *Escherichia coli* with the MIC values of 156, 156 and 78 µg/mL, respectively (Liu et al., 2019).

Strong antimycobacterial flavone glycosides: Apigenin-7-*O*-β-6″(*p*-coumaroyl)-glucopyranoside from the leaves inhibited the growth of *Mycobacterium fortuitum* (ATCC 6841), *Mycobacterium smegmatis* (ATCC 1441), and *Mycobacterium aurum* (NCTC 10437) with the MIC values of 130, 250, and 250 µg/mL, respectively (Aderogba et al., 2019). Apigenin-8-C-glucopyranoside (vitexin) from the leaves inhibited the growth of *Mycobacterium fortuitum* (ATCC 6841), *Mycobacterium smegmatis* (ATCC 1441), and *Mycobacterium aurum* (NCTC 10437) µg/mL, with the MIC values of 30, 60, and 130 µg/mL, respectively (Aderogba et al., 2019).

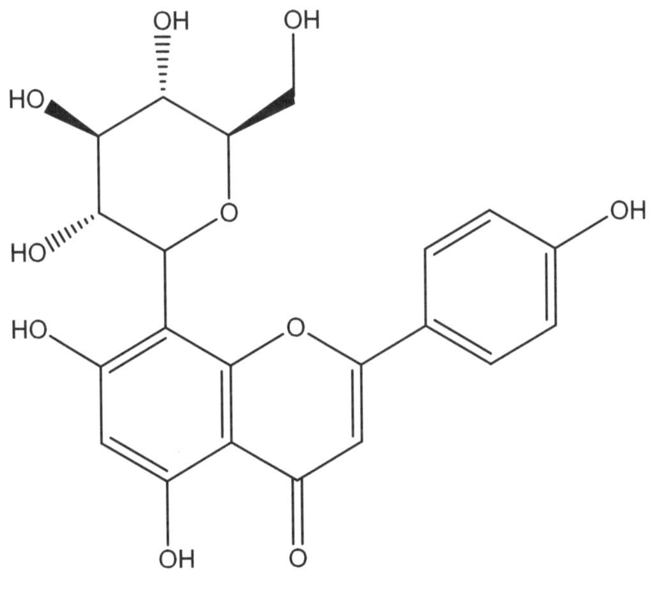

Vitexin

Weak antiviral (non-enveloped segmented linear double-stranded RNA) flavone C-glycoside: Vitexin inhibited the replication of the Rhesus rotavirus with the IC_{50} value of 129 μM (Knipping et al., 2012).

Commentaries

i. The plant synthetizes series of flavonol glycosides such as tiliroside and quercetrin (Devi & Kumar, 2018). Vitexin inhibits the growth of a broad-spectrum of microbes *in vitro.*

ii. Coronavirus encode for 3a protein which forms ion channels that become incorporated into the membrane of the host cells (Lu et al., 2006). Tiliroside at the concentration of 20 μM inhibited coronavirus ion channel 3a-mediated current by almost 100% (Schwarz et al., 2014). In this context, one can mention that Ivermectin binds to chlorine channel controlled by glutamate, typical in helminths and other parasitic worms, whence its clinical use in Human and veterinary medicine (Pandey et al., 2020). Ivermectin is used notably in India for the management of COVID-19 (Naquira et al., 1989; Behera et al. 2020). Ivermectin is a derivative of the polyketide avermectin isolated from soil *Streptomyces avermitilis* by Satoshi Ōmura and colleagues at the Antibiotics Research Group at Tokyo's Kitasato Institute forming a collaboration with the US-based Merck, Sharp and Dohme (MSD) pharmaceutical company Crump & Omura, 2011). Ivermectin needs to be given serious consideration globally as a drug to be used for COVID-19. Another molecule of interest is the synthetic antidepressant fluvoxamine which mitigates cytokine outburst during COVID-19 by interfering with σ-1 receptor (Lenze et al., 2020). Both molecules could be added to a list of first line drugs to be used in the advent of vaccine-resistant COVID-19 mutants and forthcoming coronavirus pandemics.

REFERENCES

Aderogba, M.A., Madikizela, B. and McGaw, L.J., 2019. Bioactive constituents from *Malvastrum coromandelianum* (L.) Garcke leaf extracts. *South African Journal of Botany, 126*, pp. 371–376.

Behera, P., Patro, B.K., Singh, A.K., Chandanshive, P.D., Ravikumar, S.R., Pradhan, S.K., Pentapati, S.S.K., Batmanabane, G., Padhy, B.M., Bal, S. and Singh, S.R., 2020. Role of ivermectin in the prevention of COVID-19 infection among healthcare workers in India: a matched case-control study. *MedRxiv.*

Crump, A. and Omura, S., 2011. Ivermectin, 'wonder drug' from Japan: the human use perspective. *Proceedings of the Japan Academy, Series B, 87*(2), pp. 13–28.

Devi, S. and Kumar, V., 2018. Comprehensive structural analysis of cis- and trans-tiliroside and quercetrin from *Malvastrum coromandelianum* and their antioxidant activities. *Arabian Journal of Chemistry, 13*, pp. 1720–1730.

Knipping, K., Garssen, J. and van't Land, B., 2012. An evaluation of the inhibitory effects against rotavirus infection of edible plant extracts. *Virology Journal, 9*(1), p. 137.

Lenze, E.J., Mattar, C., Zorumski, C.F., Stevens, A., Schweiger, J., Nicol, G.E., Miller, J.P., Yang, L., Yingling, M., Avidan, M.S. and Reiersen, A.M., 2020. Fluvoxamine vs placebo and clinical deterioration in outpatients with symptomatic COVID-19: a randomized clinical trial. *Jama, 324*(22), pp.2292-2300

Liu, Y., Su, X., Xiang, Y., Wang, X. and Lai, P., 2019. Chemical composition, antibacterial, cytotoxic and antioxidant activities of the essential oil of *Malvastrum coromandelianum* aerial parts. *Journal of Essential Oil Bearing Plants, 22*(4), pp. 1040–1047.

Lu, W., Zheng, B.J., Xu, K., Schwarz, W., Du, L., Wong, C.K., Chen, J., Duan, S., Deubel, V. and Sun, B., 2006. Severe acute respiratory syndrome-associated coronavirus 3a protein forms an ion channel and modulates virus release. *Proceedings of the National Academy of Sciences, 103*(33), pp. 12540–12545.

Naquira, C., Jimenez, G., Guerra, J.G., Bernal, R., Nalin, D.R., Neu, D. and Aziz, M., 1989. Ivermectin for human strongyloidiasis and other intestinal helminths. *The American journal of tropical medicine and hygiene, 40*(3), pp. 304–309.

Pandey, S., Pathak, S.K., Pandey, A., Salunke, A.A., Chawla, J., Sharma, A., Sharma, S., Thivari, P. and Ratna, H.V., 2020. Ivermectin in COVID-19: What do we know?. *Diabetes & Metabolic Syndrome, 14*(6), p. 1921.

Schwarz, S., Sauter, D., Wang, K., Zhang, R., Sun, B., Karioti, A., Bilia, A.R., Efferth, T. and Schwarz, W., 2014. Kaempferol derivatives as antiviral drugs against the 3a channel protein of coronavirus. *Planta Medica*, *80*(2/3), pp. 177–182.
Sittiwet, C., Jesadanont, S., Pongpech, P., Naenna, P. and Pongsamart, S., 2008. Antibacterial activity of *Malvastrum coromandelianum* Garcke against methicillin-sensitive and methicillin-resistant strains of *Staphylococcus aureus*. *Current Research in Bacteriology*, *1*(1), pp. 42–45.

8.4.4.11 *Pavonia zeylanica* (L.) Cav.

Synonyms: *Hibiscus zeylanicus* L.

Common names: Ceylon leadwort, Ceylon swamp mallow; bala, joli, sithamutti (India)

Habitat: Roadsides, fields, wastelands

Distribution: Pakistan, India, Sri Lanka, Bangladesh

Botanical observation: This graceful herb grows to about 1 m tall. The stems are hairy. The leaves are simple, spiral, and stipulate. The petiole is up to 5 cm long. The blade is lanceolate, 3-lobed 1.5–4.5 cm × 1–2.5 cm, hairy, cordate at base, and acute at apex. The flowers are solitary and axillary. An epicalyx is present with up to 11 lobes. The calyx includes 5 sepals which are lanceolate and up to about 8 mm long. The 5 petals are obovate, beautifully light pinkish, and about 1.5cm long. The androecium includes numerous stamens partially fused in a column. The schizocarps are about 5 mm across and contain numerous tiny and hairy seeds.

Medicinal use: Fever (India)

Broad-spectrum antibacterial halo developed by polar extract: Methanol extract of leaves (9 mm diameter disc impregnated with 10 mg/mL solution) inhibited the growth of *Bacillus cereus*, *Escherichia coli*, and *Klebsiella aerogenes* with the inhibition zone diameters ranging from 10 to 15 mm (Samy & Ignacimuthu, 2000).

REFERENCE

Samy, R.P. and Ignacimuthu, S., 2000. Antibacterial activity of some folklore medicinal plants used by tribals in Western Ghats of India. *Journal of Ethnopharmacology*, *69*(1), pp. 63–71.

8.4.4.12 *Sida acuta* Burm. f.

Synonyms: *Malvastrum carpinifolium* (Medik.) A. Grey; *Malvinda carpinifolia* Medik.; *Sida balbisiana* DC.; *Sida berlandieri* Turcz.; *Sida bodinieri* Gand.; *Sida carpinifolia* Bourg. ex Griseb.; *Sida carpinifolia* L. f.; *Sida chanetii* Gand.; *Sida disticha* Sessé & Moc.; *Sida frutescens* Cav.; *Sida garckeana* Pol.; *Sida lancea* Gand.; *Sida lanceolata* Retz.; *Sida planicaulis* Cav.; *Sida scoparia* Lour.; *Sida spiraeifolia* Link; *Sida stauntoniana* DC.; *Sida stipulata* Cav.; *Sida trivialis* Macfad.; *Sida ulmifolia* Willd.; *Sida vogelii* Hook. f.

Common names: Common wire weed, hornbeam-leaved Sida, broom weed, kureta sipsedip (Bangladesh); huang hua ren (China); chipchirip, pata (India)

Habitat: Roadsides, wastelands

Distribution: India, Nepal, Bhutan, Bangladesh, Cambodia, Laos, Vietnam, Thailand, China

Botanical observation: This erect herb grows about 1 m tall. The leaves are simple, spiral, and stipulate. The stipules are linear, 4–6 mm long, and persistent. The petiole is 4–6 mm long. The blade is oblong, lanceolate, 2–5 × 0.4–1 cm, obtuse at base, serrate, and acute at apex. The flowers are axillary on a pedicel which is 4–12 mm long and hairy. The calyx is tubular, 6 mm long, and 5-lobed. The corolla consists of 5 petals which are yellow, 6–7 mm long, and membranaceous. The androecium includes numerous stamens joined in a column which is 4 mm long. The schizocarps are globose, about 8 mm in diameter, and contain minute seeds.

Medicinal uses: Dysentery (Bangladesh, India); syphilis (Palau); fever, gonorrhea (the Philippines); influenza (Taiwan)

Antibacterial (Gram-positive) halo developed by polar extract: Ethanol extract of aerial parts inhibited the growth of *Staphylococcus aureus*, *Staphylococcus aureus* (NCTC 10788), *Streptococcus faecalis*, and *Bacillus subtilis* with inhibition zone diameter of 21.3, 16, 14, and 16 mm (25 mg/mL solution, 200 µL per well; 8 mm diameter wells) (Obah et al., 2007).

Strong broad-spectrum antibacterial alkaloidal fraction: Alkaloidal fraction of aerial parts inhibited the growth of *Staphylococcus aureus* (ATCC 25923), *Staphylococcus carmonum* (LMG 13567), *Bacillus cereus* (LMG 13569), *Listeria innocua* (LMG 13568), *Enterococcus faecalis*, *Shigella boydii*, *Shigella flexneri*, and *Escherichia coli* with the MIC/MBC values of 400/>400, 80/240, 80/>400, 16/240, 16/>400, 80/240, 80/186, and 80/240 µg/mL, respectively (Karou et al., 2006).

Strong broad-spectrum antibacterial amphiphilic indoloquinoline alkaloid: Cryptolepine (Log D = 1.4 at pH = 7.4; molecular mass = 232.2 g/mol) inhibited the growth of *Escherichia coli* (ATCC 25922), *Shigella dysenteriae* (ATCC 13313), *Salmonella typhimurium* (ATCC 43971), *Staphylococcus aureus* (ATCC 25923), *Streptococcus faecalis* (ATCC 10541), and *Vibrio cholera* (ATCC 11623) with the MIC values of 12.5, 6.2, 50, 12.5, 12.5, and <1.5 µg/mL, respectively (Paulo et al., 1994). Cryptolepine inhibited the growth of *Staphylococcus aureus* (NCTC 10788) with the MIC value of 5 µg/mL via a mechanism involving bacterial lysis (Sawer et al., 2005).

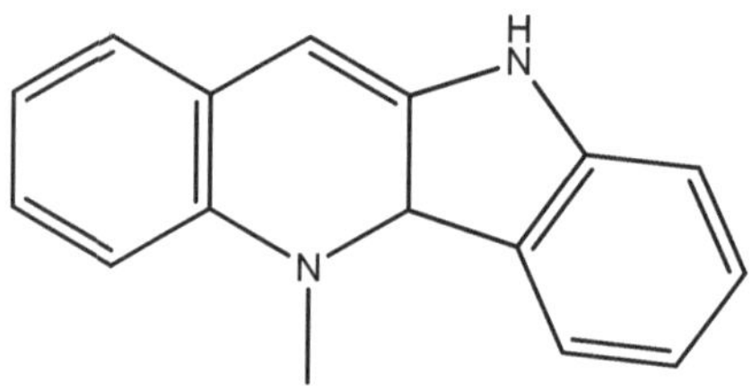

Cryptolepine

Strong broad-spectrum antifungal indoloquinoline alkaloid: Cryptolepine inhibited the *Saccharomyces cerevisiae* (NCPF 3139), *Saccharomyces cerevisiae* (NCPF 3178), and *Candida albicans* (NCPF 3242) with the MIC values of 10, 20, and 160 µg/mL, respectively (Sawer et al., 1995).

Commentaries

i. The plant produces series of indoloquinolines alkaloids of which cryptolepine and cryptolepinone as well as evofolin-A and B, the coumarin scopoletin, the megastigmanes vomifoliol, loliolid, and the lignan 4-ketopinoresinol (Jang et al., 2003; Karou et al., 2007).

ii. Compounds from medicinal plants, including cryptolepin which cause bacterial lysis, act by a number of mechanisms including protein synthesis inhibition or/and interference with cell wall synthesis (Greenwood & O'Grady, 1972). Cryptolepin is planar and is able to intercalate in DNA (Guittat et al., 2003) and could be able to inhibit bacterial topoisomerase, and may very well inhibit the replication of viruses.

iii. Asian traditional systems of medicine often use members of the genus *Sida* L. to treat infectious diseases. *Sida cordata* (Burm. f.) Borss. Waalk. is used in India (local name: *Japkhasakam*) to treat abscesses. *Sida cordifolia* L. is used to treat blenorrhoea and genital herpes in Cambodia, Laos, and Vietnam. *Sida rhombifolia* L. is used to treat congenital venereal diseases in Taiwan and fever and chicken pox and ulcers in Malaysia, and fever in the Philippines. Ethylacetate fraction of leaves of *Sida rhombifolia* L. inhibited the growth of *Bacillus subtilis, Bacillus megaterium, Staphylococcus aureus, Sarcinia lutea, Escherichia coli, Shigella shiga, Shigella dysenteriae, Shigella sonei, Shigella boydii, Pseudomonas aeruginosa*, and *Klebsiella* sp. with zones of inhibition of 17, 13, 15, 24, 15, 17, 24, 12, 14, 24, and 14 mm, respectively (200 µg/disc) (Ekramul Islam et al., 2003). In the Philippines *Sida javensis* Cav. is used to treat gonorrhea. Siddha healers in South India use the plant to treat COVID-19 (Sasikumar et al., 2020).

iv. Aqueous extract of seeds of *Sida mysorensis* Wright and Arn. at the concentration of 250 µg/mL inhibited the Human immunodeficiency virus type-1 protease by 34.9% (Xu et al., 1996).

REFERENCES

Ekramul Islam, M., Ekramul Haque, M. and Mosaddik, M.A., 2003. Cytotoxicity and antibacterial activity of *Sida rhombifolia* (Malvaceae) grown in Bangladesh. *Phytotherapy Research*, *17*(8), pp. 973–975.

Greenwood, D. and O'Grady, F., 1972. Scanning electron microscopy of *Staphylococcus aureus* exposed to some common antistaphylococcal agents. *Journal of General Microbiology*, *70*, 263–270.

Guittat, L., Alberti, P., Rosu, F., Van Miert, S., Thetiot, E., Pieters, L., Gabelica, V., De Pauw, E., Ottaviani, A., Riou, J.F. and Mergny, J.L., 2003. Interactions of cryptolepine and neocryptolepine with unusual DNA structures. *Biochimie*, *85*(5), pp. 535–547.

Jang, D.S., Park, E.J., Kang, Y.H., et al., 2003. Compounds obtained from *Sida acuta* with the potential to induce quinone reductase and to inhibit 7,12-dimethylbenz[a]anthracene-induced preneoplastic lesions in a mouse mammary organ culture model. *Archives of Pharmacal Research*, *26*, pp. 585–590.

Karou, S.D., Nadembega, W.M., Ilboudo, D.P., Ouermi, D., Gbeassor, M., De Souza, C. and Simpore, J., 2007. *Sida acuta* Burm. f.: a medicinal plant with numerous potencies. *African Journal of Biotechnology*, *6*(25), pp. 2953–2959.

Karou, D., Savadogo, A., Canini, A., Yameogo, S., Montesano, C., Simpore, J., Colizzi, V. and Traore, A.S., 2006. Antibacterial activity of alkaloids from *Sida acuta*. *African Journal of Biotechnology*, *5*(2), pp. 195–200.

Obah, I.E., Akerele, J.O. and Obasuyi, O., 2007. Antimicrobial activity of the ethanol extract of the aerial parts of *Sida acuta* burm. f. (malvaceae). *Tropical Journal of Pharmaceutical Research*, *6*(4), pp. 809–813.

Paulo, A., Duarte, A. and Gomes, E.T., 1994. In vitro antibacterial screening of *Cryptolepis sanguinolenta* alkaloids. *Journal of Ethnopharmacology*, *44*(2), pp. 127–130.

Sasikumar, R., Priya, S.D. and Jeganathan, C., 2020. A case study on domestics spread of SARS-CoV-2 pandemic in India. *International Journal of Advanced Science and Technology*, *29*(7), pp. 2570–2574.

Sawer, I.K., Berry, M.I., Brown, M.W. and Ford, J.L., 1995. The effect of cryptolepine on the morphology and survival of *Escherichia coli, Candida albicans* and *Saccharomyces cerevisiae*. *Journal of Applied Bacteriology*, *79*(3), pp. 314–321.

Sawer, I.K., Berry, M.I. and Ford, J.L., 2005. The killing effect of cryptolepine on *Staphylococcus aureus*. *Letters in Applied Microbiology*, *40*(1), pp. 24–29.

Xu, H.X., Wan, M., Loh, B.N., Kon, O.L., Chow, P.W. and Sim, K.Y., 1996. Screening of traditional medicines for their inhibitory activity against HIV-1 Protease. *Phytotherapy Research*, *10*(3), pp. 207–210.

8.4.4.13 *Thespesia populnea* (L.) Soland ex Correa

Synonyms: *Bupariti populnea* (L.) Rothm.; *Hibiscus bacciferus* Blume; *Malvaviscus populneus* (L.) Gaertn.; *Parita populnea* (L.) Scop.; *Thespesia howii* S.Y. Hu; *Thespesia populneoides* (Roxb.) Kostel.

Common names: Pacific rosewood, portia tree; parashpipul, balai (Bangladesh); tong mian (China); pahari pipal (India); baru laut (Indonesia); baru baru (Malaysia); sabu-bani (Myanmar); banalo (the Philippines); milo (Samoa, Tonga); gan sooriya (Sri Lanka); po kamat phrai (Thailand)

Habitat: Beaches, cultivated

Distribution: Tropical Asia and Pacific

Botanical observation: This graceful tree grows to a height of 6 m. The leaves are simple, spiral, and stipulate. The stipules are linear and up to 7 mm long. The petiole is up to 10 cm long. The blade is cordate, 7–18×4.5–12 cm, glossy, with about 5 pairs of secondary nerves, and caudate at apex. The flowers are solitary and terminal or axillary. The epicalyx presents 3–4 lobes which are about 1 cm long. The calyx is tubular and develops 5 minute lobes. The 5 petals are yellow, purple at base, obovate and up to 7 cm long. The androecium comprises numerous stamens merged in a column which is 2.5 cm long. The capsules are globose, up to 5 cm across, 5-parted, and contain 20 seeds angular and 6 mm across.

Medicinal uses: Abscesses (Bangladesh); cholera, fever (Indonesia); dysentery, herpes (the Philippines); colds (Solomon Islands)

Broad-spectrum antibacterial halo developed by polar extracts: Methanol extract of flowers inhibited the growth of *Rhodococcus terrae* (NCIM 5126), *Brevibacterium leuteum* (ATCC 15830), *Flavobacterium devorans* (NCIM 2581), *Shigella sonnei* (ATCC 29930), *Shigella flexneri* (NCIM 4924), *Shigella boydii* (ATCC 8700), *Streptococcus faecalis* (ATCC 8043), *Escherichia coli* (ATCC 11229), *Bacillus licheniformis* (NCIM 2468), *Proteus mirabilis* (NCIM 8268), *Klebsiella pneumoniae* (ATCC 11229), and *Micrococcus luteus* (NCIM 9341) (Saravanakumar et al., 2009). Ethanol extract of roots of inhibited the growth of *Staphylococcus aureus* (NCIM 2079), *Escherichia coli* (NCIM 2065), *Pseudomonas aeruginosa* (NCIM 2036), *Bacillus subtilis* (NCIM 2063), and *Klebsiella aerogenes* (NCIM 2098) with inhibition zone diameter of 14, 16, 14, 13, and 11 mm, respectively (50 µg/disc) (Senthil-Rajan et al., 2013).

Antibiotic potentiator polar extract: Methanol extract decreased the MIC of oxytetracyline against *Shigella sonei* (ATCC 29930), *Salmonella typhi* (NCIM 2479), *Shigella flexneri* (NCIM 4924), *Shigella boydii* (ATCC NCIM 4924), *Escherichiacoli* (ATCC 11775), *Klebsiella pneumoniae* (ATCC 13883), and *Micrococcus luteus* (ATCC 2984) (Kumar et al., 2009).

Weak broad-spectrum antifungal polar extract: Ethanol extract of roots inhibited the growth of *Candida albicans* and *Aspergillus niger* with the MIC values of 750 and 1750 µg/mL, respectively (Senthil-Rajan et al., 2013).

Strong antibacterial (Gram-positive) amphiphilic sesquiterpenes and bis-naphthalenols: Populene C, populene D, mansonone D, mansonone E, 7-hydroxycadalene (Log D = 4.7 at pH 7.4; molecular mass = 214.3 g/mol) gossypol, and (+) 6,6′-methoxygossypol isolated from the wood inhibited the growth of *Bacillus cereus* with the MIC of 4.6, 4.6, 2.3, 4.6, 0.5, 1.1, and 2.3 µg/mL, respectively (Boonsri et al., 2008). Gossypol and (+) 6,6′-methoxygossypol inhibited *Staphylococcus aureus* (TISTR517) with the MIC values of 1.1 and 4.6 µg/mL, respectively (Boonsri et al., 2008). (+) 6,6′-Methoxygossypol inhibited the growth of *Enterococcusa faecalis* with the MIC of 1.1 µg/mL (Boonsri et al., 2008).

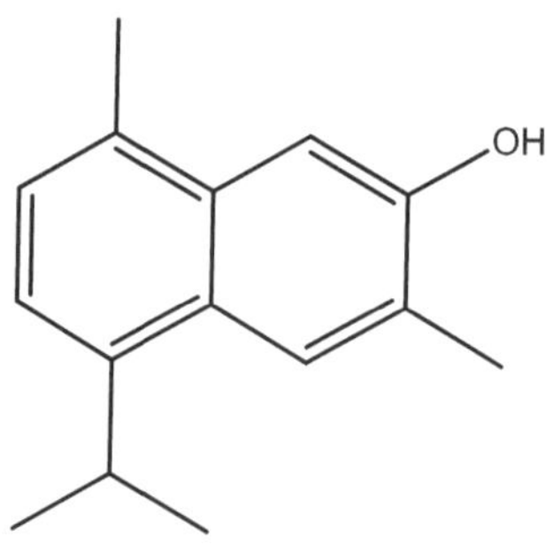

7-Hydroxycadalene

Strong antiviral (enveloped monopartite single-stranded (−) RNA) polar extract: Methanolic extract of flowers inhibited the replication of the Vesicular stomatitis virus and Respiratory syncytial viruses that is with the EC_{50} values of about 20 µg/mL (Arthanari et al., 2011).

Strong antiviral (non-enveloped monopartite single-stranded (+) RNA) polar extract: Methanolic extract of flowers inhibited the replication of the Coxsackie B4 with the EC_{50} value of about 20 µg/mL (Arthanari et al., 2011).

Commentary: What is (are) the antiviral principle(s) of this plant?

REFERENCES

Arthanari, S., Renukadevi, P., Vanitha, J., Venkateshwaran, K., Ganesh, M. and De Clercq, E., 2011. Evaluation of antiviral and cytotoxic activities of methanolic extract of *Thespesia populnea* (Malvaceae) flowers. *Journal of Herbs, Spices & Medicinal Plants, 17*(4), pp. 386–391.

Boonsri, S., Karalai, C., Ponglimanont, C., Chantrapromma, S. and Kanjana-opas, A., 2008. Cytotoxic and antibacterial sesquiterpenes from *Thespesia populnea. Journal of Natural Products, 71*(7), pp. 1173–1177.

Kumar, A.S., Venkateshwaran, K., Vanitha, J., Saravanan, V.S., Ganesh, M., Vasudevan, M. and Sivakumar, T., 2009. Synergistic activity of methanolic extract of *Thespesia populnea* (Malvaceae) flowers with oxytetracycline. *Bangladesh Journal of Pharmacology, 4*(1), pp. 13–16.

Saravanakumar, A., Venkateshwaran, K., Vanitha, J., Ganesh, M., Vasudevan, M. and Sivakumar, T., 2009. Evaluation of antibacterial activity, phenol and flavonoid contents of *Thespesia populnea* flower extracts. *Pakistan Journal of Pharmaceutical Sciences, 22*(3), pp. 282–286.

Senthil-Rajan, D., Rajkumar, M., Srinivasan, R., Kumarappan, C., Arunkumar, K., Senthilkumar, K.L. and Srikanth, M.V., 2013. Investigation on antimicrobial activity of root extracts of *Thespesia populnea* Linn. *Tropical Biomedicine, 30*(4), pp. 570–578.

8.4.4.14 *Urena lobata* L.

Synonyms: *Urena americana* L. f.; *Urena grandiflora* DC.; *Urena reticulata* Cav.; *Urena trilobata* Vell.

Common names: Aramina, Caesar weed, burr mallow; jangligagra (Bangladesh); otati (India); pulutan (Indonesia); pulut pulut (Malaysia); bito bito (Papua New Guinea); dalupang (the Philippines); po seng (Thailand)

Habitat: Roadsides, fields, wastelands

Distribution: Tropical and Subtropical Asia

Botanical observation: This beautiful herb grows to 1 m tall. The stems are terete and hairy. The leave are simple, spiral, and stipulate. The stipules are filiform, minute, and caducous. The petiole is 1–4 cm long and hairy. The blades are orbicular to ovate, 4–7×3–6.5 cm, rounded or nearly cordate at base, serrate, 3-lobed, 5–7×3–6.5 cm, and hairy. The flowers are solitary and axillary. The calyx is tubular, 5-lobed, and hairy. The 5 petals are membranaceous, about 1.5 cm long, pink, and obovate. The androecium includes numerous stamens joined into a column which is 1.5 cm long. The ovary develops 10 styles hairy and hirsute. The schizocarps are flattened globose, about 1 cm in diameter and spiny.

Medicinal uses: Cough, colds, sores, fever, ear infections, skin infections, wounds (Bangladesh)

Broad-spectrum antibacterial halo developed by polar extract: Methanol extract of roots (125 µg/mL, 6 mm diameter wells) inhibited the growth of *Staphylococcus epidermidis, Micrococcus luteus, Escherichia coli, Shigella dysenteriae*, and *Vibrio cholerae* with the inhibition zone diameter of 8, 11, 9, 8, and 9 mm, respectively (Mazumder et al., 2001).

In vivo antibacterial (Gram-positive) polar extract: Aqueous extract given at the dose of 40 g/Kg for 14 days to mice infected with *Staphylococcus aureus* evoked a decrease of colony count/lung weight (Yang et al., 2017).

Weak antibacterial (Gram-negative) essential oil: Essential oil of leaves inhibited the growth of *Escherichia coli* (EIEC 240-1) with the MIC value of 1000 µg/mL (Duarte et al., 2007).

Strong antifungal (yeast) triterpene saponin: Clematoside S isolated from the leaves inhibited the growth of *Saccharomyces cerevisiae* (ATCC 204505) with the MIC/MFC value of 0.6/2.4 µg/mL (Gao et al., 2015).

Commentary: The plant produces the dibenzylbutyrolactone lignan (–)-trachelogenin (Gao et al., 2015) which could be examined for antimicrobial effects.

REFERENCES

Duarte, M.C.T., Leme, E.E., Delarmelina, C., Soares, A.A., Figueira, G.M. and Sartoratto, A., 2007. Activity of essential oils from Brazilian medicinal plants on *Escherichia coli*. *Journal of Ethnopharmacology*, *111*(2), pp. 197–201.

Gao, X.L., Liao, Y., Wang, J., Liu, X.Y., Zhong, K., Huang, Y.N., Gao, H., Gao, B. and Xu, Z.J., 2015. Discovery of a potent anti-yeast triterpenoid saponin, clematoside-S from *Urena lobata* L. *International Journal of Molecular Sciences*, *16*(3), pp. 4731–4743.

Mazumder, U.K., Gupta, M., Manikandan, L. and Bhattacharya, S., 2001. Antibacterial activity of *Urena lobata* root. *Fitoterapia*, *72*(8), pp. 927–929.

Yang, Y., Huang, Z., Zou, X., Zhong, X., Liang, X. and Zhou, J., 2017. The antibacterial effect of *Urena lobata* L. fromv guangxi on mice with *Staphylococcus aureus* pneumonia. *African Journal of Traditional, Complementary and Alternative Medicines*, *14*(1), pp. 73–78.

8.4.5 FAMILY MUNTINGIACEAE C. BAYER, M. W. CHASE & M. F. FAY (1998)

The family Mutingiaceae includes 3 genera of shrubs or trees. The leaves are simple, alternate, and stipulate. The blade is asymmetrical at base, serrate, and hairy. The inflorescences are solitary and axillary. The calyx comprises 5 sepals and the corolla is made of 5 petals. The androecium includes numerous stamens. The gynoecium comprises a single ovary developing up to 7 stigma. The fruits are berries.

8.4.5.1 *Muntingia calabura* L.

Common names: Jamaica cherry; kakhop (Cambodia); nakkaraegu (India); khum som (Laos); kerukup siam (Malaysia); seresa (the Philippines); krop farang (Thailand); mat sam (Vietnam)

Habitat: Roadsides, wastelands

Distribution: Tropical Asia

Botanical observation: This fast growing treelet can reach a height of 10 m. It is native to South-Central America and the Caribbean and was probably brought in Asia by Portuguese or spaniard colonialists. The bark is light grey, beautifully somewhat marked with fine dark lines, and smooth. The leaves are simple, alternate, and stipulate. The petiole is up to 5 mm long. The blade is lanceolate, serrate, asymmetrical, 4–15 × 1–4 cm. The inflorescence is solitary and axillary. The calyx comprises 5 sepals. The 5 petals are pure white, spathulate, membranaceous, ephemeral and up to about 2 cm long. At first glance the flower looks very much like a flower of Rosaceae. The berries are globose, edible, up to 1.5 cm across, contain numerous tiny seeds, and looking very much like some sort of reddish-burgundy grapes.

Medicinal uses: Fever, flu (the Philippines)

Moderate antibacterial (Gram-positive) mid-polar extract: Ethylacetate extract of leaves inhibited the growth of *Staphylococcus aureus* (25923) and *Staphylococcus aureus* (33591) (Zakaria et al., 2010).

Moderate antibacterial (Gram-positive) chalcone: 2′,4′-Dihydroxychalcone from the leaves inhibited the growth of *Staphylococcus aureus* (ATCC 25293) and methicillin-resistant *Staphylococcus aureus* (ATCC 33591) with the MIC/MBC values of 50/100 and 100/200 µg/mL, respectively (Sufian et al., 2013).

Moderate anticandidal polar extract: Ethanol extract of leaves inhibited the growth of *Candida albicans* with the MIC value of 625 µg/mL (Buhian et al., 2016).

Commentary: The antiviral properties of this plant have not been apparently examined. Note that angiotensin-converting enzyme 2 inhibitors have been identified as a potential pharmacological target for the Severe acute respiratory syndrome-associated coronavirus by hampering virus attachment to host cells (Li et al., 2003) and ethylacetate extract of leaves inhibited angiotensin--converting enzyme with the IC_{50} value of 0.6 µg/mL (Lasboi & Rissyelly, 2017). The plant abounds with flavonoids (Kaneda et al., 1991) with probable antiviral potencies. 2′,4′-Dihydroxychalcone could be examined for its antiviral properties.

REFERENCES

Buhian, W.P.C., Rubio, R.O., Valle Jr, D.L. and Martin-Puzon, J.J., 2016. Bioactive metabolite profiles and antimicrobial activity of ethanolic extracts from *Muntingia calabura* L. leaves and stems. *Asian Pacific Journal of Tropical Biomedicine*, *6*(8), pp. 682–685.

Kaneda, N., Pezzuto, J.M., Soejarto, D.D., Kinghorn, A.D., Farnsworth, N.R., Santisuk, T., Tuchinda, P., Udchachon, J. and Reutrakul, V., 1991. Plant anticancer agents, XLVIII. New cytotoxic flavonoids from *Muntingia calabura* roots. *Journal of Natural Products*, *54*(1), pp. 196–206.

Lasboi, E. and Rissyelly, K.B., 2017. Angiotensin I-converting enzyme inhibitory activity, total phenolic and flavonoid content of extract and fraction of jam fruit leaves (*Muntingia calabura* L.). *Asian Journal of Pharmaceutical and Clinical Research*, *10*(Special Issue October), pp. 166–168.

Li, W., Moore, M.J., Vasilieva, N., Sui, J., Wong, S.K., Berne, M.A., Somasundaran, M., Sullivan, J.L., Luzuriaga, K., Greenough, T.C. and Choe, H., 2003. Angiotensin-converting enzyme 2 is a functional receptor for the SARS coronavirus. *Nature*, *426*(6965), pp. 450–454.

Sufian, A.S., Ramasamy, K., Ahmat, N., Zakaria, Z.A. and Yusof, M.I.M., 2013. Isolation and identification of antibacterial and cytotoxic compounds from the leaves of *Muntingia calabura* L. *Journal of Ethnopharmacology*, *146*(1), pp. 198–204.

Zakaria, Z.A., Sufian, A.S., Ramasamy, K., Ahmat, N., Sulaiman, M.R., Arifah, A.K., Zuraini, A. and Somchit, M.N., 2010. In vitro antimicrobial activity of *Muntingia calabura* extracts and fractions. *African Journal of Microbiology Research*, *4*(4), pp. 304–308.

8.4.6 FAMILY STERCULIACEAE VENTENAT (1807)

The family Sterculiaceae consists of about 70 genera and 1100 species of mostly shrubs or trees. The stems are terete, not uncommonly hairy, and fibrous. The leaves are simple, alternate, or spiral, and stipulate. The petiole is often slender, straight, and long. The blades in this family often present characteristic secondary nerve patterns and are often palmate. The inflorescences are solitary, or cymes, racemes or panicles. The calyx presents 3–5 sepals fused at base or free. Five petals are present or none. The androecium consists of 5 to numerous stamens more or less merged at base. The gynoecium consists of 2–5 fused carpels forming a plurilocular ovary, with 2 to many ovules per locules, and develops 1 or 2–5 styles. The fruits are capsules or follicles containing numerous reniform seeds. Quinolinone alkaloids, sesquiterpenes, and triterpenes of antimicrobial interest have been characterized from this family.

8.4.6.1 *Abroma augusta* (L.) L. f.

Synonyms: *Theobroma augusta* L.

Common names: Devil's cotton, Indian hemp; ulot-kumbol, udurful (Bangladesh); kapasan (Indonesia); pishaacha, ulatkarnbal (India); mway ma naing (Myanmar); rami sangat (Malaysia); rap (Palau, Yap); anabo (the Philippines); thiam dan (Thailand); bong vang (Vietnam)

Habitat: Wastelands, roadsides

Distribution: India, Southeast Asia, China, Pacific Islands

Botanical observation: This sinister shrub grows to about 3 m tall. The bark is fibrous and has been used to make ropes in India. The stems are terete and hairy. The leaves are simple, alternate, and stipulate. The stipules are narrowly triangular, about 1 cm long, and caducous. The petiole grows up to about 10 cm long. The blade is lanceolate to palmately lobed, 6.5–20×4–15 cm, hairy below, with 5–7 pairs of secondary nerves, cordate at base, and acute or acuminate at apex. The inflorescences are axillary and solitary with pendulous showy flowers. The calyx comprises 5 sepals which are lanceolate, up to 2 cm long, and hairy. The 5 petals are dull-purplish, about 2.5 cm long, and spathulate-clawed. The androecium includes 5 groups of 3 stamens. The ovary is oblong and develops a 5-branched stigma. The capsules are upright, brownish, papery, 5-partite, obconic, about 5 cm in diameter, dehiscent, hairy within, and contain minute black seeds.

Medicinal uses: Leucorrhea, diarrhea, dysentery (Bangladesh); gonorrhea (Bangladesh; Palau); leucorrhoea (India)

Broad-spectrum antibacterial halo developed by mid-polar extract: Chloroform extract (500 µg/disc) inhibited the growth of *Bacillus megaterium, Bacillus subtilis, Salmonella paratyphi, Shigella boydii, Vibrio mimicus, Salmonella typhi,* and *Escherichia coli* with the inhibition zone diameters of 29,14, 25, 23, 15, 13, and 28 mm, respectively (Zulfiker et al., 2013).

Broad-spectrum antifungal halo developed by mid-polar extract: Chloroform extract (500 µg/disc) inhibited the growth of *Aspergillus niger, Candida albicans,* and *Saccharomyces cerevisae* with the inhibition zone diameters of 15, 12, and 8 mm, respectively (Zulfiker et al., 2013).

Strong broad-spectrum antibacterial polar extract: Acetone extract of leaves inhibited the growth of *Bacillus subtilis, Bacillus megaterium, Staphylococcus aureus, Escherichia coli, Shigella dysenteriae, Shigella sonnei,* and *Salmonella typhi* with the MIC values of 16, 8, 64, 32, 64, 32, and 16 µg/mL, respectively (Saikot et al., 2012).

Strong antiviral (monopartite enveloped linear single-stranded (−) RNA) lipophilic friedelane triterpene: Friedelin (LogD = 10 at pH 7.4; molecular mass = 426.7 g/mol) inhibited the replication of the Parainfluenza virus with the IC_{50} value of 14×10^{-5}M (Jiang et al., 2002).

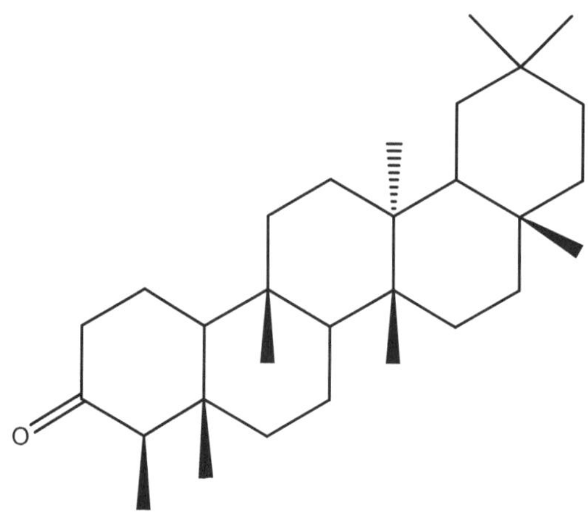

Friedelin

Commentary: The plant yields the triterpenes friedelin, abromasterol, taraxeryl acetate, and taraxerol (Dasgupta & Basu, 1970). The taraxerane-type triterpene taraxeryl acetate inhibited the growth of *Escherichia coli, Bacillus cereus, Streptococcus faecalis, Candida albicans,* and *Microsporum audouinii* with the MIC values of 78, 156, 78, 156, and 78 µg/mL, respectively (Kuete et al., 2009). One may wonder if triterpenes in plant are phytoalexins. The strong activities of taraxerol against filamentous fungi have been described earlier. An interesting fact about friedelane and taraxerane-type triterpenes is that they often protect cells against coronaviruses *in vitro.* For instance, 3β-friedelanol, 3β-acetoxy friedelane, friedelane, or epitaraxerol at the concentration of 5 µg/mL protected MRC-5 cells against the Human coronavirus (strain 229E) (Chang et al., 2012).

REFERENCES

Chang, F.R., Yen, C.T., Ei-Shazly, M., Lin, W.H., Yen, M.H., Lin, K.H. and Wu, Y.C., 2012. Anti-human coronavirus (anti-HCoV) triterpenoids from the leaves of *Euphorbia neriifolia. Natural Product Communications,* 7(11), p. 1934578X1200701103.

Dasgupta, B. and Basu, K., 1970. Chemical investigation of *Abroma augusta* Linn. identity of abromine with betaine. *Experientia*, *26*(5), pp. 477–478.
Jiang, R.W., Ma, S.C., He, Z.D., Huang, X.S., But, P.P.H., Wang, H., Chan, S.P., Ooi, V.E.C., Xu, H.X. and Mak, T.C., 2002. Molecular structures and antiviral activities of naturally occurring and modified cassane furanoditerpenoids and friedelane triterpenoids from *Caesalpinia minax*. *Bioorganic & Medicinal Chemistry*, *10*(7), pp. 2161–2170.
Kuete, V., Nana, F., Ngameni, B., Mbaveng, A.T., Keumedjio, F. and Ngadjui, B.T., 2009. Antimicrobial activity of the crude extract, fractions and compounds from stem bark of *Ficus ovata* (Moraceae). *Journal of Ethnopharmacology*, *124*(3), pp. 556–561.
Saikot, F.K., Khan, A. and Hasan, M.F., 2012. Antimicrobial and cytotoxic activities of *Abroma augusta* Linn. leaves extract. *Asian Pacific Journal of Tropical Biomedicine*, *2*(3), pp. S1418–S1422.
Zulfiker, A.H.M., Roy, P.P., Momin, M.A.M., Khan, M.S., Bulbul, I.J., Ahmed, T. and Rana, M.S., 2013. Investigation of antioxidant and antimicrobial potential of chloroform and petroleum ether extracts of selected medicinal plants of Bangladesh. *Journal of Advances in Medicine and Medical Research*, *3*, pp. 1418–1436.

8.4.6.2 *Firmiana simplex* (L.) W. Wight

Synonyms: *Firmiana platanifolia* (L. f.) Schott & Endl.; *Hibiscus simplex* L.; *Sterculia platanifolia* L. f.; *Sterculia urens* Roxb.

Common names: Chinese parasol-tree; wu tong (China)

Habitat: Cultivated

Distribution: China, Japan

Botanical observation: This handsome (but with somewhat of a dark aura) tree grows up to 15 m tall. The bark is smooth and greyish. The leaves are simple and spiral. The petiole is reddish and grows up to 30 cm long. The blade is palmately and characteristically 3–5-lobed, cordate at base, acuminate at apex, dull green, and up to 30 cm across. The panicles are terminal, showy, and about 50 cm long. The 5 sepals are linear, fleshy, whitish-yellow, linear, up to about 1 cm long, and reflexed. The perianth is absent. The 15 stamens are partially joined in a staminal column. The follicles are elliptic, somewhat leaf-like, membranaceous, up to about 10 cm long, and present up to 4 globose seeds.

Medicinal uses: Oral abscesses (China); diarrhea (Korea)

Commentary: The plant produces series of furofuran and benzofuran lignan glycosides (Woo et al., 2019) as well as isoprenylated naphthoquinone dimers firmianones A, B, and C (Bai et al., 2005). These quinones could be examined for antimicrobial activities.

REFERENCES

Bai, H., Li, S., Yin, F. and Hu, L., 2005. Isoprenylated naphthoquinone dimers firmianones A, B, and C from *Firmiana platanifolia*. *Journal of Natural Products*, *68*(8), pp. 1159–1163.
Woo, K.W., Park, J.E., Cha, J.M., Subedi, L., Kim, S.Y. and Lee, K.R., 2019. Three new lignan glycosides from the Firmiana simplex. *Chemical and Pharmaceutical Bulletin*, *67*(1), pp. 18–22.

8.4.6.3 *Helicteres angustifolia* L.

Common names: Shan zhi ma (China)

Habitat: Fields, grassy spots

Distribution: Myanmar, Cambodia, Laos, Vietnam, Thailand, Malaysia, Indonesia, the Philippines, Australia

Botanical observation: This herb grows up to 1 m tall. The stems are hairy and fibrous. The leaves are simple, spiral, and stipulate. The petiole is up to about 7 mm long. The blade is narrowly oblong, 3.5-5 cm × 1.5–2.5 cm, hairy below, rounded at base, and obtuse or acute at apex. The inflorescences are axillary cymes. The calyx is tubular, hairy, up to about 6 mm long, 5-lobed, the lobes triangular. The 5 petals are spathulate, bluish, pinkish, or purplish, and about 8 mm long.

The androecium includes 10 stamens. The ovary is globose and 5-lobed. The capsules are hairy, oblong, up to 2 cm long, and contain numerous tiny brown seeds.

Medicinal use: Abscesses (China, Taiwan)

Strong antiviral (enveloped monopartite double-stranded DNA) lipophilic lupane-type triterpene: Methyl helicterate (Log D = 10.3 at pH 7.4; molecular mass = 632.8 g/mol) inhibited the secretion of the Hepatitis B virus surface-antigen and Hepatitis B virus e-antigen with the IC_{50} values of 8.9 and 10.4 µM, respectively (Huang et al., 2012). At 15.8 µM, this pentacyclic triterpene decreased intracellular viral DNA levels (Huang et al., 2012).

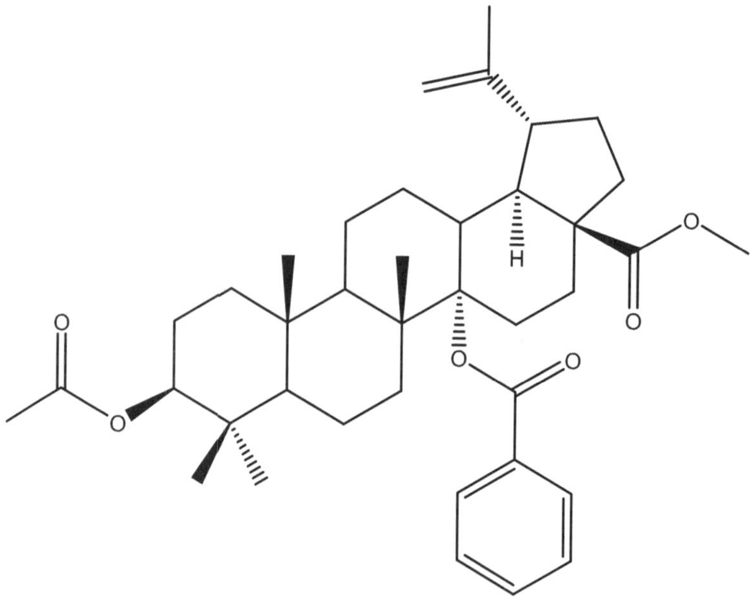

Methyl helicterate

In vivo antiviral *(enveloped circular double stranded DNA)* lupane-type triterpene: In duckling methyl helicterate given orally at 100 mg/Kg/day for 14 days reduced serum and hepatic Heptitis B virus DNA levels (Huang et al., 2012).

Commentaries

i. What is the antiviral mode of action of methyl helicterate?
ii. The plant produces series of sesquiterpene quinones of which mansonone E, F, H, and M (Chen et al., 1990). Mansonone E inhibited the growth of *Alternaria porri*, *Colletotrichum gloeosporioides*, *Fusarium oxysporum*, and *Phytophthora parasitica* with the MIC/MFC values of 250/250, 31.2/31.2, 500/1000, and 31.2/125 µg/mL, respectively (Mongkol & Chavasiri, 2016). Mansonone F (Log D = 2.5 at pH 7.4; molecular mass = 240.2 g/mol) inhibited the growth of methicillin-resistant *Staphylococcus aureus* with the MIC value of 2 µg/mL (equivalent as MIC for vancomycin), and *Bacillus subtilis* ATCC 6633, *Staphylococcus aureus* (ATCC 25923), *Staphylococcus epidermidis* (ATCC 12228), and *Klebsiella pneumoniae* (ATCC 10031) with the MIC values of 2, 1, 0.5, and 8 µg/mL, respectively (Suh et al., 2006). Are mansonones antiviral? Aromatic substances with ketone groups are always antibacterial *in vitro*.

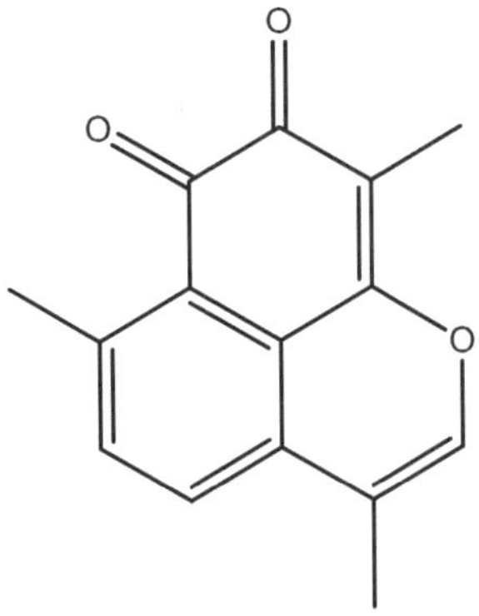

Mansonone F

iii. *Helicteres isora* L. is used in China to treat fever (local name: *Huo suo ma*). Hexane:acetone (92:8, v/v) fraction from an acetone extract of fruits of *Helicteres isora* L. at 800 μg/mL evoked phenotypic loss of gentamycin resistance by *Enterococcus faecalis*, gentamicin resistance by *Escherichia coli*, kanamycin resistance by *Bacillus cereus* via plasmid curing (loss of plasmid DNA) (Shriram et al., 2010). Butanol fraction of roots of *Helicteres isora* L. (10 mg/mL/8 mm wells) inhibited the growth of *Bacillus subtilis* (NCIM 2063), *Micrococcus luteus* (*NCIM* 2103), *Staphylococcus aureus* (NCIM 2079), *Escherichia coli* (NCIM 2068) with inhibition zone diameter of 3, 32, 24, and 10 mm, respectively (Venkatesh et al., 2007).

REFERENCES

Chen, C.M., Chen, Z.T. and Hong, Y.L., 1990. A mansonone from *Helicteres angustifolia*. *Phytochemistry*, *29*(3), pp. 980–982.

Huang, Q., Huang, R., Wei, L., Chen, Y., Lv, S., Liang, C., Zhang, X., Yin, F., Li, H., Zhuo, L. and Lin, X., 2013. Antiviral activity of methyl helicterate isolated from *Helicteres angustifolia* (Sterculiaceae) against hepatitis B virus. *Antiviral Research*, *100*(2), pp. 373–381.

Mongkol, R. and Chavasiri, W., 2016. Antimicrobial, herbicidal and antifeedant activities of mansonone E from the heartwoods of *Mansonia gagei* Drumm. *Journal of Integrative Agriculture*, *15*(12), pp. 2795–2802.

Shriram, V., Jahagirdar, S., Latha, C., Kumar, V., Dhakephalkar, P., Rojatkar, S. and Shitole, M.G., 2010. Antibacterial & antiplasmid activities of *Helicteres isora* L. *Indian Journal of Medical Research*, *132*, pp. 94–99.

Suh, Y.G., Kim, S.N., Shin, D.Y., Hyun, S.S., Lee, D.S., Min, K.H., Han, S.M., Li, F., Choi, E.C. and Choi, S.H., 2006. The structure–activity relationships of mansonone F, a potent anti-MRSA sesquiterpenoid quinone: SAR studies on the C6 and C9 analogs. *Bioorganic & Medicinal Chemistry Letters*, *16*(1), pp. 142–145.

Venkatesh, S., Sailaxmi, K., Reddy, B.M. and Ramesh, M., 2007. Antimicrobial activity of *Helicteres isora* root. *Indian Journal of Pharmaceutical Sciences*, *69*(5), p. 687.

8.4.6.4 *Heritiera littoralis* Aiton

Synonyms: *Balanopteris totila* Gaertn.; *Heritiera minor* Bojer

Common names: Looking-glass tree; sundari (Bangladesh; India); khleay (Cambodia); dungun (Indonesia, Malaysia); yin ye shu (China); pinle kasano (Myanmar); dungon late (the Philippines); khai khwai, ngon kai thale (Thailand); cui bien (Vietnam)

Habitat: Mangroves

Distribution: Tropical Asia and Pacific

Botanical observation: This tree grows to a height of about 10 m. The buttresses are massive and beautifully if not mysteriously undulate. The stems are hairy when young. The leaves are simple, spiral, and stipulate. The stipules are lanceolate and up to 2 cm long. The blade is oblong-lanceolate, elliptic, or ovate, dark green glossy above, 10–20×5–10 cm, coriaceous, hairy and somewhat glaucous below, marked with about 5–10 pairs of secondary nerves, obtuse at base and acute at apex. The inflorescences are panicles. The calyx is tubular, hairy, urceolate, pinkish, about 6 mm long and 5-lobed. The androecium includes 5 stamens merged in a column. The gynoecium is ovoid and develops a 5-lobed stigma. The follicles are coriaceous, elliptic, brown, keeled, about 6×3.5 cm, and contain a single seed.

Medicinal uses: Diarrhea, dysentery (Indonesia, Vietnam)

Viral enzyme inhibition by triterpene: Betulinic acid inhibited Severe acute respiratory syndrome-associated coronavirus 3-chymotrypsin-like protease activity with the IC_{50} value of 10.2 μM (Wen et al., 2007).

Commentaries

i. The plant assembles series of cadinane-type sesquiterpene lactones of which vallapin (Miles et al., 1991) and the simple phenol ester methyl β-orinol carboxylate (Tewtrakul et al., 2010) which could be examined for their antimicrobial effects. The plant also produces the anthraquinone physcion, and series of stigmastane-type steroids as well as the triterpenes betulinic acid and friedelin (Tewtrakul et al., 2010).

ii. In Bangladesh, *Heritiera fomes* Buch.-Ham. (local name: *Sundari*) is used for the treatment of diarrhea. Ethanol extract of bark of *Heritiera fomes* (Malvaceae) inhibited the growth of *Staphylococcus epidermidis*, *Streptococcus pyogenes*, *Escherichia coli*, *Enterococcus aerogenes*, and *Pseudomonas aeruginosa* with inhibition zone diameter of 10, 11, 10, 8, and 15 mm, respectively (0.5 mg/disc) (Uddin et al., 2008).

REFERENCES

Miles, D.H., Chittawong, V., Lho, D.S., Payne, A.M., de la Cruz, A.A., Gomez, E.D., Weeks, J.A. and Atwood, J.L., 1991. Toxicants from mangrove plants, VII. Vallapin and vallapianin, novel sesquiterpene lactones from the mangrove plant *Heritiera littoralis*. *Journal of Natural Products*, *54*(1), pp. 286–289.

Tewtrakul, S., Tansakul, P., Daengrot, C., Ponglimanont, C. and Karalai, C., 2010. Anti-inflammatory principles from *Heritiera littoralis* bark. *Phytomedicine*, *17*(11), pp. 851–855.

Uddin, S.J., Rouf, R., Shilpi, J.A., Alamgir, M., Nahar, L. and Sarker, S.D., 2008. Screening of some Bangladeshi medicinal plants for *in vitro* antibacterial activity. *Oriental Pharmacy and Experimental Medicine*, *8*(3), pp. 316–321.

Wen, C.C., Kuo, Y.H., Jan, J.T., Liang, P.H., Wang, S.Y., Liu, H.G., Lee, C.K., Chang, S.T., Kuo, C.J., Lee, S.S. and Hou, C.C., 2007. Specific plant terpenoids and lignoids possess potent antiviral activities against severe acute respiratory syndrome coronavirus. *Journal of Medicinal Chemistry*, *50*(17), pp. 4087–4095.

8.4.6.5 *Leptonychia caudata* (Wall, ex G. Don) Burrett

Synonyms: *Grewia acuminata* Bedd.; *Grewia heteroclita* Roxb.; *Leptonychia glabra* Turcz; *Turraea trichostylis* Miq.

Common names: Pakan manok (Indonesia); ceremai hantu (Malaysia); phak waan (Thailand)

Habitat: Forests

Distribution: India, Bangladesh, Andamans, Myanmar, Vietnam, Thailand, Malaysia, Indonesia

Botanical observation: This tree grows up to 12 m tall. The bark is greyish. The leaves are simple, alternate, and stipulate. The stipules are lanceolate, and up to about 6 mm long. The petiole is hairy and up to about 1 cm long. The blade is obovate, 4.5–20 × 1.5–6.5 cm, with 3–9 pairs of secondary nerves sunken above, dark green and glossy, cuneate at base, and acute to caudate at apex. The inflorescence is an axillary cyme which bears up to about 3 flowers. The 5 sepals are oblong, up to 1 cm long, whitish, and hairy. The 5 petals are greenish white, rectangular, about 3 mm long, and incised at apex. The 10 stamens are fused at base into a ring. The ovary is globose. The capsule is somewhat globose, up to about 4 cm across, velvety, and opens to release a few black, oblong, about 1 cm long, and glossy seeds in a bright red aril.

Medicinal uses: Fever, syphilis (Malaysia)

Commentary: The plant (which is most probably on its way to extinction) has apparently not been studied for its possible antimicrobial properties. Note the presence the pyrrole alkaloid brunfelsamidine and 1-β-ribofuranosylbrunfelsamidine from a member of the genus *Leptonychia* Turcz. (Yang et al., 2005). 1-β-Ribofuranosylbrunfelsamidine inhibited tryptase with the IC_{50} value of 3 μM (Yang et al., 2005). Severe acute respiratory syndrome-associated coronavirus spike glycoprotein is susceptible to airway protease cleavage enhancing virus entry in host cells (Kam et al., 2009). *Leptonychia heteroclita* Kurz is used to treat fever and syphilis in Malaysia.

REFERENCES

Kam, Y.W., Okumura, Y., Kido, H., Ng, L.F., Bruzzone, R. and Altmeyer, R., 2009. Cleavage of the SARS coronavirus spike glycoprotein by airway proteases enhances virus entry into human bronchial epithelial cells *in vitro*. *PloS One*, *4*(11), p. e7870.

Yang, L.K., Ng, S.P., Flotow, H., Yoganathan, K., Daramola, B.O., Soejarto, D.D., Buss, A.D. and Butler, M.S., 2005. Pyrrole carboxamidine tryptase inhibitors from *Leptonychia pubescens*. *Planta Medica*, *71*(11), pp. 1071–1072.

8.4.6.6 *Melochia corchorifolia* L.

Synonyms: *Riedlea corchorifolia* (L.) DC.; *Visenia corchorifolia* (L.) Spreng.

Common names: Red weed; ma song zi (China)

Habitat: Roadsides, wastelands

Distribution: Tropical Asia

Botanical observation: This herb grows up to about 60 cm tall. The stems are hairy. The leaves are simple, spiral, and stipulate. The stipules are linear and up to 4 mm long. The petiole is up to 2.5 cm long and hairy. The blade is lanceolate, 2.5–7 × 1–1.5 cm, with about 5–12 pairs of secondary nerves, hairy below, cordate of rounded at base, serrate, and acute at apex. The inflorescences are terminal or axillary cymes. The epicalyx develops 4 lobes which are linear and hairy. The calyx is tubular, 5-lobed, hairy, and about 3 mm long. The 5 petals are whitish, spathulate, and about 6 mm. The androecium comprises 5 stamens connate at base. The ovary is hairy and develops a 5-lobed stigma. The capsules are 5-lobed, globose, 5 mm across, dehiscent, hairy, and contain a few blackish seeds.

Medicinal uses: Dysentery (India); sores (Malaysia)

Broad-spectrum antibacterial halo developed polar extract: Methanol extract of aerial parts (100 μg/ 6 mm well) inhibited the growth of *Streptococcus faecalis*, *Staphylococcus epidermidis*, *Bacillus megaterium*, *Staphylococcus aureus*, *Escherichia coli*, *Pseudomonas aeruginosa*, *Salmonella typhimurium*, and *Klebsiella pneumonia* (Rao et al., 2012).

Moderate antifungal (yeasts) amide alkaloid: Frangulanine inhibited the growth of *Candida albicans* and *Saccharomyces cerevisae* (Emile et al., 2007).

Commentaries

i. The plant assembles rare series of pseudoxindole alkaloids of which melochicorine as well as cyclopeptide alkaloids of which franganine (Bhakuni et al., 1986, 1991).

ii. In Papua New Guinea, *Melochia odorata* L. f. (local name: *Ulari*) is used to treat urinary tract infection and produces the epoxide quinolinone alkaloids waltheriones A and C which protected CEM-TART cells from killing by the Human immunodeficiency virus with the EC_{50} values of 56.2 and 0.8 µM, respectively (Jadulco et al., 2014). Waltherione A inhibited the growth of *Candida albicans*, *Cryptococcus neoformans*, and *Saccharomyces cerevisae* with the minimum amounts of 50 µM (Emile et al., 2007).

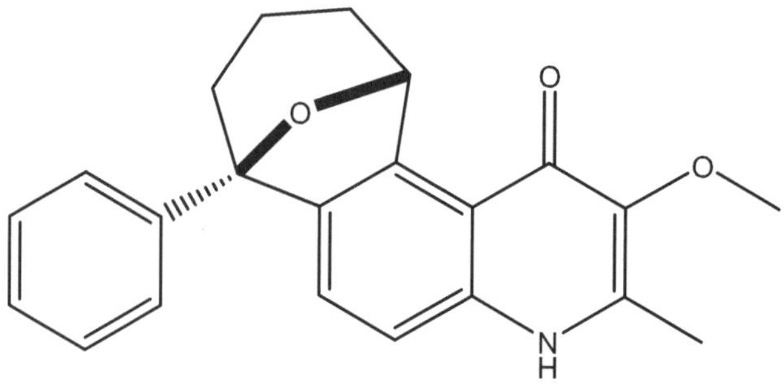

Waltherione C

Waltheria indica L. is used to treat fever and syphilis in the Philippines and produces the quinolinone alkaloids antidesmone and waltherione C (Liang et al., 2019). Antidesmone at the concentration of 50 µg/mL inhibited the growth of *Botryosphaeria dothidea*, *Colletotrichum musae*, *Pestalotipsis guepinii*, *Colletotrichum orbiculare*, *Phylophthora nicotianae*, *Pestalotiopsis longiseta*, Carbendazim-sensitive strains of *Sclerotinia sclerotiorum*, and Carbendazim-resistant strains of *Sclerotinia sclerotiorum* by 71.9, 84.6, 86.7, 75.5, 78.8, 83.4, 56.5, and 100%, respectively (Liang et al., 2019).

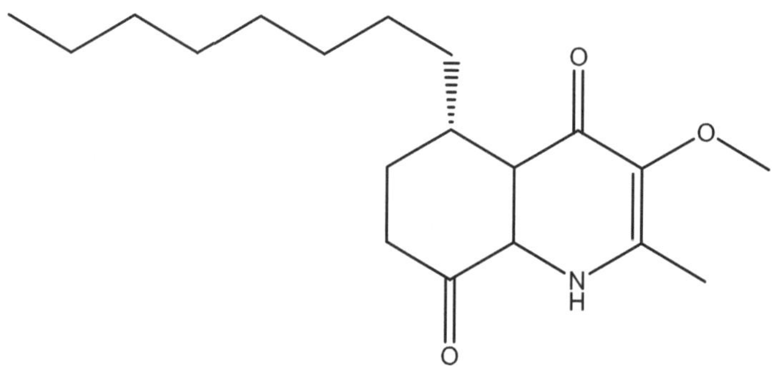

Antidesmone

Waltherione C at the concentration of 50 µg/mL inhibited the growth of *Botryosphaeria dothidea*, *Colletotrichum musae*, *Pestalotipsis guepinii*, *Colletotrichum orbiculare*, *Phylophthora nicotianae*, *Pestalotiopsis longiseta*, Carbendazim-sensitive strains of *Sclerotinia sclerotiorum*, and Carbendazim-resistant strains of *Sclerotinia sclerotiorum* by 60, 43, 71.4, 60.9, 41.6, 3.6, 79.8, and 81.8%, respectively (Liang et al., 2019). The quinolinone chamaedrone from *Melochia chamaedrys* A. St.-Hil. inhibited a broad-spectrum of bacteria and fungi (Dias et al., 2007). Quinolines with ketone groups and congeners are often antiviral such as the synthetic efavirenz used to treat AIDS via reverse transcriptase inhibition (Di Santo et al., 2006).

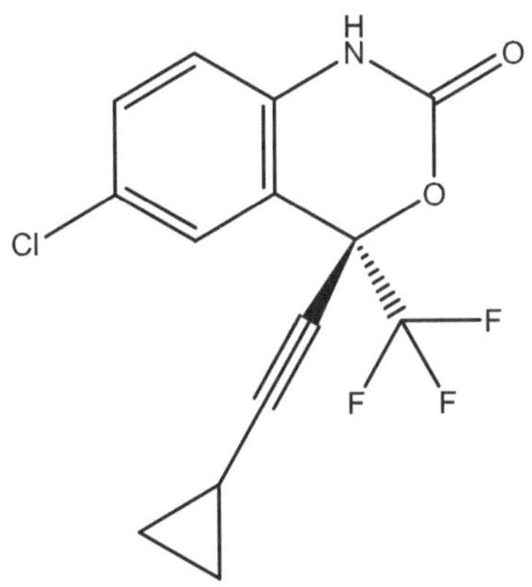

Efavirenz

REFERENCES

Bhakuni, R.S., Shukla, Y.N. and Thakur, R.S., 1986. Cyclopeptide alkaloids from *Melochia corchorifolia. Phytochemistry*, *26*(1), pp. 324–325.

Bhakuni, R.S., Shukla, Y.N. and Thakur, R.S., 1991. Melochicorine, a pseudooxindole alkaloid from *Melochia corchorifolia. Phytochemistry*, *30*(9), pp. 3159–3160.

Dias, G.O., Porto, C., Stüker, C.Z., Graessler, V., Burrow, R.A., Dalcol, I.I., Silva, U.F. and Morel, A.F., 2007. Alkaloids from *Melochia chamaedrys. Planta Medica*, *73*(3), pp. 289–292.

Di Santo, R., Costi, R., Roux, A., Artico, M., Lavecchia, A., Marinelli, L., Novellino, E., Palmisano, L., Andreotti, M., Amici, R. and Galluzzo, C.M., 2006. Novel bifunctional quinolonyl diketo acid derivatives as HIV-1 integrase inhibitors: design, synthesis, biological activities, and mechanism of action. *Journal of Medicinal Chemistry*, *49*(6), pp. 1939–1945.

Emile, A., Waikedre, J., Herrenknecht, C., Fourneau, C., Gantier, J.C., Hnawia, E., Cabalion, P., Hocquemiller, R. and Fournet, A., 2007. Bioassay-guided isolation of antifungal alkaloids from *Melochia odorata. Phytotherapy Research: An International Journal Devoted to Pharmacological and Toxicological Evaluation of Natural Product Derivatives*, *21*(4), pp. 398–400.

Jadulco, R.C., Pond, C.D., Van Wagoner, R.M., Koch, M., Gideon, O.G., Matainaho, T.K., Piskaut, P. and Barrows, L.R., 2014. 4-Quinolone alkaloids from *Melochia odorata. Journal of Natural Products*, *77*(1), pp. 183–187.

Liang, C., Yang, L., Shao, Y., Zhu, X., Zhao, H., Chen, B., Song, W., Song, X., Ding, X. and Sun, R., 2019. Broad-spectrum antifungal activity of dichloromethane extract of *Waltheria indica* stems and isolated compounds. *Industrial Crops and Products*, *142*, p. 111855.

Rao, B.G., Rao, Y.V. and Rao, T.M., 2012. Anti-bacterial activity of different extracts of *Melochia corchorifolia* and *Spilanthes acmella* aerial parts. *Journal of Pharmacy Research*, *5*(6), pp. 3022–3024.

8.4.6.7 *Scaphium macropodum* Beumee ex K. Heyne

Synonyms: *Scaphium lychnophorum* (Hance) Pierre; *Sterculia lychnophora* Hance; *Firmiana lychnophora* (Hance) K. Schum.

Common names: Malva nut; kepayang (Indonesia); kembang semangkuk jantung (Malaysia); samrong (Thailand)

Habitat: Forests

Distribution: Cambodia, Laos, Vietnam, Thailand, Malaysia, Indonesia

Botanical observation: This magnificent timber tree grows up to a height of 45 m. The bole develops large buttresses. The leaves are simple, spiral, and stipulate. The petiole is up to 15 cm long, slender, straight, and somewhat swollen at apex. The blade is broadly elliptic to oblong, 6.5–12.5 ×15–40 cm, somewhat membranaceous, cordate to round at base, acuminate at apex, and with up to about 11 pairs of secondary nerves. The panicles are axillary and showy. The calyx is yellow, tubular, 5 mm across, and 5-lobed. Corolla none. The folicles are about 20 cm long, reddish to yellowish, membranaceous and bearing a nut.

Medicinal uses: Fever, small pox (China); bronchitis (Cambodia, Laos, Vietnam)

Moderate antibacterial (Gram-positive) polar extract: Methanol extract of stem bark inhibited the growth of *Staphylococcus aureus* and *Bacillus subtilis* with the MIC/MBC values of 780/3130 and 3130/12500 μg/mL, respectively (Al Muqarrabun et al., 2014).

Weak antileptospiral lupane-type triterpene: Lupenone inhibited the growth of leptospiral strains belonging to 10 serovars with the MIC values ranging from 400 to 800 μg/mL (Punnam Chander et al., 2015).

Moderate antibacterial (Gram-positive) lupane-type triterpene: Lupeol inhibited the growth of *Enterococcus faecalis* (ATCC 29212) with the MIC_{50} of 62.5 μg/mL (Al Bonvicini et al., 2017).

Moderate antimycobacterial lupane-type triterpene: Lupeol, inhibited the growth of *Mycobacterium tuberculosis* with the MIC value of 100 μg/mL (Bibi et al., 2010).

Clinical antiviral (enveloped monopartite linear dimeric single stranded (+) RNA) polar extract: Intake of juice of seeds 250 cm^3/day for 1 month by a AIDS patient increased $CD4^+$ count by 27.4% and $CD8^+$ by 40.7% (Chaiyasit & Wiwanitkit, 2017).

Antiviral (enveloped monopartite linear double-stranded DNA) lupane-type triterpene: Lupenone exhibited viral plaque inhibitory effect against Herpes simplex virus type-1 and type-2 (Madureira et al., 2003).

Commentary: The plant produces the chromone sesquiterpene malayscaphiol (Al Muqarrabun et al., 2014) which is very probably antimicrobial owed to the presence of cyclohexenone and phenolic moieties. The plant yields the triterpenes lupeol and lupenone (Al Muqarrabun et al., 2014). Due to intense logging, deforestation, this tree and its potential antimicrobial principles may disappear soon. Emergence of COVID-19 may halt the deforestation process as the forests is one of the last sources of drugs. Is malayscaphiol antiviral?

REFERENCES

Al Bonvicini, F., Antognoni, F., Mandrone, M., Protti, M., Mercolini, L., Lianza, M., Gentilomi, G.A. and Poli, F., 2017. Phytochemical analysis and antibacterial activity towards methicillin-resistant *Staphylococcus aureus* of leaf extracts from *Argania spinosa* (L.) Skeels. *Plant Biosystems – An International Journal Dealing with all Aspects of Plant Biology*, *151*(4), pp. 649–656.

Al Muqarrabun, L.R., Ahmat, N., Aris, S.R.S., Shamsulrijal, N., Baharum, S.N., Ahmad, R., Rosandy, A.R., Suratman, M.N. and Takayama, H., 2014. A new sesquiterpenoid from *Scaphium macropodum* (Miq.) Beumee. *Natural Product Research*, *28*(9), pp. 597–605.

Bibi, N., Tanoli, S.A.K., Farheen, S., Afza, N., Siddiqi, S., Zhang, Y., Kazmi, S.U. and Malik, A., 2010. *In vitro* antituberculosis activities of the constituents isolated from *Haloxylon salicornicum*. *Bioorganic & Medicinal Chemistry Letters*, *20*(14), pp. 4173–4176.

Chaiyasit, K. and Wiwanitkit, V., 2017. $CD4^+$ count after Malva nut juice supplementation: an observation. *Annals of Tropical Medicine and Public Health*, *10*(4), pp. 1074–1075.

Madureira, A.M., Ascenso, J.R., Valdeira, L., Duarte, A., Frade, J.P., Freitas, G. and Ferreira, M.J.U., 2003. Evaluation of the antiviral and antimicrobial activities of triterpenes isolated from *Euphorbia segetalis*. *Natural Product Research*, *17*(5), pp. 375–380.

Punnam Chander, M., Vinod Kumar, K., Shriram, A.N. and Vijayachari, P., 2015. Anti-leptospiral activities of an endemic plant *Glyptopetalum calocarpum* (Kurz.) Prain used as a medicinal plant by Nicobarese of Andaman and Nicobar Islands. *Natural Product Research*, *29*(16), pp. 1575–1577.

8.4.6.8 *Sterculia foetida* L.

Synonyms: *Clompanus foetidus* (L.) Kuntze; *Clompanus molucanus* Raf.

Common names: Wild Indian Almond tree, Java olive tree; xiang ping po (China); samrong (Cambodia, Thailand); kabu-kabu, randu (Indonesia); kelumpang jari (Malaysia); kelumpang (the Philippines); trom (Vietman)

Habitat: Cultivated

Distribution: India, Sri Lanka, Bangladesh, Myanmar, Cambodia, Laos, Vietnam, Malaysia, Indonesia, the Philippines, Australia

Botanical observation: This ornamental tree grows up to 40 m tall. The bole is buttressed. The bark is greyish and the wood is smelly. The leaves are palmately lobed, spiral, and stipulate. The petiole is slender and up to about 20 cm long. The blade is 7–9 foliolate, the folioles elliptic-lanceolate, 10–15 × 3–5 cm, cuneate at base and acuminate or caudate at apex, and with 9–12 pairs of secondary nerves. The inflorescences are terminal, showy, and many flowered panicles. The calyx presents 5 purple-red recurved lobes which are up to about 1 cm long, elliptic-lanceolate, and hairy. The 12–15 stamens merge into a column. The stigma is 5-lobed. The follicles are massive, ellipsoid, reddish to orange, coriaceous, dehiscent, up to 8 cm long and contain numerous seeds which are black, glossy, and up to about 1.5 cm long.

Medicinal uses: Herpes zoster (Indonesia); fever, sore (the Philippines).

Very weak broad-spectrum antibacterial oleanane-type triterpene: β-Amyrin inhibited the growth of *Bacillus subtilis, Staphylococcus aureus, Staphylococcus epidermidis, Escherichia coli, Klebsiella pneumoniae, Micrococcus luteus, Pseudomonas aeruginisa, Salmonella typhi, Salmonella paratyphi*, and *Shigella dysenteriae* (Saeed & Sabir, 2001).

Strong antibacterial (Gram-positive) lipophilic oleanane-type triterpene: β-Amyrin (Log D 10.1 at pH 7.4; molecular mass = 426.7 g/mol) inhibited the growth of *Streptococcus mutans* (ATCC 25175) with the MIC value of 48.8 μg/mL (Zheng et al., 2011).

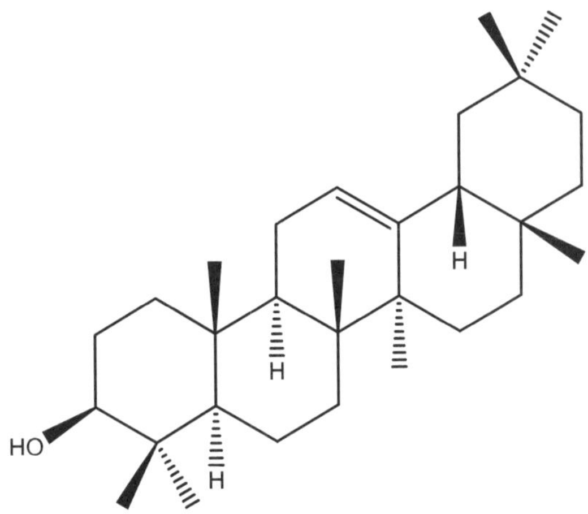

β-Amyrin

Moderate antimycobacterial lipophilic oleanane-type triterpene: β-Amyrin inhibited the growth of *Mycobacterium tuberculosis* with the MIC value of 100 μg/mL (Bibi et al., 2010).

Antibiotic-potentiator oleanane-type triterpene: Oleanolic acid inhibited NorA efflux pump in *Staphylococcus aureus* (Sharma et al., 2019).

Strong antifungal (filamentous) lipophilic oleanane-type triterpene: β-Amyrin inhibited the growth of *Trichophyton mentagrophytes* and *Trichophyton rubrum* with the MIC value of 12.5 μg/mL (Aguilar-Guadarrama et al., 2009).

Strong antiviral (enveloped monopartite linear double-stranded DNA) lipophilic olanane-type triterpene: Oleanolic acid (Log D = 5.9 at pH 7.4; molecular mass = 456.7 g/mol) from the roots inhibited the formation of plaques by the Herpes simplex virus type-1 and type-2 by with the EC_{50} of 6.8 and 7.8 μg/mL, respectively, via a mechanism involving the early period of virus multiplication (Mukherjee et al., 2013). At 13.8 μg/mL, this pentacyclic triterpene inhibited the replication of the Herpes simplex virus type-1 by 100% (Mukherjee et al., 2013). In Vero cells, oleanolic acid had a CC_{50} value of 98 μg/mL (Mukherjee et al., 2013).

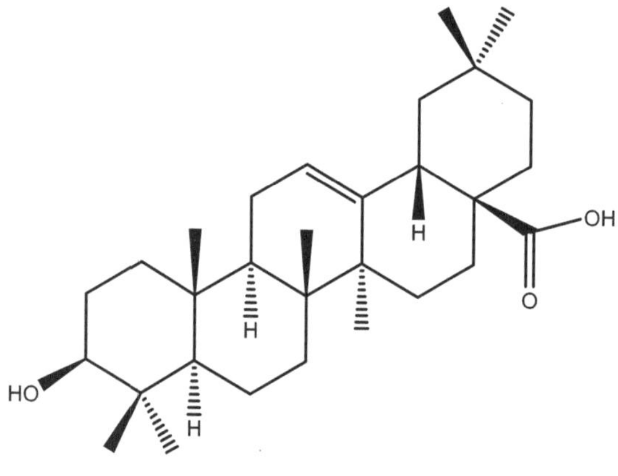

Oleanolic acid

Commentaries

i. The seeds contain cyclopropene fatty acids (Rani & Rajasekharreddy, 2010). The plant produces the oleanane-type triterpene stercufoetin A, vergatic acid, β-amyrin, oleanolic acid and maslinic acid (Pham et al., 2019), as well as flavonoid glycosides (Xia et al., 2009).
ii. An unidentified member of the genus *Sterculia* L. is used in the Solomon Islands to treat colds.
iii. The antifungal and antibacterial mode of action of pentacyclic triterpenes is not fully elucidated but evidence suggests increase of cytoplasmic membrane permeability involving impaired lipid biosynthesis in fungi (Haraguchi et al., 1999) and destruction of membrane integrity with leakage of cytoplasmic constituents in bacteria (Liu et al., 2015).

REFERENCES

Aguilar-Guadarrama, B., Navarro, V., Leon-Rivera, I. and Rios, M.Y., 2009. Active compounds against tinea pedis dermatophytes from *Ageratina pichinchensis* var. bustamenta. *Natural Product Research*, *23*(16), pp. 1559–1565.

Bibi, N., Tanoli, S.A.K., Farheen, S., Afza, N., Siddiqi, S., Zhang, Y., Kazmi, S.U. and Malik, A., 2010. *In vitro* antituberculosis activities of the constituents isolated from *Haloxylon salicornicum*. *Bioorganic & Medicinal Chemistry Letters*, *20*(14), pp. 4173–4176.

Haraguchi, H., Kataoka, S., Okamoto, S., Hanafi, M. and Shibata, K., 1999. Antimicrobial triterpenes from Ilex integra and the mechanism of antifungal action. *Phytotherapy Research: An International Journal Devoted to Pharmacological and Toxicological Evaluation of Natural Product Derivatives*, *13*(2), pp. 151–156.

Liu, W.H., Liu, T.C. and Mong, M.C., 2015. Antibacterial effects and action modes of asiatic acid. *Biomedicine (Taipei)*, *5*(3), p. 16.

Mukherjee, H., Ojha, D., Bag, P., Chandel, H.S., Bhattacharyya, S., Chatterjee, T.K., Mukherjee, P.K., Chakraborti, S. and Chattopadhyay, D., 2013. Anti-herpes virus activities of *Achyranthes aspera*: an Indian ethnomedicine, and its triterpene acid. *Microbiological Research*, *168*(4), pp. 238–244.

Pham, N.K.T., Nguyen, T.D., Doan, T.D.C., Ha, T.D., Tran, N.M.A., Tran, T.D., Mai, D.T. and Nguyen, T.P., 2019. Stercufoetin A, new oleanane-type triterpenoid from the leaves of *Sterculia foetida* L. *Natural Product Research*, *35*(7), pp. 1226–1231.

Rani, P.U. and Rajasekharreddy, P., 2010. Insecticidal activity of (2n-octylcycloprop-1-enyl)-octanoic acid (I) against three Coleopteran stored product insects from *Sterculia foetida* (L.). *Journal of Pest Science*, *83*(3), pp. 273–279.

Saeed, M.A. and Sabir, A.W., 2001. Antibacterial activity of *Caesalpinia bonducella* seeds. *Fitoterapia*, *72*(7), pp. 807–809.

Sharma, A., Gupta, V.K. and Pathania, R., 2019. Efflux pump inhibitors for bacterial pathogens: from bench to bedside. *The Indian Journal of Medical Research*, *149*(2), p. 129.
Xia, P.F., Feng, Z.M., Yang, Y.N. and Zhang, P.C., 2009. Two flavonoid glycosides and a phenylpropanoid glucose ester from the leaves of *Sterculia foetida*. *Journal of Asian natural Products Research*, *11*(8), pp. 766–771.
Zheng, Y., Huang, W., Yoo, J.G., Ebersole, J.L. and Huang, C.B., 2011. Antibacterial compounds from *Siraitia grosvenorii* leaves. *Natural Product Research*, *25*(9), pp. 890–897.

8.4.7 FAMILY TILIACEAE A. L. DE JUSSIEU (1789)

The family Tiliaceae consists of about 50 genera and 500 species of shrubs, trees, or herbs. The leaves are simple, alternate, or spiral, and stipulate. The petiole is often slender and long. The blades in this family often present characteristic secondary nerve patterns. The inflorescences are cymes. The calyx presents 4–5 sepals fused at base or free. The corolla comprises 4–5 petals or none. The androecium consists of numerous stamens. The gynoecium is made of 2–6 fused carpels forming a plurilocular ovary, with 1 to many ovules per locules, and develops a single style and a lobed stigma. The fruits are mainly drupes, capsules, or schizocarps

8.4.7.1 *Corchoropsis crenata* Siebold & Zucc.

Synonym: *Corchoropsis tomentosa* Makino

Common name: Tian ma (China)

Habitat: Sunny fields, wastelands, rocky soils, roadsides

Distribution: China, Korea, Japan

Botanical observation: This herb grows to a height of 1 m. The stems are hairy. The leaves are simple, spiral, and stipulate. The stipules are up to 4 mm long and linear. The petiole is up to about 5 cm long. The blade is lanceolate to ovate, crenate, 2–8 × 1–4.5 cm, hairy, rounded to cordate at base, marked with about 4–5 pairs of secondary nerves, and acuminate at apex. The inflorescence is solitary and axillary on pedicel which is up to 3 cm long. The flowers at first glance give an impression of somewhat those of *Chelidonium*. The 5 sepals are narrowly lanceolate, up to 8 mm long, and reflexed. The 5 petals are yellow, obovate, and up to 1 cm long. The androecium comprises up to 15 stamens arranged in 5 bundles. The capsules are terete, up to 4 cm long, hairy, splitting in 3 valves, and contain numerous tiny, ovoid, and brown seeds.

Medicinal uses: Abscesses, furuncles (Cambodia, Laos, Vietnam)

Viral inhibition by polar extract: Methanol extract of aerial part inhibited the Human immunodeficiency virus type-1 protease (Park, 2003).

Commentaries

i. The plant could be further examined for its antimicrobial and particularly antiviral constituents.
ii. Methanol extract of leaves of *Tilia amurensis* Rupr. inhibited the Human immunodeficiency type-1 protease (Park, 2003).

REFERENCE

Park, J.C., 2003. Inhibitory effects of Korean plant resources on human immunodeficiency virus Type 1 Protease activity. *Oriental Pharmacy and Experimental Medicine*, *3*(1), pp. 1–7.

8.4.7.2 *Corchorus olitorius* L.

Synonyms: *Corchorus catharticus* Blanco; *Corchorus decemangularis* Roxb. ex G. Don; *Corchorus quinquelocularis* Moench

Common names: Jew's mallow, red jute; cham ghas (Bangladesh); chang shuo huang ma, má yè (China); pat (Bangladesh; India); birbat, kaala shaaka (India); mulukhiyah (Irak); gedangan (Indonesia); moroheya (Japan); patsan (Pakistan); lumhay (the Philippines); kra chao (Thailand); day (Vietnam)

Habitat: Roadsides, wastelands, cultivated

Distribution: Tropical Asia and Pacific

Botanical observation: This herb grows up to a height of 3 m. The stems are slightly reddish and fibrous and a source of fibers. The leaves are simple, edible, spiral, and stipulate. The petiole is up to about 3.5 cm long or reddish. The blade is oblong-lanceolate, 7–10 × 2–4.5 cm, with 7–10 pairs of secondary nerves, rounded at base, serrate, and acuminate at apex. The cymes are opposite to the leaves. The 5 sepals are oblong, up to 7 mm long, and caudate at apex. The 5 petals are yellow, up to 8 mm long, and obovate. The androecium consists of numerous stamens. The ovary is cylindrical. The capsule is cylindrical, 5–6-valved, up to about 8 cm long and opening to release numerous tiny and angled seeds.

Medicinal uses: Cystitis, gonorrhea (Bangladesh); dysentery (China; India); fever (Iraq); sore throat (China)

Broad-spectrum antibacterial halo developed by polar extract: Methanol extract of seeds (100 µg/mL/6 mm well) inhibited the growth of *Staphylococcus aureus*, *Bacillus pumilus*, *Salmonella typhi*, *Salmonella typhimurium*, *Shigella boydii*, *Shigella sonnei*, *Shigella dysenteriae*, *Klebsiella pneumoniae*, *Escherichia coli*, and *Vibrio cholerae* with inhibition zone diameter of 14, 15, 12, 8, 10, 11, 13, 5, 14, and 9 mm, respectively (Pal et al., 2006).

Antifungal (filamentous) halo developed by polar extract: Methanol extract of leaves (100 mg/mL, 100 µL/ 10 mm well) inhibited the growth of *Aspergillus niger* (ATCC 10949), *Botrytis cinerea* (9492), *Fusarium graminearum*, and *Geotrichum candidum* with inhibition zone diameter of 12, 14, 12, and 18 mm, respectively (Ilhan et al., 2007).

Strong antiviral (enveloped monopartite linear single-stranded (−) RNA) polar extract: Aqueous extract of leaves inhibited the replication of the Measles virus (Schwartz Edmonston strain) with an IC_{50} value of about 25 µg/mL (Hasan & Kadhim, 2018).

Commentaries

i. One could examine the plant for activity against coronaviruses.
ii. Members of the genus *Corchorus* L. are not uncommonly used in traditional Asian medicinal practice for the treatment of microbial infections: *Corchorus capsularis* L. is used in Bangladesh (local name: *Paat gach*) to treat influenza. In Cambodia, Laos, Vietnam, and Malaysia (local name = *Kancing baju*), *Corchorus capsularis* L. affords a remedy for abscesses and dysentery, while in the Philippines it is used for fever. *Corchorus aestuans* L. is used to treat abscesses and pneumonia in China.

REFERENCES

Hasan, H.T. and Kadhim, E.J., 2018. Phytochemical investigation of *Corchorus olitorius* L. leaves cultivated in Iraq and it's *in vitro* antiviral activity. *Iraqi Journal of Pharmaceutical Sciences*, pp. 115–122 (P-ISSN: 1683-3597, E-ISSN: 2521-3512).

Ilhan, S., Savaroğlu, F. and Çolak, F., 2007. Antibacterial and antifungal activity of *Corchorus olitorius* L. (Molokhia) extracts. *International Journal of Natural & Engineering Sciences*, *1*(3), pp. 59–61.

Pal, D.K., Mandal, M., Senthilkumar, G.P. and Padhiari, A., 2006. Antibacterial activity of *Cuscuta reflexa* stem and *Corchorus olitorius* seed. *Fitoterapia*, *77*(7–8), pp. 589–591.

8.4.7.3 *Grewia hirsuta* Vahl

Synonyms: *Grewia hirsuta* Roxb.; *Grewia hirsuta* Sm.

Common names: Seta kata, seta andir (Bangladesh); cu mao bian dan gan (China); nagabala (India)

Habitat: Forests, thickets

Distribution: India, Sri Lanka, Nepal, Bangladesh, Myanmar, Laos, Cambodia, Vietnam, Malaysia

Botanical observation: This shrub grows to about 3 m tall. The stems are hairy. The leaves are simple, alternate, and stipulate. The stipules are minute. The petiole is minute and hairy. The blade is lanceolate, 5–15 × 2.5–3.5 cm, coriaceous, hairy below, with 4–5 pairs of secondary nerves, minutely serrate, round, asymmetrical, or cordate at base and acute at apex. The inflorescences are axillary cymes of few flowers (which look somewhat like those of *Citrus*). The calyx includes 5 sepals which are lanceolate, hairy, and about 1 cm long. The 5 petals are oblong, whitish, and shorter than the sepals. The androecium includes 40 stamens which are about 5 mm long. The ovary is hairy and develops a slender style and a 4-lobed stigma. The drupe is globose.

Medicinal uses: Fever, typhus, diarrhea, gonorrhea (Bangladesh)

Strong broad-spectrum antibacterial phenolics: Phenolic fraction from the fruits inhibited the growth of *Bacillus subtilis* (clinical isolate), *Staphylococcus aureus* (clinical isolate), *Staphylococcus epidermidis* (clinical isolate), and *Klebsiella pneumoniae* (clinical isolate) with the MIC values of 31.2, 15.6, 31.2, and 31.2 μg/mL, respectively (Siddiqi et al., 2011).

Commentaries: The plant could be examined for its antiviral properties (Babbar et al., 1982). Members of the genus *Grewia* L. are used for the treatment of infectious diseases in Asian systems of medicines: in Bangladesh, *Grewia serrulata* DC. (local name: *Monricca*) is used to heal wounds. In India, *Grewia sapida* Roxb. ex DC. (local name: *Barsa pakar*) and *Grewia helicterifolia* Wall. ex G. Don. (local name: *Kukurbhendi*) are used to treat diarrhea and *Grewia abutilifolia* Vent. ex Juss. (local name: *Mirga*) is used to treat dysentery. In Rajasthan, *Grewia damine* Gaertn. is used to heal boils. In Myanmar, *Grewia polygama* Roxb. is used to treat dysentery. *Grewia eriocarpa* Juss. is used for the treatment of fever in the Philippines.

REFERENCES

Babbar, O.P., Joshi, M.N. and Madan, A.R., 1982. Evaluation of plants for antiviral activity. *Indian Journal of Medical Research*, *76*(S), pp. 54–65.

Siddiqi, R., Naz, S., Ahmad, S. and Sayeed, S.A., 2011. Antimicrobial activity of the polyphenolic fractions derived from *Grewia asiatica*, *Eugenia jambolana* and *Carissa carandas*. *International Journal of Food Science & Technology*, *46*(2), pp. 250–256.

8.4.7.4 *Microcos paniculata* L.

Synonyms: *Fallopia nervosa* Lour.; *Grewia microcos* L.; *Grewia nervosa* (Lour.) Panigrahi; *Microcos nervosa* (Lour.) S.Y. Hu

Common names: Asar, Boot gach (Bangladesh); po bu ye (China); shiral (India); khom (Laos, Thailand); bung lai (Vietnam)

Habitat: Roadsides, open forests

Distribution: India, Bangladesh; China, Southeast Asia

Botanical observation: This tree grows up to about 10 m tall. The stem is hairy and somewhat zig-zag shaped at apex. The leaves are simple, somewhat pendulous, alternate, and stipulate. The stipule is linear and up to 7 mm long. The petiole is hairy and up to 1.5 cm long. The blade is ovate or oblong, 8–18 × 4–8 cm, coriaceous, dark grey glossy above, with about 4–5 pairs of secondary nerves and scalariform tertiary nerves (somewhat *Melastoma*-like), round at base, laxly crenate, and acuminate at apex. The inflorescence is a terminal and hairy panicle which is up to 10 cm long. The calyx comprises 5 sepals which are spathulate, up to 7 mm, and hairy. The 5 petals are oblong, whitish, 4 mm long and hairy at base. The androecium consists of numerous stamens. The ovary is globose and glabrous. The drupe is subglobose, green, smooth, glossy and up to 1 cm across.

Medicinal uses: Diarrhea (Bangladesh); fever, diarrhea, jaundice (China); fever and diarrhea (Indonesia)

Commentaries

i. The plant produces epicatechin (see earlier), the triterpene (–) maslinic acid, the apocarotenoid loliolide (Still, 2013), as well as series of long chain piperidine alkaloids such as *N*-methyl-6β-(deca-l′, 3′, 5′-trienyl)-3β-methoxy-2β-methylpiperidine (Bandara et al., 2000) and microgrewiapine A (Still et al., 2013).

ii. *Microcos tomentosa* Sm. is used for the treatment of cough in Cambodia, Laos, and Vietnam, fever in Malaysia, and trush in Indonesia.

REFERENCES

Bandara, K.P., Kumar, V., Jacobsson, U. and Molleyres, L.P., 2000. Insecticidal piperidine alkaloid from *Microcos paniculata* stem bark. *Phytochemistry*, *54*(1), pp. 29–32.

Still, P.C., 2013. *Cytotoxic alkaloids from Microcos paniculata with activity at neuronal nicotinic receptors* (Doctoral dissertation, The Ohio State University).

Still, P.C., Yi, B., González-Cestari, T.F., Pan, L., Pavlovicz, R.E., Chai, H.B., Ninh, T.N., Li, C., Soejarto, D.D., McKay, D.B. and Kinghorn, A.D., 2013. Alkaloids from *Microcos paniculata* with cytotoxic and nicotinic receptor antagonistic activities. *Journal of Natural Products*, *76*(2), pp. 243–249.

8.4.7.5 *Triumfetta rhomboidea* Jacq.

Synonyms: *Bartramia indica* L.; *Triumfetta angulata* Lam.; *Triumfetta bartramia* L.; *Triumfetta indica* Lam.; *Triumfetta riparia* Hochst.; *Triumfetta trilocularis* Roxb.; *Triumfetta velutina* Vahl

Common names: Burweed, Chinese burr, common burbush; ci chuo ma (China); sukupan (Indonesia); chiki (India); pulut pulut (Malaysia); gavana (Papua New Guinea); moropoto (the Philippines); poyumyu (Thailand)

Habitat: Wastelands, roadsides

Botanical observation: This herb grows up to 1.5 m tall. The stem is hairy and fibrous. The leaves are simple, spiral, and stipulate. The stipules are linear and up to 4 mm long. The petiole is up to 4 cm long. The blade is 2.5–8 × 2.5–6.5 cm, lanceolate to 3-lobed, rounded or cordate at base, serrate, and acute or acuminate at apex. The cymes are opposite the leaves. The 5 sepals are 5 mm long, lanceolate, and hairy. The 5 petals are golden-yellow, 5 mm long, and oblong to spathulate. The androecium comprises up to 15 stamens. The ovary is minute and hairy. The capsules are 5 mm across and covered with hooks.

Medicinal uses: Abscesses (China); blenorrhagia, boils (the Philippines)

Antibacterial (Gram-positive) halo developed by polar extract: Ethanol extract of leaves (200 μL of a 500 mg/mL solution in 9 mm diameter agar wells) inhibited the growth of *Staphylococcus aureus* (ATCC 25923) (Vlietinck et al., 1995).

Moderate broad-spectrum antibacterial essential oil: Essential oil (mainly containing β-caryophyllene, kessane, and caryophyllene oxide) inhibited the growth of *Escherichia coli* (CIP: 54127) and *Enterococcus hirae* (CIP: 5855) with the MIC values of 100 μg/mL (Mevy et al., 2006).

Antifungal polar extract: Ethanol extract of leaves inhibited the growth of *Epidermophyton floccosum, Microsporum canis* and *Trichophyton rubrum* (Cos et al., 2002).

Antiviral (enveloped monopartite linear double-stranded DNA) polar extract: Ethanol extract of leaves inhibited the replication of the Herpes simplex virus (Vlietinck et al., 1995).

Antiviral (non-enveloped monopartite linear single-stranded (+) RNA) polar extract: Ethanol extract of leaves inhibited the replication of the Coxsackie virus (Vlietinck et al., 1995).

Antiviral (enveloped monopartite linear single-stranded (+) RNA) polar extract: Ethanol extract of leaves inhibited the replication of the Semliki forests virus (Vlietinck et al., 1995).

Strong antiviral (enveloped monopartite linear double-stranded DNA) lipophilic sesquiterpenes: β-Caryophyllene (Log D = 6.4 at pH 7.4; molecular mass = 204.3 g/mol) and caryophyllene oxide inhibited the replication of the Herpes simplex virus type-1 with the IC_{50} values of 0.2 and 0.7 μg/mL, respectively, and selectivity indexes of 140 and 25.7, respectively (Astani et al., 2007).

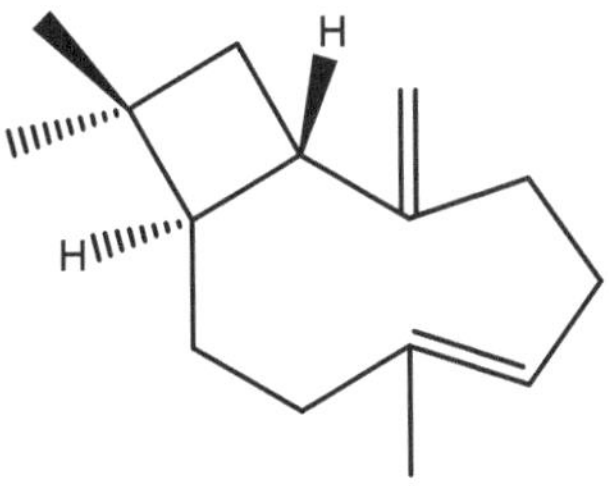

β-Caryophyllene

Commentaries

i. The plant produces flavonoid glycosides such as scutellarein 6-xyloside 7-rhamnoside (Nair et al., 1986). It may produce quinolinone alkaloids like waltherione A and E which were isolated from another member of the genus *Triumfetta* L. (Jang et al., 2015).
ii. Ethanol extract being active against the Semliki forests virus *in vitro* could be active against coronaviruses because both coronaviruses and Semliki forests virus are enveloped, monopartite, linear, single-stranded (+) RNA viruses.

REFERENCES

Astani, A., Reichling, J. and Schnitzler, P., 2011. Screening for antiviral activities of isolated compounds from essential oils. *Evidence-Based Complementary and Alternative Medicine, 2011.* doi:10.1093/ecam/nep187.

Cos, P., Hermans, N., De Bruyne, T., Apers, S., Sindambiwe, J.B., Berghe, D.V., Pieters, L. and Vlietinck, A.J., 2002. Further evaluation of Rwandan medicinal plant extracts for their antimicrobial and antiviral activities. *Journal of Ethnopharmacology, 79*(2), pp. 155–163.

Jang, J.Y., Le Dang, Q., Choi, Y.H., Choi, G.J., Jang, K.S., Cha, B., Luu, N.H. and Kim, J.C., 2015. Nematicidal activities of 4-quinolone alkaloids isolated from the aerial part of *Triumfetta grandidens* against *Meloidogyne incognita. Journal of Agricultural and Food Chemistry, 63*(1), pp. 68–74.

Mevy, J.P., Bessiere, J.M., Rabier, J., Dherbomez, M., Ruzzier, M., Millogo, J. and Viano, J., 2006. Composition and antimicrobial activities of the essential oil of *Triumfetta rhomboidea* Jacq. *Flavour and Fragrance Journal, 21*(1), pp. 80–83.

Nair, A.R., Seetharaman, T.R., Voirin, B. and Favre-Bonvin, J., 1986. True structure of triumboidin, a flavone glycoside from *Triumfetta rhomboidea. Phytochemistry, 25*(3), pp. 768–769.

Vlietinck, A.J., Van Hoof, L., Totte, J., Lasure, A., Berghe, D.V., Rwangabo, P.C. and Mvukiyumwami, J., 1995. Screening of hundred Rwandese medicinal plants for antimicrobial and antiviral properties. *Journal of Ethnopharmacology, 46*(1), pp. 31–47.

8.4.5 FAMILY THYMELEACEAE A. L. DE JUSSIEU (1789)

The family Thymeleaceae consists of about 50 genera and 650 species of herbs, shrubs, or trees not uncommonly resinous. The leaves are simple, opposite or alternate, and exstipulate. The inflorescences are racemes, spikes, umbels, or fascicles. The calyx is tubular and develops 4–5 lobes. Corolla none or very much reduced. The androecium comprises 2 to many stamens. A disc is present. The gynoecium comprises an ovary which is 1 to 2-loculed with 1 pendulous ovule per locules and a slender style. The fruit is a capsule or drupe.

8.4.5.1 *Aquilaria sinensis* (Lour.) Spreng

Synonyms: *Agallochum sinense* (Lour.) Kuntze; *Aquilaria grandiflora* Benth.; *Aquilaria ophispermum* Poir.; *Ophispermum sinense* Lour.

Common name: Tu chen xiang (China)

Habitat: Forests

Distribution: China

Botanical observation: This resinous tree grows up to a height of 15 m. The bark is grey. The leaves are simple, alternate, and exstipulate. The petiole is about 5–7 mm long. The blade is elliptic to oblong, coriaceous, glossy, dark green, cuneate at base, acuminate at apex, 5–9×2.5–6 cm, and marked with 15–20 pairs of unconspicuous secondary nerves. The panicles of umbels are axillary or terminal. The calyx is yellowish-green, tubular, about 5 mm long, and develops 5 lobes. The perianth is reduced to scales. The flower altogether looks like a tiny *Rafflesia*. The androecium includes 10 stamens. The ovary is ovoid. The capsule is obovoid, glaucous, hairy, and shelters dark brown seeds, which are about 1 cm long.

Medicinal use: Cuts (China)

Commentaries

i. The plant produces chromanes, agarofuran-type and agarospirane-type sesquiterpenes, abietane-type diterpenes, tirucallane-type triterpenes (Wang et al., 2015) which could be examined for their antimicrobial properties.
ii. It is used for the treatment of COVID-19 in China (Wang et al., 2020).
iii. In the Philippines, *Aquilaria cumingiana* (Decne.) Ridley (local name= *Binukat*) is used to heal wounds.

REFERENCES

Wang, S.X., Wang, Y., Lu, Y.B., Li, J.Y., Song, Y.J., Nyamgerelt, M. and Wang, X.X., 2020. Diagnosis and treatment of novel coronavirus pneumonia based on the theory of traditional Chinese medicine. *Journal of Integrative Medicine*, *18*(4), pp. 275–283.

Wang, S.-C., Wang, F., and Yue, C.-H., 2015. Chemical constituents from the petioles and leaves of *Aquilaria sinensis. Biochemical Systematics and Ecology*, *61*, pp. 458–461.

8.4.5.2 *Stellera chamaejasme* L.

Synonyms: *Chamaejasme stelleriana* Kuntze; *Passerina chamaejasme* Fisch. ex Meisn.; *Passerina dichotoma* (Fisch. ex Sweet) Steud.; *Passerina stelleri* Wikstr.; *Stellera bodinieri* H. Lév.; *Wikstroemia chamaejasme* (L.) Domke

Common name: Lang du, rui xiang lang du, ri jia aa (China)

Habitat: Sunny and rocky hill slopes, sandy soils

Distribution: Bhutan, Nepal, China

Botanical observation: This heavenly and diffuse herbs grow from a rhizome up to a height of about 50 cm. The stems are woody at base. The leaves are simple, subsessile, alternate, opposite, or whorled. The blade is lanceolate to oblong, 1–2.5×0.5–1 cm, rounded to cuneate at base, and acuminate, or acute at apex. The umbels are globose and terminal. The calyx is tubular, white, yellow, or reddish purple, about 1 cm long, and develops 5 ovate-oblong, about 5 mm long, lobes. Petals are none. The 10 stamens are inserted on the calyx tube. A disc is present. The ovary develops a short style and a capitate stigma. The drupes are about 5 mm long and enclosed in persistent calyx.

Medicinal uses: Tuberculosis, ulcers (China)

Strong antiviral (enveloped, monopartite, double-stranded DNA) amphiphilic flavonol: Quercetin (molecular mass = 320.2 g/mol) isolated from the roots at the concentration of 0.2 μM inhibited the secretion of the Hepatits B virus surface-antigen secretion by 64.3% (Yang & Chen, 2008).

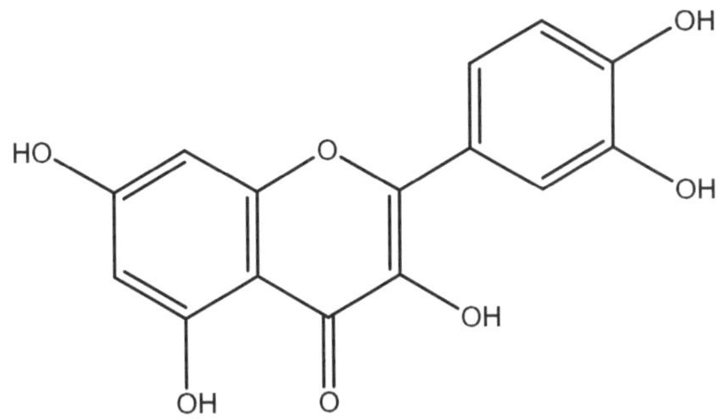

Quercetin

Strong antiviral (enveloped, monopartite, double-stranded DNA) biflavanol: Sikokianin A isolated from the roots at the concentration of 0.2 μM inhibited the secretion of the Hepatits B virus by 71.9% (Yang & Chen, 2008).

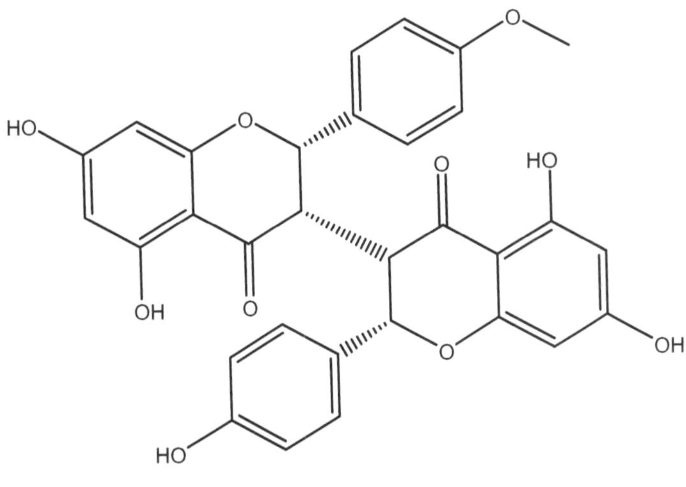

Sikokianin A

Moderate antiviral (enveloped, monopartite, double-stranded DNA) chromone: Chamaechromone from the roots at the concentration of 0.2 μM inhibited the secretion of the Hepatits B virus by 34% (Yang & Chen, 2008).

Strong antiviral (monopartite, linear, dimeric, single stranded (+) RNA) lipophilic tigliane-type diterpenes: Stelleralide A (LogD = 6.9 at pH 7.4; molecular mass = 712.8 g/mol) and gnidimacrin inhibited the replication of the Human immunodeficiency virus in MT4 cells with the EC_{90} value around 0.4 nM (CC_{50} = 4.3 μM) (Asada et al., 2013).

Strong antiviral (monopartite linear dimeric single stranded (+) RNA) biflavonoid: Chamaeflavone A inhibited the replication of the Human immunodeficiency virus (NL4-3 strain) virus in MT4 cells with the EC_{90} value of 4.2 µM (CC_{50} = 17.5 µM) (Asada et al., 2013).

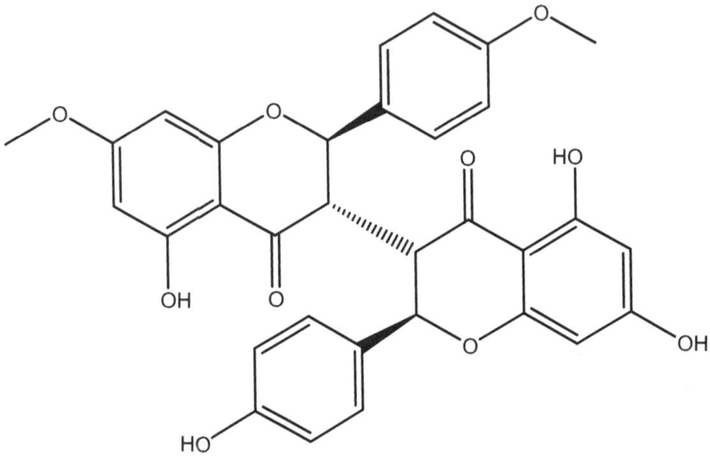

Chamaeflavone A

Strong antiviral (monopartite linear dimeric single stranded (+) RNA) tigliane-type diterpenes: Stellerarin, 12-*O*-benzoylphorbol 13-octanoate, stelleracin A, stelleracin B, stelleracin C, stelleracin D, and stelleracin E, inhibited the replication of the Human immunodeficiency virus (NL4-3 strain) virus in MT4 cells with the EC_{90} values of 4.2, 0.0005, 0.0005, 0.0006, 0.003, 0.4, and 1.5 µM (Asada et al., 2013). Stellerarin and 12-*O*-benzoylphorbol 13-octanoate were cytotoxic for MT-4 cells with the IC_{50} value of 5.1 and 4.4 µM, respectively (Asada et al., 2013).

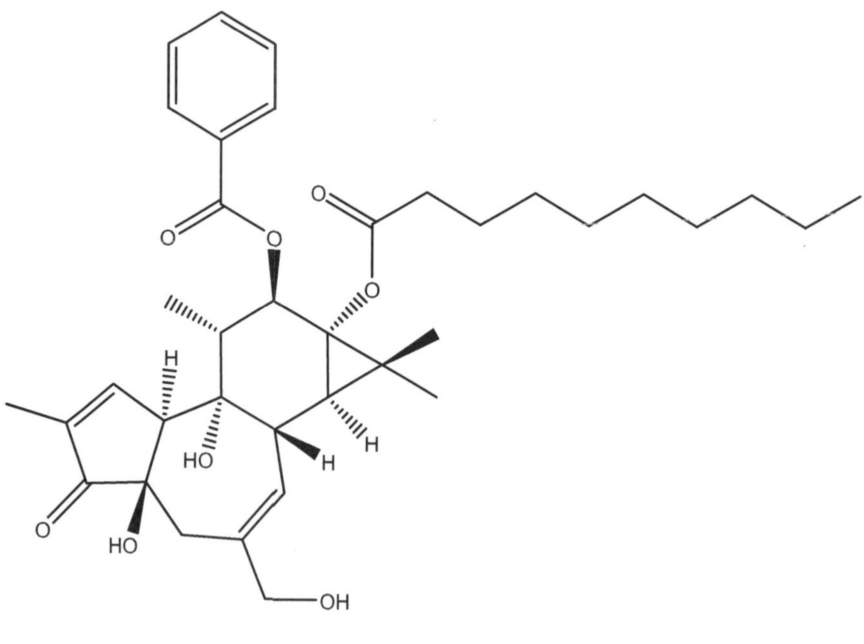

Stellerarin

Commentaries

i. Herbs growing in rocky soils or in fissures of stones are not uncommonly woody at base.

ii. Tigliane-type diterpenes (which are common in members of the Euphorbiaceae) are toxic for humans and of no therapeutic use but their exact antiviral mechanism could be a mean to indentify novel cellular pharmacological targets. Anti-Human immunodeficiency virus compounds are not uncommonly active against coronaviruses.

iii. In the Euphorbiaceae, the tigliane-type diterpene 4-deoxy-20-oxophorbol 12-tiglyl 13-acetate isolated from *Croton tiglium* L. inhibited the growth of *Mycobacterium tuberculosis* with the IC_{50} value of 19.5 µM (Zhao et al., 2016). Thus, the antitubercular medicinal use of *Stellera chamaejasme* L. could be owed to tigliane-type diterpene yet to be isolated. One could wonder why some plants in different families produce the same type of characteristic natural products? Is there a link between Euphorbiaceae and Thymeleaceae?

iv. In Indonesia, *Phaleria macrocarpa* (Scheff.) Boerl. (local name = *Mahkota dewa*) is used for the treatment of pharyngitis and diarrhea. Ethanol extract of fruits inhibited the growth of *Enterococcus faecalis* with the MIC value of 512 µg/mL (Romulo et al., 2018). Ethylacetate fraction of the fruits (edible Indonesia) of this shrub inhibited the replication of the Herpes simple virus type-1 in Vero cells with the EC_{50} value of 150 µg/mL (Ismaeel et al., 2018).

REFERENCES

Asada, Y., Sukemori, A., Watanabe, T., Malla, K.J., Yoshikawa, T., Li, W., Kuang, X., Koike, K., Chen, C.H., Akiyama, T. and Qian, K., 2013. Isolation, structure determination, and anti-HIV evaluation of tigliane-type diterpenes and biflavonoid from *Stellera chamaejasme*. *Journal of Natural Products*, *76*(5), pp. 852–857.

Ismaeel, M.A.H.M.U.D., Dyari, H.R., Nor, N.S., Yaacob, W.A. and Ibrahim, N.A.Z.L.I.N.A., 2018. Anti-human herpesvirus type-1 activity of *Phaleria macrocarpa* fruits methanol extract and fractions. *Malaysian Applied Biology*, *47*(5), pp. 31–40.

Romulo, A., Zuhud, E.A., Rondevaldova, J. and Kokoska, L., 2018. Screening of *in vitro* antimicrobial activity of plants used in traditional Indonesian medicine. *Pharmaceutical Biology*, *56*(1), pp. 287–293.

Yang, G. and Chen, D., 2008. Biflavanones, flavonoids, and coumarins from the roots of *Stellera chamaejasme* and their antiviral effect on hepatitis B virus. *Chemistry & Biodiversity*, *5*(7), pp. 1419–1424.

Zhao, B.Q., Peng, S., He, W.J., Liu, Y.H., Wang, J.F. and Zhou, X.J., 2016. Antitubercular and cytotoxic tigliane-type diterpenoids from *Croton tiglium*. *Bioorganic & Medicinal Chemistry Letters*, *26*(20), pp. 4996–4999.

8.5 Order Sapindales Juss. ex Bercht. & J. Presl (1820)

8.5.1 FAMILY ANACARDIACEAE R. BROWN (1818)

The family Anacardiaceae consists of about 60 genera and 600 species of resinous or sappy trees, shrub, or climbers. The sap can induce anaphylactic shock by simple skin contact. The leaves are mostly alternate, simple or compound, coriaceous, with somewhat showy and characteristic secondary nerves. Stipules are absent. The inflorescences are panicles. The flowers are minute, whitish to green, with 5 sepals, 5 petals and 10 stamens and present a well-developed 5-lobed nectary disc. The gynoecium comprises 1–5 carpels fused into a 1–3 locular ovary, each locule containing a single ovule. The fruits are drupes. Members in this family produce antimicrobial hydrolysable tannins, biflavonoids, and vesicant alkyl phenols.

8.5.1.1 *Anacardium occidentale* L.

Synonyms: *Acajuba occidentalis* (L.) Gaertn.; *Anacardium microcarpum* Ducke; *Cassuvium pomiferum* Lam.

Common names: Cashew-nut tree; kajul; hijli badam (Bangladesh); svai chanti (Cambodia); yao guo (China); jambu monyet (Indonesia); gajus, jambu golok, keterek (Malaysia); thiho-thayet, shitkale, mak-mong-sang-yip (Myanmar); balubad (the Philippines); yaruang (Thailand); cây diêù (Vietnam)

Habitat: Cultivated

Distribution: Tropical Asia

Botanical observation: This beautiful tree grows to a height of 10 m leaves. The bole is seldom straight and the bark exudes upon incision a resinous and acrid sap which blackens on exposure to air. The wood is very valuable and exudes a yellow gum, which can raise blisters. The leaves are simple, spiral, and exstipulate. The petiole is about 2 cm long. The blade is coriaceous, obovate, 5–15 × 3–10 cm, tapering at base, round or blunt and often notched at apex, and shows 10–18 pairs of secondary nerves. The flowers are minute, with a sickly sweet odor and arranged in terminal panicles. The calyx develops 5 sepals which are about 1 cm long. The 5 petals are greenish – white to pink with red longitudical lines. The androecium includes up to 10 stamens. The style is about 1 cm long. The drupes are 2–3 cm long, greyish brown, asymmetrical, somewhat boxing gloves-shaped, up to 3 cm long, and sheltering a curved seed, which is edible after being roasted. The fruit pedicels are swollen, succulent, palatable, cushion-shaped, glossy, yellow, pink or red, pulpy, fragrant, and 3–6 cm long.

Medicinal uses: Leprosy, warts, syphilis (Bangladesh)

Strong antibacterial (Gram-positive) lipophilic alkyl-phenol: Anacardic acid (also known as 6-[8(Z), 11(Z), 14-pentadecatrienyl] salicylic acid; Log D = 6.4 at pH 7.4; molecular mass = 348.5 g/mol) inhibited the growth of *Bacillus subtilis* (ATCC 9372), *Brevibacterium ammoniagenes* (ATCC 6872), *Staphylococcus aureus* (ATCC 12598), *Streptococcus mutans* (ATCC 25175), and *Propionibacterium acnes* (ATCC 11827) with the MIC values of 3.1, 3.1, 6.2, 1.5, and 0.7 μg/mL, respectively (Kubo et al., 1993). Anacardic acid inhibited the growth of methicillin-resistant *Staphylococcus aureus* (ATCC 33591) and penicillin-resistant *Staphylococcus aureus* (ATCC 11632) with the MIC values of 6.2 and 3.1 μg/mL, respectively (Muroi & Kubo, 1996).

Moderate antifungal akyl-phenol: Anacardic acid at the concentration of 75 μM abrogated the spore germination of *Magnaporthe oryzae* (Muzaffar et al., 2016).

DOI: 10.1201/9781003180371-2

Antibiotic potentiator alkyl-phenol: Anacardic acid decreased the resistance of *Staphylococcus aureus* (ATCC 33591) to methicillin (Muroi & Kubo, 1996).

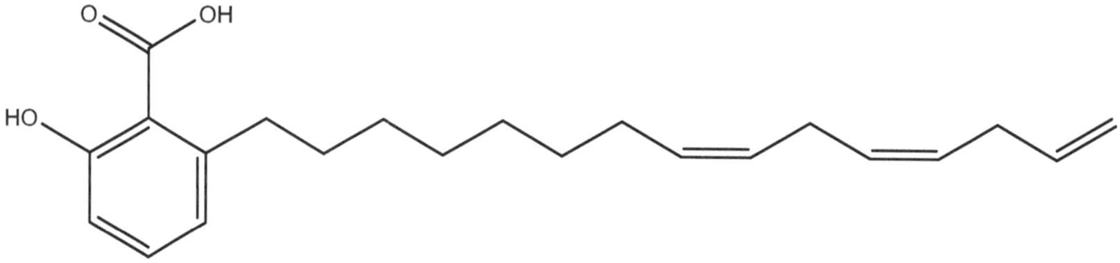

Anacardic acid

Strong antiviral (non-enveloped, segmented, linear, double-stranded RNA) polar extract: Aqueous extract of leaves at the concentration of 4 µg/mL inhibited the replication of the Human rotavirus and the Simian rotavirus in African Rhesus monkey kidney cells (MA-104) by 25.9 and 84.5%, respectively (Gonçalves et al., 2005).

Strong antiviral (enveloped, segmented, linear, single stranded (−) RNA) hydrophilic bisflavone: Agathisflavone (Log D = −0.9 at pH 7.4; molecular mass = 538.4 g/mol) inhibited the replication of the Influenza A virus in Madin-Darby canine kidney with the EC_{50} value of 1.3 µM and the selectivity index of 268.4 (De Freitas et al., 2020). Agathisflavone inhibited Influenza A H3N2 wild type and H1N1 wild type with the IC_{50} values of 20.4 and 28.6 µM, respectively (De Freitas et al., 2020).

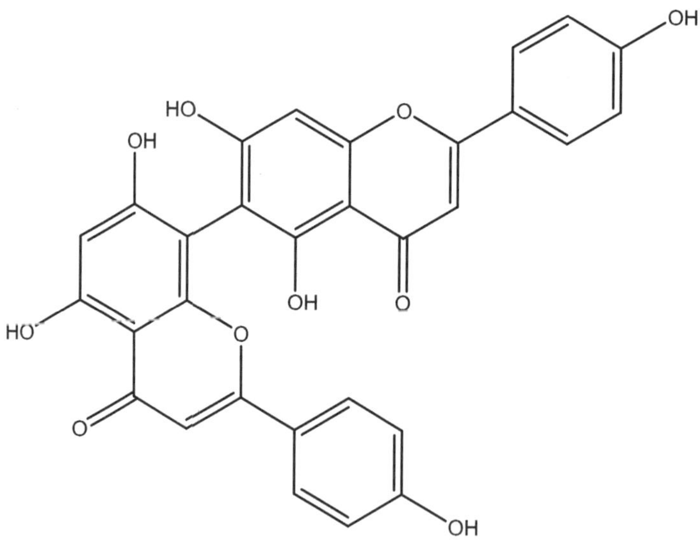

Agathisflavone

Commentaries

i. Long chain in anacardic acid may contribute to bacterial cystoplasmic membrane penetration.
ii. Malays eat young fresh leaves as protective medicine or "*ulam.*"
iii. Natural products inhibiting Influenza viruses *in vitro* are often hydrophilic flavonoids.

REFERENCES

De Freitas, C.S., Rocha, M.E., Sacramento, C.Q., et al., 2020. Agathisflavone, a biflavonoid from *Anacardium occidentale* L., inhibits influenza virus neuraminidase. *Current Topics in Medicinal Chemistry*, *20*(2), pp. 111–120.

Gonçalves, J.L.S., Lopes, R.C., Oliveira, D.B., et al., 2005. *In vitro* anti-rotavirus activity of some medicinal plants used in Brazil against diarrhea. *Journal of Ethnopharmacology*, *99*(3), pp. 403–407.

Kubo, I., Muroi, H., Himejima, M., Yamagiwa, Y., Mera, H., Tokushima, K., Ohta, S. and Kamikawa, T., 1993. Structure-antibacterial activity relationships of anacardic acids. *Journal of Agricultural and Food Chemistry*, *41*(6), pp. 1016–1019.

Muroi, H. and Kubo, I., 1996. Antibacterial activity of anacardic acid and totarol, alone and in combination with methicillin, against methicillin resistant *Staphylococcus aureus*. *Journal of Applied Bacteriology*, *80*(4), pp. 387–394.

Muzaffar, S., Bose, C., Banerji, A., et al., 2016. Anacardic acid induces apoptosis-like cell death in the rice blast fungus *Magnaporthe oryzae*. *Applied Microbiology and Biotechnology*, *100*(1), pp. 323–335.

8.5.1.2 *Dracontomelon dao* (Blanco) Merr. & Rolfe

Synonym: *Dracontomelon duperreanum* Pierre; *Dracontomelon mangiferum* Blume; *Dracontomelon sinensis* Stapf; *Paliurus dao* Blanco

Common names: New Guinea walnut, Papua New Guinea walnut, Pacific walnut; dahu (Indonesia); sengkuang (Malaysia); nga-bauk (Myanmar); mon (Papua New Guinea); dao, lamyo (the Philippines); phrachao ha phra ong (Thailand); sau (Vietnam)

Habitat: River banks and swampy areas in rainforest, cultivated

Distribution: From India to the Solomon Islands

Botanical observation: This majestic tree grows to about 40 m tall. The bole is straight and buttressed. The bark is greyish brown, the inner bark is pink, and a resin is exuded upon incision. The stems are covered with a few rusty hairs at apex. The leaves are spiral, imparipinnate, and exstipulate. The rachis is 30–45 cm long and presents 5–8 pairs of folioles which are 4–22.5 × 2.5–7.5 cm. The folioles are oblong, glossy, marked with 10–15 pairs of secondary nerves, round to somewhat asymmetrical at base, and acuminate at apex. The panicles are about 60 cm long, lax, and present numerous fragrant flowers. The calyx is minute and 5-lobed. The 5 petals are about 7 mm long. The androecium includes 10 stamens. A disc is present. The gynoecium includes 5 carpels which are partially free. The drupes are globose, up to 4 cm across, green turning yellow to brownish, edible, and with 5 oval markings on the upper side when dry.

Medicinal use: Dysentery (Malaysia, Philippines); wounds (the Philippines); dental infections (Vietnam)

Broad-spectrum antibacterial halo developed by mid-polar extract: Dichloromethane extract of leaves inhibited the growth of *Bacillus cereus*, *Bacillus subtilis*, *Micrococcus luteus*, *Staphylococcus aureus*, *Streptococcus pneumoniae*, *Citrobacter freundii*, *Klebsiella pneumoniae*, *Neisseria gonorrhea*, *Proteus mirabilis*, *Pseudomonas aeruginosa*, and *Salmonella typhi* with diameter inhibition zones of 16, 20, 18, 20, 20, 20, 20, 16, 20, 20, and 20 mm, respectively (4 mg/disc) (Khan & Omoloso, 2002).

Broad-spectrum antifungal halo developed by mid-polar extract: Dichloromethane extract of leaves inhibited the growth of *Aspergillus versicolor*, *Candida albicans*, *Cladosporium cladoporoides*, and *Trichophyton tonsurum* (4 mg/disc) (Khan & Omoloso, 2002).

Commentary: The plant builds cardol and anacardic acid (Ragasa et al., 2017) (see previous plant). Not much is known about possible antiviral activities.

REFERENCES

Khan, M.R. and Omoloso, A.D., 2002. Antibacterial and antifungal activities of *Dracontomelon dao*. *Fitoterapia*, *73*(4), pp. 327–330.

Ragasa, C.Y., Batarra, T.C., Vivar, J.L.A., Mariquit, M. and Shen, C.C., 2017. Chemical constituents of *Dracontomelon dao* (Blanco) Merr. et Rolfe. *Pharmacognosy Journal*, *9*(5). doi:10.5530/pj.2017.5.103.

8.5.1.3 *Lannea coromandelica* (Houtt.) Merr.

Synonyms: *Calesium grande* Kuntze; *Dialium coromandelicum* Houtt.; *Haberlia grandis* Dennst.; *Lannea grandis* (Dennst.) Engl.; *Lannea wodier* (Roxb.) Parker; *Odina pinnata* Rottler; *Odina wodier* Roxb.; *Rhus odina* Buch.-Ham. ex Wall.

Common names: Jiwul (Bangladesh); hou pi shu (China); kakausil (India); baing (Myanmar); kamlai, kembal (Pakistan)

Habitats: Forests

Distribution: Pakistan to Malaysia

Botanical observation: This resinous tree grows to about 10 m tall. The leaves are imparipinnate, spiral, and exstipulate. The petiole is hairy. The blade includes 3–4 pairs of oblong folioles which are membranaceous, 5.5–14 × 2.5-4 cm, sessile, asymmetrical, marked with 5–10 pairs of secondary nerves, and acuminate at apex. The racemes are terminal, up to 30 cm long, and present numerous minute flowers. The calyx develops 4 ciliate lobes. The 4 petals are yellow to purplish, 2 mm long, and elliptic to lanceolate. A lobed disc is present. The 8 stamens are as long as the petals and arranged in 4 pairs. The ovary is ovoid and develops 3–4 styles. The drupes are coriaceous, about 1 cm long, and purplish red.

Medicinal uses: Dysentery (Bangladesh; India); boils, mumps, ulcers, wounds (Bangladesh)

Broad-spectrum antibacterial halo developed by polar extract: Ethanol extract of barks (10 µL of a 100 mg/mL solution/6 mm disc) inhibited the growth of *Bacillus cereus* (MTCC 430), *Staphylococcus aureus* (MTCC 96), *Escherichia coli* (MTCC 1771), and *Proteus vulgaris* (MTCC 1771) with the inhibition zone diameters of 20, 26, 13, and 12 mm, respectively (Satish et al., 2010). Ethanol extract of bark inhibited the growth of *Staphylococcus aureus* (MTCC 3160) and *Streptococcus pyogenes* (MTCC 1928) with inhibition zone diameters of 19.2 and 18.1 mm, respectively (200 mg/mL/6 mm wells) (Kaur et al., 2013). Ethanol extract of bark inhibited the growth of *Escherichia coli* (Panda et al., 2017).

Broad-spectrum antifungal halo developed by polar extract: Ethanol extract of barks (10 µL of a 100 mg/mL solution/6 mm disc) inhibited the growth of *Aspergillus niger* (MTCC 872) with the inhibition zone diameters of 21 mm (Sathish et al., 2010). Ethanol extract of bark inhibited the growth of *Candida albicans* (MTCC 183) with the inhibition zone diameter of 13.2 mm (200 mg/mL/6 mm well) (Kaur et al., 2013).

Strong antifungal extract: Methanol extract of stem bark at the concentration of 5 µg/mL inhibited the growth of *Aphanomyces cochlioides* by 80% (Islam et al., 2002).

Commentary: The plant generates flavanols such as (2*R*, 3*S*)-(+) 3′, 5-dihydroxy-4′, 7-dimethoxydihydroflavonol, (2*R*, 3*R*)-(+) 4′, 5,7-trimethoxydihydroflavonol, (2*R*, 3*R*)-(+) 4′, 7-di-*O*-methyldihydroquercetin, and (2*R*, 3*R*)-(+) 4′, 7-di-*O* methyldihydrokaempferol, and (2*R*, 3*R*)-(+) 4′-*O*-methyldihydroquercetin (Islam & Tahara, 2000). Consider that flavanols are often antiviral *in vitro*.

REFERENCES

Islam, M.T. and Tahara, S., 2000. Dihydroflavonols from *Lannea coromandelica*. *Phytochemistry*, *54*(8), pp. 901–907.

Islam, M.T., Ito, T., Sakasai, M. and Tahara, S., 2002. Zoosporicidal activity of polyflavonoid tannin identified in *Lannea coromandelica* stem bark against phytopathogenic oomycete *Aphanomyces cochlioides*. *Journal of Agricultural and Food Chemistry*, *50*(23), pp. 6697–6703.

Kaur, R., Jaiswal, M.L. and Jain, V., 2013. Protective effect of *Lannea coromandelica* Houtt. Merrill. against three common pathogens. *Journal of Ayurveda and Integrative Medicine*, *4*(4), p. 224.

Panda, S.K., Padhi, L., Leyssen, P., Liu, M., Neyts, J. and Luyten, W., 2017. Antimicrobial, anthelmintic, and antiviral activity of plants traditionally used for treating infectious disease in the Similipal Biosphere Reserve, Odisha, India. *Frontiers in Pharmacology*, *8*, p. 658.

Sathish, R., Ahmed, M.H., Natarajan, K. and Lalitha, K.G., 2010. Evaluation of wound healing and antimicrobial activity of *Lannea coromandelica* (Houtt) Merr. *Journal of Pharmacy Research*, *3*(6), pp. 1225–1228.

8.5.1.4 *Mangifera indica* L.

Synonyms: *Mangifera austroyunnanensis* H.H. Hu; *Rhus laurina* Nutt.

Common name: Mango tree; svai (Cambodia); mang guo (China): mempelan, buah mangga (Indonesia, Malaysia); aam (India, Pakistan, Bangladesh); thayet (Myanmar); paho (The Philippines); mamuang (Thailand); xoai (Vietnam)

Habitat: Cultivated

Distribution: Tropical Asia and Pacific

Botanical observation: This handsome tree grows to a height of 15 m and is sacred both for Hindus and Buddhists. It has a positive aura of life force and offers a fresh shade. Hindus believe that cutting the tree will render the offender sterile. The wood is yellow. The bark is greyish and fissured and exudes a sticky resin upon incision. The leaves are simple, spiral, and exstipulate. The petiole is up to about 6 cm long and swollen at apices. The blade is lanceolate, somewhat coriaceous, 12–30 × 4–9 cm, typically narrowly elliptic or lanceolate, pointed, at apex, wavy, and marked with about 10–25 pairs of secondary nerves. Young leaves can be beautifully dark-burgundy red and glossy. The flowers are minute, greenish-yellow or white, fragrant, and arranged in showy yellowish-pink panicles attracting insects. The drupes are more or less kidney-shaped, green first then ripening yellow to someone pinking and smooth. The pulp is delicious and yellow to orange, and the seed is elongated, fibrous, and flattened. Mangoes from Cebu are particularly delicious.

Medicinal uses: Diarrhea, dysentery, leucorrhea (Bangladesh, Pakistan); septicaemia (Palau).

Broad-spectrum antibacterial halo developed by polar extract: Acetone extract of leaves inhibited the growth of *Salmonella typhi* (50 strains), multidrug-resistant *Salmonella typhi* (30 strains), antibiotic-sensitive *Salmonella typhi* (20 strains), *Staphylococcus aureus* (ATCC 25923), *Salmonella typhimurium* (ATCC 4028) (Hannan et al., 2013).

Moderate antibacterial (Gram-positive) polar extract: Aqueous ethanol extract of stem bark inhibited the growth of *Micrococcus luteus* (MTCC 106), *Bacillus subtilis* (MTCC 121), *Bacillus cereus* (MTCC 430), and *Streptococcus pneumonia* (MTCC 2672) with the MIC values of 300, 600, 600, and 600 µg/mL, respectively (Singh et al., 2010).

Antibacterial (Gram-positive) gallotannins: Penta-*O*-galloyl-glucose and hexa-*O*-galloyl-glucose isolated from the seeds inhibited the growth of *Bacillus subtilis*, *Listeria monocytogenes*, and *Pediococcus acidilactici* (Engels et al., 2009).

Viral enzyme inhibition by gallotannins: Penta-*O*-galloyl-glucose and tetra-*O*-galloyl-glucose inhibited Influenza neuraminidase with IC_{50} values of 11.9 and 9.2 µM, respectively (Abdel-Mageed et al., 2014). Penta-*O*-galloyl-glucose and tetra-*O*-galloyl-glucose inhibited Coxsackie virus 3C protease with IC_{50} values of 1.1 and 2 µM, respectively (Abdel-Mageed et al., 2014).

Commentaries

i. The antibacterial mechanism of action of hydrolyzable tannins *in vitro* includes at least in part: interaction with proteins, inhibition of enzymatic activities, damage of lipid bilayer membranes, and complexation of metal ions (Engels et al., 2009).
ii. The plant produces mangiferin (see earlier) which being active against both (+) and (–) RNA viruses could be examined for its activity against coronaviruses including the Severe acquired respiratory syndrome-associated coronavirus 2.

REFERENCES

Abdel-Mageed, W.M., Bayoumi, S.A., Chen, C., et al., 2014. Benzophenone C-glucosides and gallotannins from mango tree stem bark with broad-spectrum anti-viral activity. *Bioorganic & Medicinal Chemistry*, *22*(7), pp. 2236–2243.

Engels, C., Knodler, M.A.T.T.H.I.A.S., Zhao, Y.Y., Carle, R., Ganzle, M.G. and Schieber, A., 2009. Antimicrobial activity of gallotannins isolated from mango (*Mangifera indica* L.) kernels. *Journal of Agricultural and Food Chemistry*, *57*(17), pp. 7712–7718.

Hannan, A., Asghar, S., Naeem, T., Ullah, M.I., Ahmed, I., Aneela, S. and Hussain, S., 2013. Antibacterial effect of mango (*Mangifera indica* Linn.) leaf extract against antibiotic sensitive and multi-drug resistant *Salmonella typhi*. *Pakistan Journal of Pharmaceutical Sciences*, *26*(4), pp. 715–719.
Singh, M., Khatoon, S., Singh, S., Kumar, V., Rawat, A.K.S. and Mehrotra, S., 2010. Antimicrobial screening of ethnobotanically important stem bark of medicinal plants. *Pharmacognosy Research*, *2*(4), p. 254.

8.5.1.5 *Semecarpus anacardium* L.f.

Common name: Indian marking nut; bela (Bangladesh); bhilama, nall geedi (India); bhalaayo (Nepal)

Habitat: Forests

Distribution: India, Bhutan, Bangladesh, Myanmar

Botanical observation: This tree grows up to 10 m tall. Upon incision of the bole and stem, a vesicant sap is secreted. The leaves are simple, spiral, and exstipulate. The petiole is 2–5 cm long. The blade is broadly elliptic, obovate, coriaceous, slightly glossy, wedge-shaped at base, round at apex, 5.5–25 × 10–45 cm, and presents about 10–15 pairs of showy secondary nerves. The inflorescence is a terminal or axillary hairy panicle. The flowers are minute and yellowish. The calyx is minutely 5-lobed. The corolla is made of 5 yellowish and triangular petals. A disc is present. The androecium is made of 5 stamens. The gynoecium consists of three styles. The drupe is dark green to black, flat, ovoid, somewhat glans-like, glossy, up to 3 cm long on a fleshy, cupular, and orange stalk (edible after preparation).

Medicinal uses: Leprosy, syphilis (Bangladesh); cough, dysentery, diarrhea, fever (Nepal); putrefied wounds (India).

Weak antibacterial (Gram-positive) polar extract: Aqueous ethanol extract of stem bark inhibited the growth of *Bacillus subtilis* (MTCC 121) and *Enterobacter aerogenes* (MTCC 111) with the MIC values of 0.1 and 1.2 mg/mL, respectively (Singh et al., 2010).

Strong broad-spectrum antibacterial alkyl phenols: Catechol derivative I isolated from the seeds inhibited the growth of *Bacillus subtilis* (MTCC 441), *Staphylococcus aureus* (MTCC 96), *Escherichia coli* (MTCC 1689), and *Proteus mirabilis* (MTCC 742) (Sundaram et al., 2014). Catechol derivative IV isolated from the seeds inhibited the growth of *Bacillus subtilis* (MTCC 441), *Staphylococcus aureus* (MTCC 96), *Escherichia coli* (MTCC 1689), and *Proteus mirabilis* (MTCC 742) (Sundaram et al., 2014).

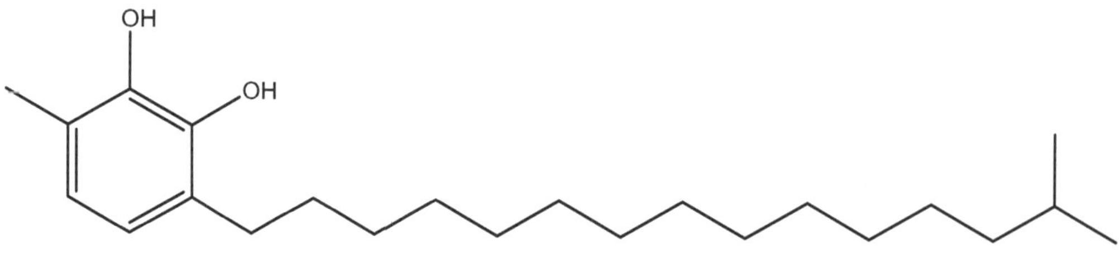

Catechol derivative I

Strong antifungal (filamentous) polar extract: Methanol extract of seeds inhibited the mycelial growth of *Rhizoctonia solanii*, *Alternaria* sp., *Sclerotium rolfsii*, and *Fusarium oxysporum* with the IC_{50} value of 8.1, 31.2, 30, and 62.5 μg/mL, respectively (Jain et al., 2014).

Strong antiviral (enveloped monopartite linear single stranded (−) RNA) polar extract: Methanol extract of seed inhibited the replication of the Mumps and Measles viruses in Vero cells at the dilutions of 1:128 and 1:256, respectively (Pa et al., 2012).

Commentaries

i. The plant brings to being the biflavanone semecarpetin (Murthy, 1988). Antiviral agents probably exits in this tree.
ii. In Myanmar, *Semecarpus albescens* Kurzis is used to treat leprosy.

REFERENCES

Jain, P., Singh, S.K., Sharma, H.P. and Basri, F., 2014. Phytochemical screening and antifungal activity of *Semecarpus anacardium* L. (An anti-cancer plant). *International Journal of Pharmaceutical Sciences and Research*, *5*(5), pp. 1884–1891.

Murthy, S.S.N., 1988. Semecarpetin, a biflavanone from *Semecarpus anacardium*. *Phytochemistry*, *27*(9), pp. 3020–3022.

Pa, S., Mangalaa, S.D., Sophiea, A.M. and Lathab, C.P., 2012. Antiviral and antioxidant activities of two medicinal plants. *International Journal of Current Science*, pp. 256–261.

Singh, M., Khatoon, S., Singh, S., Kumar, V., Rawat, A.K.S. and Mehrotra, S., 2010. Antimicrobial screening of ethnobotanically important stem bark of medicinal plants. *Pharmacognosy Research*, *2*(4), p. 254.

Sundaram, R., Muthu, K., Nagaraj, S., Shanthi, P. and Sachdanandam, P., 2014. Isolation and characterization of catechol derivatives from *Semecarpus anacardium* seeds and their antibacterial potential in *in vitro*. *Biomedicine & Preventive Nutrition*, *4*(2), pp. 177–180.

8.5.1.6 *Spondias dulcis* Parkinson

Synonyms: *Evia dulcis* (Parkinson) Comm. ex Bl.; *Poupartia dulcis* (Parkinson) Bl.; *Spondias cytherea* Sonn.; *Spondias mombin* L.; *Spondias purpurea* L.

Common names: Hog plum; amra, bilati amra (Bangladesh); mokak (Cambodia); kuk van (Laos); kedondong (Indonesia, Malaysia); cemcem (Indonesia); gway (Myanmar); akanang (Papua New Guinea); hevi (the Philippines); makok-farang (Thailand); coc (Vietnam)

Habitat: Cultivated

Distribution: Tropical Asia and Pacific

Botanical observation: This tree grows to 40 m tall. The bark is grayish and smooth to fissured. The leaves are compound with 4–10 pairs of folioles on a 10–20 cm long rachis. The foliole is ovate or lanceolate, 7.5–15 × 3–5 cm, somewhat coriaceous, pointed at apex, cuneate at base, and marked with 14–24 pairs of secondary nerves forming an intramarginal nerve. The panicle is terminal and up to 35 cm long. The flowers are cream or white and small. The calyx is 5-lobed. The corolla consists of 5 petals which are pure white, lanceolate, and about 2 mm long. A disc is present. The androecium includes 10 stamens. The gynoecium develops 5 white styles. The drupe is edible, ellipsoid, 4–10 × 3–8 cm, green to bright orange, and 5-celled.

Medicinal uses: Dysentery (Bangladesh); treat infection in infants (the Philippines); sores (Papua New Guinea)

Broad-spectrum antibacterial halo developed by polar extract: Methanol extract of leaves inhibited the growth of *Bacillus subtilis*, *Staphylococcus aureus*, *Escherichia coli*, *Salmonella typhi*, *Shigella dysenteriae*, *Vibrio mimicus*, and *Pseudomonas aeruginosa* with inhibition zone diameters of 15, 10, 9, 14, 13, 11, and 15 mm, respectively (Islam et al., 2013).

Antifungal (yeasts) halo developed by polar extract: Methanol extract of leaves inhibited the growth of *Candida albicans* and *Saccharomyces cerevisiae* with inhibition zone diameters of 10 and 7 mm, respectively (800 μg/disc) (Islam et al., 2013).

Strong antimycobacterial polar extract: Methanol extract stem bark at the concentration of 40 μg/mL inhibited the growth of *Mycobacterium tuberculosis* (H37Rv) by 92.8% (Olugbuyiro et al., 2013).

Antiviral (enveloped monopartite linear double-stranded DNA) ellagitannins: Geraniin and galloylgeraniin inhibited the replication of the Herpes simplex type-1 (Corthout et al., 1992).

Antiviral (non-enveloped monopartite linear single-stranded (+) RNA) ellagitannins: Geraniin and galloylgeraniin inhibited the replication of the Coxscackie virus B4 (Corthout et al., 1992).

Antiviral (enveloped monopartite linear double-stranded DNA) hydroxycinnamic acid derivative: Chlorogenic acid butyl ester inhibited the growth of the Herpes simplex 1 at the maximal nontoxic dose 100 µg/mL (Corthout et al., 1992).

Antiviral (non-enveloped monopartite linear single-stranded (+) RNA) hydroxycinnamic acid derivative: 2-*O*-caffeoyl-(+)-allohydroxycitric acid Coxsackie B virus the maximal nontoxic dose 100 µg/mL (Corthout et al., 1992).

Commentaries

i. The plant produces ellagic acid (da Silva et al., 2012) which is antimicrobial (see earlier).
ii. Is 2-*O*-caffeoyl-(+)-allohydroxycitric acid active against coronaviruses?

REFERENCES

Corthout, J., Pieters, L., Claeys, M., Berghe, D.V. and Vlietinck, A., 1992. Antiviral caffeoyl esters from *Spondias mombin*. *Phytochemistry*, *31*(6), pp. 1979–1981.

Islam, S.M.A., Ahmed, K.T., Manik, M.K., Wahid, M.A. and Kamal, C.S.I., 2013. A comparative study of the antioxidant, antimicrobial, cytotoxic and thrombolytic potential of the fruits and leaves of *Spondias dulcis*. *Asian Pacific Journal of Tropical Biomedicine*, *3*(9), p. 682.

da Silva, A.R.A., de Morais, S.M., Mendes Marques, M.M., de Oliveira, D.F., Barros, C.C., de Almeida, R.R., Vieira, Í.G.P. and Guedes, M.I.F., 2012. Chemical composition, antioxidant and antibacterial activities of two Spondias species from Northeastern Brazil. *Pharmaceutical Biology*, *50*(6), pp. 740–746.

Olugbuyiro, J.A.O., Moody, J.O. and Hamann, M.T., 2013. Phytosterols from *Spondias mombin* Linn with antimycobacterial activities. *African Journal of Biomedical Research*, *16*(1), pp. 19–24.

8.5.1.7 *Spondias pinnata* (L. f.) Kurz

Synonyms: *Mangifera pinnata* L.f., *Poupartia pinnata* (L.f.) Blco., *Spondias acuminata* Roxb., *Spondias bivenomarginalis* K.M. Feng & Mao, *Spondias mangifera* Willd., *Tetrastigma megalocarpum* W.T. Wang

Common names: Wild mango, Andaman mombin, Indian hog plum, Indian mombin; puen si phlae, pon, mkak Prei (Cambodia); umrah (Malaysia); kywae (Myanmar); amrataka (India), amara (Nepal); emberella (Sri Lanka), libas (the Philippines); titimml (Yap)

Habitat: Cultivated

Distribution: Myanmar, Indonesia, China, Thailand, Malaysia, Cambodia, Laos, Vietnam, Indonesia, the Philippines, the Pacific island.

Botanical observation: This resinous tree grows to 10 m tall. The bark is light grey with some vertical cracks. The leaves are compound, spiral, and exstipulate. The rachis is 10–20 cm long and smooth and bears 3–5 pairs of folioles plus a terminal one. The petiolule is 0.5–0.6 cm long. The foliole is dull dark greenish above, 7.5–15 × 3–5 cm, oblong, acuminate at apex with 14–25 pairs of secondary nerves which merge into an intramarginal nerve. The inflorescence is a terminal panicle growing up to 35 cm long. The flowers are cream-white. The calyx shows minute triangular lobes. The 5 petals are 3 mm long, and oblong. A disc is present. The androecium shows 10 stamens smaller than the petals. The drupe is ovoid, soily green to pale yellowish, edible, and up to about 4 cm long.

Medicinal use: Dysentery (Cambodia, India, Yap)

Broad-spectrum antibacterial polar extract: Ethanol extract (500 µg/disc) inhibited the growth of *Shigella boydii* (ATCC 9361), *Shigella flexneri* (ATCC 12022), *Shigella sonnei* (ATCC 25931), *Salmonella aeruginosa* (ATCC 25619), *Vibrio cholerae* (ATCC 14035), *Staphylococcus epidermidis* (ATCC 12228), *Streptococcus pyogenes* (ATCC), *Staphylococcus saprophyticus* (ATCC 15305), and *Staphylococcus aureus* (ATCC 6538) with the inhibition zone diameters of 7, 10, 10, 12, 9, 15, 8, 10, and 12 mm, respectively (Muhammad et al., 2011).

Moderate antibacterial (Gram-negative) essential oil: Essential oil of fruit peels inhibited the growth of *Staphylococcus aureus*, *Acinetobacter baumannii*, and *Pseudomonas aeruginosa* (Li et al., 2020).

Strong antifungal (filamentous) essential oil: Essential oil of fruit peels inhibited the growth of *Aspergillus fumigatus* and *Candida albicans* with the MIC/MBC values of 16/32 and 128/256 µg/mL, respectively (Li et al., 2020).

Commentaries

i. Results from Gas Chromatography-Mass Spectrometry (GC-MS) essential oil analysis are often presented in antimicrobial reports. However, GC-MS does not provide structure identification of each compounds in the sample analysed but gives an overall first glance of what could be present base on retention times and database available. It is like looking from a plane to a rainforest, the trees can be seen but are not fully identifiable (Butler M., personal communication).

ii. The dammarane-type triterpene pistagremic acid isolated from the galls of *Pistacia integerrima* J. Stewart inhibited the growth of *Candida albicans*, *Microsporum canis*, *Fusarium salani*, *Candida glabrata* and *Klebsiella pneumoniae* (Rauf et al., 2014). Polyphenolic fraction of raw shelled fruits of *Pistacia vera* L. inhibited the growth of *Enterococcus faecium* (ATCC 8459), *Bacillus subtilis* (ATCC 6633), *Listeria monocytogenes* (ATCC 7644), *Listeria monocytogenes* (ATCC 1392), *Staphylococcus aureus* (ATCC 43300), *Staphylococcus aureus* (ATCC 6538P) with the MIC/MFC values of 500/>2000, 500/>2000, 15.6/1000, 31.2/2000, 62.5/>2000, and 15.6/2000 µg/mL, respectively (Bisignano et al., 2013).

iii. The oleanane-type triterpene moronic acid (Log D = 5.5 at pH 7.4; molecular mass = 454.6 g/mol) isolated from the galls of *Rhus javanica* L., a plant used in Nepal (local name = *Bhakamilo*) to treat dysentery and diarrhea (synonym: *Rhus chinensis* Mill.) inhibited the replication of the Herpes simplex virus type-1 (wild-type), Herpes simplex virus type-1 (AP^r), and Herpes simplex virus type-2 (wild-type) with EC_{50} values of 3.9, 1.6, and 5.8 µg/mL, respectively (Kurokawa et al., 1999).

The lupane-type triterpene betulonic acid (Log D = 4.8 at pH 7.4; molecular mass = 454.6 g/mol) isolated from the galls of *Rhus javanica* L. inhibited the replication of the Human immunodeficiency virus type-1 (wild-type) with EC_{50} values of 2.6 µg/mL (Kurokawa et al., 1999). In mice infected with Herpes simplex virus type-1 (wild-type), moronic acid given orally once at 8 hours and daily for 7 days after viral inoculation at 10 mg/Kg delayed the development of skin lesions induced and increased survival rate (Kurokawa et al., 1999).

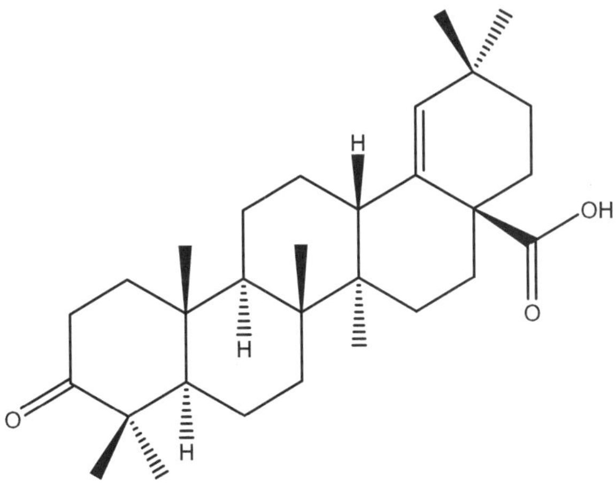

Moronic acid

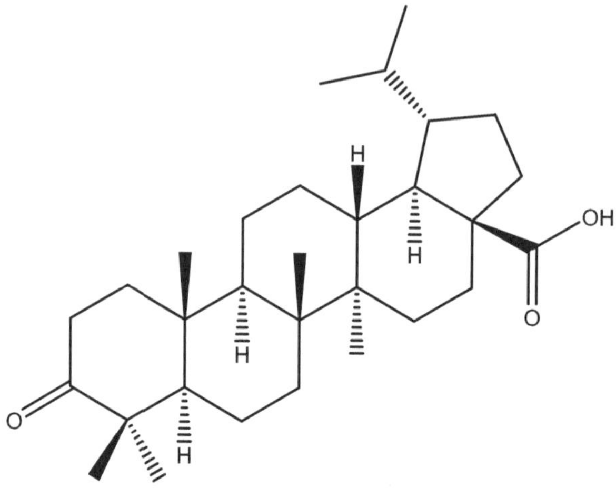

Betulonic acid

Aqueous extract of galls at the concentration of 250 µg/mL inhibited the replication of the Human immunodeficiency virus type-1 by 100% (Xu et al., 1996). Methyl gallate (Log D = 1.1 at pH 7.4; molecular mass = 184.1 g/mol) isolated from the galls inhibited the growth of *Salmonella gallinarum* (ATCC 9184), *Salmonella typhi* (ATCC 19943), and *Salmonella enteridis* with the MIC values of 31.2, 3.9, and 31.2 µg/mL, respectively. The mechanism of action did not involve membrane destabilization but ATPase inhibition (Choi et al., 2014). Methyl gallate given orally at a dose of 50 mg/Kg/day increased the survival rate of mice orally infected by *Salmonella typhimurium* by more than 70% during 6 days (Choi et al., 2014). Methylgallate inhibited the growth of *Vibrio cholerae* (Sánchez et al., 2013).

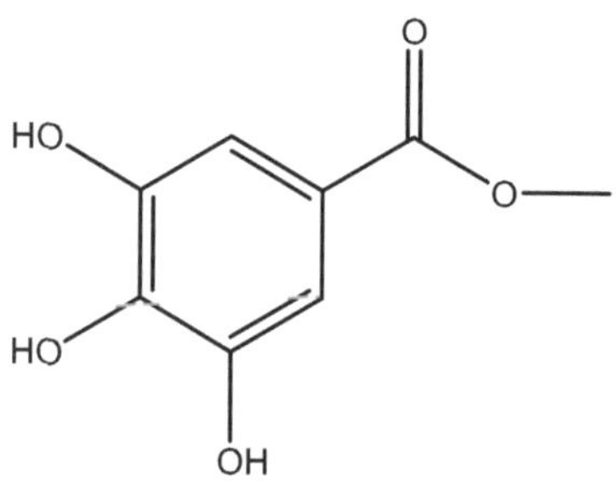

Methyl gallate

Luteolin and tetra-*O*-galloyl-β-D-glucose isolated from *Rhus javanica* L. inhibited the replication of the Severe acute respiratory syndrome-associated coronavirus with the IC_{50} values of 10.6 and 4.5 µg/mL by attachment with viral spikes and preventing entry (Yi et al., 2004). The spike S protein at the surface of coronaviruses comprises the amino-terminal S1 subunit and a carboxyl-terminal S2 subunit. The S1 subunit is responsible for virus–host cell receptor binding, whereas the S2 subunit is involved in virus–host membrane fusion (Wang et al., 2020). Aqueous extract of leaves of leaves of *Rhus typhina* L. containing 3,6-bis-*O*-di-*O*-galloyl-1,2,4-tri-*O*-galloyl-β-D-glucose (74.05%) inhibited the growth of *Bacillus cereus* (ATCC 10987) with the increase in cytoplasmic rigidity (Olchowik-Grabarek et al., 2014).

Rhus succedanea L.

Rhus coriaria L. is used in Iran (local name = *Somagh*) to treat jaundice. In Nepal, *Rhus succedanea* L. (local name = *Kagbhalayo*) is given for dysentery and diarrhea. Methanol extract of gall of *Rhus succedanea* L. inhibited the growth of *Bacillus cereus, Listeria monocytogenes, Staphylococcus aureus, Escherichia coli*, and *Salmonella anatum* (Shan et al., 2007). Hinokiflavone (Log D = 0.9 at pH 7.4; molecular mass = 538.4 g/mol) and agathisflavone isolated from *Rhus succedanea* L. inhibited the replication of the Human immunodeficiency virus type-1 (strain LAV-1) with the EC_{50} values of 4.1 and 33.6 µM, respectively (Lin et al., 1997). Hinokiflavone and agathisflavone inhibited Human immunodeficiency virus type-1 reverse transcriptase activity with IC_{50} values of 62 and 100 µM, respectively (Lin et al., 1997).

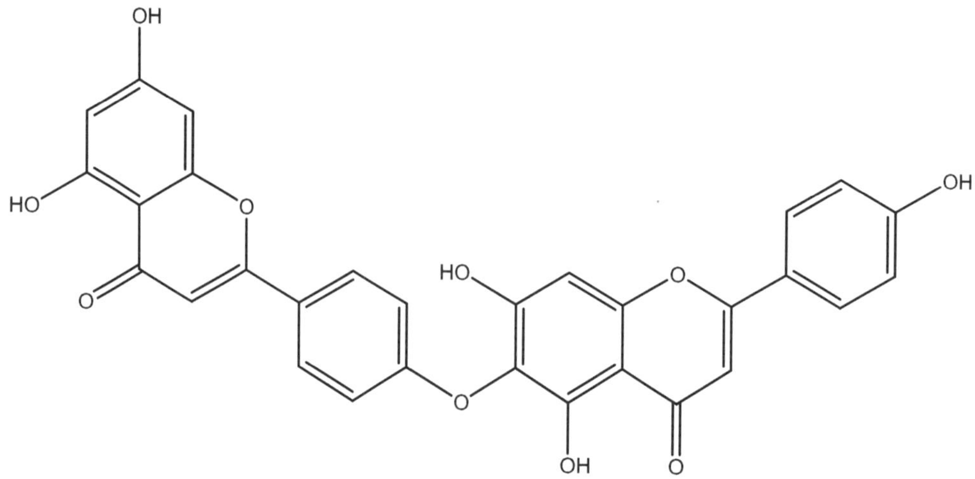

Hinokiflavone

Acetone extract of young shoots of *Cotinus coggygria* Scop. inhibited the growth of *Staphylococcus aureus* (ATCC 25923), *Staphylococcus epidermidis* (ATCC 12228), *Micrococcus luteus* (ATCC 9341), *Enterococcus faecalis* (ATCC 29212), *Pseudomonas aeruginosa* (ATCC 27853), *Escherichia coli* (ATCC 27853), *Klebsiella pneumoniae* (NCIMB 9111), *Candida albicans* (ATCC 10259), and *Candida albicans* (ATCC 24433) (Marčetić et al., 2013).

REFERENCES

Bisignano, C., Filocamo, A., Faulks, R.M. and Mandalari, G., 2013. *In vitro* antimicrobial activity of pistachio (*Pistacia vera* L.) polyphenols. *FEMS Microbiology Letters, 341*(1), pp. 62–67.

Choi, J.G., Mun, S.H., Chahar, H.S., Bharaj, P., Kang, O.H., Kim, S.G., Shin, D.W. and Kwon, D.Y., 2014. Methyl gallate from *Galla rhois* successfully controls clinical isolates of Salmonella infection in both *in vitro* and *in vivo* systems. *PloS One, 9*(7), p. e102697.

Li, R., Yang, J.J., Song, X.Z., Wang, Y.F., Corlett, R.T., Xu, Y.K. and Hu, H.B., 2020. Chemical composition and the cytotoxic, antimicrobial, and anti-inflammatory activities of the fruit peel essential oil from *Spondias pinnata* (Anacardiaceae) in Xishuangbanna, Southwest China. *Molecules, 25*(2), p. 343.

Lin, Y.M., Anderson, H., Flavin, M.T., Pai, Y.H.S., Mata-Greenwood, E., Pengsuparp, T., Pezzuto, J.M., Schinazi, R.F., Hughes, S.H. and Chen, F.C., 1997. *In vitro* anti-HIV activity of biflavonoids isolated from *Rhus succedanea* and *Garcinia multiflora*. *Journal of Natural Products, 60*(9), pp. 884–888.

Marčetić, M., Božić, D., Milenković, M., Malešević, N., Radulović, S. and Kovačević, N., 2013. Antimicrobial, antioxidant and anti-inflammatory activity of young shoots of the smoke tree, *Cotinus coggygria* Scop. *Phytotherapy Research, 27*(11), pp. 1658–1663.

Muhammad, A., Rahman, M., Kabir, A.N.M., Kabir, S. and Hossain, M., 2011. Antibacterial and cytotoxic activities of *Spondias pinnata* (Linn. f.) Kurz fruit extract. *Indian Journal of Natural Products and Resources, 2*, pp. 265–267.

Olchowik-Grabarek, E., Swiecicka, I., Andreeva-Kovaleskaya, Z., Solonin, A., Bonarska-Kujawa, D., Kleszczyńska, H., Mavlyanov, S. and Zamaraeva, M., 2014. Role of structural changes induced in biological membranes by hydrolysable tannins from sumac leaves (*Rhus typhina* L.) in their antihemolytic and antibacterial effects. *The Journal of Membrane Biology*, *247*(6), pp. 533–540.

Rauf, A., Uddin, G., Latif, A. and Muhammad, N., 2014. Pistagremic acid, a novel antimicrobial and antioxidant isolated from *Pistacia integerrima*. *Chemistry of Natural Compounds*, *50*(1), pp. 97–99.

Sánchez, E., Heredia, N., Camacho-Corona, M.D.R. and García, S., 2013. Isolation, characterization and mode of antimicrobial action against *Vibrio cholerae* of methyl gallate isolated from *Acacia farnesiana*. *Journal of Applied Microbiology*, *115*(6), pp. 1307–1316.

Shan, B., Cai, Y.Z., Brooks, J.D. and Corke, H., 2007. The *in vitro* antibacterial activity of dietary spice and medicinal herb extracts. *International Journal of Food Microbiology*, *117*(1), pp. 112–119.

Wang, N., Shang, J., Jiang, S. and Du, L., 2020. Subunit vaccines against emerging pathogenic human coronaviruses. *Frontiers in Microbiology*, *11*, p. 298.

Xu, H.X., Wan, M., Loh, B.N., Kon, O.L., Chow, P.W. and Sim, K.Y., 1996. Screening of traditional medicines for their inhibitory activity against HIV-1 protease. *Phytotherapy Research*, *10*(3), pp. 207–210.

Yi, L., Li, Z., Yuan, K., et al., 2004. Small molecules blocking the entry of severe acute respiratory syndrome coronavirus into host cells. *Journal of Virology*, *78*, pp. 11334–11339.

8.5.2 FAMILY BURSERACEAE KUNTH (1824)

The family Burseraceae consists of about 20 genera and 550 species of resinous trees and shrubs. The leaves are compound, mainly alternate, and stipulate or exstipulate. The folioles are often serrate or crenate. The inflorescences are racemes or panicles of minute flowers. The calyx presents 3–6 sepals and the corolla is made of 3–6 petals. A disc is present. The androecium includes 3–12 stamens. The gynoecium consists of a 2–5 locular ovary, each locule sheltering mainly a pair of ovules on axile placentas. The fruits are somewhat characteristic oblong and smooth drupes seated on persistent, cupular, and woody calyx. The main antimicrobial constituents in this family are triterpenes and monoterpenes mixed in the resins.

8.5.2.1 *Boswellia serrata* Roxb. ex Colebr.

Common names: Indian Frankincense, Indian olibanum; ganda barosa (Bangladesh); tau ju (China); kundur zachir, salar, shallaki (India); ganda ferozeh (Iran); salai guggul (Malaysia)

Habitat: Forests

Distribution: Pakistan, India, Bangladesh

Botanical observation: This tree grows up to about 15 tall and yields after incision an aromatic translucent and somewhat orangish-amber tear-like gum which is used in religious ceremonies in India. The tree has an aura of dryness. The leaves are imparipinnate, alternate, and exstipulate. The blade can reach about 30 cm long and presents about 15–30 pairs of folioles which are oblong to lancelate, sessile, 1–9 × 0.5–3.5 cm, coriaceous, somewhat crenate, glaucous below, and marked with 8–14 pairs of unconspicuous secondary nerves. The panicles are about 30 cm long and bear numerous minute pinkish flowers. The calyx includes 5–7 sepals. The 5–7 petals are about 8 mm long and oblong. A disc is present. The androecium includes 10 stamens. The ovary develops a minute style and a trilobed stigma. The drupe is greenish-brown, ellipsoid, and about 3 cm long.

Medicinal uses: Boils, bronchitis, diarrhea, dysentery, fever, gonorrhea, ringworm, ulcers, urethritis, wounds (India)

Antibacterial (Gram-positive) halo developed by essential oil: Essential oil (10%) inhibited the growth of *Propionibacterium acnes* (MTCC 1951) and clindamycin-resistant *Propionibacterium acnes* (CCARM 9010) with inhibition zone diameters of 30 mm (Sadhasivam et al., 2016).

Strong antibacterial (Gram-positive) amphiphilic ursane-type triterpenes: 11-Keto-β-boswellic acid (Log D = 3.7 at pH 7.4; molecular mass = 470.6 g/mol) isolated from the gum resin inhibited the growth of *Staphylococcus aureus* (ATCC 29213), methicillin-resistant *Staphylococcus aureus* (ATCC 33591), *Enterococcus faecalis* (ATCC 29212), *Enterococcus faecium* (ATCC 8042), *Staphylococcus epidermidis* (ATCC 12228), vancomycin resistant *Enterococcus faecalis* (ATCC 12228) with the MIC values of ranging from 8 to 16 µg/mL (Raja et al., 2011).

3-Acetyl-11-keto-β-boswellic acid isolated from the gum resin inhibited the growth of *Staphylococcus aureus* (ATCC 29213), methicillin-resistant *Staphylococcus aureus* (ATCC 33591), *Enterococcus faecalis* (ATCC 29212), *Enterococcus faecium* (ATCC 8042), *Staphylococcus epidermidis* (ATCC 12228), and vancomycin resistant *Enterococcus faecalis* (ATCC 12228) with the MIC values of 2, 2, 4, 4, 4, and 2 µg/mL, respectively (Raja et al., 2011). 3-Acetyl-11-keto-β-boswellic acid altered the membrane of *Staphylococcus aureus* at 64 µg/mL (Raja et al., 2011).

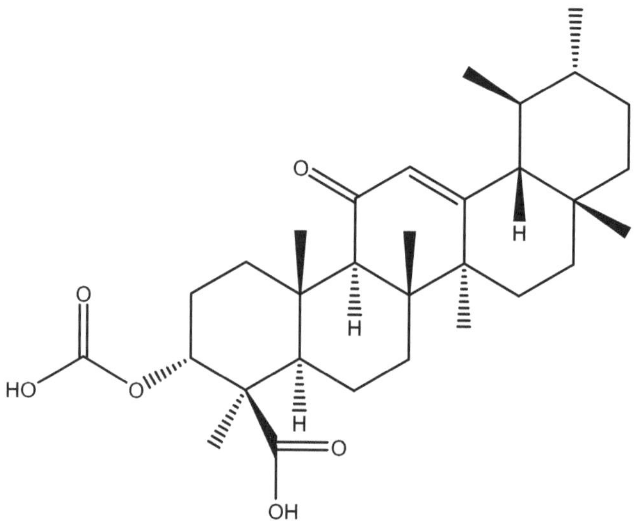

3-Acetyl-11-keto-β-boswellic acid

Broad-spectrum antifungal halo developed by essential oil: Essential oil inhibited *Candida albicans* (MTCC 3017), *Candida albicans* (MTCC 227), *Malassezia furfur* (MTCC 1374), *Malassezia globosa* (CBS 7707), *Trichophyton rubrum* (ATCC 28133), and *Trichophyton mentagrophytes* (MTCC 7687)

Antifungal potentiating essential oil: Essential oil decreased the resistance of *Candida albicans* (MTCC 3017) to ketoconazole and fluconazole (Sadhasivam et al., 2016).

Commentaries

i. Lipopolysaccharides covering the surface of Gram-negative bacteria form a polar and ionized barrier impermeable to antibacterial lipophilic and amphiphilic substances (Raja et al., 2011).
ii. The gum has been used as incense for religious ceremonies from time immemorial and was imported to Rome and known of Dioscorides (Hillson, 1988).
iii. Consider that burning incense was considered as a mean to purify air in time of pestilence by our peers the 9th century Arab physician Ishāq bin Ali al-Rohawi in his "*Adab al-tabib*" and the 10th century Persian physician Ali ibn al-'Abbas al-Majusi in his "*Kimil al-sini'a al-tibbiya*" (Conrad, 1982). "*Schola Medica Salernitana*" or medical school of Salerno where European and Arab physicians shared knowledge and worked together for the well-being of all composed the "*Regimen sanitatis Salernitanum*" used up to the 19th century by European physicians.
iv. *Boswellia carteri* Birdw. is used in China for the treatment of COVID-19 in (Luo, 2020; Wang et al., 2020).

REFERENCES

Conrad, L.I., 1982. Tāʿūn and Wabāʾ Conceptions of Plague and Pestilence in Early Islam. *Journal of the Economic and Social History of the Orient/Journal de l'histoire economique et sociale de l'Orient*, pp. 268–307.

Hillson, R.M., 1988. Gold, frankincense and myrrh. *Journal of the Royal Society of Medicine*, *81*(9), pp. 542–543.

Raja, A.F., Ali, F., Khan, I.A., Shawl, A.S., Arora, D.S., Shah, B.A. and Taneja, S.C., 2011. Antistaphylococcal and biofilm inhibitory activities of acetyl-11-keto-β-boswellic acid from *Boswellia serrata*. *BMC Microbiology*, *11*(1), p. 54.

Sadhasivam, S., Palanivel, S. and Ghosh, S., 2016. Synergistic antimicrobial activity of *Boswellia serrata* Roxb. ex Colebr. (Burseraceae) essential oil with various azoles against pathogens associated with skin, scalp and nail infections. *Letters in Applied Microbiology*, *63*(6), pp. 495–501.

Wang, S.X., Wang, Y., Lu, Y.B., Li, J.Y., Song, Y.J., Nyamgerelt, M. and Wang, X.X., 2020. Diagnosis and treatment of novel coronavirus pneumonia based on the theory of traditional Chinese medicine. *Journal of Integrative Medicine*, *18*, pp. 275–283.

8.5.2.2 *Canarium indicum* L.

Synonyms: *Canarium aboinense* Hochr.; *Canarium moluccanum* Bl.

Common names: Red Canarium; kenari Ambon (Indonesia); lawele, ngali (Papua New Guinea)

Habitat: Forests, cultivated

Distribution: Indonesia; Papua New Guinea, Pacific Islands

Botanical observation: This magnificent timber tree grows up to 40 m tall. The bole is straight and develops large butresses. The bark is greyish to brownish yellow and yields an aromatic resin after incision. The leaves are paripinnate, spiral, and stipulate. The stipules are dentate and up to 6 cm long. The blade is about 30 cm long and includes 3–7 pairs of folioles. The folioles are asymmetrical, glossy, coriaceous, oblong, marked with 10–15 pairs of conspicuous secondary nerves, and acuminate at apex. The panicles are terminal, lax, and bear minute flowers. The calyx is cupular and 3-lobed. The corolla presents 3 petals. The androecium includes 6 stamens. The drupe is about 6 cm long, ovoid, somewhat trigonal, smooth, purplish black, and seated on a persistent woody calyx. The seeds are edible and have been used as currency in Papua New Guinea.

Medicinal use: Sores (Solomon Islands)

Commentary: This plant has apparently not been examined for antimicrobial activities.

8.5.2.3 *Canarium oleosum* (Lamk.) Engler

Synonyms: *Amyris oleosa* Lam.

Common names: Grey Canarium; kanari minyak (Indonesia)

Habitat: Forests

Distribution: Indonesia, Papua New Guinea

Botanical observation: This magnificent timber tree grows up to 50 m tall. The bole is straight and present large buttresses. The bark is greyish black and yields an aromatic resin after incision. The leaves are paripinnate, spiral, and exstipulate. The rachis is about 25 cm long and presents 4–5 pairs of folioles. The folioles are about 7 × 2.5 cm, asymmetrical, glossy, coriaceous, oblong, with 10–15 pairs of conspicuous secondary nerves, and acuminate at apex. The panicles are axillary, up to 20 cm long, lax, and bear 7 mm long flowers. The calyx is cupular and 3-lobed. The corolla presents 3 petals. The androecium includes 6 stamens. The drupe is about 2 cm long, ovoid, and purplish.

Medicinal uses: Wounds (Indonesia; Papua New Guinea)

Commentary: The plant has apparently not been examined for antimicrobial effects. The coumarins scopoletin and scoparone isolated from *Canarium pentatinervium* Miq. (Malay name: *Kedondong krut*) inhibited the growth of *Staphylococcus aureus* (ATCC11632) with the MIC values of 25 and 50 µg/mL and the MBC values of 50 and 100 µg/mL, respectively (Mogana et al., 2020). Scopoletin and isocorilagin inhibited the replication of the Influenza A virus with the IC_{50} values of 22.9 and 5.4 µg/mL, respectively (Yang et al., 2018).

From *Canarium album* Leenh. methyl brevifolincarboxylate (Log D = 0.5 at pH 7.4; molecular mass = 306.22 g/mol) inhibited the replication of the Influenza virus A/Puerto Rico/8/34 (H1N1) and A/Aichi/2/68 (H3N2) in Madin–Darby canine kidney cells with IC_{50} values of 27.1 (selectivity index > 36.8) and 33.4 µM, respectively (Chen et al., 2020b). This compound could be tested for activity against coronaviruses and is probably antibacterial and antifungal *in vitro*. From the same plant, the ellagitannin isocorilagin inhibited the replication of the Influenza A/Puerto Rico/8/34 (H1N1), A/Aichi/2/68 (H3N2), and NA-H274Y (H1N1) viruses with IC_{50} value of 9.1, 23.7, and 4.6 µM, respectively, via inhibition of neuraminidase activity (Chen et al., 2020a).

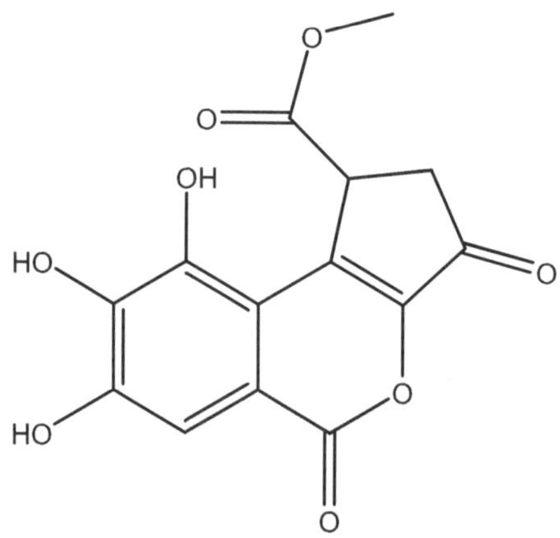

Methyl brevifolincarboxylate

REFERENCES

Chen, F., Yang, L., Huang, Y., Chen, Y., Sang, H., Duan, W. and Yang, J., 2020a. Isocorilagin, isolated from *Canarium album* (Lour.) Raeusch, as a potent neuraminidase inhibitor against influenza A virus. *Biochemical and Biophysical Research Communications*, *523*(1), pp. 183–189.

Chen, F., Yang, L., Zhai, L., Huang, Y., Chen, F., Duan, W. and Yang, J., 2020b. Methyl brevifolincarboxylate, a novel influenza virus PB2 inhibitor from *Canarium Album* (Lour.) Raeusch. *Chemical Biology & Drug Design*, *96*(5), pp. 1280–1291.

Mogana, R., Adhikari, A., Tzar, M.N., Ramliza, R. and Wiart, C., 2020. Antibacterial activities of the extracts, fractions and isolated compounds from *Canarium patentinervium* Miq. against bacterial clinical isolates. *BMC Complementary Medicine and Therapies*, *20*(1), pp. 1–11.

Yang, L.-P., Gu, X.-L., Chen, J.X., Yang, J., Tan, S.Y. and Duan, W.J., 2018. Chemical constituents from *Canarium album* Raeusch and their anti-influenza A virus activities. *Journal of Natural Medicines*, *72*(3), pp. 808–815.

8.5.2.4 *Commiphora wightii* (Arn.) Bhandari

Synonyms: *Balsamodendrum wightii* Arn.; *Commiphora mukul* (Hook. ex Stocks) Engl.; *Commiphora roxburghii* (Stocks) Engl.

Common names: Indian bdellium, gum guggul; guggul (Bangladesh; India); bu-e-Jahudaan, guggal, guggula, palankasa (India); muql-i-azraq (Iran)

Habitat: Deserts, savanna, cultivated

Distribution: Pakistan, India

Botanical observation: This shrub grows up to 4 m tall. The bark is papery and yields upon incision of gum. The stems are thorny. The leaves are sessile, simple, or compound, alternate or fascicled, and exstipulate. The petiole is up to about 1 cm long. The blade is obovate to elliptic, 1–5 × 0.5–2.5 cm, coriaceous, glossy, marked with about 5 pairs of secondary nerves, serrate or not, acute at base, and rounded to notched at apex. The flower fascicles are axillary. The calyx is tubular, minute, and develops 4 triangular lobes hairy within. The 4 petals are reflexed, acute, 3–5 mm long. A disc is present. The androecium includes 8 stamens. The ovary is ovoid and develops a bifid stigma. The drupes are up to 1 cm long, red to dull burgundy, smooth, and pointed at apex.

Medicinal uses: Bronchitis, fever, pharyngitis, tonsilitis, tuberculosis (India)

Broad-spectrum antibacterial halo developed by oleo-gum resin: Gum inhibited the growth of *Bacillus subtilis* (ATCC 6633), *Escherichia coli* (ATCC 8739), *Sarcina lutea* (ATCC 9341), *Staphylococcus aureus* (ATCC 6536), *Staphylococcus epidermidis*, *Klebsiella pneumoniae*, *Micrococcus luteus*, *Proteus vulgaris*, *Salmonella typhi*, and *Shigella dysenteria* (Saeed & Sabir, 2004).

Weak broad-spectrum antibacterial polar extract: Methanol extract of gum inhibited the growth of *Staphylococcus aureus*, *Bacillus cereus*, *Bacillus subtilis* with the MIC values of 512, 1024, 2048 µg/mL respectively (Goyal et al., 2010).

Strong antimycobacterial polar extract: Methanol extract inhibited the growth of *Mycobacterium avium* with the MIC value of 62.5 µg/mL (Newton et al., 2000).

Broad-spectrum antibacterial pregnane-type steroid: (*E*)-Guggulsterone (also known as 17E-pregnan-4,17-diene-3,16-dione) (100 µL of a 1 mg/mL solution/6 mm well) inhibited the growth of *Bacillus subtilis*, *Staphylococcus aureus*, and *Pseudomonas aeruginosa* with the inhibition zone diameters of 14, 14, and 11 mm, respectively (Choudhary et al., 2005).

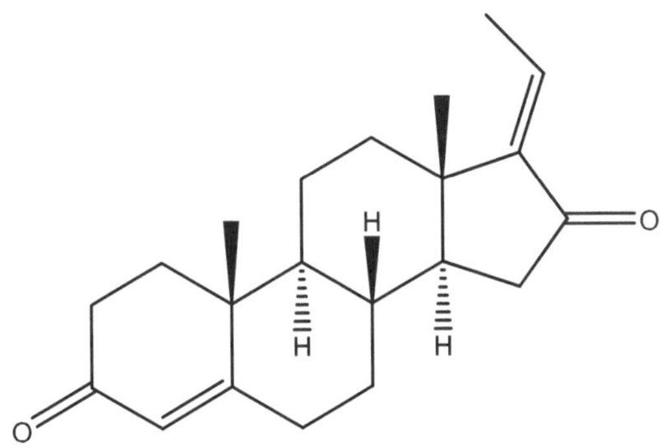

(E)-Guggulsterone

Antibacterial germacrane-type sesquiterpenes: 3-Methoxy-10-methylenefurano-germacr-1-en-6-one and 2-methoxy-4,5-dihydrofurano-dien-6-one from the gum inhibited the growth of *Bacillus subtilis* (ATCC 6633), *Escherichia coli* (ATCC 8739), *Sarcina lutea* (ATCC 9341), *Staphylococcus aureus* (ATCC 6536) *Staphylococcus epidermidis*, *Klebsiella pneumoniae*, *Micrococcus luteus*, *Proteus vulgaris*, *Salmonella typhi*, and *Shigella dysenteria* (Saeed & Sabir, 2004).

Weak anticandidal antifungal prenylated flavanone: Muscanone inhibited the growth of *Candida albicans* with the MIC value of 250 µg/mL (Fatope et al., 2003).

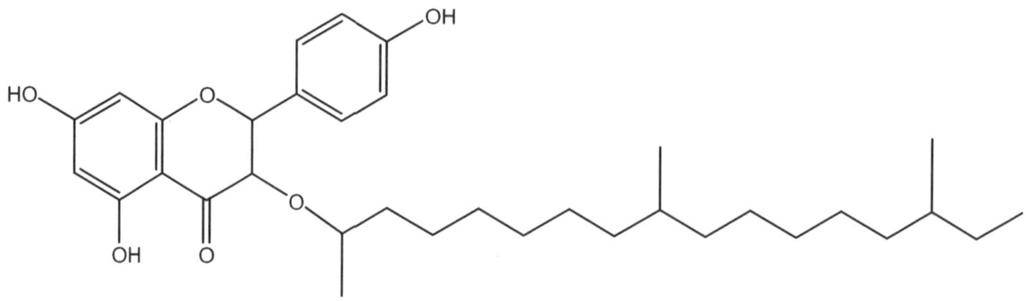

Muscanone

Commentaries

i. *Commiphora molmol* (Engl.) Engl. ex Tschirch is used in India for the treatment of boils and gingivitis (local name: *Bola*) and the source of myrrh. The octanordammarane-type triterpene 3,4-*seco*-mansumbinoic acid isolated from the oleo-resin from of this tree inhibited multi drug-effluxing strain *Staphylococcus aureus* (1199B) with the MIC value of 4 μg/mL (Rahman et al., 2008).

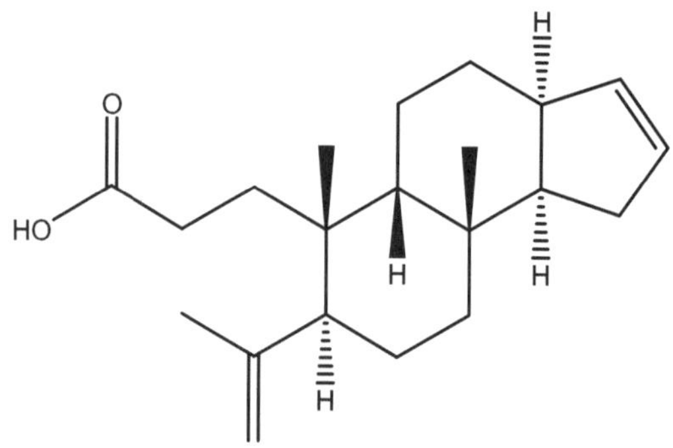

3,4-*seco*-Mansumbinoic acid

3,4-*seco*-Mansumbinoic acid evoked a 2-fold potentiation of ciprofloxacin and tetracycline against *Salmonella typhimurium* (strain SL 1344) and against *Salmonella typhimurium* (strain L10) (Rahman et al., 2008). Myrrh oil from this tree (containing mainly furanoeudesma 1,3 diene and menthofuran) inhibited the growth of *Trichophyton mentagrophytes* (PTCC 5054), *Trichophyton rubrum* (PTCC 5143), *Microsporum canis* (PTCC 5069), *Microsporum gypseum* (PTCC 5070), and *Trichophyton verrucosum* (PTCC 5056) with the MIC/MFC values of 100/200, 25/25, 25/50, 50/100, and 50/100 μg/mL, respectively (Mahboubi & Mohammad 2016). Menthofuran inhibited *Candida albicans* with the MIC value of 100 ppm (Tampieri et al., 2005). In India, *Commiphora caudata* (Wight & Arn.) Engl. (local name: *Kilimaram*) is used to treat mouth ulcers and fever. In Iran, *Commiphora opobalsamum* (L.) Engl. (local name = *Habb-i-balsan*) is used to treat colds.

ii. Methanol extract of a member of the genus *Commiphora* Jacq. inhibited the replication of the Herpes simplex virus type-2 and Respiratory syncytial virus type B with the

IC_{50} values of 21.3 and 23.3 μg/mL, respectively, and selectivity indexes of 11.1 and 10.2 (Bouslama et al., 2019).

iii. Muscanone having a long isoprenic chain could be examined for its antimycobacterial properties and ability to inhibit bacterial and/or fungal efflux pumps.

iv. Our peer, the medieval 14th Spanish Physician Abraham Ben Solomon Hen in his "*Tractatulus de pestilentia*" advocated the burning of myrrh and frankinscense to prevent Black Death pestilence. Pestilence in air was considered as "cold" and "humid" and to be fought by daily fumigations with "hot" and "dry" ingredients like myrrh, frankinscence, clove, aloe, ginger, or laurel (Bos & Mensching, 2011).

REFERENCES

Bos, G. and Mensching, G., 2011. The Black Death in Hebrew Literature: Abraham Ben Solomon Hen's "Tractatulus de pestilential". *Jewish Studies Quarterly*, *18*(1), pp. 32–63.

Bouslama, L., Kouidhi, B., Alqurashi, Y. M., Chaieb, K. and Papetti, A., 2019. Virucidal effect of guggulsterone isolated from *Commiphora gileadensis*. *Planta Medica*, *85*(16), pp. 1225–1232.

Choudhary, M.I., Shah, S.A.A., Sami, A., Ajaz, A. and Shaheen, F., 2005. Fungal metabolites of (*E*)-guggulsterone and their antibacterial and radical-scavenging activities. *Chemistry & Biodiversity*, *2*(4), pp. 516–524.

Fatope, M.O., Al-Burtomani, S.K.S., Ochei, J.O., Abdulnour, A.O., Al-Kindy, S.M. and Takeda, Y., 2003. Muscanone: a 3-O-(1″, 8″, 14″-trimethylhexadecanyl) naringenin from *Commiphora wightii*. *Phytochemistry*, *62*(8), pp. 1251–1255.

Goyal, P., Chauhan, A. and Kaushik, P., 2010. Assessment of *Commiphora wightii* (Arn.) Bhandari (Guggul) as potential source for antibacterial agent. *Journal of Medicine and Medical Sciences*, *1*(3), pp. 71–75.

Mahboubi, M. and Mohammad Taghizadeh Kashani, L., 2016. The anti-dermatophyte activity of *Commiphora molmol*. *Pharmaceutical Biology*, *54*(4), pp. 720–725.

Newton, S.M., Lau, C. and Wright, C.W., 2000. A review of antimicrobial natural products. *Phytotherapy Research*, *14*, pp. 303–322.

Rahman, M.M., Garvey, M., Piddock, L.J. and Gibbons, S., 2008. Antibacterial terpenes from the oleo-resin of *Commiphora molmol* (Engl.). *Phytotherapy Research: An International Journal Devoted to Pharmacological and Toxicological Evaluation of Natural Product Derivatives*, *22*(10), pp. 1356–1360.

Saeed, M.A. and Sabir, A.W., 2004. Antibacterial activities of some constituents from oleo-gum-resin of *Commiphora mukul*. *Fitoterapia*, *75*(2), pp. 204–208.

Tampieri, M.P., Galuppi, R., Macchioni, F., Carelle, M.S., Falcioni, L., Cioni, P.L. and Morelli, I., 2005. The inhibition of *Candida albicans* by selected essential oils and their major components. *Mycopathologia*, *159*(3), pp. 339–345.

8.4.2.5 *Garuga pinnata* Roxb.

Common names: Garuga; dhadhabe, jum, laljigachal (Bangladesh); yu ye bai tou shu (China); annakkara (India); chinyok (Myanmar); kharpat (Pakistan)

Habitat: Forests

Distribution: Pakistan, India, Bangladesh, Myanmar, Cambodia, Laos, Vietnam, Thailand, Malaysia, the Philippines

Botanical observation: This tree grows up to 15 m tall. The bark is grey and rough. The stems are hairy at apex. The leaves are compound, spiral, and exstipulate. The blade reaches 45 cm long and comprises 4–11 pairs of folioles and a terminal one and the rachis is hairy. The folioles are oblong-lanceolate, 5–11 × 2–3 cm, sessile, crenate, base rounded, shortly acuminate at apex, and present 6–15 pairs of secondary nerves. The panicles are axillary or terminal, whitish, up to about 20 cm long and hairy. The calyx is urceolate, hairy, white, 5-lobed, the lobes about 5 mm long. The 5 petals are white, triangular, about 5 mm long, and 5-lobed. A lobed disc is present. The 10 stamens are about 8 mm long. The ovary is oblong, hairy, and produces a style with 5-lobed stigma. The drupes are globose, dull light green, edible (Pakistan), and about 1.5 cm in across.

Medicinal uses: Conjunctivitis (Bangladesh; India); dysentery (Bangladesh); cough, diarrhea (India)

Broad-spectrum antibacterial halo developed by phenolic: Garuganin V (100 µg/disc) isolated from the stem bark inhibited the growth of *Bacillus cereus, Bacillus megaterium, Bacillus subtilis, Staphylococcus aureus, Sarcinia lutea, Escherichia coli, Pseudomonas aeruginosa, Salmonella paratyphi, Shigella boydii, Shigella dysenteriae, Vibrio mimicus,* and *Vibrio parahaemolyticus* with the inhibition zone diameters of 40, 34, 40, 35, 35, 32, 34, 40, 36, 32, 36, and 36 mm, respectively (Ara et al., 2012).

Broad-spectrum antifungal halo developed by phenolic: Garuganin V (100 µg/disc) from the stem bark inhibited the growth of *Aspergillus niger, Candida albicans, Saccharomyces cerevisae* with the inhibition zone diameters of 31, 35, and 33 mm, respectively (Ara et al., 2012).

Commentary: The plant produces series of euphane-type and lanostane-type triterpenes (Venkatraman et al., 1992, 1994) as well as diarylheptanoids (Ara et al., 2006).

REFERENCES

Ara, K., Kaisar, M.A., Rahman, M.S., Chowdhury, S.R., Islam, F. and Rashid, M.A., 2012. Antimicrobial constituents from *Garuga pinnata* Roxb. *Latin American Journal of Pharmacy, 31*(7), pp. 1071–1073.

Ara, K., Rahman, A.H.M.M., Hasan, C.M., Iskander, M.N., Asakawa, Y., Quang, D.N. and Rashid, M.A., 2006. Macrocyclic diarylheptanoids from *Garuga pinnata. Phytochemistry, 67*(24), pp. 2659–2662.

Venkatraman, G., Thombare, P.S. and Sabata, B.K., 1992. Euphane triterpenoid from *Garuga pinnata. Phytochemistry, 32*(1), pp. 161–163.

Venkatraman, G., Thombare, P.S. and Sabata, B.K., 1994. A tetracyclic triterpenoid from *Garuga pinnata. Phytochemistry, 36*(2), pp. 417–419.

8.4.2.6 *Protium javanicum* Burm.f.

Common names: Ketimis, loa (Indonesia); katos (Malaysia)

Habitat: Forests

Distribution: Malaysia, Indonesia

Botanical observation: This magnificent tree grows up to 10 m tall and exudes a resin upon incision. The stems are spiny. The leaves are compound, spiral, and exstipulate. The folioles are ovate to elliptic, 3.5–10 × 1.5–4.5 cm, coriaceous, crenate, acuminate at apex, and marked with 8–13 pairs of secondary nerves. The panicles are lax, axillary, and up to 25 cm long. The calyx develops 5-minute lobes. The corolla is made of 5 petals which are 2 mm long. A disc is present. The androecium comprises 10 stamens. The drupes are somewhat globose to obcordate, edible (Indonesia), and about 1.5 cm long.

Medicinal use: Dysentery (Indonesia)

Commentary: The plant has apparently not been examined for antimicrobial effects. It brings to being 6-desacetylnimbin, quercitrin, myricitrin, and scopoletin (Adfa et al., 2010). This tree and its antimicrobial principles (which could be of clinical value) are on their ways to disappearance owed to palm oil and logging businesses.

REFERENCE

Adfa, M., Yoshimura, T., Komura, K. and Koketsu, M., 2010. Antitermite activities of coumarin derivatives and scopoletin from *Protium javanicum* Burm. f. *Journal of Chemical Ecology, 36*(7), pp. 720–726.

8.5.3 FAMILY MELIACEAE A. L. DE JUSSIEU (1789)

The family Meliaceae consists of about 50 genera and 650 species of trees, shrubs, or herbs. Bark and leaves are bitter. The leaves are simple, compound, spiral, and exstipulate. The inflorescences are cymes and racemes. The calyx includes 3–6 sepals. The corolla presents 4 or 5 petals. The androecium consists of three more stamens fused in a characteristic staminal tube. A disc is present or not. The gynoecium consists of a plurilocular ovary, with one to many ovules per locule, a single style and a disciform or capitate stigma. The fruits are berries, capsules, or drupes containing arillate and winged seeds. Members in this family produce a bewildering array of triterpenes including oxygenated triterpene derivatives known as limonoids.

8.5.3.1 *Aglaia edulis* (Roxb.) Wall.

Synonym: *Aglaia motleyana* Stapf ex Ridl.; *Aglaia sulingi* Bl.; *Aglaia testicularis* C.Y. Wu; *Milnea edulis* Roxb.

Common name: Ma shen guo (China); gumi (India); balik balik (Indonesia); late mahwa (Nepal); lasaging (the Philippines); khangkhao (Thailand)

Habitat: Forests by seashores

Distribution: India, Bhutan, Myanmar, Cambodia, China, Vietnam, Thailand, Malaysia, Indonesia, the Philippines

Botanical observation: This tree grows up to 10 m tall. The bole presents massive buttresses. The bark is reddish to yellowish-brown. The leaves are compound, spiral, and exstipulate. The blade is up to 30 cm long. The rachis presents 5–9 pairs of folioles which are marked with 5–16 pairs of secondary nerves, rounded to cuneate at base, acute and acuminate at apex, 5–10 × 1.5–4 cm, and hairy below. The cymes are axillary and up to 15 cm long. The calyx is 5-lobed. The corolla presents 5 petals which are ovate. The staminal tube is globose and develops 5 anthers. The berries are about 5 cm long, yellowish, and shelter about 3 seeds in an edible aril (India).

Medicinal use: Diarrhea (India)

Moderate antifungal (filamentous) cyclopentabenzofuran furan: Pannellin inhibited the growth of *Pyricularia grisea* with the EC_{50} value of 50 µg/mL (Engelmeier et al., 2000).

Weak antifungal (filamentous) cyclopentabenzofuran furan: Aglaroxin A inhibited the growth of *Pyricularia grisea* with the EC_{50} value of 100 µg/mL (Engelmeier et al., 2000).

Antiviral (enveloped monopartite linear double-stranded DNA) sulfur amide alkaloids: Aglaithioduline and aglaidithioduline from the leaves inhibited the replication of the Herpes simplex virus type-1 and type-2 at the concentration of 20 µg/mL (Saifa et al., 1999).

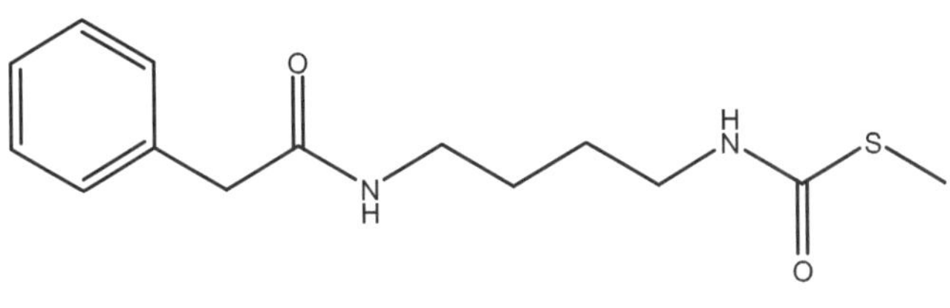

Aglaithioduline

Commentaries

i. Members of the genus *Aglaia* Lour. bring to being antimicrobial lignans and triterpenes.
ii. *Aglaia elliptica* Bl. is used to heal wounds in Southeast Asia. In China, the flowers of *Aglaia odorata* Lour. are used to make aromatic teas (local name: *San yeh la*), and in Indonesia, the plant is used to treat venereal diseases (local name = *Pacar cina*) and fever. This tree produces the cyclopentabenzofuran-type lignans rocaglaol, aglafolin, rocaglamide, and desmethyl rocaglamide which inhibited *Pyricularia grisea* (VIAM MA1627) with the MIC values of 1.6, 3, 25, and 100 µg/mL, respectively (Engelmeier et al., 2000). Rocaglaol inhibited *Fusarium avenaceum* (VIAM MA1512) and *Alternaria citri* (VIAM MA1512) with the MIC values of 25 and 50 µg/mL, respectively (Engelmeier et al., 2000).

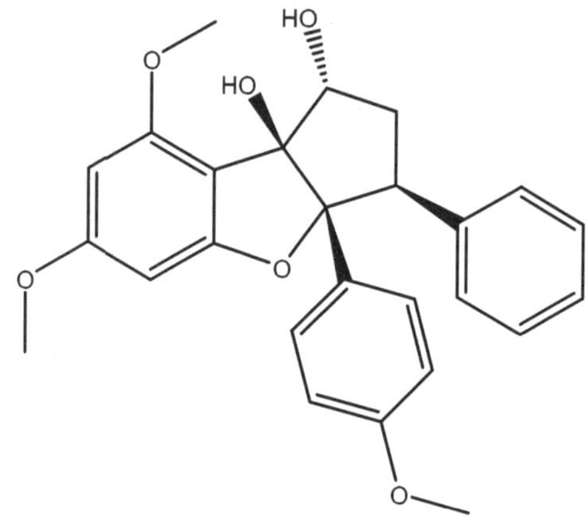

Rocaglaol

Methanol extract of *Aglaia odorata* Lour. at the concentration of 200 μg/mL inhibited the Human immunodeficiency virus type-1 reverse transcriptase by 9% (Tan et al., 1991). The cyclopentabenzofuran-type lignan silvestrol (Log D = 2.1 at pH 7.4; molecular mass = 654.6 g/mol) found in the genus *Aglaia* Lour. inhibited the replication of the Human coronavirus 229E, Middle East Respiratory syndrome coronavirus, Human rhinovirus A1 (HRV A1), and poliovirus type-1 with the IC_{50} values of 3, 1.3, 100, and 20 nM, respectively, and selectivity indexes above 100 in MRC-5 cells (Müller et al., 2018). Silvestrol at the concentration of 50 nM inhibited the replication of the Hepatitis E virus (Todt et al., 2018). Silvestrol inhibited the replication of the Chikungunya virus with the IC_{50} values of 1.8 and 5 nM in 293T and NIH3T3 cells, respectively (Henss et al., 2018). Silvestrol inhibited the replication of the Zika virus in A549 cells at the concentration of 50 nM (Elgner et al., 2018) and Ebola virus at 10 nM in primary human macrophages (Elgner et al., 2018). The antiviral mode of action of silvestrol is not fully understood but being a broad-spectrum antiviral lignan one could deduce interference with genetic material replication. Silverstrol is cytotoxic in mammalian cells (Cencic et al., 2009; Yang et al., 2013). Note that lignans are decomposed by gut bacteria into "enterolignans" (Clavel et al., 2006) and the oral bioavailability of silvestrol could be low. However, being a broad-spectrum antiviral molecule, it may have to usefulness against COVID-19 and other pandemic viruses.

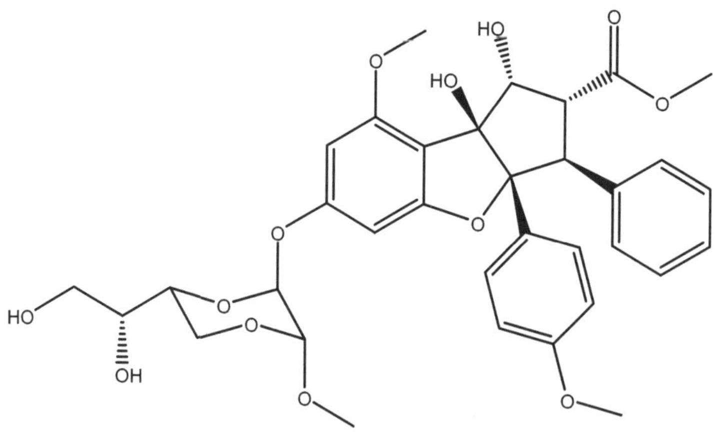

Silvestrol

iii. *Aglaia cucullata* Pellegr. (local names: *Uanra* (India); *Daeng nam* (Thailand); *Dai ngu ai nioc* (Vietnam)) is used for the treatment of dysentery in India. The lanostane-type triterpenes niloticin and piscidol A isolated from the stems of this tree of inhibited plaque formation in Hep-2 cells by respiratory syncytial virus (Long strain) with IC_{50} values of 15.8 and 12.8 µg/mL and the selectivity indexes of 4.2 and 1.2 (Esimone et al., 2008).

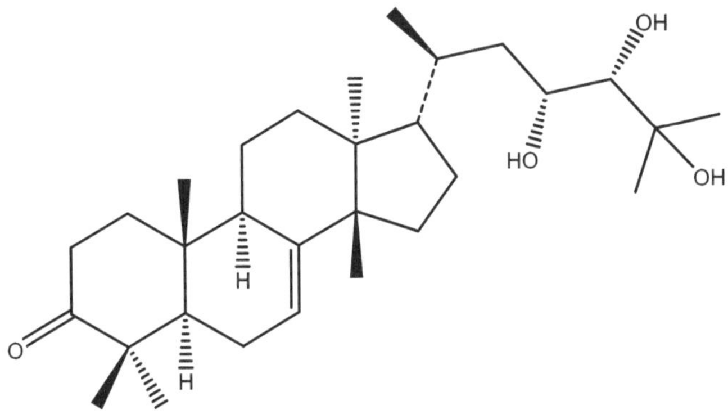

Piscidol A

iv. The dammarane-type triterpenes aglaiol, 24,25-epoxy-dammar-20-ene-3-one, and 24,25-dihydroxy-5α-dammar-20-ene-3-one isolated from the leaves of *Aglaia duperreana* Pierre inhibited plaque formation in Hep-2 cells by Respiratory syncytial virus (Long strain) with IC_{50} values of 11.8, 21.7, and 25.5 µg/mL, respectively (Esimone et al., 2008). Aglaiol inhibited Respiratory syncytial virus replication at a post entry stage (Esimone et al., 2008).
v. 4α, 14-Dimethyl-9,19-cyclocholestan-3β, and 24α, 25-triol, 24,25-epoxy-cycloartan-3-ol, lupeol, and lupeone isolated from *Aglaia tsangi* Merr. inhibited plaque formation in Hep-2 cells by Respiratory syncytial virus (Long strain) (Esimone et al., 2008).
vi. The *seco*-dammarane-type triterpene dammarenolic acid isolated from the bark of *Aglaia ignea* Valeton ex K. Heyne inhibited plaque formation in Hep-2 cells by Respiratory syncytial virus (Long strain) with the IC_{50} value of 0.1 µg/mL and the selectivity index of 29 2 (Esimone et al., 2008). This triterpene inhibited Respiratory syncytial virus replication at a postentry stage (Esimone et al., 2008). Dammarenolic acid isolated from the bark at a concentration of 15 µg/mL nullified the Human immunodeficiency virus type-1 infectivity by inhibiting a postentry step in the virion replication cycle (Esimone et al., 2010).

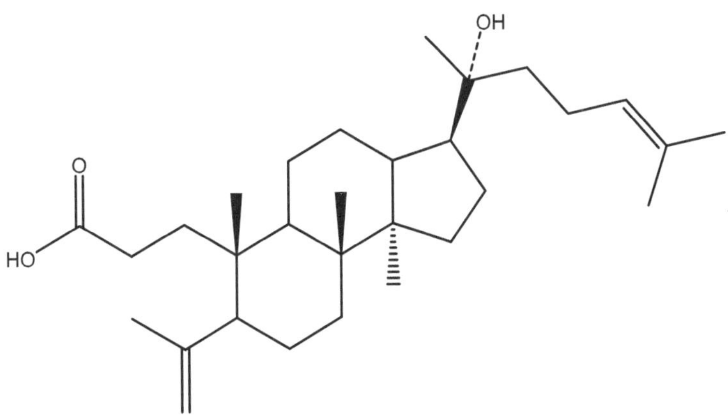

Dammarenolic acid

vii. *Aglaia roxburgiana* (Wight & Arn.) Miq. (local name: *Priyangu* (Bangladesh; India) is used in India for the treatment of urinary tract infections.

REFERENCES

Cencic, R., Carrier, M., Galicia-Vázquez, G., Bordeleau, M.E., Sukarieh, R., Bourdeau, A., Brem, B., Teodoro, J.G., Greger, H., Tremblay, M.L. and Porco Jr, J.A., 2009. Antitumor activity and mechanism of action of the cyclopenta [b] benzofuran, silvestrol. *PloS One*, *4*(4), p. e5223.

Clavel, T., Doré, J. and Blaut, M., 2006. Bioavailability of lignans in human subjects. *Nutrition Research Reviews*, *19*(2), pp. 187–196.

Elgner, F., Sabino, C., Basic, M., Ploen, D., Grünweller, A. and Hildt, E., 2018. Inhibition of Zika virus replication by silvestrol. *Viruses*, *10*(4), p. 149.

Engelmeier, D., Hadacek, F., Pacher, T., Vajrodaya, S. and Greger, H., 2000. Cyclopenta [b] benzofurans from Aglaia species with pronounced antifungal activity against rice blast fungus (*Pyricularia grisea*). *Journal of Agricultural and Food Chemistry*, *48*(4), pp. 1400–1404.

Esimone, C.O., Eck, G., Duong, T.N., Überla, K., Proksch, P. and Grunwald, T., 2008. Potential anti-respiratory syncytial virus lead compounds from Aglaia species. *Die Pharmazie-An International Journal of Pharmaceutical Sciences*, *63*(10), pp. 768–773.

Esimone, C.O., Eck, G., Nworu, C.S., Hoffmann, D., Überla, K. and Proksch, P., 2010. Dammarenolic acid, a secodammarane triterpenoid from Aglaia sp. shows potent anti-retroviral activity *in vitro*. *Phytomedicine*, *17*(7), pp. 540–547.

Henss, L., Scholz, T., Grünweller, A. and Schnierle, B.S., 2018. Silvestrol inhibits chikungunya virus replication. *Viruses*, *10*(11), p. 592.

Kato-Noguchi, H., Suzuki, M., Noguchi, K., Ohno, O., Suenaga, K. and Laosinwattana, C., 2016. A potent phytotoxic substance in *Aglaia odorata* Lour. *Chemistry & Biodiversity*, *13*(5), pp. 549–554.

Müller, C., Schulte, F.W., Lange-Grünweller, K., Obermann, W., Madhugiri, R., Pleschka, S., Ziebuhr, J., Hartmann, R.K. and Grünweller, A., 2018. Broad-spectrum antiviral activity of the eIF4A inhibitor silvestrol against corona-and picornaviruses. *Antiviral Research*, *150*, pp. 123–129.

Saifah, E., Suttisri, R., Shamsub, S., Pengsuparp, T. and Lipipun, V., 1999. Bisamides from *Aglaia edulis*. *Phytochemistry*, *52*(6), pp. 1085–1088.

Tan, G.T., Pezzuto, J.M., Kinghorn, A.D. and Hughes, S.H., 1991. Evaluation of natural products as inhibitors of human immunodeficiency virus type 1 (HIV-1) reverse transcriptase. *Journal of Natural Products*, *54*(1), pp. 143–154.

Todt, D., Moeller, N., Praditya, D., et al., 2018. The natural compound silvestrol inhibits hepatitis E virus (HEV) replication *in vitro* and *in vivo*. *Antiviral Research*, *157*, pp. 151–158.

Yang, J., 2013. *Natural Product Anticancer Drug Discovery and Mechanistic Studies on Hapolosin and Silvestrol* (Doctoral dissertation, University of Illinois at Chicago).

8.5.3.2 *Amoora rohituka* (Roxb.) Wight & Arn.

Synonyms: *Aphanamixis grandifolia* Bl.; *Aphanamyxis rohituka* (Roxb.) Pierre; *Aphanamixis polystachya* (Wall.) R. Parker

Common names: Tikhta raj (Bangladesh); chemarum, khana, rohitaka (India); chaya kaya, khayo (Myanmar)

Habitat: Forests

Distribution: India, Sri Lanka, Bangladesh Myanmar, Laos, Malaysia, Indonesia, Papua New Guinea, Philippines Thailand, Vietnam, Pacific islands

Botanical observation: This handsome tree grows up to 30 m tall. The leaves are compound, spiral, and exstipulate. The blade grows up to 90 cm long and presents about 8–10 pairs of folioles. The folioles are oblong, elliptic, or ovate, 10–25 × 5–10 cm, marked with 11–20 pairs of secondary nerves, asymmetrical, and acute to caudate at apex. The spikes are axillary, up to 30 cm long, and support a few flowers which are about 6 mm long. The 5 sepals are minute. The 5 petals are about 7 mm long, spoon-shaped, and white. The staminal tube is globose and develops 5 anthers. The capsules are globose to obovoid, up to 3 cm long, orange, and shelter 3 seeds.

Medicinal uses: Astringent (Myanmar); cold (Bangladesh; Indonesia); jaundice, leucorrhea (India)

Broad-spectrum antibacterial non-polar extract: Petroleum ether extract of stem bark (500 μg/disc) inhibited the growth of *Bacillus cereus* (QL 29), *Bacillus subtilis* (QL 40), *Escherichia coli* (ATCC 25922), *Klebsiella* sp., *Pseudomonas aeruginosa* (ATCC 27853), *Salmonella typhi* (AM 16406), *Sarcina lutea* (QL 166), *Shigella dysenteriae* (ATCC 26131), *Staphylococcus aureus* (ATCC 25923), *Vibrio parahemolyticus* (AM 16362), and *Shigella flexneri* (Y 976) with inhibition zone diameters of 11.5, 11.5. 14.5, 13, 16.5, 8.5, 14, 12, 10.5, 9.5, and 17 mm, respectively (Chowdhury et al., 2003).

Broad-spectrum antibacterial halo developed by essential oil: Essential oil of leaves (20 μg/disc) inhibited the growth of *Bacillus subtilis* (NCTC 6276), *Staphylococcus aureus* (NCIMP 95181), *Streptococcus faecalis* (NCTC 7551), *Micrococcus flavoroscus* (NCTC 75281), *Propionibacterium acnes* (NCTC 7371), *Listeria monocytogenes* (M 79731), *Alcaligenes faecalis* (NCTC 6551), *Klebsiella aerogenes* (NCTC 50551), *Salmonella enterica* subsp. *enterica* serovar *Abony* (NCTC 60171), *Staphylococcus epidermidis* (NCTC 11047), *Escherichia coli* (NCTC 90021), and *Pseudomonas aeruginosa* (NCTC 104211) (Aboutabl et al., 2000).

Strong antibacterial (Gram-positive) cycloartane-type triterpenes: Aphagrandinoid A isolated from the leaves and stems inhibited the growth of *Staphylococcus aureus*, *Pseudomonas aeruginosa*, methicillin-resistant *Staphylococcus aureus* (92#), and methicillin-resistant *Staphylococcus aureus* (98#) (Wang et al., 2013). (20*R*)-3β-Hydroxy-24,25,26,27-tetranor-5α-cycloartan-23,21-olide isolated from the leaves and stems inhibited the growth of *Staphylococcus aureus*, *Pseudomonas aeruginosa*, methicillin-resistant *Staphylococcus aureus* (92#), and methicillin-resistant *Staphylococcus aureus* (98#) with the MIC values of 25, 25, 1.5, and 1.5 μg/mL, respectively (Wang et al., 2013).

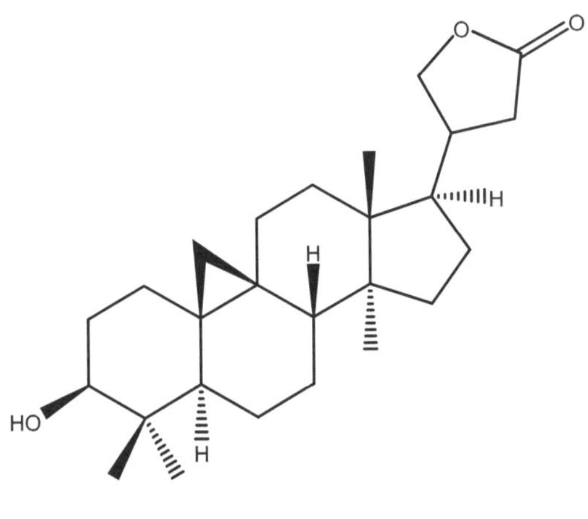

(20*R*)-3*β*-Hydroxy-24,25,26,27-tetranor-5*α*-cycloartan-23,21-olide

Anticandidal halo developed by essential oil: Essential oil of leaves (20 μg/disc) inhibited the growth of *Candida albicans* (NCTC 31791) with the inhibition zone diameter of 24 mm (Aboutabl et al., 2000).

Commentaries

i. The hemisynthetic alkaloid flavopiridol developed from the piperidine chromone alkaloid rohitukine abrogated the replication of the Human immunodeficiency virus type-1 (NL4-3 strain) in Jurkat cells at the concentration of 26 nM via a mechanism involving inhibition of RNA polymerase II by inhibition of the cyclin-dependent kinase P-TEFb I (Chao et al., 2000). Since the Severe acute respiratory syndrome-associated coronavirus replication involves host cell cyclin-dependent kinase (Gargouri et al., 2021), it is possible that flavopiridol or rohitukine could be of value for the treatment of COVID-19, as well as other pandemic coronaviruses. Rohitukine and flavopiridol can be added to the list of first line natural products to be tested in face of the upcoming pandemics.

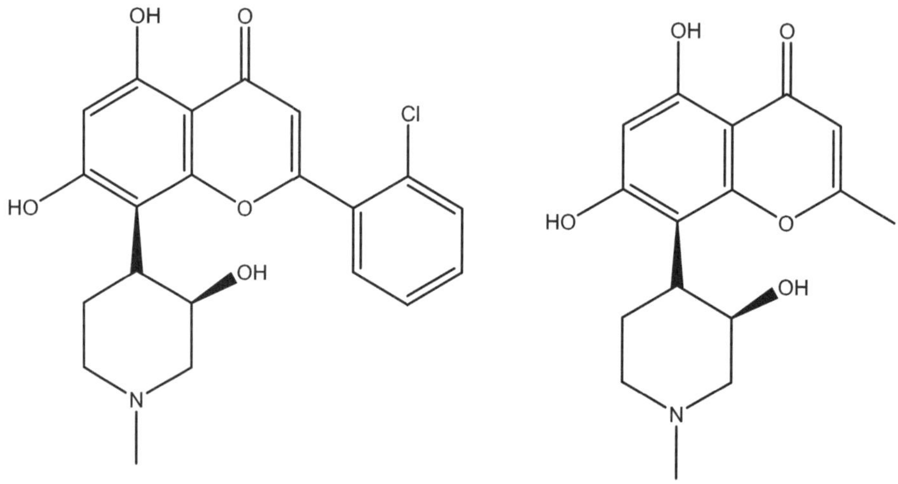

Flavopiridol Rohitukine

ii. The plant generates rare series of nornemoralisin-type diterpenoids (Wu et al., 2020) which could be examined for their antimicrobial properties.

REFERENCES

Aboutabl, E.A., El-Sakhawy, F.S., Fathy, M.M. and Megid, R.M.A., 2000. Composition and antimicrobial activity of the leaf and fruit oils from *Amoora rohituka* Wigth. et Arn. *Journal of Essential Oil Research*, *12*(5), pp. 635–638.

Chao, S.H., Fujinaga, K., Marion, J.E., Taube, R., Sausville, E.A., Senderowicz, A.M., Peterlin, B.M. and Price, D.H., 2000. Flavopiridol inhibits P-TEFb and blocks HIV-1 replication. *Journal of Biological Chemistry*, *275*(37), pp. 28345–28348.

Chowdhury, R., Hasan, C.M. and Rashid, M.A., 2003. Antimicrobial activity of *Toona ciliata* and *Amoora rohituka*. *Fitoterapia*, *74*(1), pp. 155–158.

Gargouri, M., Alzwi, A. and Abobaker, A., 2021. Cyclin dependent kinase inhibitors as a new potential therapeutic option in management of COVID-19. *Medical Hypotheses*, *146*, p. 110380.

Wang, X.Y., Tang, G.H., Yuan, C.M., Zhang, Y., Zou, T., Yu, C., Zhao, Q., Hao, X.J. and He, H.P., 2013. Aphagrandinoids A–D, cycloartane triterpenoids with antibacterial activities from *Aphanamixis grandifolia*. *Fitoterapia*, *85*, pp. 64–68.

Wu, X.Z., Fang, F.H., Huang, W.J., Shi, Y.Y., Pan, H.Q., Ning, L. and Yuan, C.S., 2020. Two novel nornemoralisin-type diterpenoids from *Aphanamixis polystachya* (Wall.) R. Parker. *Fitoterapia*, *140*, p. 104431.

8.5.3.3 *Azadirachta indica* L.

Synonyms: *Melia azadirachta* L.; *Melia indica* (A. Juss.) Brandis

Common names: Neem tree; neem (Bangladesh; India; Nepal); nimba (India); tama (Myanmar); dhrek (Pakistan)

Botanical observation: This magnificent tree grows to about 10 m and has a positive aura. It is often grown with mango trees in front of houses as good omen. The bark is smooth and reddish-brown. The leaves are imparipinnate, spiral, and exstipulate. The rachis is slender. The blade comprises about 5–7 pairs of folioles which are asymmetrical, 2.5–7 × 1.5–4 cm, falcate, glossy, and serrate. The panicles are axillary, lax, light yellowish green, pendulous and present few white, sweet-scented, and ephemeral flowers. The 5 sepals are obovate and minute. The 5 petals are about 6 mm long, oblong to spathulate, and heavenly white. The staminal tube is about 5 mm long and develops 10 anthers. The ovary is subglobose and develops a linar style and a trifid stigma. The drupes are olive-shaped to oblong, light yellowish green, 1.3–2 cm long, and smooth.

Medicinal uses: Chicken pox, leprosy, measles, pimples, small pox, wounds, syphilis, ringworms (Bangladesh); pustular eruptions, sores, tuberculosis, ulcers (India); boils (Myanmar; Pakistan); fever (Bangladesh; India; Nepal); cough (Nepal; Pakistan); mouthwash, urinary tract infection (Myanmar)

Antibacterial (Gram-positive) halo developed by polar extract: Ethanol extract of leaves (50 µL of a 10 mg/mL solution/8 mm disc) inhibited the growth of *Listeria monocytogenes* (ATCC 49594) and *Staphylococcus aureus* (JCM 2151) with the inhibition zone diameters of 20 and 13 mm, respectively (Mahfuzul Hoque et al., 2007).

Antibacterial (Gram-negative) halo developed by polar extract: Methanol extract of leaves inhibited *Vibrio cholerae* (SG24 O139), *Vibrio cholerae* (NB2 O1), and *Vibrio cholerae* (PC4 (nonO1, nonO139) with the zones of inhibition of 12.7, 12.8, and 12.7 mm, respectively (200 µL/well; 50 mg/mL) (Thakurta et al., 2997).

In vivo antibacterial (Gram-positive) limonoid: Azadirachtin (40% aqueous solution) was as effective as 0.2% chlorhexidine mouthrinse in reducing the *Streptococcus mutans* count in dental plaque (Kankariya et al., 2016).

Antifungal limonoids: Azadiradione inhibited the growth of *Puccinia arachidis* (Govindachari et al., 2000).Gedunin inhibited the growth of *Polyporus* sp. (Sundarasivarao et al., 1977).

Strong antiviral (non-enveloped monopartite linear single-stranded (+) RNA) polar extract: Ethanol extract of leaves inhibited the replication of the Enterovirus 71 the IC_{50} value of 5.5 µg/mL and a selectivity index above 57.7 in Vero cells (Choi et al., 2010).

Strong antiviral (enveloped monopartite double-stranded DNA) hydrophilic limonoid: Azadirachtin (Log D = 0.9 at pH 7.4; molecular mass = 720 g/mol) at the concentration of 10 µg/mL inhibited the secretion of Hepatitis B virus surface-antigen and Hepatitis B virus e-antigen by HepG2.2.15 by about 60 and 75% (Parvez et al., 2019).

Strong antiviral (monopartite linear dimeric single-stranded (+) RNA) hydrophilic limonoid: Azadirachtin at the concentration of 10 µM inhibited the replication of the Human immunodeficiency virus by more than 50% (David et al., 2017).

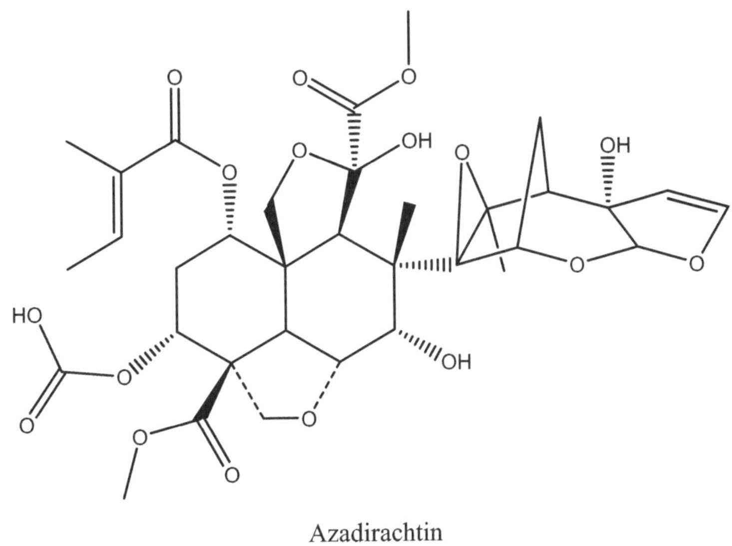

Azadirachtin

Very weak antiviral (non-enveloped monopartite linear single-stranded (+) RNA) polar extract: Methanol extract of leaves inhibited the replication of the Coxsackie B viruses at the concentration of 1000 µg/mL (Badam et al., 1999).

Commentaries

i. Being active against the Human immunodeficiency virus, azadirachtin or aqueous extracts of leaves could be active against the Severe acute respiratory-acquired coronavirus. In Ayurvedic medicine the plant is recommended for COVID-19 (Soni et al., 2020).
ii. Note that neuroactive natural products are often able to inhibit bacterial efflux pumps and one could anticipate that insect antifeedants could inhibit efflux pumps. Insecticidal substances like azadirachtin (Jennifer Mordue et al., 1998) are neuroactive and antifeedant in insects (Jennifer Mordue et al., 1998) and the antibiotic potentiating effects of azadiachtin is probable.
iii. Limonoids (oxygenated tetranortriterpenes) as often antiretroviral by mechanisms yet to be understood: *Cipadessa baccifera* (Roth.) Miq. is used in Nepal (local name: *Kaligeri*) for the treatment of cough and cold. The limonoids ciparasin B and P isolated from this plant inhibited the replication of the Human immunodeficiency virus type-1 (NL4-3 strain) in MT-4 cells with the EC_{50} values of 5.5 (selectivity index>7.2) and 6.1 µM (selectivity index>6.5), respectively (Yu et al., 2015). Trichiconin B and C isolated from the fruits of the magnificent rainforest tree *Trichilia connaroides* (Wight & Arn.) Bentv. (a plant used to treat cholera in China) inhibited the replication of the Human immunodeficiency virus type-1 (stain NL 43) in MT4 cells with the EC_{50} values of 5.9 (selectivity index>4.2) and 3.6 µM (selectivity index = 3.5), respectively (Liu et al., 2014).
iv. *Walsura trifoliolata* (A. Juss.) Harms in the tribe Melieae, Subfamily Melioideae, is used in India to treat cough (local name: *Kanjiram*). Ethanol extract of *Walsura rubusta* Roxb. inhibited the growth of 35 clinical isolates of methicillin-resistant *Staphylococcus aureus* and *Staphylococcus aureus* (Voravuthikunchai & Kitpipit, 2005). Ethanol extract of wood of *Walsura robusta* Roxb. inhibited the growth of *Escherichia coli* (O157:H7) (Voravuthikunchai & Limsuwan, S., 2006). Methanol extract of *Walsura trijuga* (Roxb. ex Sims) Kurz at the concentration of 200 µg/mL inhibited the Human immunodeficiency virus type-1 reverse transcriptase by 12% (Tan et al., 1991).

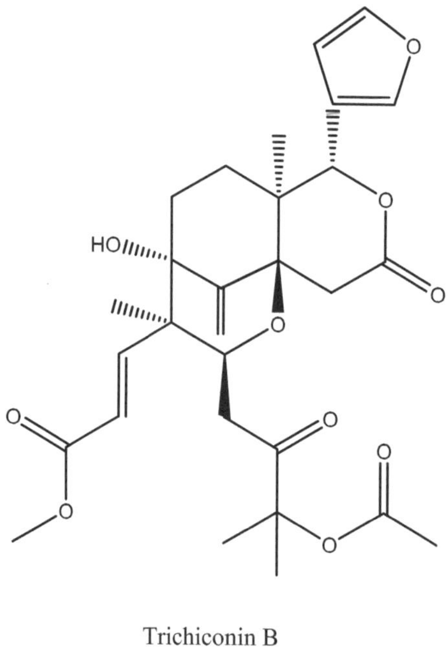

Trichiconin B

v. The plant was known of our peer the Ayurvedic physician Sushruta (800 BC) who describes cholera and warns against physical contact or touch, use of contaminated cloths, ornaments, and cosmetics, and avoiding sitting, sleeping, and taking meal together with the infected person in his *Sushruta Samhita* translated in Arab and Persian by the 10th century (Arora, 2020; Joshi & Dudhamal, 2020)

REFERENCES

Arora, N., 2020. Pandemic COVID-19 Diseases and Rajyakshma in Charak Samhita: A conceptual study. *Journal of Ayurveda*, *14*(4), p. 28.

Badam, L., Joshi, S.P. and Bedekar, S.S., 1999. '*In vitro*' antiviral activity of neem (*Azadirachta indica*. A. Juss) leaf extract against group B coxsackieviruses. *The Journal of Communicable Diseases*, *31*(2), pp. 79–90.

Choi, H.J., Song, J.H., Park, K.S. and Baek, S.H., 2010. *In vitro* anti-enterovirus 71 activity of gallic acid from *Woodfordia fruticosa* flowers. *Letters in Applied Microbiology*, *50*(4), pp. 438–440.

David, P.E., Benjamín, S.G.L., Delia, E.Á.E.A., Patricia, N.H.M. and Carmen, V.M.M.D., 2017. HIV-1 infection inhibition by neem (*Azadirachta indica* A. Juss.) leaf extracts and Azadirachtin. *Indian Journal of Traditional Knowledge*, *16*, pp. 437–441.

Govindachari, T.R., Suresh, G., Gopalakrishnan, G., Masilamani, S. and Banumathi, B., 2000. Antifungal activity of some tetranortriterpenoids. *Fitoterapia*, *71*(3), pp. 317–320.

Jennifer Mordue, A., Simmonds, M.S., Ley, S.V., Blaney, W.M., Mordue, W., Nasiruddin, M. and Nisbet, A.J., 1998. Actions of azadirachtin, a plant allelochemical, against insects. *Pesticide Science*, *54*(3), pp. 277–284.

Joshi, F. and Dudhamal, T., 2020. Pandemic COVID-19 in relation to Ayurveda: A brief review through Sushruta Samhita. *Journal of Ayurveda*, *14*(4), p. 58.

Kankariya, A.R., Patel, A.R. and Kunte, S.S., 2016. The effect of different concentrations of water soluble azadirachtin (neem metabolite) on *Streptococcus mutans* compared with chlorhexidine. *Journal of Indian Society of Pedodontics and Preventive Dentistry*, *34*(2), p. 105.

Liu, C.P., Xu, J.B., Han, Y.S., Wainberg, M.A. and Yue, J.M., 2014. Trichiconins A–C, limonoids with new carbon skeletons from *Trichilia connaroides*. *Organic Letters*, *16*(20), pp. 5478–5481.

Mahfuzul Hoque, M.D., Bari, M.L., Inatsu, Y., Juneja, V.K. and Kawamoto, S., 2007. Antibacterial activity of guava (*Psidium guajava* L.) and neem (*Azadirachta indica* A. Juss.) extracts against foodborne pathogens and spoilage bacteria. *Foodborne Pathogens and Disease*, *4*(4), pp. 481–488.

Parvez, M.K., Rehman, M.T., Alam, P., Al-Dosari, M.S., Alqasoumi, S.I. and Alajmi, M.F., 2019. Plant-derived antiviral drugs as novel hepatitis B virus inhibitors: cell culture and molecular docking study. *Saudi Pharmaceutical Journal*, *27*(3), pp. 389–400.

Siddiqui, S., Faizi, S. and Siddiqui, B.S., 1992. Constituents of *Azadirachta indica*: isolation and structure elucidation of a new antibacterial tetranortriterpenoid, mahmoodin, and a new protolimonoid, naheedin. *Journal of Natural Products*, *55*(3), pp. 303–310.

Soni, H., Sharma, S. and Malik, J.K., 2020. Role of Indian herbs against COVID-19: a review. *International Journal of Traditional and Complementary Medicine*, *3*(9), pp. 169–176.

Sundarasivarao, B. and Madhusudhanarao, NJ., 1977. Antifungal activity of gedunin. *Current Science*, *46*(20), pp. 714–716.

Tan, G.T., Pezzuto, J.M., Kinghorn, A.D. and Hughes, S.H., 1991. Evaluation of natural products as inhibitors of human immunodeficiency virus type 1 (HIV-1) reverse transcriptase. *Journal of Natural Products*, *54*(1), pp. 143–154.

Thakurta, P., Bhowmik, P., Mukherjee, S., Hajra, T.K., Patra, A. and Bag, P.K., 2007. Antibacterial, antisecretory and antihemorrhagic activity of *Azadirachta indica* used to treat cholera and diarrhea in India. *Journal of Ethnopharmacology*, *111*(3), pp. 607–612.

Voravuthikunchai, S.P. and Kitpipit, L., 2005. Activity of medicinal plant extracts against hospital isolates of methicillin-resistant *Staphylococcus aureus*. *Clinical Microbiology and Infection*, *11*(6), pp. 510–512.

Voravuthikunchai, S.P. and Limsuwan, S., 2006. Medicinal plant extracts as anti–*Escherichia coli* O157: H7 agents and their effects on bacterial cell aggregation. *Journal of Food Protection*, *69*(10), pp. 2336–2341.

Yu, J.H., Wang, G.C., Han, Y.S., Wu, Y., Wainberg, M.A. and Yue, J.M., 2015. Limonoids with anti-HIV activity from *Cipadessa cinerascens*. *Journal of Natural Products*, *78*(6), pp. 1243–1252.

8.5.3.4 *Chisocheton penduliflorus* Planch. ex Hiern

Habitat: Forests

Distribution: Malaysia, Thailand

Botanical observation: This handsome tree grows up to 10 m tall. The bark is blackish. The inner bark is pale fawn. The stems are velvety. The leaves are imparipinnate, spiral, and exstipulate. The rachis is velvety, 40 cm long, and shows to 8 pairs of folioles pubescent below. The folioles are 10–14 × 5–6 cm, obovate, and marked with 11–22 pairs of secondary nerves. The inflorescences are 15–23.5 cm long with tiny and dull red flowers grouped at apex. The corolla consists of 4–5 petals. The staminal tube shows 3–5 apical forked lobes and develops 3–5 anthers. The nectary disc is obscure to cup shaped. The ovary is discoid to rounded. The berries are 3.5 × 1.2 cm, beaked when young to 5 cm long, splitting into 3, containing 3 glossy seeds which with a partial red aril.

Medicinal use: The fruits are eaten by the Malays as medicinal food or "*ulam*".

Moderate antibacterial aromadendrane-type sesquiterpenes: Allo-aromadendrane-10β, 14-diol, allo-aromadendrane-10α, 14-diol, and allo-aromadendrane-10β, 13, 14-triol inhibited the growth of *Mycobacterium tuberculosis* H_{37}Ra with the MIC values of 50, 100, and 50 μg/mL, respectively (Phongmaykin et al., 2008).

Strong antibacterial dammarane-type triterpenes: Cabraleadiol, eichlerialactone, cabraleahydroxylactone, and cabralealactone inhibited the growth of *Mycobacterium tuberculosis* H_{37}Ra with the MIC values of 50, 25, 50, and 50 μg/mL, respectively (Phongmaykin et al., 2008).

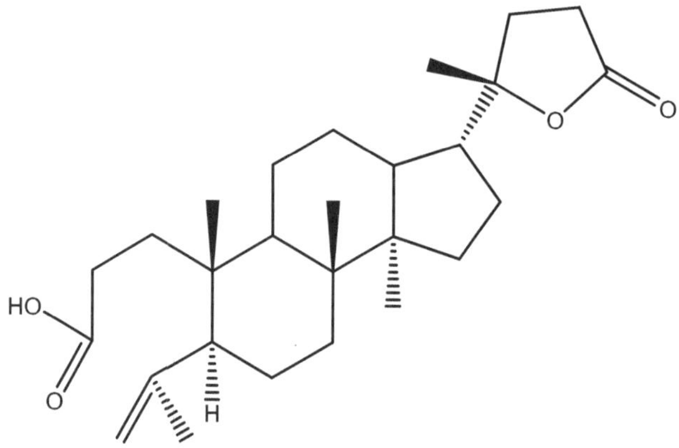

Eichlerialactone

Commentary: Members of the genus *Chisocheton* bring to being interesting series of limonoids (Yang et al., 2009) which could be examined for antimicrobial properties. Consider that within less than five decades the whole primary rain forest of Southeast Asia and its unique reserve of drugs should be gone owed to the rampage of palm oil and logging businesses.

REFERENCES

Phongmaykin, J., Kumamoto, T., Ishikawa, T., Suttisri, R. and Saifah, E., 2008. A new sesquiterpene and other terpenoid constituents of *Chisocheton penduliflorus*. *Archives of Pharmacal Research*, *31*(1), pp. 21–27.

Yang, M.H., Wang, J.S., Luo, J.G., Wang, X.B. and Kong, L.Y., 2009. Tetranortriterpenoids from *Chisocheton paniculatus*. *Journal of Natural Products*, *72*(11), pp. 2014–2018.

8.5.3.5 *Chukrasia tabularis* A. Juss.

Synonyms: *Chukrasia velutina* Wight & Arn.; *Dysoxylum esquirolii* H. Lév

Common names: Bastard cedar, Chittagong wood, golden mahogany; ma lian (China); aglay, chikrassy (India); kin thabut gyi (Myanmar); nhom (Laos); lát (Vietnam).

Habitat: Forests

Distribution: Pakistan, India, Nepal, Sri Lanka, Bhutan, Andamans, Myanmar, Thailand, Vietnam, China, Laos, Malaysia, Indonesia

Botanical observation: This tree grows up to 30 m tall. The bole is buttressed. The bark is brown and the inner bark is reddish. The leaves are compound, alternate, and exstipulate. The petiole is 4–7 cm long and swollen at base. The blade grows up about 50 cm long and presents 7–16 pairs of folioles. The folioles are alternate or subopposite, ovate, oblong, or lanceolate, 6–15 × 3–6 cm, asymmetrical at base, hairy below, acuminate, glossy, and marked with 10–15 pairs of secondary nerves.. The panicles are axillary, lax, and bear numerous flowers. The calyx is minute, cupular, slightly 5-lobed, and hairy. The 5 petals are free, about 1.5 cm long, spathulate, yellow, whitish, purplish, or red. The staminal tube is cylindrical, yellow, and develops 10 anthers. The ovary is bottle-shaped and hairy. The capsules are somewhat globose, up to about 5 cm long, coriaceous, verrucose, dark brown, 2–4.5 × 2–3.5 cm, and opening by 3 valves to release numerous seeds which are 3 mm long, winged, and about 1 cm long.

Medicinal use: Diarrhea (India, Myanmar, Cambodia, Laos, Vietnam); fever (India)

Broad-spectrum antibacterial essential oil: Essential oil of leaves (6 mm disc impregnated with the oil) inhibited the growth of *Bacillus subtilis* (MTTC 121), *Staphylococcus aureus* (MTTC 737), *Proteus vulgaris* (MTCC 1771), *Klebsiella pneumonoae* (MTCC 109), with inhibition zone diameters of 12, 20, 23, and 14 mm (Nagalakshmi et al., 2003).

Broad-spectrum antifungal essential oil: Essential oil of leaves (6 mm disc impregnated with the oil) inhibited the growth of *Aspergillus fumigatus* (MTCC 1811), *Aspergillus niger* (MTCC 1344), and *Fusarium oxysporum* (MTCC 1755) with inhibition zone diameters of 13, 7, and 17 mm (Nagalakshmi et al., 2003).

Commentary: The plant brings to being series of limonoids (Luo et al., 2010) which could be examined for antimicrobial properties. Limonoids are antibacterial. The limonoid 3-deacetylkhivorin isolated from *Khaya senegalensis* (Desr.) A. Juss. inhibited the growth of *Staphylococcus aureus*, *Pseudomonas aeruginosa*, methicillin-resistant *Staphylococcus aureus* (clinical strain 92#), and methicillin-resistant *Staphylococcus aureus* (clinical strain 98#) with the MIC values of 25, 12.5, 12.5, and 12.5 µg/mL, respectively (Yuan et al., 2012). Swietmanin B isolated from *Khaya senegalensis* (Desr.) A. Juss. inhibited the growth of *Staphylococcus aureus*, *Pseudomonas aeruginosa*, methicillin-resistant *Staphylococcus aureus* (clinical strain 92#), and methicillin-resistant *Staphylococcus aureus* (clinical strain 98#) with the MIC values of 25, 12.5, 50, and 12.5 µg/mL, respectively (Yuan et al., 2012). $\Delta^{8(30)}$-3β-Acetoxy-1-oxo-methylmeliacate isolated from *Khaya senegalensis* (Desr.) A. Juss. inhibited the growth of *Staphylococcus aureus*, *Pseudomonas aeruginosa*, methicillin-resistant *Staphylococcus aureus* (clinical strain 92#), and methicillin-resistant *Staphylococcus aureus* (clinical strain 98#) with the MIC values of 25, 12.5, 12.5, and 12.5 µg/mL, respectively (Yuan et al., 2012).

REFERENCES

Nagalakshmi, M.A.H., Thangadurai, D. and Pullaiah, T., 2003. *In vitro* antimicrobial efficacy of leaf essential oils of *Chukrasia tabularis* Adr. Juss. and *Melia dubia* Cav. (Meliaceae). *Phytotherapy Research: An International Journal Devoted to Pharmacological and Toxicological Evaluation of Natural Product Derivatives*, *17*(4), pp. 414–416.

Yuan, C.M., Zhang, Y., Tang, G.H., et al., 2012. Khayseneganins A–H, limonoids from *Khaya senegalensis*. *Journal of Natural Products*, *76*(3), pp. 327–333.

8.5.3.6 *Dysoxylum alliaceum* (Blume) Blume

Synonyms: *Guarea alliacea* Bl.; *Prasoxylon alliaceum* (Blume) M. Roem.

Common names: Kayu bawang (Indonesia) kulim burong (Malaysia); paluahan (the Philippines); ta suea khao (Thailand); chặc khế (Vietnam)

Habitat: Forests

Distribution: Andaman Islands, Thailand, Malaysia, Indonesia, the Philippines, Papua New Guinea, Australia, Pacific Islands

Botanical observation: This handsome tree grows to about 40m. The bole is straight and buttressed. The bark is thin and the inner bark reddish with a somewhat onion-like smell (from where the local name *kayu bawang* = wood onion). The leaves are compound, spiral, and exstipulate. The petiole is 5–15cm long and enlarged at base. The blade grows up to 1m long and presents 6–8 pairs of folioles. The folioles are 7.5–25 × 2.5–7.5 cm, elliptic, asymmetrical at base, dark green and glossy above, marked with 8–12 pairs of secondary nerves sunken above, and acuminate at apex. The panicles are axillary and up to 40 cm long. The calyx is minute, cupular, and 4-lobed. The 4 or 5 petals are 5–8mm long, linear, white, or pinkish. The staminal tube develops 8 sessile anthers. The ovary is hairy. The capsule is about 7.5cm across, brownish, obovate, somewhat velvety, and sheltering numerous smelly seeds.

Medicinal uses: Postpartum protective remedy (Malaysia)

Strong antibacterial (Gram-positive) sesquiterpene: (+) 8-Hydroxy calamenene isolated from the seeds inhibited the growth of *Staphylococcus aureus* with an MIC value below 20ppm.

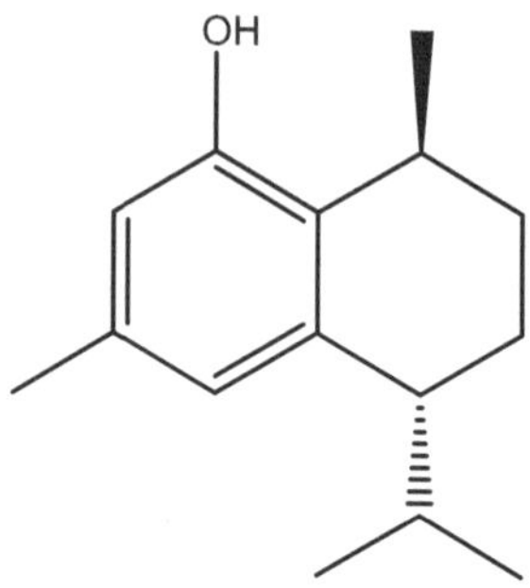

(+) 8-Hydroxycalmenene

Strong broad-spectrum antifungal sesquiterpene: (+) 8-hydroxy calamenene inhibited the growth of *Candida albicans* and *Trichophyton mentagrophytes* with the MIC values ranging from 5 to 20ppm (MIC).

Commentaries

i. The pungent smell is most probably owed to sulfur compounds. Consider that the aliphatic sulfur dysoxysulfone isolated from *Dysoxylum richii* C DC. is antibacterial (Jogia et al., 1989).

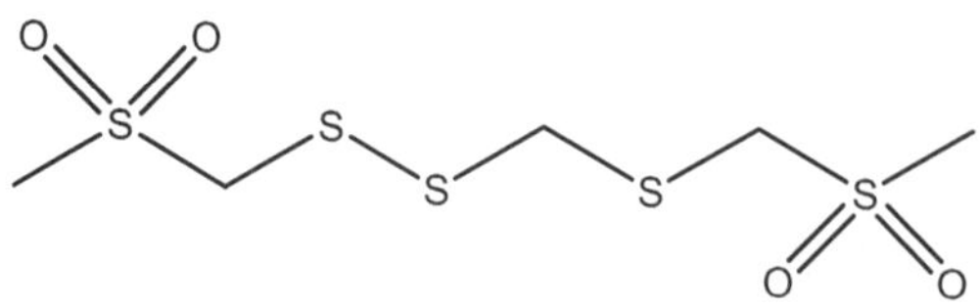

Dysoxysulfone

ii. Seco-oleanane type triterpenes in members of the genus *Dysoxylum* Bl. inhibit the growth of Gram-positive bacteria. Dysoxyhainic acid G isolated from the stems and leaves of *Dysoxylum hainanense* Merr. inhibited the growth of *Staphylococcus epidermidis*, *Micrococcus luteus*, and *Bacillus subtilis* (He et al., 2011). Dysoxyhainic acid H isolated from the stems and leaves of *Dysoxylum hainanense* Merr. inhibited the growth of *Micrococcus luteus* and *Bacillus subtilis* (He et al., 2011). Dysoxyhainic acid I isolated from the stems and leaves of *Dysoxylum hainanense* Merr. inhibited the growth of *Staphylococcus epidermidis* and *Bacillus subtilis* with the MIC values of 12.5 and 1.5 μg/mL, respectively (He et al., 2011). Koetjapic acid isolated from the stems and leaves of *Dysoxylum hainanense* Merr. inhibited *Staphylococcus epidermidis* and *Bacillus subtilis* with the MIC values of 12.5 and 6.2 μg/mL, respectively (He et al., 2011).

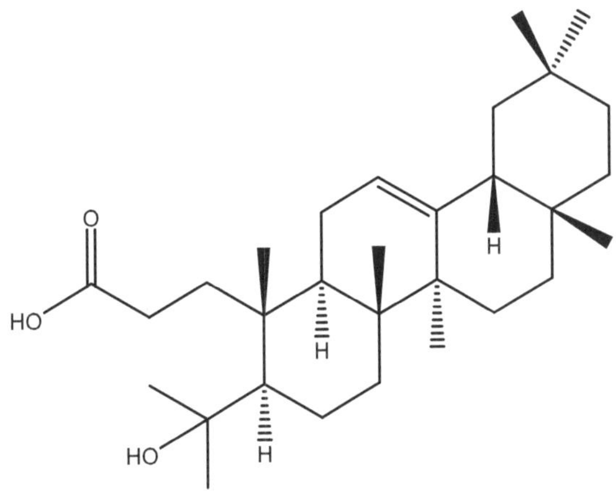

Dysoxyhainic acid I

iii. Members of the genus *Dysoxylum* Bl. produce interesting series of antibacterial and antifungal sesquiterpenes such as dysoxyphenol and 7*R*, 10*S*-2-hydroxycalamenene isolated from the seeds of *Dysoxylum densiflorum* (Blume) Miq. inhibited the growth of *Bacillus subtilis* with the MIC value of 28 μM (Dharmayani et al., 2020).
iv. The limonoids dysoxylins A, B, and D isolated from the bark of *Dysoxylum gaudichaudianum* Miq. inhibited the replication of the Respiratory syncitial virus (strain A2) in MA-104 cells with the EC_{50} values of 2, 4, and 1 μg/mL, respectively (Chen et al., 2007).

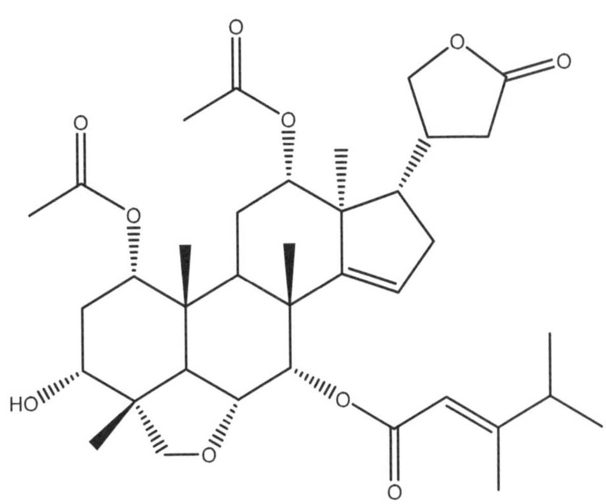

Dysoxylin D

v. Members of the genus *Dysoxylum* Bl. synthethize chromone alkaloids including rohitukine (see earlier) (Ismail et al., 2009) which could be examined for their anticoronavirus properties.

REFERENCES

Chen, J.L., Kernan, M.R., Jolad, S.D., Stoddart, C.A., Bogan, M. and Cooper, R., 2007. Dysoxylins A–D, Tetranortriterpenoids with potent anti-RSV activity from *Dysoxylum gaudichaudianum*. *Journal of Natural Products*, *70*(2), pp. 312–315.

Dharmayani, N.K.T., Yoshimura, T., Hermawati, E., Juliawaty, L.D. and Syah, Y.M., 2020. Antibacterial and antifungal two phenolic sesquiterpenes from *Dysoxylum densiflorum*. *Zeitschrift für Naturforschung C*, *75*(1–2), pp. 1–5.

He, X.F., Wang, X.N., Yin, S., Dong, L. and Yue, J.M., 2011. Ring A-seco triterpenoids with antibacterial activity from *Dysoxylum hainanense*. *Bioorganic & Medicinal Chemistry Letters*, *21*(1), pp. 125–129.

Ismail, I.S., Nagakura, Y., Hirasawa, Y., Hosoya, T., Lazim, M.I.M., Lajis, N.H., Shiro, M. and Morita, H., 2009. Chrotacumines A–D, chromone alkaloids from *Dysoxylum acutangulum*. *Journal of Natural Products*, *72*(10), pp. 1879–1883.

Jogia, M.K., Andersen, R.J., Mantus, E.K. and Clardy, J., 1989. Dysoxysulfone, a sulfur rich metabolite from the Fijian medicinal plant dysoxylum richii. *Tetrahedron Letters*, *30*(37), pp. 4919–4920.

Nishizawa, M., Inoue, A., Sastrapradja, S., & Hayashi, Y. (1983). (+) 8-Hydroxycalmenene: a fish–poison principle of *Dysoxylum acutangulum* and *D. alliaceum*. *Phytochemistry*, *22*, 2083–2085.

8.5.3.7 *Lansium domesticum* Corrêa

Synonyms: *Aglaia domestica* (Correa) Pell., *Aglaia dookoo* Griff.

Common names: Langsat; duku (India); kokosan (Indonesia), duku langsat (Malaysia), boocan, buahan, lanzones (the Philippines), longkong (Thailand), bon bon (Vietnam)

Habitat: Cultivated

Distribution: India, Bangladesh, Southeast Asia to Papua New Guinea

Botanical observation: This handsome tree grows to 15 m tall. The bark is smooth and brownish and the bole presents buttresses. The leaves are pinnate, alternate, up to 50 cm long, and exstipulate. The petiole is up to 7 cm long. The rachis bears 5–9 alternate folioles which are somewhat coriaceous, oblong. 7–20 × 5–10 cm, dark green above and glossy, asymmetrical at base, marked with 1–14 pairs of secondary nerves, and acuminate at apex. The fascicles are cauliflorous and up to about 30 cm long. The calyx is 5-lobed and cupular. The 5 petals are yellowish, ovate, 5 mm long. The staminal tube develops 10 anthers. The berries are yellowish, globose, about 3 cm across, arranged in showy bunches, with a soft pericarp and containing up to five bitter seeds in a whitish, somewhat translucent, and delicious aril.

Medicinal uses: Diarrhea (India); dysentery (Malaysia); fever (India; the Philippines)

Broad-spectrum antibacterial halo developed by onoceranoid-type triterpenes: Lansiolic acid (60 µg/10 mm well) isolated from the epicarp inhibited the growth of *Staphylococcus aureus* (UPCC 1143), *Escherichia coli* (UPCC 1195), *Pseudomonas aeruginosa* (UPCC 1244), and *Bacillus subtilis* (UPCC 1149) with the inhibition zones of 12, 11, 12, and 13 mm, respectively (Ragasa al., 2006). Lansioside C (60 µg/10 mm well) isolated from the epicarp inhibited the growth of *Staphylococcus aureus* (UPCC 1143), *Escherichia coli* (UPCC 1195), *Pseudomonas aeruginosa* (UPCC 1244), and *Bacillus subtilis* (UPCC 1149) with the inhibition zones of 19, 12, 12, and 26 mm, respectively (Ragasa et al., 2006).

Broad-spectrum antifungal halo developed by onoceranoid-type triterpenes: Lansiolic acid (60 µg/10 mm well) inhibited the growth of *Candida albicans* (UPCC 2168), *Aspergillus niger*

(UPCC 3701), and *Trichophyton mentagrophytes* (UPCC 4193) with the inhibition zones of 14, 14, and 14 mm, respectively (Ragasa al., 2006). Lansioside C (60 µg/10 mm well) inhibited the growth of *Candida albicans* (UPCC 2168), *Aspergillus niger* (UPCC 3701), and *Trichophyton mentagrophytes* (UPCC 4193) with the inhibition zones of 13, 14, and 20 mm, respectively (Ragasa al., 2006).

Strong antibacterial (Gram-positive) hydrophilic onoceranoid-type triterpenes: Lamesticumin A (Log D = 4.6 at pH 7.4; molecular mass = 502.4 g/mol) isolated from the stems inhibited the growth of *Staphylococcus aureus* (ATCC 25923), *Staphylococcus epidermidis* (ATCC 12228), *Micrococcus luteus* (ATCC 9341), *Bacillus subtilis* ATCC 6633, *Micrococcus pyogenes* (clinical isolate), and *Bacillus cereus* (clinical isolate) with the MIC values of 6.2, 12.5, 6.2, 3.1, 3.1, and 3.1 µg/mL, respectively (Dong et al., 2011).

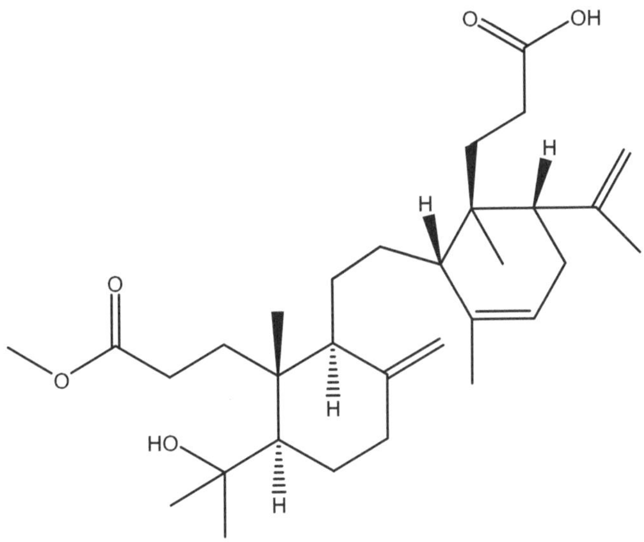

Lamesticumin A

Lamesticumin B isolated from the stems inhibited the growth of *Staphylococcus aureus* (ATCC 25923), *Staphylococcus epidermidis* (ATCC 12228), *Micrococcus luteus* (ATCC 9341), *Bacillus subtilis* (ATCC 6633), *Micrococcus pyogenes* (clinical isolate), and *Bacillus cereus* (clinical isolate) with the MIC values of 6.2, 12.5, 6.2, 3.1, 3.1, and 3.1 µg/mL, respectively (Dong et al., 2011).

Ethyl lansiolate isolated from the stems inhibited the growth of *Staphylococcus aureus* (ATCC 25923), *Staphylococcus epidermidis* (ATCC 12228), *Micrococcus luteus* (ATCC 9341), *Bacillus subtilis* (ATCC 6633), *Micrococcus pyogenes* (clinical isolate), and *Bacillus cereus* (clinical isolate) with the MIC values of 6.2, 12.5, 6.2, 3.1, 3.1, and 3.1 µg/mL, respectively (Dong et al., 2011).

Commentary: Are onoceranoid-type triterpenes in this plant antiviral?

REFERENCE

Dong, S.H., Zhang, C.R., Dong, L., Wu, Y. and Yue, J.M., 2011. Onoceranoid-type triterpenoids from Lansium domesticum. *Journal of Natural Products*, *74*(5), pp. 1042–1048.

Lansium domesticum Corrêa

8.5.3.8 *Melia azedarach* L.

Synonyms: *Melia dubia* Cav.; *Melia orientalis* M. Roem.; *Melia toosendan* Siebold & Zucc.

Common names: Bead tree, China berry, Indian lilac; bokain, goraneem, sadi rayssia (Bangladesh); sdau khmaôch (Cambodia); ku lien shu (China); bakayan, daraink, mahaanimba (India); mindi (Indonesia); kadau s'angz (Laos); mindi kecil (Malaysia); bakaino (Nepal); bakain, dreik (Pakistan); bagalunga (the Philippines); mahanimba (Sri Lanka); lian (Thailand); sâu dông (Vietnam)

Habitat: Forests, cultivated

Distribution: Tropical Asia and Pacific

Botanical observation: This noble tree, probably native to India, grows to about 10 m tall. The stems are hairy at apex. The leaves are bipinnate, spiral, and exstipulate. The petiole is up to 30 cm long and swollen at base. The blade grows up to about 50 cm long and presents 3–7 pairs of pinnae each of them bearing 3–7 pairs of folioles. The folioles are opposite, elliptic, 2.5–5 × 0.5–1.9 cm, asymmetrical, serrate, acuminate at apex, somewhat dark-green, and glossy. The panicles are axillary, up to 20 cm long, and present numerous beautiful and fragrant flowers. The calyx is 5–6-lobed and minute. The 5 petals are up to about 1 cm long, spathulate to oblong, and purplish-white. The staminal tube is 6–7 mm long and develops 10 anthers. An annular disc is present. The ovary produces a style which is 4–5 mm long and a capitate stigma. The berries are green to orangish, glossy, ovoid, up to 4 cm long, and contain up to 5 seeds.

Medicinal uses: Cystitis (China); abscess, infection, for brushing teeth, jaundice, skin diseases (Bangladesh); leprosy (Bangladesh; Nepal); pox (Bangladesh; India); ulcers (Pakistan); boils, wounds (India); fever (India; Pakistan; Sri Lanka)

Broad-spectrum antibacterial essential oil: Essential oil of leaves (impregnated 6mm disc) inhibited the growth of *Bacillus subtilis* (MTTC 121), *Staphylococcus aureus* (MTTC 737), *Pseudomonas aeruginosa* (MTCC 1688), *Escherichia coli* (MTCC 1687), and *Klebsiella pneumoniae* (MTCC 109) with inhibition zone diameters of 27, 18, 32, 40, and 18 mm, respectively (Nagalakshmi et al., 2003).

Strong antibacterial (Gram-positive) limonoids: 7-Cinnamoyltoosendanin isolated from the seeds inhibited the growth of *Micrococcus luteus* (ATCC 9341) and *Bacillus subtilis* (ATCC 6633) with the MIC values of 6.2 and 25 µg/mL, respectively (Liu et al, 2011). 21β-Methylmelianodiol isolated from the seeds inhibited the growth of *Micrococcus luteus* (ATCC 9341) and *Bacillus subtilis* (ATCC 6633) with the MIC values of 12.5 and 25 µg/mL, respectively (Liu et al, 2011).

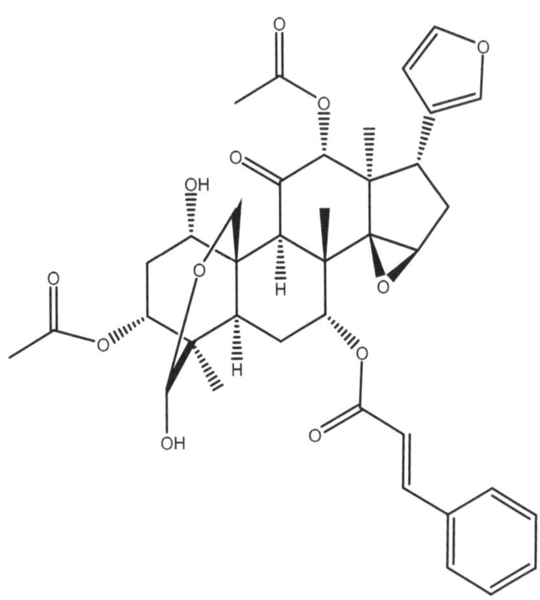

7-Cinnamoyltoosendanin

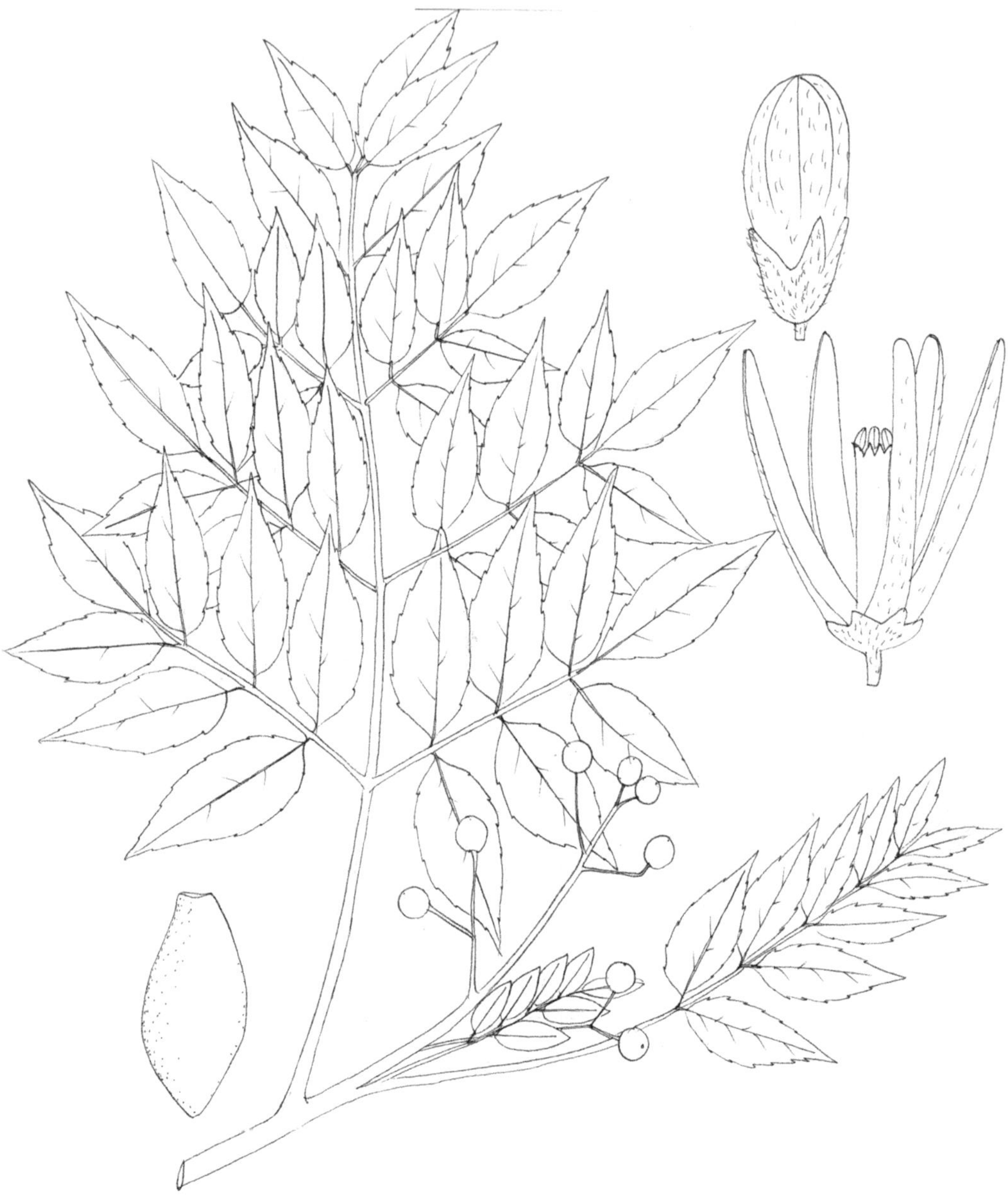

Melia azedarach L.

Strong antibacterial (Gram-negative) limonoids: 1α, 7α-Ditigloyloxy-3α-acetoxyl-12α-ethoxylnimbolinin isolated from the fruits inhibited *Porphyromonas gingivalis* (ATCC 33277) with an MIC value of 15.2 μg/mL (Zhang et al., 2016). 1α-Tigloyloxy-3α-acetoxyl-7α-hydroxyl-12β-ethoxylnimbolinin isolated from the fruits inhibited *Porphyromonas gingivalis* (ATCC 33277) with an MIC value of 31.2 μg/mL (Zhang et al., 2016). 12-Ethoxynimbolinin C, 1-cinnamoyltrichilinin, and trichilinin B isolated from the fruits inhibited the growth of *Porphyromonas gingivalis* (ATCC 33277) with the MIC values of 15.6, 31.3, and 35.5 μg/mL, respectively (Zhang et al., 2007).

Moderate antimycobacterial limonoid: 3-α-Tigloyl-melianol inhibited the growth of *Mycobacterium tuberculosis* (H37Ra ATCC 25177) with the MIC value of 29 μM (Sanna et al., 2015).

In vivo antibacterial (Gram-positive) polar extract: Methanol extract of flowers healed wounds infected with *Staphyloccocus aureus* (Saleem et al., 2002).

Strong antiviral (enveloped monopartite linear double-stranded DNA) limonoid: 1-Cinnamoyl-3,11-dihydroxymeliacarpin from the leaves inhibited the replication of the Herpes simplex virus type-1 (strain F) in Vero cells with the IC_{50} value of 20 μM (Alché et al., 2003). In a subsequent study, 1-cinnamoyl-3,11-dihydroxymeliacarpin isolated from the leaves inhibited the replication of the Herpes simplex virus type-1 with the EC_{50} value of 0.7 μM in human corneal-limbal epithelial cells by affecting the trafficking of viral gD glycoprotein and nuclear factor B activation in host cells (Bueno et al., 2009).

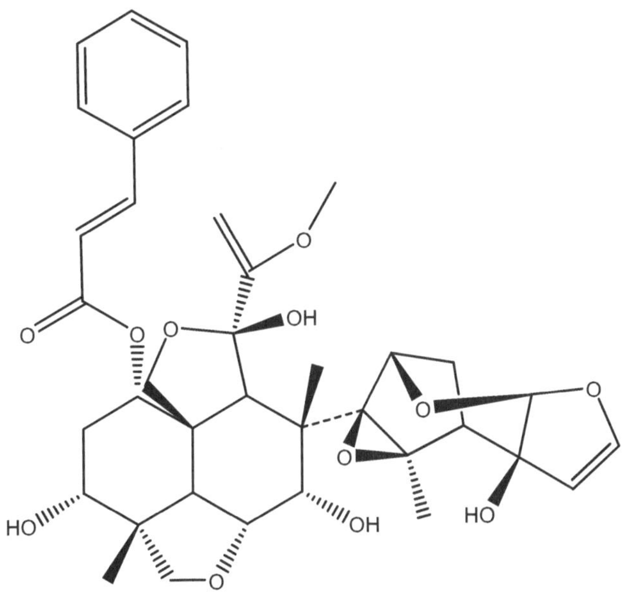

1-Cinnamoyl-3,11-dihydroxymeliacarpin

Strong antiviral (enveloped monopartite single-stranded (−) RNA) limonoid: 1-Cinnamoyl-3,11-dihydroxymeliacarpin from the leaves inhibited the replication of the Vesicular stomatitis virus in Vero cells with the IC_{50} value of 6 μM (Alché et al., 2003).

Strong antiviral (enveloped monopartite linear single-stranded (+) RNA) tirucallane-type triter-penes: 3-α-Tigloyl-melianol and melianone isolated from the fruits inhibited the replication of the Yellow fever virus (strain 17-D vaccine) with the EC_{50} values of 7 and 3μM, respectively (Sanna et al., 2015). 3-α-Tigloyl-melianol and melianone inhibited Dengue virus-2 (Clinical isolate) with EC_{50} values of 3 and 12 μM and West Nile virus with the EC_{50} values of 3 and 11 μM (Sanna et al., 2015).

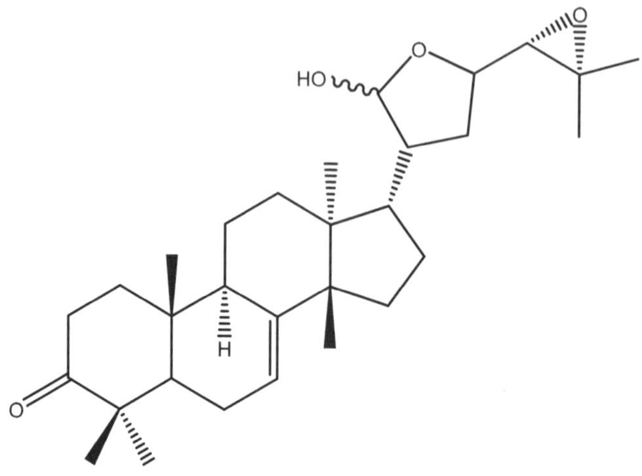

Melianone

Broad-spectrum antifungal essential oil: Essential oil of leaves (impregnated 6 mm) disc inhibited the growth of *Candida albicans* (MTCC 183) and *Fusarium oxysporum* (MTCC 1755) with inhibition zone diameters of 26 and 19 mm, respectively (Nagalakshmi et al., 2003).

Weak antifungal (filamentous) polar extract: Ethanol extract of seeds inhibited the growth of *Aspergillus flavus, Diaporthe phaseolorum* var *meridionale, Fusarium oxysporum, Fusarium solani, Fusarium verticilloides,* and *Sclerotinia sclerotiorum* with the minimum MFC values of 5000, 500, 500, 500, 2500, and 2500 µg/mL, respectively (Carpinella et al., 2003).

Weak antifungal (filamentous) simple phenolic: Vanillin isolated from the seeds inhibited the growth of *Fusarium verticilloides* with the MIC value of 400 µg/mL (Carpinella et al., 2003).

Weak antifungal (filamentous) hydroxycinnamic acid: 4-Hydroxy-3-methoxycinnamaldehyde inhibited the growth of *Fusarium verticilloides* with the MIC values of 400 µg/mL (Carpinella et al., 2003).

Weak antifungal (filamentous) furofuran-type lignan: Pinoresinol inhibited the growth of *Fusarium verticilloides* with the MIC values of 1000 µg/mL (Carpinella et al., 2003).

Commentaries

i. The antiviral activity of a natural product tested *in vitro* is influenced by the type of host cells (Bueno et al., 2009).

ii. When vanillin, 4-hydroxy-3-methoxycinnamaldehyde and pinoresinol were combined at concentration below MIC values, growth inhibition of *Fusarium verticilloides* was as high as positive standard drugs (Carpinella et al., 2003), suggesting that plants synthesize series of natural products belonging to different groups working synergistically to inhibit the growth of microbes. These compounds may help one another and have specific and complementary roles against microbes (several targets, facilitating entry, inhibiting efflux) to defend the plant against phytopathogenic fungi.

iii. The antiviral mechanism of action of limonoid is yet not understood. Consider that limonoids are cytotoxic *in vitro* (Priyadarsini, et al., 2010). Being active against enveloped monopartite linear single-stranded (+) RNA viruses, melianone and 3-α-tigloyl-melianol could be tested against the Severe acute respiratory syndrome associated coronavirus 2.

iv. In the Subfamily Melioideae, Tribe Turraeeae, *Munronia delavayi* Franch. generates the limonoid mulavanin D and 2α, 3α, 15β-trihydroxy20(*S*)-tigloyl-pregnane which inhibited the growth of *Microsporum gypseum* with the MIC values of 25 and 50 µg/mL, respectively, and *Trichophyton rubrum* with the MIC values of 25 and 50 µg/mL (Lin et al., 2010).

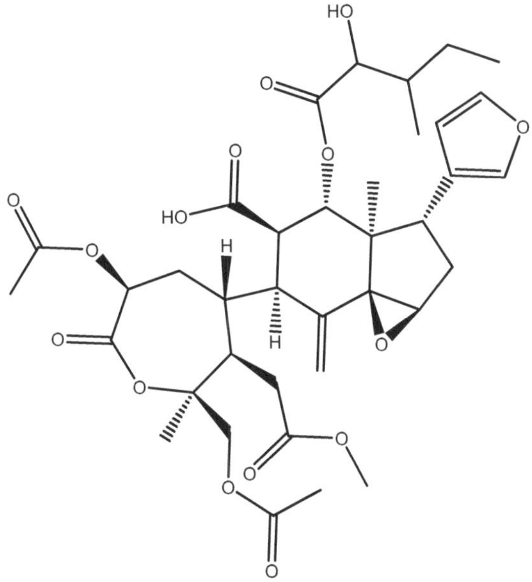

Mulavanin D

Munronia henryi Harms produces the limonoid munronin Q which inhibited the Tobacco mosaic virus (U1 strain) with an IC_{50} value of 14.8 µg/mL (Yan et al., 2015). Munronoid K isolated from *Munronia unifolia* Oliv. at 30 µg/mL inhibited by 67.2% Tobacco mosaic virus (U1 strain) (Ge et al., 2012).

Stem hexane extract of *Munronia pumila* Wight (1 mg/ 6 mm disc) inhibited the growth of *Bacillus subtilis* and *Escherichia coli* as well as *Saccharomyces cerevisiae* with inhibition zone diameters of 10, 11, and 14 mm, respectively (Jayasinghe et al., 2002).

v. The Yellow fever virus belongs to the genus Flavivirus in the family Flaviviridae transmitted by mosquitoes and native to tropical Africa which incure lethal bleedings and liver failure. This virus accounted to the Philadelphia Yellow fever of 1793 (Gum, 2010). Global warming means more mosquitoes and Yellow fever virus could strike in overpopulated and unprepared warm countries (Wilder-Smith et al., 2019). Vaccines to prevent the Yellow fever virus exist but concerns are raised over side effects (de Menezes Martins et al. 2015). There is a need to develop drugs to be taken orally for the prevention of zoonotic pandemic Flaviviruses and the medicinal plants in Asia-Pacific offer a vast source of potential candidates. Again, the solution to zoonotic pandemic looks very much like the use of both Western medicines and herbals or natural products from medicinal plants.

REFERENCES

Alché, L.E., Ferek, G.A., Meo, M., Coto, C.E. and Maier, M.S., 2003. An antiviral meliacarpin from leaves of *Melia azedarach* L. *Zeitschrift für Naturforschung C*, *58*(3–4), pp. 215–219.

Bueno, C.A., Barquero, A.A., Di Cónsoli, H., Maier, M.S. and Alché, L.E., 2009. A natural tetranortriterpenoid with immunomodulating properties as a potential anti-HSV agent. *Virus Research*, *141*(1), pp. 47–54.

Carpinella, M.C., Giorda, L.M., Ferrayoli, C.G. and Palacios, S.M., 2003. Antifungal effects of different organic extracts from *Melia azedarach* L. on phytopathogenic fungi and their isolated active components. *Journal of Agricultural and Food Chemistry*, *51*(9), pp. 2506–2511.

de Menezes Martins, R., da Luz Fernandes Leal, M. and Homma, A., 2015. Serious adverse events associated with yellow fever vaccine. *Human vaccines & immunotherapeutics*, *11*(9), pp. 2183–2187.

Ge, Y.H., Liu, K.X., Zhang, J.X., Mu, S.Z. and Hao, X.J., 2012. The limonoids and their antitobacco mosaic virus (TMV) activities from *Munronia unifoliolata* Oliv. *Journal of Agricultural and Food Chemistry*, *60*(17), pp. 4289–4295.
Gum, S.A., 2010. *Philadelphia under siege: the yellow fever of 1793*. Pennsylvania Center for the Book, USA.
Jayasinghe, U.L.B., Jayasooriya, C.P., Bandara, B.M.R., Ekanayake, S.P., Merlini, L. and Assante, G.M., 2002. Antimicrobial activity of some Sri Lankan Rubiaceae and Meliaceae. *Fitoterapia*, *73*, pp. 424–427.
Liu, H.B., Zhang, C.R., Dong, S.H., Dong, L., Wu, Y. and Yue, J.M., 2011. Limonoids and triterpenoids from the seeds of *Melia azedarach*. *Chemical and Pharmaceutical Bulletin*, *59*(8), pp. 1003–1007.
Nagalakshmi, M.A.H., Thangadurai, D. and Pullaiah, T., 2003. *In vitro* antimicrobial efficacy of leaf essential oils of *Chukrasia tabularis* Adr. Juss. and *Melia dubia* Cav. (Meliaceae). *Phytotherapy Research: An International Journal Devoted to Pharmacological and Toxicological Evaluation of Natural Product Derivatives*, *17*(4), pp. 414–416.
Priyadarsini, R.V., Murugan, R.S., Sripriya, P., Karunagaran, D. and Nagini, S., 2010. The neem limonoids azadirachtin and nimbolide induce cell cycle arrest and mitochondria-mediated apoptosis in human cervical cancer (HeLa) cells. *Free Radical Research*, *44*(6), pp. 624–634.
Saleem, R., Ahmed, S.I., Shamim, S.M., Faizi, S. and Siddiqui, B.S., 2002. Antibacterial effect of *Melia azedarach* flowers on rabbits. *Phytotherapy Research: An International Journal Devoted to Pharmacological and Toxicological Evaluation of Natural Product Derivatives*, *16*(8), pp. 762–764.
Sanna, G., Madeddu, S., Giliberti, G., Ntalli, N.G., Cottiglia, F., De Logu, A., Agus, E. and Caboni, P., 2015. Limonoids from *Melia azedarach* fruits as inhibitors of flaviviruses and *Mycobacterium tubercolosis*. *PloS One*, *10*(10), p. e0141272.
Wilder-Smith, A., Lee, V. and Gubler, D.J., 2019. Yellow fever: is Asia prepared for an epidemic?. *The Lancet Infectious Diseases*, *19*(3), pp. 241–242.
Yan, Y., Yuan, C.M., Di, Y.T., Huang, T., Fan, Y.M., Ma, Y., Zhang, J.X. and Hao, X.J., 2015. Limonoids from *Munronia henryi* and their anti-tobacco mosaic virus activity. *Fitoterapia*, *107*, pp. 29–35.
Zhang, Q., Shi, Y., Liu, X.T., Liang, J.Y., Ip, N.Y. and Min, Z.D., 2007. Minor limonoids from *Melia toosendan* and their antibacterial activity. *Planta Medica*, *73*(12), pp. 1298–1303.
Zhang, Q., Zheng, Q.H., Liang, J.Y., Qing-Shan, L.I. and Zhi-Da, M.I.N., 2016. Two new limonoids isolated from the fuits of *Melia toosendan*. *Chinese Journal of Natural Medicines*, *14*(9), pp. 692–696.

8.5.3.9 *Sandoricum koetjape* (Burm. f.) Merr.

Synonyms: *Melia koetjape* Burm. f.; *Sandoricum indicum* Cav.; *Sandoricum maingayi* Hiern., *Sandoricum nervosum* Bl.; *Sandoricum vidalii* Merr

Common names: Lolly fruit tree, santol; kom ping rech (Cambodia); sevai (India); kecapi, sentul (Indonesia); sentul hutan (Malaysia); tung (Laos); thittoo (Myanmar); malasantol (Philippines); kra thon (Thailand); sau (Vietnam)

Habitat: Forests, cultivated

Distribution: India, Bangladesh, Southeast Asia

Botanical description: This beautiful tree grows to a height of 45 m. The bole is straight and buttressed, the buttresses to 5 m tall. The bark is pale pinkish brown, smooth, and lenticelled. The inner bark is pink. The leaves are trifoliolate, hairy, arranged in terminal spirals, and to 40 cm long. The petiole is 10 cm long and somehow flat at apex. The petiolules are 5 mm long. The rachis is 15 cm long and supports 3 folioles which are elliptic, 8–14.6 × 6–8 cm, somewhat coriaceous, dark green, glossy, marked with 10–12 pairs of secondary nerves sunken above, and acute at apex. The flowers are fragrant, in axillary panicles which are 15 cm long and pendulous. The calyx is truncate. The 5 petals are yellowish and free. The staminal tube is whitish, cylindrical, ribbed above, tipped with 10 short appendages, and develops 10 anthers. The ovary is 4–5 lobed as well as the stigma. The drupes are up to about 10 cm across, rough, of a unique type of soft orangish-brown, velvety, yellow – red, laticiferous, and contain 5 seeds in a white and delicious and bitter aril. The fruits from Cebu are particularly delicious.

Medicinal uses: Diarrhea (India; Indonesia; Myanmar); dysentery (Indonesia; Myanmar); fever (the Philippines); ringworm (Indonesia)

Broad-spectrum antibacterial halo developed by polar extract: Ethanol extract of leaves inhibited the growth of *Escherichia coli* (0157:H7), methicillin-resistant *Staphylococcus aureus* (ATCC 43300), *Listeria monocytogenes, Pseudomonas aeruginosa, Staphylococcus epidermidis, Staphylococcus aureus, Bacillus cereus*, and *Salmonella typhimurium* (Nogodula et al., 2013).

Strong antibacterial (Gram-negative) lipophilic multiflorane-type triterpene: Bryononic acid (also known as 3-oxomultiflora-8-en-29-oic acid; Log D = 5.7 at pH 7.4; molecular mass = 454.6 g/mol) from the fruit peels inhibited the growth of *Salmonella enterica* with the MIC value of 6 µg/mL (Heliawati et al., 2019).

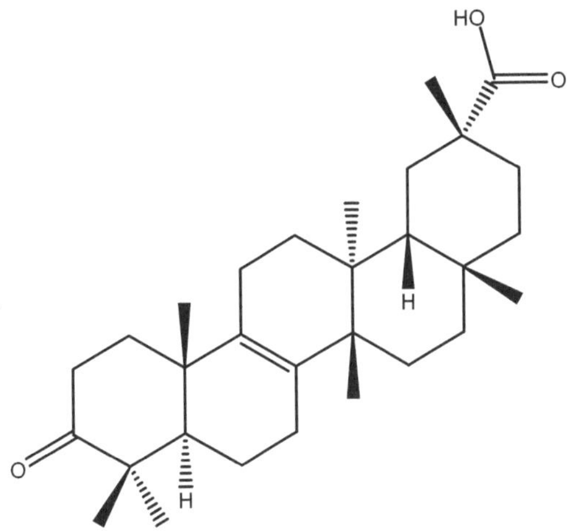

Bryononic acid

Anticandidal halo developed mid-polar extract: Dichloromethane extract of stem bark (100 µg/TLC) inhibited the growth of *Candida albicans* (Cavin et al., 1999).

Viral enzyme inhibition by polar extract: Methanol extract at the concentration of 200 µg/mL inhibited the Human immunodeficiency virus type-1 reverse transcriptase by 9% (Tan et al., 1991).

Commentaries

i. The plant yields limonoids such as sandrapins A, B, and C (Ismail et al., 2003) as well as secomultiflorane-type triterpenes acids (Kosela et al., 1995) which could be examined for their antimicrobial properties. Triterpenes from Meliaceae are often active against Gram-negative bacteria; why? Koetjapic acid from this plant (Ismail et al., 2003) is antibacterial (see earlier)

ii. The oleanane-type triterpenes 3-oxo-olean-12-en-29-oic acid, katonic acid, and koetjapic acid inhibited DNA polymerase β with IC_{50} values of 22, 36, and 20 µM, respectively (Sun et al., 1999). DNA polymerase β repairs DNA (Sun et al., 1999) and one could suggest that such agents could attenuate resistance of bacteria to DNA-targeting antibiotics and the resistance of fungi to DNA-targeting antifungal agents. Lentiviruses, such as the Human immunodeficiency virus, utilize DNA polymerase β for the insertion of their DNA into host cell chromosome (Goetze et al., 2017) and since antiretroviral principles are often active against coronaviruses, the anti-Severe acute respiratory syndrome - associated coronavirus activities of these triterpenes could be examined. Remdesivir, which targets RNA-dependent RNA polymerase, has been approved by the FDA for the prevention/treatment of COVID-19 (Cao et al., 2020; Lamb, 2020).

REFERENCES

Cao, Y.C., Deng, Q.X. and Dai, S.X., 2020. Remdesivir for severe acute respiratory syndrome coronavirus 2 causing COVID-19: an evaluation of the evidence. *Travel Medicine and Infectious Disease*, *35*, p. 101647.

Cavin, A., Dyatmyko, W. and Hostettmann, K., 1999. Screening of Indonesian plants for antifungal and free radical scavenging activities. *Pharmaceutical Biology*, *37*(4), pp. 260–268.

Goetze, R.W., Kim, D.H., Schinazi, R.F. and Kim, B., 2017. A CRISPR/Cas9 approach reveals that the polymerase activity of DNA polymerase β is dispensable for HIV-1 infection in dividing and nondividing cells. *Journal of Biological Chemistry*, *292*(34), pp. 14016–14025.

Heliawati, L., Syah, Y.M. and Bumi, M.B., 2019. Bryononic acid: antibacterial compound from fruit hulls of *S. koetjape* Merr extract. *Journal of Chemical and Pharmaceutical Sciences*, *12*. ISSN (Print 0974-2115) (Online 2349-8552).

Ismail, I.S., Ito, H., Hatano, T., Taniguchi, S. and Yoshida, T., 2003. Modified limonoids from the leaves of *Sandoricum koetjape*. *Phytochemistry*, *64*(8), pp. 1345–1349.

Kosela, S., Yulizar, Y., Tori, M. and Asakawa, Y., 1995. Secomultiflorane-type triterpenoid acids from stem bark of *Sandoricum koetjape*. *Phytochemistry*, *38*(3), pp. 691–694.

Lamb, Y.N., 2020. Remdesivir: first approval. *Drugs*, *80*, pp. 1355–1363.

Nogodula, J.N., Aguhar, B.A., Gardose, D.L. and Juntilla, A.A., 2013. *In vitro* antibacterial screening of Santol (*Sandoricum koetjape* Merr.) leaf crude extract. *Root Gatherers*, *4*, pp. 67–67.

Sun, D.A., Starck, S.R., Locke, E.P. and Hecht, S.M., 1999. DNA polymerase β inhibitors from *Sandoricum k oetjape*. *Journal of Natural Products*, *62*(8), pp. 1110–1113.

Tan, G.T., Pezzuto, J.M., Kinghorn, A.D. and Hughes, S.H., 1991. Evaluation of natural products as inhibitors of human immunodeficiency virus type 1 (HIV-1) reverse transcriptase. *Journal of Natural Products*, *54*(1), pp. 143–154.

8.5.3.10 *Swietenia mahagoni* (L.) Jacq.

Synonyms: *Cedrela mahagoni* L.;*Swietenia mahogoni* Lam.

Common names: Mahogany tree; mehagoni (Bangladesh); tao hua xin mu (China); ciminukku (India)

Habitat: Cultivated

Distribution: Tropical Asia and Pacific

Botanical observation: This noble tree probably native to South America grows to about 40 m tall. The bole is buttressed, the bark is rough and scaly, and the wood somewhat brownish and durable. The leaves are compound, alternate, and exstipulate. The blade presents 8–12 pairs of folioles which are ovate to lanceolate, 10–16 × 4–6 cm, coriaceous, dark green, glossy, asymmetrical, marked with about 10 pairs of secondary nerves, and acuminate at apex. The panicles are up to 15 cm long. The calyx is cupular and 5-lobed. The 5 petals are whitish, obovate, up to 4 mm long, and glabrous. A disc is present. The staminal tube develops 10 anthers. The ovary is somewhat conical and develops a disciform stigma. The capsules are brownish, somewhat ovoid, up to about 18 cm long, woody, dehiscent in 5 parts (like somewhat an umbrella) and contain numerous beautiful smooth, brown, woody, closely packed, glossy and winged seeds which can reach about 7 cm long.

Medicinal uses: Diarrhea (India; the Philippines); dysentery, fever (India)

Broad-spectrum antibacterial halo developed by mid-polar extract: Chloroform extract of bark (500 µg/5 mm disc) inhibited the growth of *Bacillus subtilis*, *Staphylococcus aureus*, *Escherichia coli*, *Salmonella typhi*, *Salmonella paratyphi*, *Vibrio parahemolyticus*, *Shigella boydii*, *Shigella dysenteriae*, and *Pseudomonas aeruginosa* with the diameter zones of inhibition of 12, 13, 13, 12, 28, 11.5, 14, 16, and 15 mm, respectively (Haque et al., 2009).

Broad-spectrum antibacterial halo developed by limonoid: Swietenolide (100 µg/disc) isolated from the seeds inhibited the growth of multidrug resistant clinical isolates *Streptococcus aureus* (Group A β-hemolytic), *Staphylococcus aureus*, *Streptococcus pneumoniae*, *Haemophilus influenzae*, *Escherichia coli*, *Klebsiella pneumoniae*, *Salmonella typhi* and *Salmonella paratyphi* with inhibition zone diameters of 14, 13, 19, 29, 12, 9, 27, and 29 mm, respectively (Rahman et al., 2009).

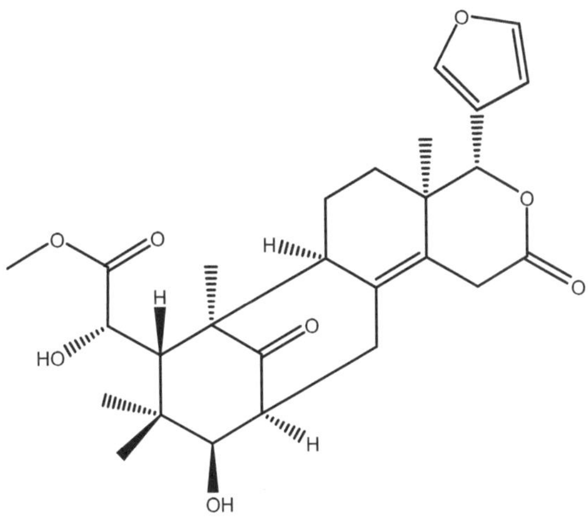

Swietenolide

2-Hydroxy-3-*O*-tigloylswietenolide isolated from the seeds inhibited the growth of multidrug-resistant clinical isolates *Streptococcus aureus* (Group A β-hemolytic), *Staphylococcus aureus*, *Streptococcus pneumoniae*, *Haemophilus influenzae*, *Escherichia coli*, *Klebsiella pneumoniae*, *Salmonella typhi*, and *Salmonella paratyphi* with inhibition zone diameters of 15, 23, 35, 24, 15, 10, 30, and 31 mm, respectively (100 μg/disc) (Rahman et al., 2009).

Strong antibacterial (Gram-positive) limonoid: 2-Hydroxy-3-*O*-isobutyrylproceranolide and 2-hydroxyfissinolide isolated from the fruits inhibited the growth of *Micrococcus luteus* (ATCC 9341) with the MIC values of 50 and 12.5 μg/mL, respectively (Lin et al., 2009).

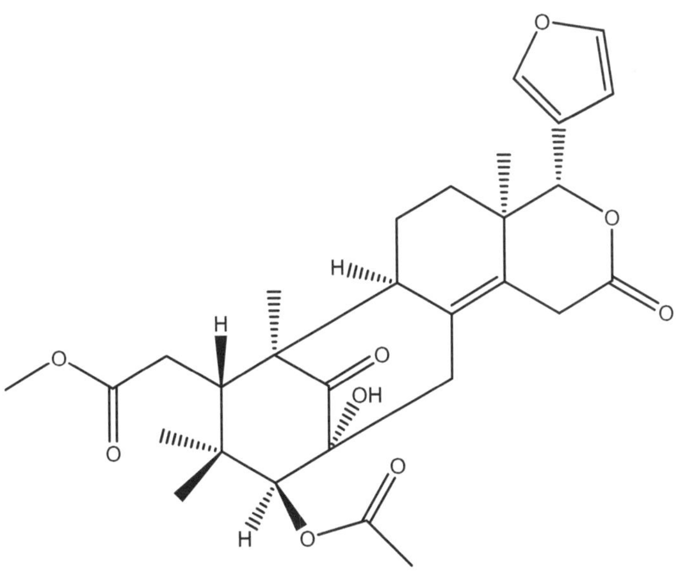

2-Hydroxyfissinolide

Moderate antimycobacterial polar extract: Ethanol extract of stems at the concentration of 100 μg/mL inhibited the growth of *Mycobacterium tuberculosis* by 83% (Antoun et al., 2001).

Commentaries

i. 3-Hydroxy caruilignan C isolated from the stems inhibited the Hepatitis C virus RNA levels by 50% at a concentration of about 10.5 μM (Wu et al., 2012). Swielimonoid B, swietenolide, swietenine acetate, and methyl angolensate isolated from the seeds of *Swietenia macrophylla* King inhibited the Dengue virus-2 replication in Huh-7 cells with the EC_{50} values of 7.2, 3.5, 6.3, and 4.3 μM, respectively, and selectivity indexes of >27.7, 19.4, 13.2, and 27, respectively (Cheng et al., 2014).

ii. *Swietenia febrifuga* Roxb. also known as rohun bark (local name: *Rohan*) is used to treat fever, dysentery, and diarrhea in India.

REFERENCES

Antoun, M.D., Ramos, Z., Vazques, J., Oquendo, I., Proctor, G.R., Gerena, L. and Franzblau, S.G., 2001. Evaluation of the flora of Puerto Rico for *in vitro* antiplasmodial and antimycobacterial activities. *Phytotherapy Research*, *15*, pp. 638–642.

Cheng, Y.B., Chien, Y.T., Lee, J.C., Tseng, C.K., Wang, H.C., Lo, I.W., Wu, Y.H., Wang, S.Y., Wu, Y.C. and Chang, F.R., 2014. Limonoids from the seeds of *Swietenia macrophylla* with inhibitory activity against dengue virus 2. *Journal of Natural Products*, *77*(11), pp. 2367–2374.

Haque, M., Ullah, M.O. and Nahar, K., 2009. *In vitro* antibacterial and cytotoxic activities of different parts of plant *Swietenia mahagony*. *Pakistan Journal of Biological Sciences*, *12*(7), p. 599.

Lin, B.D., Yuan, T., Zhang, C.R., Dong, L., Zhang, B., Wu, Y. and Yue, J.M., 2009. Structurally diverse limonoids from the fruits of *Swietenia mahagoni*. *Journal of Natural Products*, *72*(12), pp. 2084–2090.

Rahman, A.S., Chowdhury, A.A., Ali, H.A., Raihan, S.Z., Ali, M.S., Nahar, L. and Sarker, S.D., 2009. Antibacterial activity of two limonoids from *Swietenia mahagoni* against multiple-drug-resistant (MDR) bacterial strains. *Journal of Natural Medicines*, *63*(1), pp. 41–45.

Wu, S.F., Lin, C.K., Chuang, Y.S., Chang, F.R., Tseng, C.K., Wu, Y.C. and Lee, J.C., 2012. Anti-hepatitis C virus activity of 3-hydroxy caruilignan C from *Swietenia macrophylla* stems. *Journal of Viral Hepatitis*, *19*(5), pp. 364–370.

8.5.3.11 *Toona sureni* (Bl.) Merr.

Synonyms: *Cedrela febrifuga* Bl., *Cedrela surenii* (Bl.) Burk.; *Toona febrifuga* Roem.

Common names: Red cedar, Suren, Indian mahogany, toon tree; toon (Bangladesh); zi chun (China); loodh, toona (India); surian wangi, suryan (Malaysia); thit kado (Myanmar)

Habitat: Forests

Distribution: Himalaya. India, China, Papua New Guinea

Botanical description: This noble tree grows to a height of 30 m and a girth of 2 m. The bole is butteressed, the buttresses up to 3 m tall. The bark is scaling, the scales elongate, fissured regularly longitudinally, and greyish. The inner bark is fibrous, orange red, or pink and turns rapidly orange-brown, and evokes a pleasant cedary fragrance. The sapwood is yellowish white. The wood is sweet scented and peppery. The leaves are paripinnate with 4–5 pairs of folioles. The rachis is velvety and 38–50 cm long. The petiolules are 4 mm long. The folioles are 13–17 × 3.5–5 cm and show 11 pairs of secondary nerves. The flowers are 4 mm long, somewhat smelly, white, and arranged in 40 cm long panicles. The 5 sepals are triangular and minute. The corolla is white or flushed pink, 3 mm long and consists of 5 petals. A red disc is present. The androecium consists of 5 stamens. The ovary is hairy. The capsules are 1.5 cm × 8 mm, rusty lenticeled, and opening to release several flat seeds which are winged at both ends.

Medicinal uses: Gonorrhea, measles (China); dysentery, ulcers (India); fever (India; Indonesia)

Broad-spectrum antibacterial halo developed by polar extract: Ethanol extract of stem bark inhibited *Micrococcus luteus* (MTCC 106), *Enterobacter aerogenes* (MTCC 111), *Escherichia coli* (MTCC 443), *Klebsiella pneumonia* (MTCC 109), *Salmonella typhimurium* (MTCC 98),

Streptococcus pneumonia (MTCC 2672) with the MIC values of 300, 600, 300, 600, 300, and 300 µg/mL (Singh et al., 2010).

Antifungal limonoid: Gedunin displayed antifungal effects (Wu et al., 2008).

Strong antiviral (enveloped monopartite linear double-stranded DNA) polar extract: Methanol extract of leaves inhibited the replication of the Herpes simplex type-1 virus (strain 7401H) in Vero cells by 100% at the concentration of 100 µg/mL (Nawawi et al., 1999).

Strong antiviral (enveloped monopartite single-stranded (+) RNA) mid-polar extract: Ethylacetate extract inhibited the replication of the Hepatitis C virus (J6/JFH1) in Huh-7.5 cells with the IC_{50} value of 1.7 µg/mL (Hafid et al., 2018).

Strong antiviral (enveloped monopartite single-stranded (+) RNA) hydrophilic phenolic: Gallic acid (Log D = –2.3; molecular mass = 170.1 g/mol) inhibited the replication of the Hepatitis C virus (J6/JFH1) in Huh-7.5 cells with the IC_{50} value of 15.9 µg/mL (Hafid et al., 2018).

Strong antiviral (enveloped monopartite linear double-stranded DNA) hydrophilic phenolic: Gallic acid methyl ester isolated from the leaves inhibited the replication of the Herpes simplex type-1 virus (strain 7401H) in Vero cells with the IC_{50} value of 20 µg/mL (Nawawi et al., 1999).

Strong antiviral (enveloped monopartite linear double-stranded DNA) gallotannin: 1,2,3,4,6-penta-*O*-galloyl-β-glucose isolated from the leaves inhibited the replication of the Herpes simplex type-1 virus (strain 7401H) in Vero cells with the IC_{50} value of 16 µg/mL (Nawawi et al., 1999).

Commentaries

i. The plant produces series of tetranortriterpenoids of which surenone and surenin (Kraus & Kypke, 1979) which may have some levels of antimicrobial properties, but this remains to be examined. Note that tetranortriterpenoids produced by members of the genus *Toona* (Endl.) M. Roem. have the tendency to be antitrypanosomal (Githua et al., 2011). Antitrypanosomal natural products are often antifungal (Fukuda et al., 2010) and one could examine the antifungal properties of tetranortriterpenoids.

ii. The flowers are used to make a red dye in India.

iii. Members of the genus *Toona* (Endl.) M. Roem. could be examined for their antiviral principles. *Toona sinensis* (A. Juss.) M. Roem. is used in China (local name = *Xiang chun*) for the treatment of measles. Aqueous extract of young leaves inhibited the replication of the Severe acute respiratory syndrome-associated coronavirus (strain FFM1) in Vero cells with the EC_{50} value of 30 µg/mL and selectivity index of 17 (Chen et al., 2008). Essential oil of leaves inhibited the growth of *Staphylococcus aureus* (ATCC 25923) with the MIC value of 500 µg/mL (Wu et al., 2014).

iv. *Toona ciliata* M. Roem. (synonym: *Cedrela toona* Roxb. ex Rottl.) is used in Nepal (local name = *Tuni*) to treat dysentery as well as in India (local name: *Tuunikaa*). In Indonesia, this tree is a remedy for boils. In the Himalayas it is used to treat bronchitis. Petroleum ether extract of stem bark of *Toona ciliata* M. Roem. (500 µg/disc) inhibited the growth of *Bacillus cereus* (QL 29), *Bacillus subtilis* (QL 40), *Escherichia coli* (ATCC 25922), *Klebsiella* sp., *Pseudomonas aeruginosa* (ATCC 27853), *Salmonella typhi* (AM 16406), *Sarcina lutea* (QL 166), *Shigella dysenteriae* (ATCC 26131), *Staphylococcus aureus* (ATCC 25923), *Vibrio parahemolyticus* (AM 16362), and *Shigella flexneri* (Y 976) with inhibition zone diameters of 11.5, 11.5. 14.5, 13, 16.5, 8.5, 14, 12, 10.5, 9.5, and 17 mm, respectively (Chowdhury et al., 2003). Siderin (also known as 4,7-dimethoxy-5-methyl-coumarin) from this plant (50 µg/disc) inhibited the growth of *Bacillus cereus* (QL 29), *Escherichia coli* (ATCC 25922), and *Shigella flexneri* (Y 976) with inhibition zone diameters of 7, 8.5, and 7.5 mm, respectively (Chowdhury et al., 2003). Petroleum ether extract of stem bark of *Toona ciliata* M. Roem. inhibited at the concentration of 360 µg/mL the growth of *Macrophomina phaseolina*, *Botryodiplodia theobromae*, and *Curvularia lunata* by 19.7%, 35.9%, and 34.2%, respectively (Chowdhury et al., 2003). The pimarane-type diterpene toonaciliatin M (also known as 8-hydroxypimar-15-en-18-oic acid; Log D = 2.1 at pH 7.4; molecular mass = 320.4 g/mol) from *Toona ciliata* M. Roem. inhibited the growth of *Trichophyton rubrum* with the MIC value of 12.5 μg/mL (Chen et al., 2009).

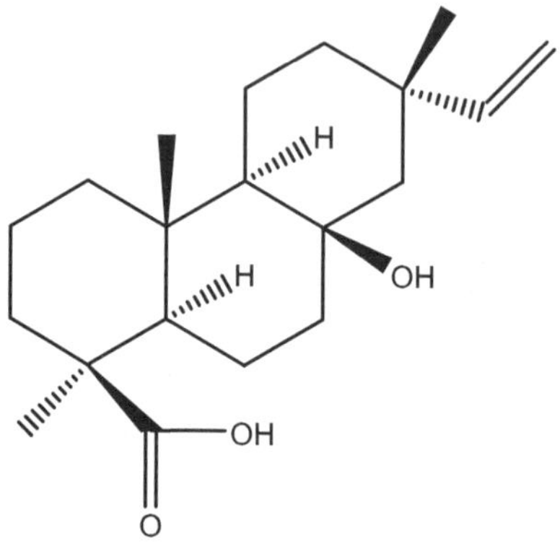

Toonaciliatin M

The protolimonoid nourjotinoline A from *Toona ciliata* M. Roem. inhibited the growth of *Staphylococcus epidermidis* (ATCC 12228) with an MIC value of 6.2 μg/mL (Chen et al., 2009).

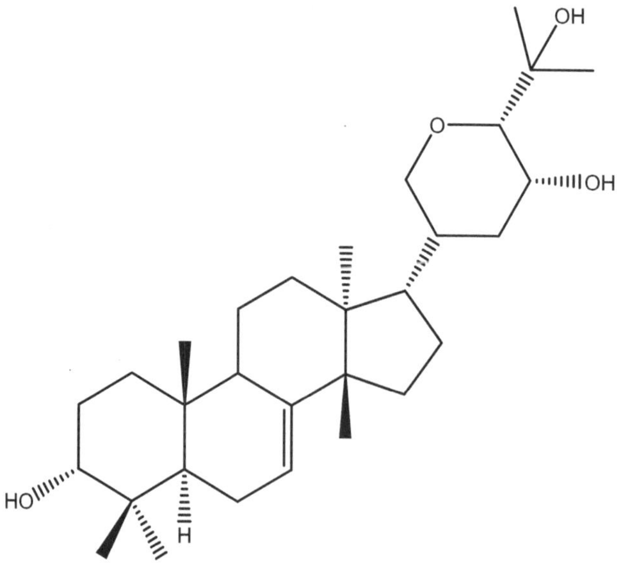

Bourjotinoline A

v. *Toona calantas* Merr. & Rolfe (Philippine cedar; local name: *Kalantas*) is used in the Philippines to cleanse wounds.

REFERENCES

Chen, C.J., Michaelis, M., Hsu, H.K., Tsai, C.C., Yang, K.D., Wu, Y.C., Cinatl Jr, J. and Doerr, H.W., 2008. *Toona sinensis* Roem tender leaf extract inhibits SARS coronavirus replication. *Journal of Ethnopharmacology*, *120*(1), pp. 108–111.

Chen, H.D., Yang, S.P., Wu, Y., Dong, L. and Yue, J.M., 2009. Terpenoids from *Toona ciliata*. *Journal of Natural Products*, *72*(4), pp. 685–689.

Chowdhury, R., Hasan, C.M. and Rashid, M.A., 2003. Antimicrobial activity of *Toona ciliata* and *Amoora rohituka*. *Fitoterapia*, *74*(1), pp. 155–158.

Fukuda, K., Tamura, T., Ito, H., Yamamoto, S., Ochi, K. and Inagaki, K., 2010. Production improvement of antifungal, antitrypanosomal nucleoside sinefungin by rpoB mutation and optimization of resting cell system of *Streptomyces incarnatus* NRRL 8089. *Journal of Bioscience and Bioengineering*, *109*(5), pp. 459–465.

Hafid, A.F., Wahyuni, T.S., Tumewu, L., et al., 2018. Antihepatitis C virus activity of Indonesian Mahogany (*Toona sureni*). *Asian Journal of Pharmaceutical and Clinical Research*, *11*(2), pp. 87–90.

Kraus, W. and Kypke, K., 1979. Surenone and surenin, two novel tetranortriterpenoids from *Toona sureni* [blume] merrill. *Tetrahedron Letters*, *20*(29), pp. 2715–2716.

Nawawi, A.A., Nakamura, N., Hattori, M., Kurokawa, M. and Shiraki, K., 1999. Inhibitory effects of Indonesian medicinal plants on the infection of herpes simplex virus type 1. *Phytotherapy Research: An International Journal Devoted to Pharmacological and Toxicological Evaluation of Natural Product Derivatives*, *13*(1), pp. 37–41.

Singh, M., Khatoon, S., Singh, S., Kumar, V., Rawat, A.K.S. and Mehrotra, S., 2010. Antimicrobial screening of ethnobotanically important stem bark of medicinal plants. *Pharmacognosy Research*, *2*(4), p. 254.

Wu, J., Xiao, Q., Xu, J., Li, M.Y., Pan, J.Y. and Yang, M.H., 2008. Natural products from true mangrove flora: source, chemistry and bioactivities. *Natural Product Reports*, *25*(5), pp. 955–981.

Wu, J.G., Peng, W., Yi, J., Wu, Y.B., Chen, T.Q., Wong, K.H. and Wu, J.Z., 2014. Chemical composition, antimicrobial activity against *Staphylococcus aureus* and a pro-apoptotic effect in SGC-7901 of the essential oil from *Toona sinensis* (A. Juss.) Roem. leaves. *Journal of Ethnopharmacology*, *154*(1), pp. 198–205.

8.5.3.12 *Xylocarpus granatum* Koenig.

Synonyms: *Carapa granatum* (J. Koenig) Alston; *Carapa obovata* Bl.; *Xylocarpus obovatus* Juss.

Common names: Cannonball mangrove; tbun (Cambodia); mu guo lian (China); nyireh bunga (Indonesia; Malaysia); dhundul, pussur (India); pinle on (Myanmar); tabigi (the Philippines); ta bun (Thailand); dang dinh (Vietnam); emogul (Yap)

Habitat: Mangrove, beaches

Distribution: India, Bangladesh, Southeast Asia, Papua New Guinea, Australia, Pacific Islands

Botanical description: This magnificent tree grows to a height of 15 m. The bole is crooked and develops eery snake – like buttresses. The bark is smooth to peeling. The leaves are compound, spiral, and exstipulate. The folioles are elliptic to obovate, 10.5–18.5 × 4–6 cm, marked with about 9 pairs of secondary nerves, glossy, and rounded at apex. The cymes are axillary. The calyx is minute and includes 4 lobes. The corolla presents 4 white to greenish petals which are fleshy and about 5 mm long. A red disc is present. The staminal tube develops 8 anthers. The berries are globose, green at first then brownish and woody, dehiscing into parts, up to about 15 cm across, heavy, smooth, and containing about 10 angular seeds.

Medicinal uses: Dysentery (India; Indonesia; Yap); diarrhea, fever (India); cholera (India; Indonesia)

Broad-spectrum antibacterial halo developed by polar extract: Ethanol extract of bark (500 μg/disc) inhibited the growth of *Staphylococcus aureus*, *Staphylococcus epidermidis*, *Streptococcus pneumoniae*, *Shigella boydi*, *Salmonella typhimurium*, and *Escherichia coli* with the inhibition zone diameters of 19, 18, 16, 20, 18, and 20 mm, respectively (Uddin et al., 2008).

Xylocarpus granatum J. Koenig

Strong antifungal dihydrofuranone: 3-(1-Hydroxyethyl)-4,4-dimethyl-4- butyrolactone from the leaves at concentration of 20 μg/mL inhibited the growth of *Blumeria graminis* by 67.4% (Du et al., 2009).

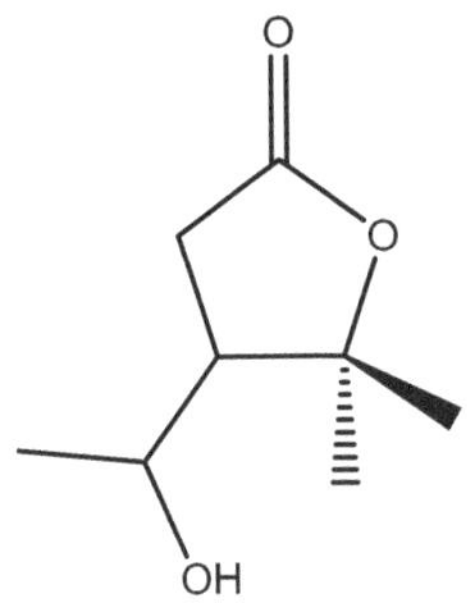

3-(1-Hydroxyethyl)-4,4-dimethyl-4- butyrolactone

Moderate antiviral (monopartite linear dimeric single-stranded (+) RNA) limonoid: Sundarbanxylogranin B isolated from the seeds inhibited the replication of the Human immunodeficiency virus type-1 in 293 T cells with the IC_{50} value of 23.1 μM (Dai et al., 2017).

Viral enzyme inhibition by polar extract: Methanol extract at the concentration of 200 μg/mL inhibited the Human immunodeficiency virus type-1 reverse transcriptase by 11% (Tan et al., 1991).

Commentaries: The plant brings to being a fascinating array of limonoids including gedunin and photogedunin. Is Sundarbanxylogranin B active against coronaviruses?

REFERENCES

Dai, Y.G., Wu, J., Padmakumar, K.P. and Shen, L., 2017. Sundarbanxylogranins A–E, five new limonoids from the *Sundarban mangrove, Xylocarpus granatum. Fitoterapia, 122*, pp. 85–89.

Du, S., Wang, M., Zhu, W. and Qin, Z., 2009. A new fungicidal lactone from *Xylocarpus granatum* (Meliaceae). *Natural Product Research, 23*(14), pp. 1316–1321.

Tan, G.T., Pezzuto, J.M., Kinghorn, A.D. and Hughes, S.H., 1991. Evaluation of natural products as inhibitors of human immunodeficiency virus type 1 (HIV-1) reverse transcriptase. *Journal of Natural Products, 54*(1), pp. 143–154.

Uddin, S.J., Rouf, R., Shilpi, J.A., Alamgir, M., Nahar, L. and Sarker, S.D., 2008. Screening of some Bangladeshi medicinal plants for *in vitro* antibacterial activity. *Oriental Pharmacy and Experimental Medicine, 8*(3), pp. 316–321.

8.5.3.13 *Xylocarpus moluccensis* (Lam.) Roem.

Synonyms: *Carapa moluccensis* Lam.; *Xylocarpus australasicus* Ridley; *Xylocarpus gangeticus* (Prain) CE. Parkinson; *Xylocarpus mekongensis* Pierre

Common names: Cedar mangrove, nyireh batu (Indonesia; Malaysia); kyana (Myanmar); piagau (Philippines); ta buun dam (Thailand); xu'o'ng cá (Vietnam)

Habitat: Mangrove

Distribution: India, Bangladesh, Southeast Asia, Papua New Guinea, Australia, and Pacific Islands

Botanical observation: This magnificent tree grows to a height of 20 m. The plant presents some pneumatophores. The bark is strongly fissured and brownish-yellow. The leaves are paripinnate, spiral and exstipulate. The rachis is up to 10 cm long and bears 3–4 pairs of folioles. The petiolule is about 5 mm long. The foliole is elliptic-ovate, up to 12 × 5 cm, wedge-shaped at base, glossy, smooth, and acute at apex. The inflorescence is a lax raceme of cymes at apex of stems. The calyx is about 3 mm long with 4 sepals which are rounded. The corolla comprises 4 petals which are about 5 mm long, fleshy, whitish, obovate or elliptical. The staminal tube is about 5 mm long, fleshy, and develops 8 stamens. The berries are about 20 cm in diameter, woody, globose, and with several angled seeds which are 4–8 cm long.

Medicinal uses: Cholera, diarrhea (Myanmar); dysentery (Indonesia; Myanmar; the Philippines)

Weak antiviral (enveloped segmented linear single-stranded (−) RNA) khayanolides: Thaixylomolin I, K, and M isolated from the seeds inhibited the replication of the Influenza A virus (subtype H1N1) with IC_{50} values of 77.1, 113.5, and 121.5 µM (Li et al., 2015).

Commentary: The plant produces limononoids of which deacetylgenudin and 2-hydroxyfissinolide (Ravangpai et al., 2011). These could be examined for antiviral properties.

REFERENCES

Li, W., Jiang, Z., Shen, L., Pedpradab, P., Bruhn, T., Wu, J. and Bringmann, G., 2015. Antiviral limonoids including khayanolides from the Trang mangrove plant *Xylocarpus moluccensis*. *Journal of Natural Products*, *78*(7), pp. 1570–1578.

Ravangpai, W., Sommit, D., Teerawatananond, T., Sinpranee, N., Palaga, T., Pengpreecha, S., Muangsin, N. and Pudhom, K., 2011. Limonoids from seeds of Thai Xylocarpus moluccensis. *Bioorganic & medicinal chemistry letters*, *21*(15), pp. 4485–4489.

8.5.4 FAMILY RUTACEAE A.L DE JUSSIEU (1789)

The family Rutaceae comprises 160 genera and 1800 species of trees, shrubs, or climbers. The stems are often thorny. The leaves are simple or compound, alternate or opposite, and exstipulate. The petiole can be winged. The blade is characteristically more or less marked with somewhat tiny and translucent oil glands. The inflorescences are solitary, axillary or terminal cymes, panicles, or racemes. The calyx presents 4–5 free or basally united sepals. The 4–5 petals are characteristically white, somewhat fleshy, and ephemeral. The stamens are equal to or twice as numerous as the number of petals or more, filaments free or more or less united at base. The ovary is 4–5-locular, each locule sheltering ovules on axille placentas, and develops a characteristic style and capitate and humid stigma. The fruits are capsules, hesperidia, drupes, or schizocarps. Medicinal plants in this family generate antimicrobial coumarins, acridone and carbazole alkaloids, flavones, flavanones, and monoterpenes.

8.5.4.1 *Aegle marmelos* (L.) Corrêa

Synonyms: *Belou marmelos* (L.) A. Lyons, *Crataeva marmelos* L.; *Crateva marmelos* L.; *Cydonia indica* Spach.; *Feronia pellucida* Roth

Common names: Bael, Bengal quince, stone apple; bel, shepalpubaong, singadar, srifal (Bangladesh); haigakok (Bangladesh; India); belpatra, belthei, bill, bilva, koovalam, shriphala (India); bilak (Malaysia); hpun ja, ki -bok, mak phyn, okshit (Myanmar); matum (Thailand)

Habitat: Cultivated

Distribution: India, Bangladesh, Southeast Asia

Botanical observation: This graceful tree grows up to 10 m tall. The plant is sacred in India and often cultivated in temples. The stems present straight woody thorns which are up to 3 cm long. The leaves are spiral, exstipulate, and 3-foliolate. The folioles are broadly lanceolate, somewhat asymmetrical at base, 4–12 × 2–6 cm, beautifully crenate, soft, dull green, reddish and glossy when very young, and acuminate. The calyx comprises 5 lobes which are minute. The 4–5 petals are whitish to yellow, somewhat fleshy, marked with 4–10 pairs of secondary nerves, elliptical, and up to 1.5 cm long. The stamens are numerous with elongated anthers. The gynoecium is somewhat bowling pin-shaped and about 6 mm long. The capsule is greenish yellow, up to about 15 cm in diameter, somewhat globose, woody, and contains numerous, flat, and about 8 mm long seeds in an edible orangish flesh. Fruits from Bangladesh are particularly tasty.

Medicinal uses: Boils, cold, cholera, cuts, fever, jaundice, ophthalmia, wounds (India); leucorrhea (Bangladesh); tuberculosis (India; Pakistan); bronchitis, sores (Myanmar); abscesses (Bangladesh; India); cough (India; Myanmar); diarrhea, dysentery, fever (Bangladesh, India, Myanmar).

Broad-spectrum antibacterial halo developed by mid-polar extract: Ethylacetate extract of leaves (1 mg/disc) inhibited the growth of *Bacillus subtilis*, *Staphylococcus aureus*, *Enterococcus faecalis*, *Escherichia coli*, *Klebsiella pneumoniae*, *Pseudomonas aeruginosa*, and *Salmonella typhi* with zones of inhibition of 20, 15, 10, 25, 15, 21, and 22 mm, respectively (Rejiniemon et al., 2014).

Strong broad-spectrum antibacterial amphiphilic prenylated furanocoumarin: Imperatonin (also known as marmelosin) isolated from the fruits and at the concentration of 100 µg/mL inhibited the growth *Escherichia coli* and *Bacillus subtilis* by 64.6 and 74.9%, respectively (Chinchansure et al., 2015).

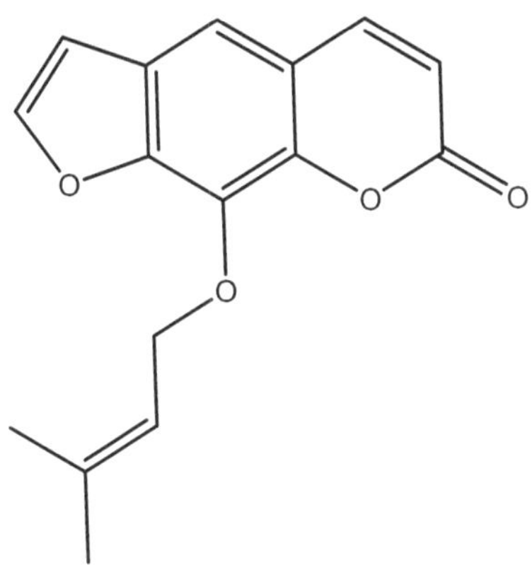

Imperatonin

Strong antibacterial (Gram-positive) hydrophilic furanocoumarin: Xanthoarnol (Log D = 1.1 at pH 7.4; molecular mass = 262.2 g/mol) isolated from the fruits inhibited the growth of *Enterococcus faecalis* (TISTR459) with the MIC value of 18.7 µg/mL (Chakthong et al., 2012).

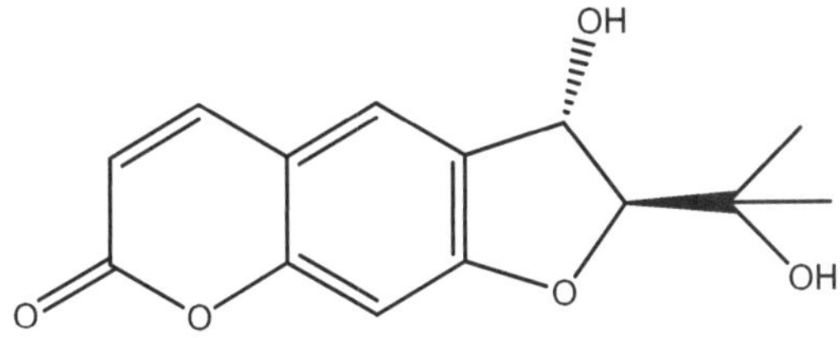

Xanthoarnol

Strong antimycobacterial hydrophilic prenylated coumarins: Imperatonin and marmin (Log D = 2.6 at pH 7.4; molecular mass = 332.3 g/mol) isolated from the fruits inhibited the growth of *Mycobacterium tuberculosis* (H37Ra) with the IC_{50} values of 12.4 and 4.3 µg/mL respectively and at the concentration of 100 μg/mL, inhibited the growth of *Mycobacterium bovis* by 62.5 and 82.4%, respectively (Chinchansure et al., 2015).

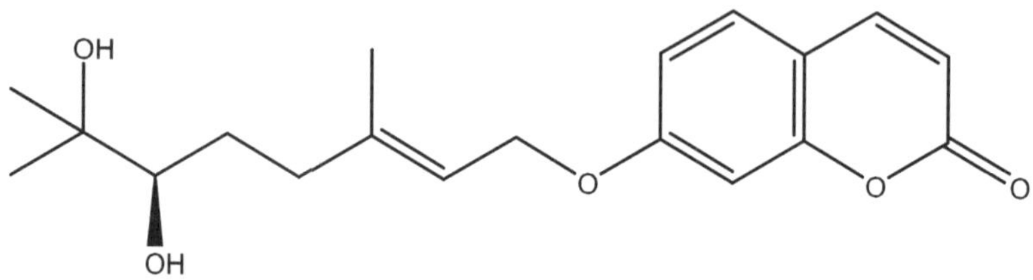

Marmin

Antibiotic potentiator coumarin: Imperatonin increased the sensitivity of *Staphylococcus aureus* (expressing efflux pump for tetracyclin), *Staphylococcus aureus* (expressing efflux pump for macrolides), *Staphylococcus aureus* (expressing the NorA efflux pump) (Madeiro et al., 2017). This prenylated coumarin evoked an 8-fold reduction in the MIC of ethidium bromide suggesting that the increase of activity of norfloxacin in *Staphylococcus aureus* may have occurred as a result of the inhibition of NorA efflux pump system (Madeiro et al., 2017).

Strong antifungal hydrophilic coumarin: 6-Formylumbelliferone (Log D = 0.7 at pH 7.4; molecular mass = 190.1 g/mol) isolated from the fruits inhibited *Rhizoctonia solani* (NPRC750), *Rigidoporus microporus* (PR720), and *Sclerotium oryzae* (NPRC760), with the MIC/MFC values of 15.6/31.2, 31.2/62.5, and 31.2/250 μg/mL, respectively (Chakthong et al., 2012).

Strong antifungal amphiphilic coumarin: 8-[(3″-Methyl-2″-oxo-3″-buten-1″-yl)oxy]-7*H*-furo[3,2-*g*]benzopyran-2-one isolated from the fruits inhibited *Rhizoctonia solani* (NPRC750), *Rigidoporus microporus* (PR720), and *Sclerotium oryzae* (NPRC760), with the MIC/MFC values of 7.8/7.8, 15.6/15.6, and 7.8/7.8 μg/mL, respectively (Chakthong et al., 2012).

Strong antiviral (monopartite linear dimeric single stranded (+) RNA) polar extract: Methanol extract of fruits inhibited the replication of the Human immunodeficiency virus type-1 (strain NL4.3) in human CD4+ T cell line by 44% at a concentration of 5 μg/mL (Sabde et al., 2011).

Strong antiviral (monopartite linear dimeric single stranded (+) RNA) amphiphilic prenylated coumarin: Imperatonin (Log D = 3.1 at pH 7.4; molecular mass = 270.2 g/mol) inhibited the replication of the Human immunodeficiency virus type-1 (strain NL4.3) in human CD4+ T cell line by 59.7% at the maximum noncytotoxic concentration of 8 μg/mL (Sabde et al., 2011).

Strong antiviral (monopartite linear dimeric single stranded (+) RNA) lipophilic prenylated coumarin: Aurapten (Log D = 5.1 at pH 7.4; molecular mass = 298.1 g/mol) inhibited the replication of the Human immunodeficiency virus type-1 (strain NL4.3) CD4+ T cell line by 30.8% at the maximum noncytotoxic concentration of 20 μg/mL (Sabde et al., 2011).

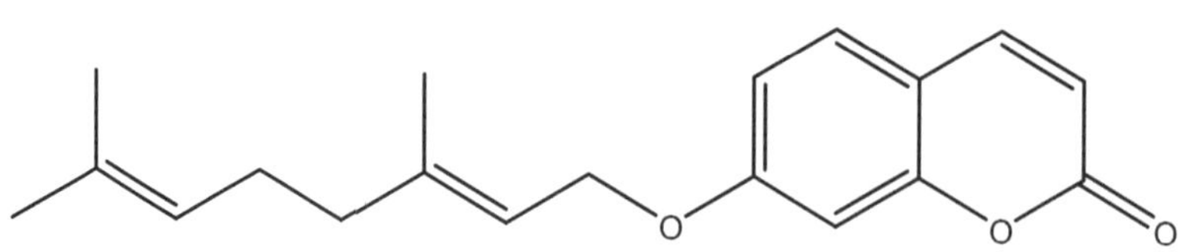

Aurapten

Commentaries

i. *Aegle marmelos* (L.) Corrêa was known of our Ayurvedic peers including Sushruta as well as Siddha practitioners as a cure for numerous illnesses (Rakulini & Kalaichelvi, 2019). Siddha practitioners use *Aegle marmelos* (L.) Corrêa to treat Chikungunya virus infection (Viswanathan et al., 2008). Chikungunya virus is like coronavirus an enveloped monopartite linear single-stranded (+) RNA virus and could be of value for coronavirus infections including COVID-19. In brief, the pulp in the fruits could be seen as a material to add on a list of first line phytomedication in case of viral pandemics. The pulp affords a delicious drink (Bangladesh).

ii. In the Subfamily Aurantioideae, Tribe Clauseneae, *Luvunga scandens* (Roxb.) Buch-Hams is used as protective postpartum remedy in Malaysia.

iii. Prenylated and geranylated natural products are often able to inhibit bacterial efflux pumps.

REFERENCES

Chakthong, S., Weaaryee, P., Puangphet, P., Mahabusarakam, W., Plodpai, P., Voravuthikunchai, S.P. and Kanjana-Opas, A., 2012. Alkaloid and coumarins from the green fruits of *Aegle marmelos*. *Phytochemistry*, *75*, pp. 108–113.

Chinchansure, A.A., Shamnani, N.H., Arkile, M., Sarkar, D. and Joshi, S.P., 2015. Antimycobacterium activity of coumarins from fruit pulp of *Aegle marmelos* (l.) Correa. *International Journal of Basic and Applied Chemical Sciences*, *5*(3), pp. 39–44.

Madeiro, S.A., Borges, N.H., Souto, A.L., de Figueiredo, P.T., Siqueira-Junior, J.P. and Tavares, J.F., 2017. Modulation of the antibiotic activity against multidrug resistant strains of coumarins isolated from Rutaceae species. *Microbial Pathogenesis*, *104*, pp. 151–154.

Rakulini, R. and Kalaichelvi, S., 2019. A review of anti-diarrheal activity of *Aegle marmelos*. *Journal of Complementary and Alternative Medical Research*, *7*, pp. 1–10.

Rejiniemon, T.S., Arasu, M.V., Duraipandiyan, V., Ponmurugan, K., Al-Dhabi, N.A., Arokiyaraj, S., Agastian, P. and Choi, K.C., 2014. *In-vitro* antimicrobial, antibiofilm, cytotoxic, antifeedant and larvicidal properties of novel quinone isolated from *Aegle marmelos* (Linn.) Correa. *Annals of Clinical Microbiology and Antimicrobials*, *13*(1), p. 48.

Sabde, S., Bodiwala, H.S., Karmase, A., Deshpande, P.J., Kaur, A., Ahmed, N., Chauthe, S.K., Brahmbhatt, K.G., Phadke, R.U., Mitra, D. and Bhutani, K.K., 2011. Anti-HIV activity of Indian medicinal plants. *Journal of Natural Medicines*, *65*(3–4), pp. 662–669.

Viswanathan, M.V., Raja, D.K. and Khanna, S.D., 2008. Siddha way to cure Chikungunya. *Indian Journal of Traditional Knowledge*, *7*(2), pp. 345–346.

8.5.4.2 *Boenninghausenia albiflora* (Hook.) Rchb. ex Meisn.

Synonyms: *Boenninghausenia brevipes* (Franch.) H. Lév., *Boenninghausenia japonica* Siebold. ex Miq., *Boenninghausenia schizocarpa* S.Y, Hu, *Boenninghausenia sessilicarpa* H. Lév., *Bodinieria thalictrifolia* H. Lév., *Bodinieria japonica* Sib. ex Miq; *Bodinieria. schizocarpa* S. Y. Hu; *Bodinieria. sessilicarpa* H. Lév; *Podostaurus thalictroides* Jungh., *Ruta albiflora* Hook.

Common names: Chou jie cao (China); dampate, makhamar (Nepal)

Habitat: Moist soils in forests

Distribution: Pakistan, India, Bhutan, Kashmir, Nepal, Myanmar, Cambodia, Laos, Vietnam, China, Thailand, the Philippines, Japan

Botanical observation: This foetid and slender herb grows to about 1 m tall. The stems are smooth, terete, and green to brownish-red. The leaves are 3–5 pinnate (somewhat *Ruta graveolens* L.-like), alternate, and exstipulate. The folioles are obcordate to spathulate, dull dark green above, marked with oil glands, 0.6–1.8 × 0.5–1.2 cm, and with 3 pairs of secondary nerves. The cymes are purplish. The calyx is cupular, minute, and 4–5 lobed. The 4–5 petals are elliptic to oblong, pure white, oil gland-dotted, and about 6 mm long. A disk is present. The 6–8 stamens are longer than the petals, the anthers are purplish and the filaments are pure white, of unequal length, and somewhat fleshy. A minute pale light greenish gynophore is present and sustains a 3–5-lobed ovary (somewhat *Euphorbia*-like). The style and stigma are light green. The follicles are about 8 mm long, oil gland-dotted, and each follicle contains a kidney- shaped, black, and muricate seed.

Medicinal uses: Cuts, fever, wounds (Nepal)

Broad-spectrum antibacterial halo developed by polar extract: Methanol extract of aerial parts (5 mm disc impregnated with 1 mg/mL solution) inhibited the growth of *Escherichia coli* (MTCC No. 40), *Bacillus subtilis* (MTCC No. 121), *Proteus vulgaris* (MTCC No. 426), *Staphylococcus aureus* (MTCC No. 87), *Agrobacterium tumefaciens* (MTCC No. 609), *Xanthomonas campestris* (MTCC No. 2286), *Erwinia crysanthmi*, and *Xanthomonas phaseoli* (Khulbe & Satti, 2009).

Antifungal (filamentous) halo developed by acridone: 1-Hydroxy-10-methylacridone (Log D = 3.3 at pH 7.4; molecular mass = 225.2 g/mol) (10 µg/TLC) inhibited the growth of *Cladosporium cucumerinum* (Ahua et al., 2004).

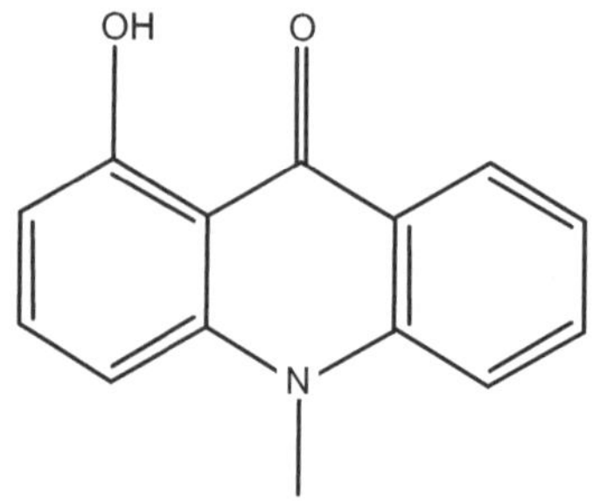

1-hydroxy-10-methylacridone

Strong antiviral (enveloped monopartite linear single-stranded (+) RNA) extract: Methanol extract inhibited the replication of the Yellow fever virus in Human hepatoma cells with the EC_{50} value of 36.1 µg/mL and the selectivity index of 2.7 (Joshi et al., 2020).

Strong antiviral (enveloped monopartite linear single-stranded (+) RNA) hydrophilic coumarin: Leptodactylone (Log D = 1.1 at pH 7.4; molecular mass = 222.1 g/mol) inhibited the replication of the Severe acute respiratory syndrome-associated coronavirus in Vero cells by 27.2% at the concentration of 20 µg/mL and 60% at the concentration of 100 µg/mL, respectively (Yang et al., 2007).

Strong antiviral (enveloped monopartite linear single-stranded (+) RNA) amphiphilic furanocoumarin: Rutamarin (Log D = 4.9 at pH 7.4; molecular mass = 356.4 g/mol) inhibited the replication of the Severe acute respiratory syndrome-associated coronavirus in Vero cells by 27.2% at the concentration of 20 µg/mL (Yang et al., 2007).

Strong antiviral (enveloped monopartite linear double-stranded DNA) amphiphilic furanocoumarin: Rutamarin at the concentration of 20 µM inhibited both viral DNA synthesis and Human herpes virus production and inhibited the ATPase activity of host cells topisomerase IIα with the EC_{50} value of 28.2 µM (Xu et al., 2014).

Commentaries

i. The plant brings to being a bewildering array of furanocoumarins of which xanthotoxol, psoralen, bergapten, isopimpinellin as well acridone-type alkaloids (Chaya et al., 2004) such as 1-hydroxy-10-methylacridone also produced by *Ruta graveolens* L. (Ahua et al., 2004).
ii. The Human herpes virus 8 accounts for the Kaposi's sarcoma developed by untreated AIDS patients (Krentz et al., 2005).

REFERENCES

Ahua, K.M., Ioset, J.R., Ransijn, A., Mauël, J., Mavi, S. and Hostettmann, K., 2004. Antileishmanial and antifungal acridone derivatives from the roots of *Thamnosma rhodesica*. *Phytochemistry*, *65*(7), pp. 963–968.

Chaya, N., Terauchi, K., Yamagata, Y., Kinjo, J. and Okabe, H., 2004. Antiproliferative constituents in Plants 14. Coumarins and acridone alkaloids from *Boenninghausenia japonica* N AKAI. *Biological and Pharmaceutical Bulletin*, *27*(8), pp. 1312–1316.

Joshi, B., Panda, S.K., Jouneghani, R.S., Liu, M., Parajuli, N., Leyssen, P., Neyts, J. and Luyten, W., 2020. Antibacterial, antifungal, antiviral, and anthelmintic activities of medicinal plants of Nepal selected based on ethnobotanical evidence. *Evidence-Based Complementary and Alternative Medicine*, *2020*. doi:10.1155/2020/1043471.

Khulbe, K. and Sati, S.C., 2009. Antibacterial activity of *Boenninghausenia albiflora* reichb. (Rutaceae). *African journal of Biotechnology*, *8*(22), pp. 6346–6348.

Krentz, H.B., Kliewer, G. and Gill, M.J., 2005. Changing mortality rates and causes of death for HIV-infected individuals living in Southern Alberta, Canada from 1984 to 2003. *HIV Medicine*, *6*, pp. 99–106.

Xu, B., Wang, L., González-Molleda, L., Wang, Y., Xu, J. and Yuan, Y., 2014. Antiviral activity of (+) rutamarin against Kaposi's sarcoma-associated herpesvirus by inhibition of the catalytic activity of human topoisomerase II. *Antimicrobial Agents and Chemotherapy*, *58*(1), pp. 563–573.

Yang, Q.Y., Tian, X.Y. and Fang, W.S., 2007. Bioactive coumarins from *Boenninghausenia sessilicarpa*. *Journal of Asian Natural Products Research*, *9*(1), pp. 59–65.

8.5.4.3 *Citrus sinensis* (L.) Osbeck

Synonyms: *Citrus aurantium* var. *sinensis* L.

Common names: Orange; krôôch pôôsat (Cambodia); jeruk manis (Indonesia); naranj (Iraq); kièngz (Laos); limau manis (Malaysia); tungchin-thi (Myanmar); malta (Pakistan); kahel (The Philippines); somkliang (Thailand); cam (Vietnam)

Habitat: Cultivated

Distribution: Native to Myanmar, Vietnam, China

Botanical observation: This beautiful tree grows to 10 m tall. The stems are angled at apex and harmed with axillary spines which are about 3 mm long. The leaves are 1-foliolate, alternate, and exstipulate. The petiole is narrowly winged, up to about 1.5 cm long, and articulated at apex. The foliole is dark green and glossy, gland oil-dotted, cuneate at base, about 7.5–15 × 2–8 cm, oblong to ovate or obovate, crenulate, and acute to obtuse or retuse at apex. The flowers are single and axillary. The calyx is 6 mm across, cupular, and develops 5 lobes. The 5 petals are pure white, about 1.5 cm long, gland oil-dotted, ephemeral, aromatic, fleshy, and reflexed. A disc is present. The androecium includes about 20–25 stamens with white, fleshy, and partially united filaments. The style is stout and the stigma capitate and moist. The hesperidium is oblate to globose, 6–12 cm across, yellow to orange-red, and contains a delicious orange-yellow or reddish pulp.

Medicinal use: Dysentery (Iraq)

Antibacterial halo developed by essential oil: Essential inhibited the growth of *Legionella pneumophila* (strain GC-4, serogroup 5/8), *Legionella pneumophila* (strain Strain GC-5, serogroup 4), *Legionella pneumophila* (strain GC-6 strain, serogroup 8), *Legionella pneumophila* (strain GC 6, serogroup 8), *Legionella pneumophila* (strain GC13, serogroup 6) and *Legionella pneumophila* (strain Paris, serogroup 1) with the MIC values of ranging from 100 to 900 µg/mL (Chaftar et al., 2015).

Moderate antibacterial (Gram-positive) flavanone: Naringenin inhibited the growth of *Streptococcus pyogenes* with the MIC of 50 µg/mL and MBC > 100 µg/mL (Macé et al., 2017).

Weak antibacterial (Gram-negative) lipophilic flavones: Nobiletin and tangeretin disrupted the cytoplasmic membrane of *Pseudomonas aeruginosa* and inhibited the protein synthesis (Yao et al., 2012).

Strong antiviral (enveloped monopartite linear single stranded (+) RNA) flavanone: Hesperitin (also known as 3',5,7-trihydroxy-4'-methoxyflavanone; LogD = 2.2 at pH 7.4; molecular mass = 302.2 g/mol) inhibited the replication of the Japanese encephalitis virus in TE-671 cells with the IC_{50} value of 3.6 µg/mL and selectivity index above 159 (Lin et al., 2008).

Strong (enveloped monopartite linear single-stranded (-) RNA) amphiphilic flavones: Tangeretin (Log D = 2.5 at pH 7.4; molecular mass = 372.3 g/mol) inhibited the replication of the Respiratory syncytial virus (A2 strain), Respiratory syncytial virus (Long strain), and Respiratory syncytial virus (B strain) with the IC_{50} values of 8.3, 6.7, and 6.7 µM, respectively, in Hep-2 cells and selectivity indexes of 50.3, 62.3, and 62.3, respectively (Xu et al., 2014). Tangeretin inhibited virion entry into Hep-2 cells and viral replication of (Xu et al., 2014).

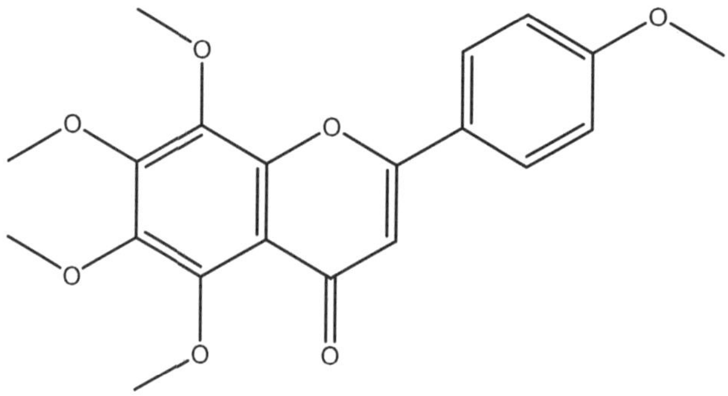

Tangeretin

Nobiletin (Log D = 2.3 at pH 7.4; molecular mass = 402.3 g/mol) inhibited the replication of the Respiratory syncytial virus (A2 strain), Respiratory syncytial virus (Long strain), and Respiratory syncytial virus (B strain) with IC_{50} values of 15.4, 10.9, and 11.2 µM, respectively, and selectivity indexes of 16.4, 23.1, and 15.6, respectively (Xu et al., 2014).

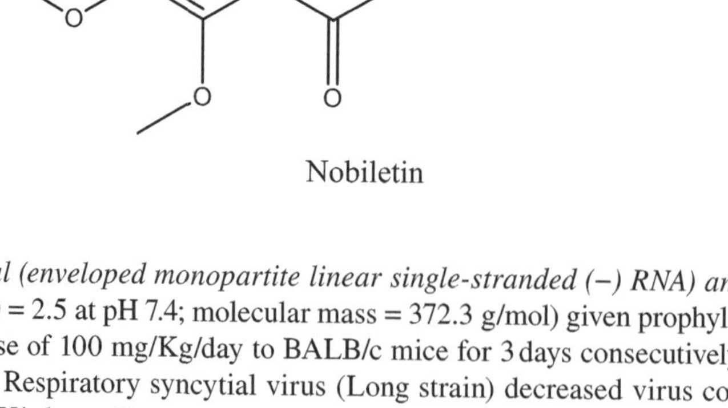

Nobiletin

In vivo antiviral (enveloped monopartite linear single-stranded (−) RNA) amphiphilic flavone: Tangeretin (Log D = 2.5 at pH 7.4; molecular mass = 372.3 g/mol) given prophylactically and intragastrically at a dose of 100 mg/Kg/day to BALB/c mice for 3 days consecutively before intranasal infection with the Respiratory syncytial virus (Long strain) decreased virus count in lung tissues from 3.3 to 3.0 PFU/g $\log_{10}$ (Xu et al., 2015).

Viral enzyme inhibition by flavanone: Hesperitin inhibited the Severe acute respiratory syndrome-associated coronavirus 3CL protease with the IC_{50} value of 18.1 µg/mL (Lin et al., 2005).

Commentaries

i. Members of the genus *Citrus* L. produce antibacterial and antifungal coumarins (Stanley & Jurd, 1971). Antibiotics with coumarin moieties (such as novobiocin) inhibit the ATPase activity of topoisomerase II (also known as DNA gyrase) and thus the supercoiling of DNA (Gormley et al., 1996).

ii. Essential oil of members of the genus *Citrus* L. is broadly antibacterial and antifungal: *Citrus aurantifolia* (Christm.) Swingle is used in Bangladesh (local name: *Kagoji lebu*) for cough, fever, influenza, ringworm, and cholera in India (local name: *Chuluk mazzapa*), and in Myanmar (local name: *Thanbayar*) for cough, fever, sore throat, and toothache. Essential oil peel of *Citrus aurantifolia* (Christm.) Swingle (15 µL/disc) inhibited the growth of *Salmonella typhi*, *Streptococcus enteriditis*, *Escherichia coli*, *Clostridium perfringens*, and *Campylobacter jejuni* with inhibition zone diameters of 16.5, 10, 10.3, 20.5, and 11.8 mm, respectively (Wannissorn et al., 2005). Essential oil of *Citrus aurantifolia* Swingle inhibited the growth of *Escherichia coli*, *Staphylococcus aureus*, *Streptococcus mutans*, and *Candida albicans* with the MIC of 2.5, 0.4, 0.6, and 0.3 µL/mL, respectively (Miller et al., 2015).

Citrus aurantium L. is used in India (local name: *Khatta*) for bronchitis, fever, and cough. Essential oil of *Citrus aurantium* L. inhibited the growth of *Staphylococcus aureus*, *Streptococcus mutans*, and *Candida albicans* with MIC values of 2.9, 2.9, and 0.3 µL/mL, respectively (Miller et al., 2015). 5-Hydroxy-6,7,3′, 4′-tetramethoxyflavone and the limonoid limonexic acid (Log D = −0.7 at pH 7.4; molecular mass = 502.5 g/mol) isolated from the flowers of *Citrus aurantium* var. *amara* L. inhibited the expression of the Hepatitis B virus surface-antigen with the IC_{50} values of 59.5 and 39.2 µg/mL, respectively (Zhao et al., 2012).

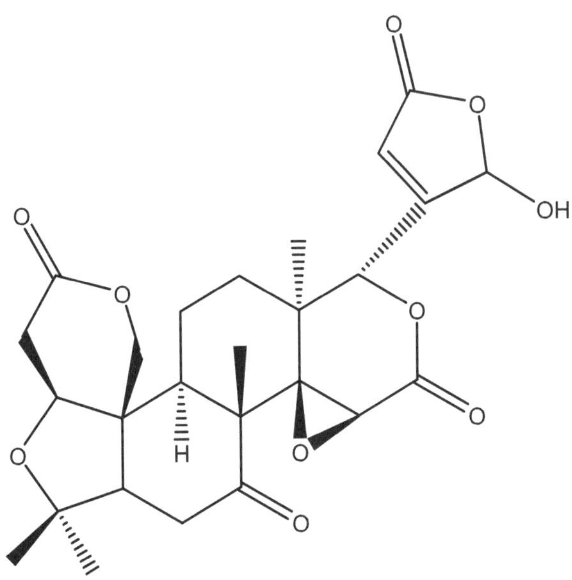

Limonexic acid

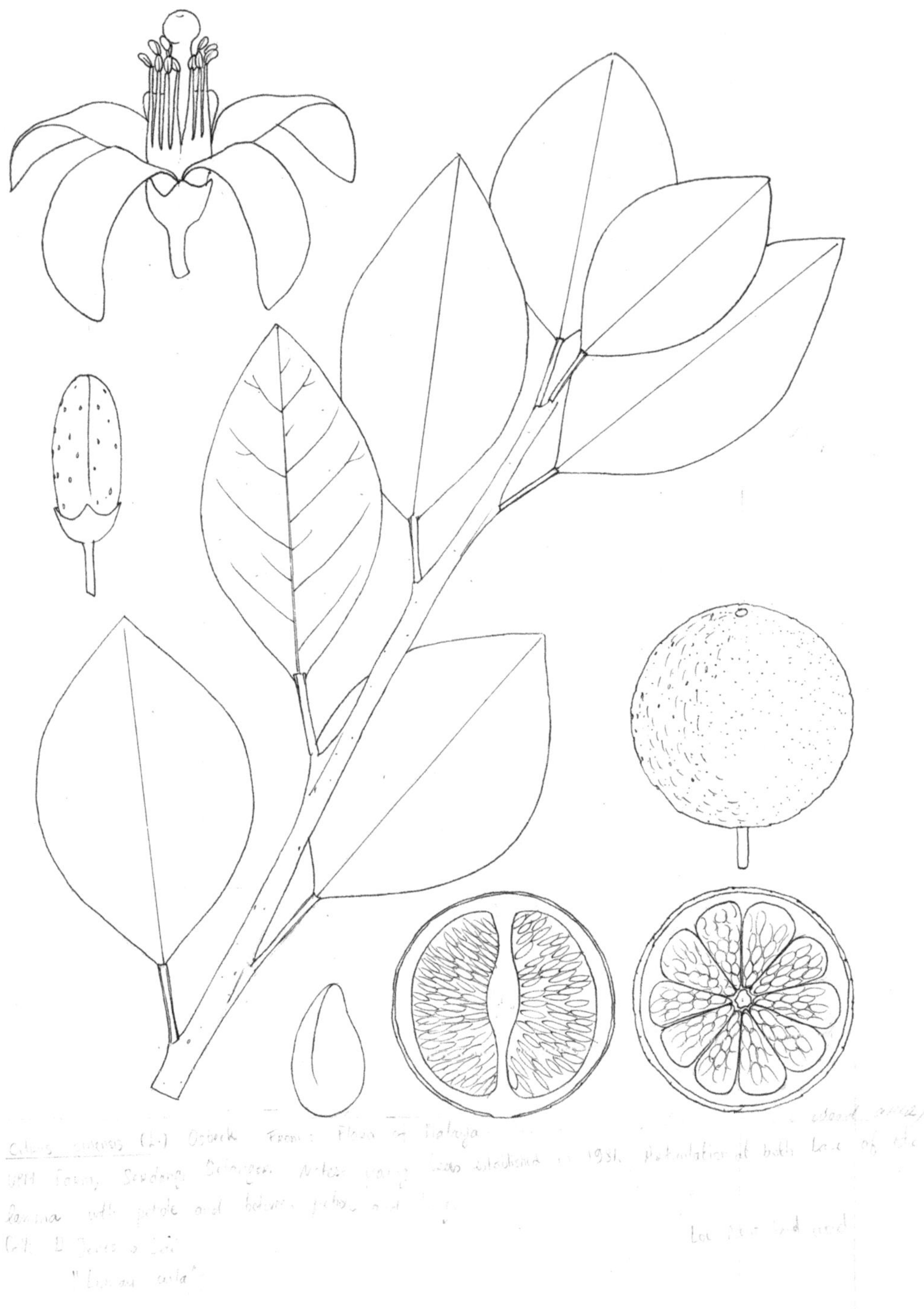

Citrus sinensis (L.) Osbeck

5-Hydroxy-6,7,3′,4′-tetramethoxyflavone and limonexic acid inhibited the expression of Hepatitis B virus e-antigen in HepG2.2.15 cells with IC_{50} values of 43.5 and 68.3 μg/mL, respectively (Zhao et al., 2012).

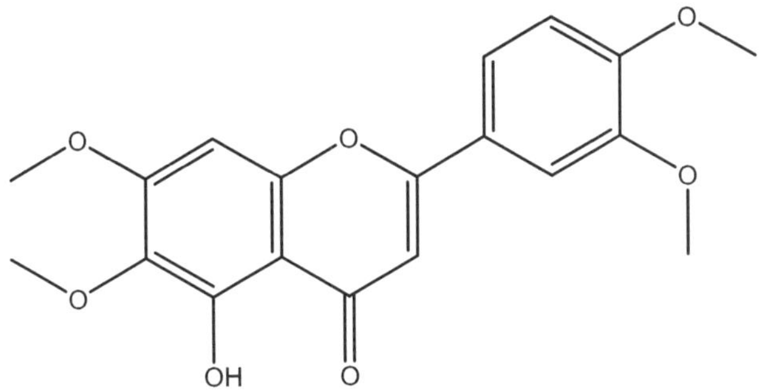

5-Hydroxy-6,7,3',4'-tetramethoxyflavone

The acridone alkaloid citrusinine I (Log D = 2.4 at pH 7.4; molecular mass = 301.2 g/mol) from *Citrus aurantium* L. inhibited the Herpes simplex virus type-1 and type-2 with the IC_{50} values of 0.5 and 0.7 μg/mL, respectively, probably via inhibition of ribonucleotide reductase (Yamamoto et al., 1989). Citrusinine-1, when combined with Acyclovir or Ganciclovir, synergistically potentiated the antiherpetic activity of these drugs (Yamamoto et al., 1989).

Citrus hystrix DC. is used in Indonesia for cough, leucorrhea, and gonorrhea. Essential oil of leaves of *Citrus hystrix* DC. (15 μL/disc) inhibited the growth of *Salmonella typhi*, *Escherichia coli*, *Clostridium perfringens*, and *Campylobacter* (Wannissorn et al., 2005).

Citrus limon (L.) Burm.f. (synonym: *Citrus medica* var. *limon* L.; *Citrus limonum* Risso; also known as lemon) is used in Turkey (local name = *Limon*) for fever, in India (local name = *Nimbu*) to treat dysentery, fever, jaundice, ringworm, sore throat, thrush, and dandruff, in Myanmar (local name: *Than-bayo*) for bronchitis, cough, sore throat, and in Palau (local name: *Debe el*) for dysentery. Acetone extract of peels of *Citrus limon* (L.) Osbeck (2 mg/8.5 mm well) inhibited *Klebsiella pneumoniae* and *Salmonella typhimurium* with inhibition zone diameters of 13 and 24.5 mm, respectively (Rakholiya et al., 2014).

Essential oil of peel of *Citrus maxima* (Burm.) Merr. (15 μL/disc) inhibited the growth of *Salmonella typhi*, *Streptococcus enteriditis*, *Escherichia coli*, *Clostridium perfringens*, and *Campylobacter jejuni* with inhibition zone diameters of 21, 20, 21.5, 20, and 34.5 mm, respectively (Wannissorn et al., 2005). Essential oil of *Citrus maxima* (Burm.) Merr. inhibited the growth of *Staphylococcus aureus* and *Streptococcus mutans* with MIC values of 3.3 and 1.6 μL/mL, respectively (Miller et al., 2015).

Citrus reticulata Blanco is used to treat fever in India. Essential oil from the fruits of *Citrus reticulata* Blanco at 2.5 μL/mL inhibited the growth of *Penicillium digitatum* and *Penicillium italicum* by 100% via membrane disruption (Tao et al., 2014). Hexane extract of peels of *Citrus reticulata* Blanco inhibited the growth of *Bacillus cereus*, *Bacillus coagulans*, *Bacillus subtilis*, *Staphylococcus aureus*, *Escherichia coli*, and *Pseudomonas aeruginosa* (Jayaprakasha et al., 2000). *Citrus reticulata* Blanco is used in China for the treatment of COVID-19 in combination with Oseltamivir phosphate (Luo, 2020; Yang et al., 2020; Wang et al., 2020).

iii. Methanol extract of bark of *Citrus unshiu* Marcov. at the concentration of 100 μg/mL inhibited RNAdependent DNA polymerase activities of Human immunodeficiency virus-1 reverse transcriptase and human immunodeficiency virus-1 protease by 14.6 and 18.3%, respectively (Min et al., 2001). Methanol extract of *Citrus bergamia* Risso & Poit. inhibited the replication of the Human T-cell leukemia virus type-1 and Human immunodeficiency virus type-1 with IC_{50} values of 3.1 and 1 μg/mL, respectively (Balestrieri et al., 2011). Limonin from the seeds of *Citrus bergamia* Risso & Poit. inhibited the replication of the Human T-cell leukemia virus type-1 and human immunodeficiency virus type-1 with IC_{50} values of 1 and 0.9 μg/mL, respectively (Balestrieri et al., 2011). Nomilin (Log D = 2.6 at pH 7.4; molecular mass = 514.5 g/mol) from the seeds of *Citrus bergamia* Risso & Poit. inhibited Human T-cell leukemia virus type-1 and human immunodeficiency virus type-1 with the IC_{50} values of 3.9 and 13.0 μg/mL, respectively (Balestrieri et al., 2011). Limonin and nomilin inhibited the Human immunodeficiency virus reverse transcriptase with the IC_{50} values of 4.2 and 3.4 μg/mL, respectively (Balestrieri et al., 2011).

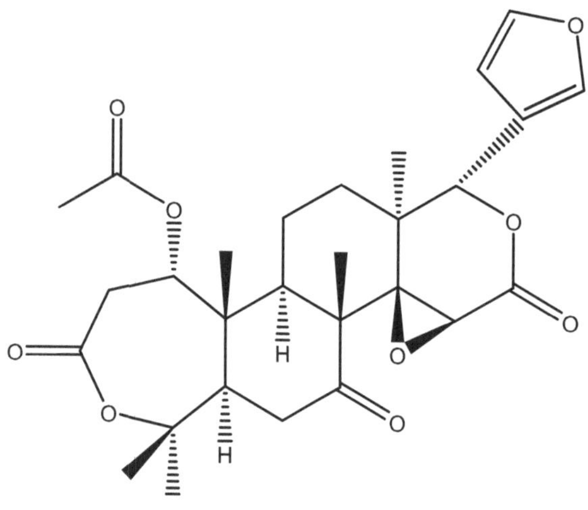

Nomilin

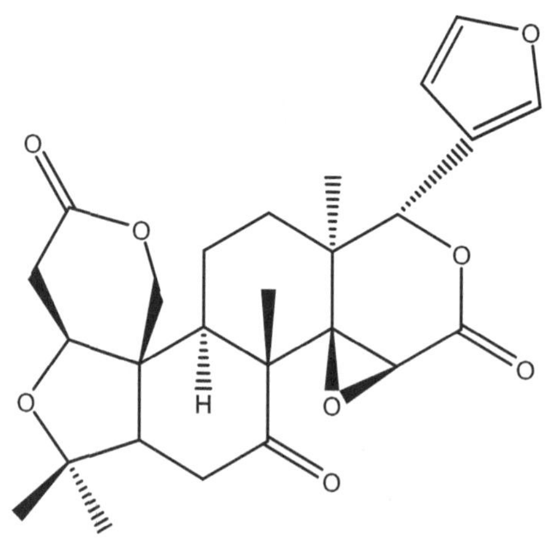

Limonin

iv. *Citrus maxima* (Burm.) Merr. is used in India (local name: *Maha nimbu*) to treat cough and cholera. Bergamottin epoxide from this plant, at a concentration of 35.7 μg/mL and 7-{[(E)-5-(3,3-dimethyl-2-oxiranyl)-3-methyl-2-pentenyl]oxy}-2H-2-chromenone at a concentration of 30 μg/mL reduced the MIC of norfloxacin on methicillin-resistant *Staphylococcus aureus* (strain 16565) and methicillin-resistant *Staphylococcus aureus* (strain 9543) (Abulrob et al., 2004).

v. Lemon probably originates from India and was known of ancient Greeks and Romans. The word lemon comes from the Unani *Limun* from the Sanskrit *Nimbuka*.

vi. In India, *Citrus macroptera* Montrouz. (local name: *Halkora*) is used to treat fever and *Citrus paradisi* Macfad. is used to treat cold and influenza. In the Philippines, *Citrus microcarpa* Bunge (local name: *Lemonsito*) is used to treat cough and to prevent COVID-19.

vii. In the Subfamily Aurantioideae and Tribe Aurantieae are the genera *Poncirus* Raf, *Citrus* L., *Burkillanthus* Swingle, *Feronia* Corrêa, *Fortunella* Swingle, *Limonia* L., and *Triphasia* Lour. The chalcone glucoside phlorizin (also known as phloretin-2'-*O*-β-glucoside) from *Fortunella japonica* (Thunb.) Swingle inhibited the growth of *Streptococcus pyogenes* with the MIC of 100 μg/mL and MBC > 100 μg/mL (Macé et al., 2017) and inhibited the growth of *Staphylococcus aureus* with the MIC value of 500 μg/mL (Barreca et al., 2014). Phlorizin inhibited the growth of *Escherichia coli* with the MIC value of 1500 μg/mL (Barreca et al., 2014). Hydrolyzation of phlorizin gives birth to the strong broad-spectrum antibacterial (MacDonald & Bishop, 1952) and antifungal (Shim et al., 2010) dihydrochalcone phloretin which is a "phytoalexin" (Gosh et al., 2010). For instance, phloretin (also known as dihydronaringenin; Log D = 2.5 at pH 7.4; molecular mass = 274.2 g/mol) inhibited the growth of *Staphylococcus aureus* (ATCC 6538), *Listeria monocytogenes* (ATCC 13311), *Salmonella typhimurium* (ATCC 13311), and methicillin-resistant *Staphylococcus aureus* (clinical) with the MIC values of 7.8, 62.5, 125, and 15.6 μg/mL, respectively (Barreca et al., 2014). Phloretin did not evoke the release of lactate dehydrogenase by *Staphylococcus aureus*, *Listeria monocytogenes*, or *Salmonella typhimurium* indicating a mechanism of antibacterial action independent of cytoplasmic membrane leakage (Barreca et al., 2014).

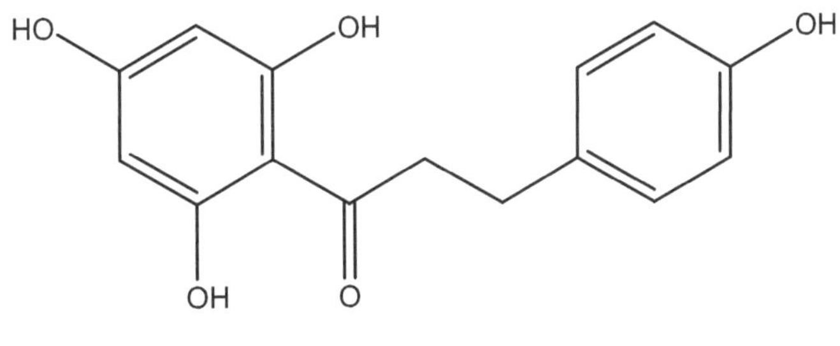

Phloretin

viii. *Citrus trifoliata* L (synomym: *Poncirus trifoliata* (L.) Raf.) is used in China to treat COVID-19 (Lem et al., 2020). From this plant, (−) 5,4′-dihydroxy-7, 8-[(3″,4″-cisdihydroxy-3″,4″-dihydro)-2″,2″-dimethylpyrano]-flavone, (−) 5,4′-dihydroxy-7, 8- [(3″-hydroxy-4″-one)-2″,2″-dimethylpyrano]-flavone, (−) 5,4″-dihydroxy-7,8-[(cis3″-hydroxy-4″-ethoxy-3″,4″-dihydro)-2″,2″-dimethylpyrano]-flavone, and poncitrin inhibited the replication of the Human immunodeficiency virus type-1 (IIIB strain) with EC_{50} values of 8.6, 0.9, 6.7, and 9.5 μM, respectively (Feng et al., 2010).

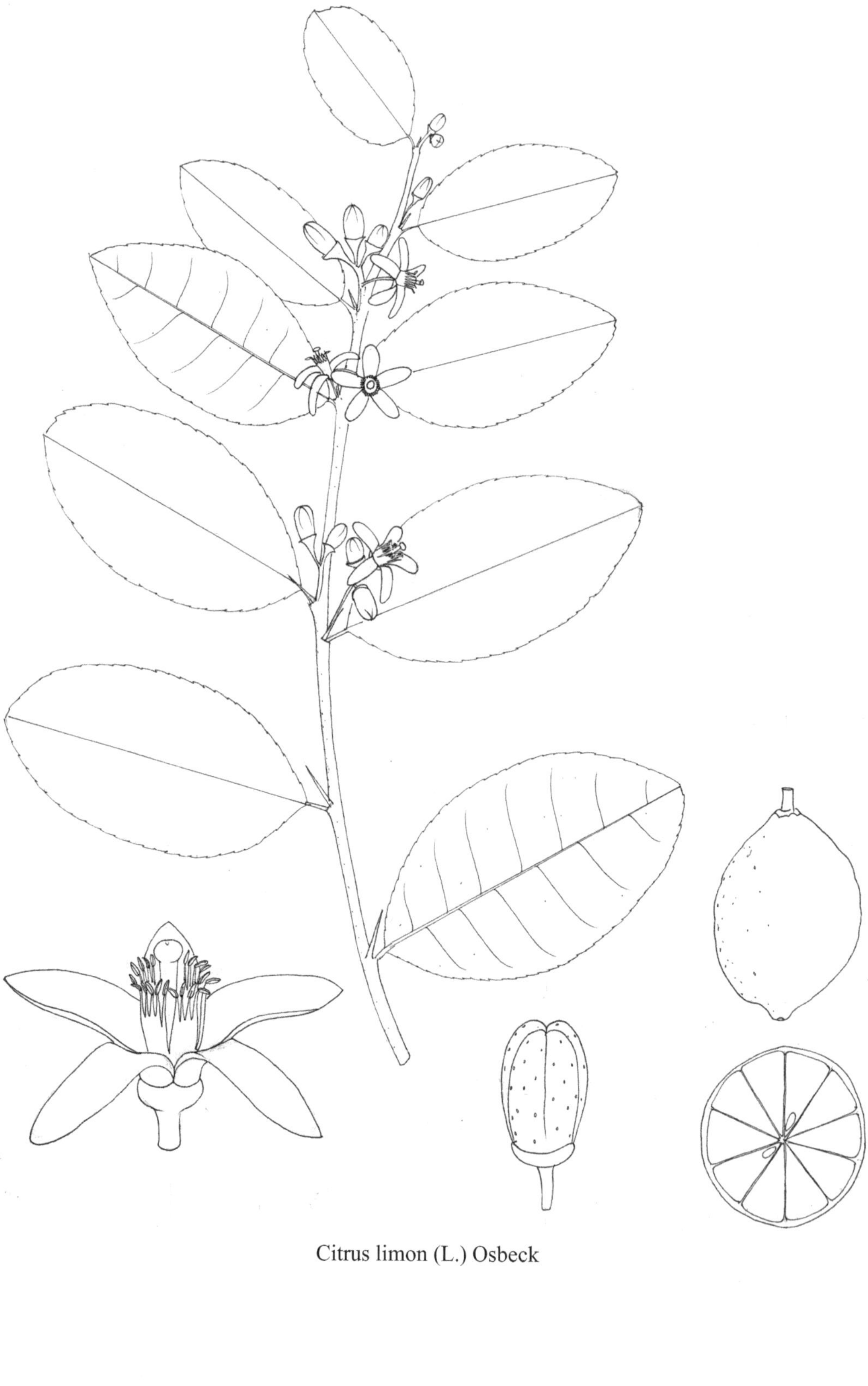

Citrus limon (L.) Osbeck

Ethylacetate fraction of *Citrus trifoliata* L inhibited the growth of methicillin-resistant *Staphylococcus aureus* (KCCM 40511) with the MIC value of 256 μg/mL. A subfraction isolated from the ethylacetate fraction at 128 μg/mL decreased the expression of MecA and penicillin-binding protein 2a (PBP2a) in methicillin-resistant *Staphylococcus aureus* (KCCM 40511) and increased the sensitivity of methicillin-resistant *Staphylococcus aureus* to ampicillin, penicillin, and oxacillin (Eom et al., 2016).

REFERENCES

Abulrob, A.N., Suller, M.T., Gumbleton, M., Simons, C. and Russell, A.D., 2004. Identification and biological evaluation of grapefruit oil components as potential novel efflux pump modulators in methicillin-resistant Staphylococcus aureus bacterial strains. *Phytochemistry*, *65*(22), pp. 3021–3027.

Balestrieri, E., Pizzimenti, F., Ferlazzo, A., Giofrè, S.V., Iannazzo, D., Piperno, A., Romeo, R., Chiacchio, M.A., Mastino, A. and Macchi, B., 2011. Antiviral activity of seed extract from *Citrus bergamia* towards human retroviruses. *Bioorganic & Medicinal Chemistry*, *19*(6), pp. 2084–2089.

Barreca, D., Bellocco, E., Laganà, G., Ginestra, G. and Bisignano, C., 2014. Biochemical and antimicrobial activity of phloretin and its glycosilated derivatives present in apple and kumquat. *Food Chemistry*, *160*, pp. 292–297.

Chaftar, N., Girardot, M., Quellard, N., Labanowski, J., Ghrairi, T., Hani, K., Frère, J. and Imbert, C., 2015. Activity of six essential oils extracted from Tunisian plants against Legionella pneumophila. *Chemistry & Biodiversity*, *12*(10), pp. 1565–1574.

Eom, S.H., Jung, Y.J., Lee, D.S., et al., 2016. Studies on antimicrobial activity of *Poncirus trifoliata* ethyl extract fraction against methicillin-resistant *Staphylococcus aureus* and to elucidate its antibacterial mechanism. *Journal of Environmental Biology*, *37*(1), p. 129.

Feng, T., Wang, R.R., Cai, X.H., Zheng, Y.T. and Luo, X.D., 2010. Anti-human immunodeficiency virus-1 constituents of the bark of *Poncirus trifoliata*. *Chemical and Pharmaceutical Bulletin*, *58*(7), pp. 971–975.

Gormley, N.A., Orphanides, G., Meyer, A., Cullis, P.M. and Maxwell, A., 1996. The interaction of coumarin antibiotics with fragments of the DNA gyrase B protein. *Biochemistry*, *35*(15), pp. 5083–5092.

Jayaprakasha, G.K., Negi, P.S., Sikder, S., Mohanrao, L.J. and Sakariah, K.K., 2000. Antibacterial activity of *Citrus reticulata* peel extracts. *Zeitschrift für Naturforschung C*, *55*(11–12), pp. 1030–1034.

Lem, F.F., Opook, F., Herng, D.L.J., Na, C.S., Lawson, F.P. and Tyng, C.F., 2020. Molecular mechanism of action of repurposed drugs and traditional Chinese medicine used for the treatment of patients infected with COVID-19: a systematic review. *medRxiv*. doi:10.1101/2020.04.10.20060376.

Lin, C.W., Tsai, F.J., Tsai, C.H., Lai, C.C., Wan, L., Ho, T.Y., Hsieh, C.C. and Chao, P.D.L., 2005. Anti-SARS coronavirus 3C-like protease effects of Isatis indigotica root and plant-derived phenolic compounds. *Antiviral Research*, *68*(1), pp. 36–42.

Lin, C.W., Wu, C.F., Hsiao, N.W., Chang, C.Y., Li, S.W., Wan, L., Lin, Y.J. and Lin, W.Y., 2008. Aloe-emodin is an interferon-inducing agent with antiviral activity against Japanese encephalitis virus and enterovirus 71. *International Journal of Antimicrobial Agents*, *32*(4), pp. 355–359.

Luo, A., 2020. Positive SARS-Cov-2 test in a woman with COVID-19 at 22 days after hospital discharge: a case report. *Journal of Traditional Chinese Medical Sciences*, *7*, pp. 413–417.

Macé, S., Truelstrup Hansen, L. and Rupasinghe, H.P., 2017. Anti-bacterial activity of phenolic compounds against *Streptococcus pyogenes*. *Medicines*, *4*(2), p. 25.

Miller, A.B., Cates, R.G., Lawrence, M., Soria, J.A.F., Espinoza, L.V., Martinez, J.V. and Arbizú, D.A., 2015. The antibacterial and antifungal activity of essential oils extracted from Guatemalan medicinal plants. *Pharmaceutical Biology*, *53*(4), pp. 548–554.

Min, B.S., Kim, Y.H., Tomiyama, M., Nakamura, N., Miyashiro, H., Otake, T. and Hattori, M., 2001. Inhibitory effects of Korean plants on HIV-1 activities. *Phytotherapy Research*, *15*(6), pp. 481–486.

Rakholiya, K., Kaneria, M. and Chanda, S., 2014. Inhibition of microbial pathogens using fruit and vegetable peel extracts. *International Journal of Food Sciences and Nutrition*, *65*(6), pp. 733–739.

Stanley, W.L. and Jurd, L., 1971. Citrus coumarins. *Journal of Agricultural and Food Chemistry*, *19*(6), pp. 1106–1110.

Tao, N., Jia, L. and Zhou, H., 2014. Anti-fungal activity of *Citrus reticulata* Blanco essential oil against *Penicillium italicum* and *Penicillium digitatum*. *Food Chemistry*, *153*, pp. 265–271.

Wang, S.X., Wang, Y., Lu, Y.B., Li, J.Y., Song, Y.J., Nyamgerelt, M. and Wang, X.X., 2020. Diagnosis and treatment of novel coronavirus pneumonia based on the theory of traditional Chinese medicine. *Journal of Integrative Medicine*, *18*, pp. 275–283.

Xu, J.J., Liu, Z., Tang, W., Wang, G.C., Chung, H.Y., Liu, Q.Y., Zhuang, L., Li, M.M. and Li, Y.L., 2015. Tangeretin from Citrus reticulate inhibits respiratory syncytial virus replication and associated inflammation *in vivo*. *Journal of agricultural and food chemistry*, *63*(43), pp.9520–9527.

Xu, J.J., Wu, X., Li, M.M., Li, G.Q., Yang, Y.T., Luo, H.J., Huang, W.H., Chung, H.Y., Ye, W.C., Wang, G.C. and Li, Y.L., 2014. Antiviral activity of polymethoxylated flavones from "Guangchenpi", the edible and medicinal pericarps of *Citrus reticulata* 'Chachi'. *Journal of Agricultural and Food Chemistry*, *62*(10), pp. 2182–2189.

Yamamoto, N., Furukawa, H., Ito, Y., Yoshida, S., Maeno, K. and Nishiyama, Y., 1989. Anti-herpesvirus activity of citrusinine-I, a new acridone alkaloid, and related compounds. *Antiviral Research*, *12*(1), pp. 21–36.

Yang, Q.X., Zhao, T.H., Sun, C.Z., Wu, L.M., Dai, Q., Wang, S.D. and Tian, H., 2020. New thinking in the treatment of 2019 novel coronavirus pneumonia. *Complementary Therapies in Clinical Practice*, *39*, p. 101131.

Yao, X., Zhu, X., Pan, S., Fang, Y., Jiang, F., Phillips, G.O. and Xu, X., 2012. Antimicrobial activity of nobiletin and tangeretin against Pseudomonas. *Food Chemistry*, *132*(4), pp. 1883–1890.

Zhao, H.Y., Yang, L., Wei, J., Huang, M. and Jiang, J.G., 2012. Bioactivity evaluations of ingredients extracted from the flowers of *Citrus aurantium* L. var. amara Engl. *Food Chemistry*, *135*(4), pp. 2175–2181.

8.5.4.4 *Clausena excavata* Burm. f.

Synonyms: *Amyris punctata* Roxb.; *Clausena forbesii* Engl.; *Clausena lunulata* Hayata; *Clausena moningerae* Merr.; *Clausena punctata* (Roxb.) Wight & Arn.; *Clausena tetramera* Hayata

Common names: Sanitrok damrey (Cambodia); gǎo yíng (China); tikusan (Indonesia); agnijhal (India); tcho kou nhia (Laos); pokok kemantu (Malaysia); daw-hke (Myanmar); buringit (the Philippines); huat mon (Thailand); chùm hôi (Vietnam)

Habitat: Forests edges, villages, gardens

Distribution: From India to the Philippines

Botanical observation: This treelet grows up to about 4 m tall. The bole is not straight. The stems are hairy at apex. The leaves are compound, spiral, and exstipulate. The rachis is up to about 30 cm long. The blade comprises 5–31 pairs of folioles. The folioles are asymmetrical, serrate, 2.4–5 × 1–1.7 cm, oil gland-dotted, dull green, and marked with about 9 pairs of secondary nerves. The flowers are very small and arranged in about 15 cm terminal and long panicles. The 5 petals are whitish, oblong, and minute. The androecium includes 8 stamens. The ovary develops a stout style. The drupes are up to 1.5 cm long, light-pink, somewhat glossy, and subglobose to ellipsoid.

Medicinal uses: Fever (India, Indonesia); leprosy (Myanmar); post-partum, sores, syphilis (Malaysia); boil, dysentery, sore throat (India); AIDS, colds (Thailand)

Moderate antimycobacterial carbazole-type alkaloids: 3-Formylcarbazole, mukonal, 3-methoxycarbonylcarbazole, 2-hydroxy-3-formyl-7-methoxycarbazole, and clauszoline J inhibited the growth of *Mycobacterium* sp. with the MIC values of 100, 200, 50, 100 and 100 µg/mL, respectively (Sunthitikawinsakul et al., 2003a).

Moderate antimycobacterial coumarins: Dentatin, nor-dentatin, and clausenidin inhibited the growth of *Mycobacterium* sp. with the MIC values of 50, 100, and 200 µg/mL, respectively (Sunthitikawinsakul et al., 2003a).

Moderate broad-spectrum antifungal coumarin: 7((2E)-4(4,5-Dihydro-3-methylene-2-oxo-5-furanyl)-3-methylbut-2-enyloxy) coumarin isolated from the leaves inhibited the growth of *Aspergillus niger, Aspergillus fumigatus, Candida albicans, Candida tropicalis, Cryptococcus curvatus, Filobasidiela neoformans, Mucor circinelloides,* and *Trichosporum cutaneum* with the MIC values of 600, 600, 70, 0.03, 70, 70, 70, and 70 μg/mL, respectively (Kumar et al., 2012).

Strong antiviral (monopartite linear dimeric single stranded (+) RNA) amphiphilic limonoid: Clausenolide-1-ethyl ether (molecular mass = 488.6 g/mol) isolated from the roots inhibited the replication of the Human immunodeficiency virus type-1 in 1A2 cells with the EC_{50} value of 3.4×10^{-5}M and a selectivity index of 16 (Sunthitikawinsakul et al., 2003b).

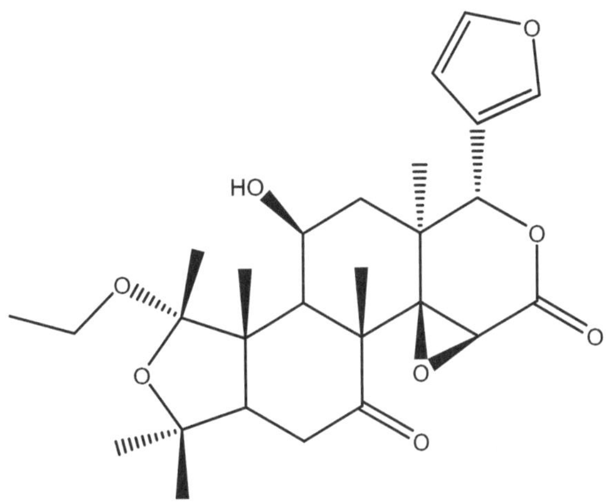

Clausenolide-1-ethyl ether

Strong antiviral (monopartite linear dimeric single stranded (+) RNA) amphiphilic carbazole-type alkaloids: O-Methylmukonal, 3-formyl-2,7-dimethoxycarbazole, and clauszoline J isolated from the roots inhibited the replication of the Human immunodeficiency virus type-1 in 1A2 cells with the EC_{50} values of 12, 29.1, and 34.2 μM and selectivity indexes of 56.7, 8, and 1.6 (Sunthitikawinsakul et al., 2003b).

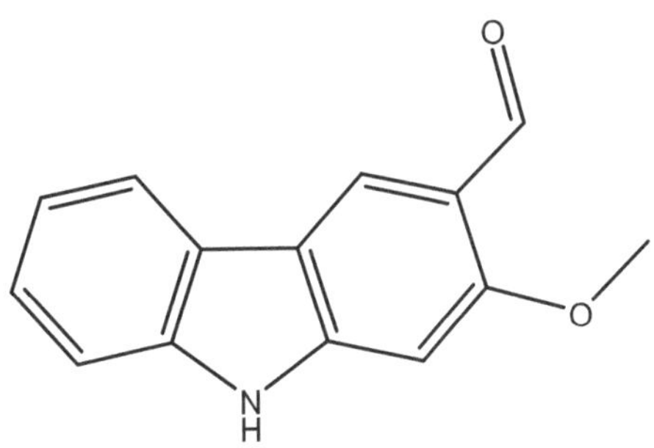

O-Methylmukonal

Strong antiviral (monopartite linear dimeric single stranded (+) RNA) amphiphilic pyranocoumarin: Clausenidin (Log D = 2.3 at pH 7.4; molecular mass = 328.3 g/mol) isolated from the roots inhibited the replication of the Human immunodeficiency virus type-1 in 1A2 cells with the EC_{50} value of 5.3 µM and selectivity index of 7 (Sunthitikawinsakul et al., 2003b).

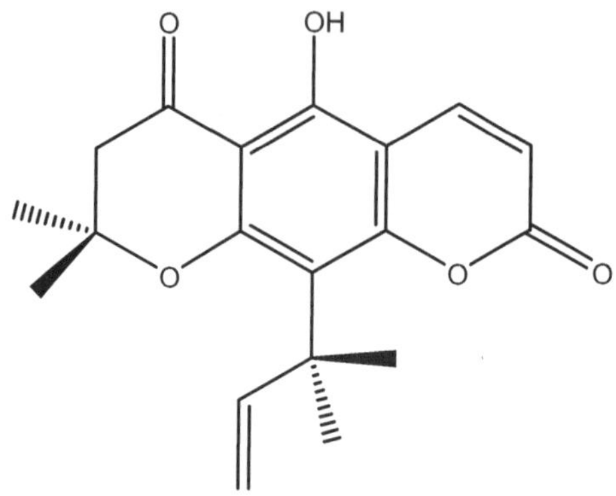

Clausenidin

Strong antiviral (enveloped monopartite double-stranded DNA) pyranocoumarins: Clausenidin and nordentatin inhibited the expression of the Hepatitis B virus surface antigen in HepA2 cells with the IC_{50} value of 1.8 and 6.3 µM, respectively (Su et al., 2009).

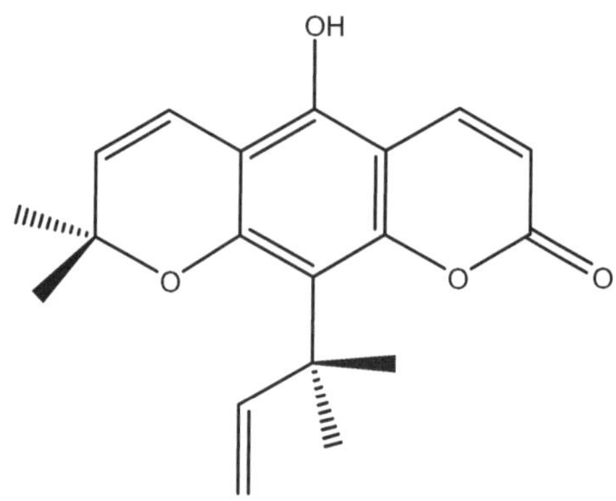

Nordentatin

Commentaries

i. When testing natural products against a virus, bacteria, or fungi, it is not rare to observe that compounds isolated from a given medicinal plant and belonging to different classes of natural products are all active.
ii. Carbazole-type alkaloids often inhibit the growth of a broad-spectrum of viruses of which the Human immunodeficiency virus, Human cytomegalovirus, Hepatitis C virus, Herpes simplex virus, and Human papilloma virus via mechanisms yet not fully delinealated

(Caruso et al., 2019). These alkaloids are planar and aromatic and thus could interfere with virion genetic material (Tanious et al., 1997) and thus impede viral replication. Natural products from plants targeting the genetic material of microbes are often broad-spectrum antibacterial, antifungal, and antiviral.

iii. The carbazole alkaloids clausine L, clausine K, and the limonoid zapoterin isolated from the roots of *Clausena harmandiana* (Pierre) Guillaumin inhibited the growth of *Pythium insidiosum* with inhibition zone diameter of 10.2, 16.2, and 11.8 mm, respectively (20 µL/6 mm disc) (Sriphana et al., 2013). The carbazole clausamine A (Log D = 3.9 at pH 7.4; molecular mass = 293.3 g/mol) isolated from the stems of *Clausena harmandiana* inhibited the growth of methicillin-resistant *Staphylococcus aureus* (SK1), *Staphylococcus aureus* (TISTR 1466), *Escherichia coli* (TISTR 780), and *Salmonella typhimurium* (TISTR 292) with the MIC values of 8, 32, 128, and 128 µg/mL, respectively (Maneerat et al., 2012). Clausamine B (Log D = 4.5) at pH 7.4; molecular mass = 307.3 g/mol) isolated from the stems inhibited the growth of methicillin-resistant *Staphylococcus aureus* (SK1) and *Salmonella typhimurium* (TISTR 292) with the MIC values of 0.2 and 128 µg/mL, respectively (Maneerat et al., 2012).

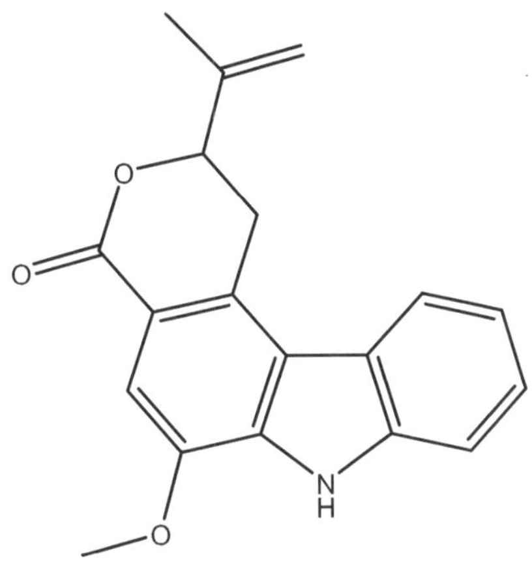

Clausamine B

Clauszoline N isolated from the stems inhibited the growth of methicillin-resistant *Staphylococcus aureus* (SK1), *Escherichia coli* (TISTR 780), and *Salmonella typhimurium* (TISTR 292) with the MIC values of 16, 64, and 128 µg/mL, respectively (Maneerat et al., 2012). Clausine F isolated from the stems methicillin-resistant *Staphylococcus aureus* (SK1), *Staphylococus aureus* (TISTR 1466), *Escherichia coli* (TISTR 780), and *Salmonella typhimurium* (TISTR 292) with the MIC values of 4, 4, 128 and 64 µg/mL, respectively (Maneerat et al., 2012). One could sense that low to medium molecular mass planar or planar aromatic alkaloids with the presence of a few hydroxy or ketone groups almost always target DNA of RNA in microbes.

iv. *Clausena pentaphylla* (Roxb.) DC. is used in India to treat wounds. *Clausena wampi* (Blanco) Oliv. is used in India for bronchitis and Thailand for cold. *Clausena heptaphylla* (Roxb.) Hook. f. is used for fever in Bangladesh (local name: *Alkatra*).

v. Aqueous extract of leaves of *Clausena anisata* (Willd.) Hook. f. ex Benth. inhibited the replication of the Human immunodeficiency virus type-1 (IIIB strain) in Molt-4 clone 8 with the EC_{50} of 700 µg/mL and a selectivity index above 2. This extract inhibited Human immunodeficiency virus type-2 (strain GH1) with EC_{50} value of 100 µg/mL and selectivity index of 12.7 (Ayisi & Nyadedzor, 2003).

REFERENCES

Ayisi, N.K. and Nyadedzor, C., 2003. Comparative *in vitro* effects of AZT and extracts of *Ocimum gratissimum*, *Ficus polita*, *Clausena anisata*, *Alchornea cordifolia*, and *Elaeophorbia drupifera* against HIV-1 and HIV-2 infections. *Antiviral Research*, *58*(1), pp. 25–33.

Caruso, A., Ceramella, J., Iacopetta, D., Saturnino, C., Mauro, M.V., Bruno, R., Aquaro, S. and Sinicropi, M.S., 2019. Carbazole derivatives as antiviral agents: an overview. *Molecules*, *24*(10), p. 1912.

Kumar, R., Saha, A. and Saha, D., 2012. A new antifungal coumarin from *Clausena excavata*. *Fitoterapia*, *83*(1), pp. 230–233.

Maneerat, W., Phakhodee, W., Ritthiwigrom, T., Cheenpracha, S., Promgool, T., Yossathera, K., Deachathai, S. and Laphookhieo, S., 2012. Antibacterial carbazole alkaloids from *Clausena harmandiana* twigs. *Fitoterapia*, *83*(6), pp. 1110–1114.

Sriphana, U., Thongsri, Y., Prariyachatigul, C., Pakawatchai, C. and Yenjai, C., 2013. Clauraila E from the roots of *Clausena harmandiana* and antifungal activity against *Pythium insidiosum*. *Archives of Pharmacal Research*, *36*(9), pp. 1078–1083.

Sunthitikawinsakul, A., Kongkathip, N., Kongkathip, B., Phonnakhu, S., Daly, J.W., Spande, T.F., Nimit, Y. and Rochanaruangrai, S., 2003a. Coumarins and carbazoles from *Clausena excavata* exhibited antimycobacterial and antifungal activities. *Planta Medica*, *69*(2), pp. 155–157.

Sunthitikawinsakul, A., Kongkathip, N., Kongkathip, B., Phonnakhu, S., Daly, J.W., Spande, T.F., Nimit, Y., Napaswat, C., Kasisit, J. and Yoosook, C., 2003b. Anti-HIV-1 limonoid: first isolation from Clausena excavata. *Phytotherapy Research: An International Journal Devoted to Pharmacological and Toxicological Evaluation of Natural Product Derivatives*, *17*(9), pp. 1101–1103.

Tanious, F.A., Ding, D., Patrick, D.A., Tidwell, R.R. and Wilson, W.D., 1997. A new type of DNA minor-groove complex: carbazole dication–DNA interactions. *Biochemistry*, *36*(49), pp. 15315–15325.

8.5.4.5 *Dictamnus albus* L.

Synonyms: *Dictamnus himalayanus* Royle; *Dictamnus fraxinella* Pers.

Common names: Burning-bush; dittany, fraxinella, gas-plant (India)

Habitat: Forests

Distribution: Temperate Asia

Botanical observation: This magnificent erect herb grows up to a height of 1 m. The stems are hairy at apex and glandular. The leaves are compound, sessile, alternate, and exstipulate. The blade presents 2–6 pairs of folioles plus a terminal one. The folioles are dotted with oil glands, 3.5–10 × 1–3.5 cm, lanceolate, serrulate, and hairy below. The inflorescence is a terminal raceme of large flowers. The 5 sepals are basally fused and 1 cm long. The 5 petals are clawed, light pink and beautifully pencilled with fine, somewhat capillary-like dark pink lines, elliptic to oblong. Four petals are above and 1 petal below. A disc is present. The 10 stamens are free, with light pink filaments, somewhat grayish-blue anthers, and curved upward (with a vague look of *Bauhinia*). The ovary is 5-lobed and develops a slender style. The fruit consists of 5 folicles which are woody, dehiscent, glandular, and shelter 2–3 black seeds.

Medicinal uses: Fever (India)

Strong antibacterial (Gram-positive) amphiphilic furanoquinoline alkaloid: Dictamine (Log D = 2.6 at pH 7.4; molecular mass = 199.2 g/mol) inhibited the growth of *Micrococcus luteus*

(TISTR 884), *Bacillus cereus* (TISTR 688), *Bacillus subtilis* (TISTR 008) with the MIC values of 26, 64, and 128 μg/mL, respectively (Tantapakul et al., 2014).

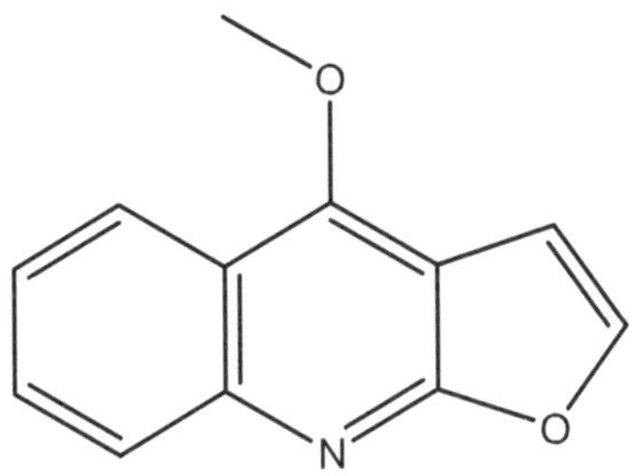

Dictamine

Strong antibacterial (Gram-positive) hydrophilic furanoquinoline alkaloid: Robustine (Log D = 0.02 at pH 7.4; molecular mass = 215.2 g/mol) inhibited the growth of *Micrococcus luteus* (TISTR 884), *Bacillus cereus* (TISTR 688), *Bacillus subtilis* (TISTR 008) (Tantapakul et al., 2014).

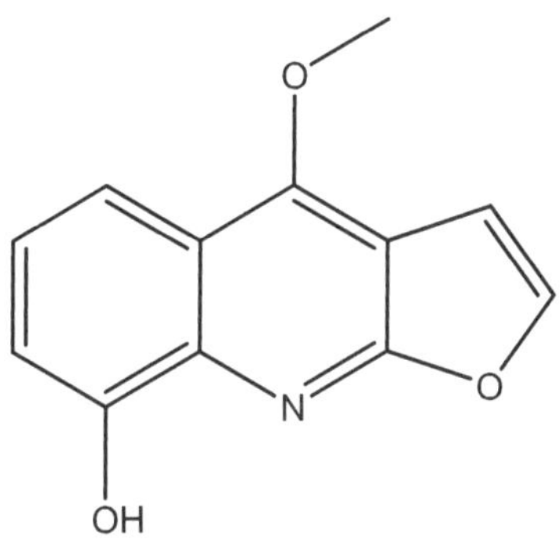

Robustine

Moderate broad-spectrum antibacterial furanoquinoline alkaloid: γ-Fagarine inhibited the growth of *Staphylococcus aureus* (ATCC 6538), *Streptococcus pyogenes* (ATCC 19615), *Klebsiella pneumoniae* (ATCC 13883), and *Escherichia coli* (ATCC 25922) with the MIC/MBC values of 50/>100, 50/>100, 50/>100, and 100/>100 μg/mL, respectively (Tavares et al., 2014).

Moderate broad-spectrum antifungal furanoquinoline alkaloid: γ-Fagarine inhibited the growth of *Candida albicans* (ATCC 10231), *Saccharomyces cerevisae* (ATCC 2601), and *Cryptococcus neoformans* (ATCC 28952) with the MIC/MBC values of 100/>100, 100/>100, and 50/>100 μg/mL, respectively (Tavares et al., 2014).

Strong broad-spectrum antibacterial benzofuranone: Fraxinellone (Log D = 2.9 at pH 7.4; molecular mass = 232.2 g/mol) at 5 ppm inhibited the growth of *Escherichia coli*, *Staphylococcus aurens*, and *Bacillus megathrium* by 100, 99.8, and 99.8%, respectively (Yuan et al., 2006).

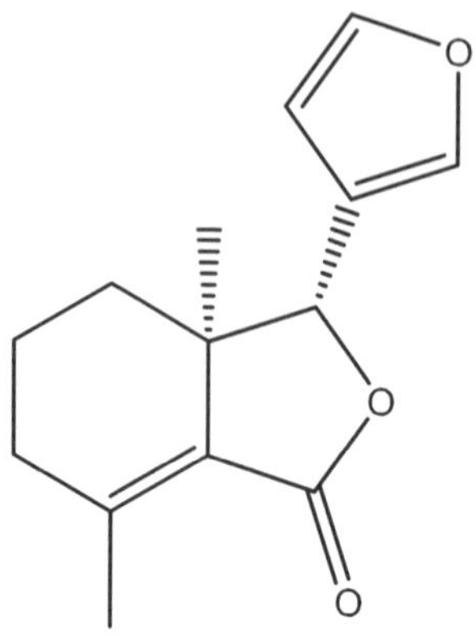

Fraxinellone

Strong antimycobacterial amphiphilic furanoquinoline alkaloids: Dictamnine and γ-fagarine (Log D = 2.9 at pH 7.4; molecular mass = 229.3 g/mol) inhibited the growth of *Mycobacterium tuberculosis* (H37Rv) with the MIC values of 30 and 30 μg/mL, respectively (Huang et al., 2008).

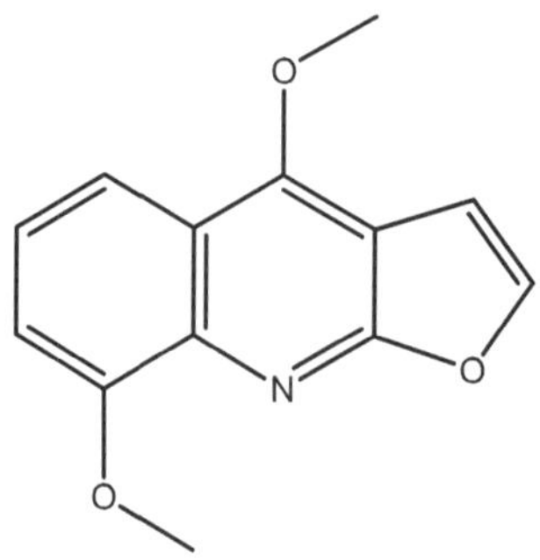

γ-Fagarine

Strong antifungal (filamentous) amphiphilic furanoquinoline: Dictamine inhibited the growth of *Cladosporium cucumerinum* with minimum amounts of 0.6 μg/TLC plates (Zhao et al., 1998).

Strong antifungal (filamentous) amphiphilic benzofuranone: Fraxinellone at 50 ppm inhibited the growth of *Rhizotonia cerealis* and *Colletrichurm lagenarium* by 100% and 72%, respectively (Yang & Chen, 2008).

Strong antiviral (enveloped monopartite double-stranded DNA) amphiphilic furanoquinoline: Dictamine (Log D = 2.6 at pH 7.4; molecular mass = 199.2 g/mol) inhibited the secretion of the Hepatits B virus surface-antigen and Hepatitis B virus e-antigen at 0.2 μM by 18.6% and 6.8%, respectively, in HepG2 2.215 cells (Yang & Chen, 2008).

Strong antiviral (enveloped monopartite single-stranded (+) RNA) amphiphilic furanoquinoline: γ-Fagarine inhibited the replication of the Hepatitis C virus (J6/JFH1-P47 strain) with the IC_{50} value of 20.4 μg/mL, respectively, and selectivity above 1.5 (Wahyuni et al., 2014).

Commentaries

i. The rootbark contains furanoquinoline alkaloids of which dictamine, γ-fagarine, robustine, haplopine, and skimmianine (Woo & Kang, 1985).
ii. Essential oils from the roots of *Dictamnus angustifolius* G. Don ex Sweet (containing mainly tetramethylenecyclobutane and fraxinellone) inhibited the growth of *Staphylococcus*

aureus (ATCC 6538), *Escherichia coli* (ATCC 8739), *Monilia albicans* (ATCC 10231), and *Aspergillus niger* (ATCC 16404) with the MIC values of 109, 225, 96, and 680 μg/mL, respectively (Sun et al., 2015).

iii. 6β-Hydroxyfraxinellone (also known as dasycarpol), fraxinellone, isofraxinellone, calodendrolide, dictamine, and haplopine isolated from the root bark of *Dictamnus dasycarpus* Turcz. (synonym: *Dictamnus albus* subsp. *dasycarpus* (Turcz.) L. Winte) inhibited the growth of *Cladosporium cucumerinum* with minimum amounts of 5, 5, 2, 5, 0.6, and 10 μg/TLC plates (Homans & Fuchs, 1970; Zhao et al., 1998). Aqueous extract of root bark of *Dictamnus dasycarpus* Turcz. at the concentration of 250 μg/mL inhibited Human immunodeficiency virus type-1 protease by 11.4% (Xu et al., 1996).

iv. The plant is used as tonic in Arabian and Persian medicines. In his "*Hieronymi Fracastorii syphilidis, sive morbis gallici*", our peer the Italian physician Girolamo Fracastoro (1484–1553) advises *Dictamnus albus* L. as a mean to treat syphilis (Riddell, 1926). In China, *Dictamnus dasycarpus* Turcz. (local name in China: *Bai xian* pi) was used for syphilis as recorded by our peer the French physician Jean Astruc (1684–1768) in his "*De morbis venereis libri novem*."

v. Furanoquinolines are planar and thus are prone to intercalate into DNA (Wink, 2020). Amphiphilic low to medium molecular weight natural products intercalating DNA are broad-spectrum antibacterial, antifungal, and antiviral effect as well as cytotoxic and mutagenic (Mohammed et al., 2016). γ-Fagarine could have activity against the Severe acute respiratory syndrome-associated coronavirus and other members of the Coronaviridae.

REFERENCES

Huang, H.Y., Ishikawa, T., Peng, C.F., Tsai, I.L. and Chen, I.S., 2008. Constituents of the root wood of *Zanthoxylum wutaiense* with antitubercular activity. *Journal of Natural Products*, *71*(7), pp. 1146–1151.

Mohammed, M.M., Ibrahim, N.A., El-Sakhawy, F.S., Mohamed, K.M. and Deabes, D.A.H., 2016. Two new cytotoxic furoquinoline alkaloids isolated from *Aegle marmelos* (Linn.) Correa. *Natural Product Research*, *30*(22), pp. 2559–2566.

Riddell, W.R., 1926. Hieronymi Fracastorii Syphilidis Sive Morbi Gallici. *The Public Health Journal*, *17*(10), pp. 471–486.

Sun, J., Wang, X., Wang, P., Li, L., Qu, W. and Liang, J., 2015. Antimicrobial, antioxidant and cytotoxic properties of essential oil from *Dictamnus angustifolius*. *Journal of Ethnopharmacology*, *159*, pp. 296–300.

Tantapakul, C., Sripisut, T., Maneerat, W., Ritthiwigrom, T. and Laphookhieo, S., 2014. Antibacterial compounds from *Glycosmis puberula* twigs. *Natural Product Communications*, *9*(12), p. 1934578X1400901210.

Tavares, L.D.C., Zanon, G., Weber, A.D., Neto, A.T., Mostardeiro, C.P., Da Cruz, I.B., Oliveira, R.M., Ilha, V., Dalcol, I.I. and Morel, A.F., 2014. Structure-activity relationship of benzophenanthridine alkaloids from *Zanthoxylum rhoifolium* having antimicrobial activity. *PloS One*, *9*(5), p. e97000.

Wahyuni, T.S., Widyawaruyanti, A., Lusida, M.I., Fuad, A., Fuchino, H., Kawahara, N., Hayashi, Y., Aoki, C. and Hotta, H., 2014. Inhibition of hepatitis C virus replication by chalepin and pseudane IX isolated from *Ruta angustifolia* leaves. *Fitoterapia*, *99*, pp. 276–283.

Woo, W.S. and Kang, S.S., 1985. Furoquinoline alkaloids in *Dictamus albus* Root Bark. *Korean Journal of Pharmacognosy*, *16*(3), pp. 125–128.

Xu, H.X., Wan, M., Loh, B.N., Kon, O.L., Chow, P.W. and Sim, K.Y., 1996. Screening of traditional medicines for their inhibitory activity against HIV-1 Protease. *Phytotherapy Research*, *10*(3), pp. 207–210.

Yang, G. and Chen, D., 2008. Alkaloids from the roots of *Zanthoxylum nitidum* and their antiviral and antifungal effects. *Chemistry & Biodiversity*, *5*(9), pp. 1718–1722.

Yuan, C., Wang, X. and Yang, D., 2006. Antimicrobial and pesticidal activity of fraxinellone from *Dictamnus dasycarpus*. *Mortality*, *24*(48h), p. 72h.

Zhao, W., Wolfender, J.L., Hostettmann, K., Xu, R. and Qin, G., 1998. Antifungal alkaloids and limonoid derivatives from *Dictamnus dasycarpus*. *Phytochemistry*, *47*(1), pp. 7–11.

8.5.4.6 *Euodia hortensis* J.R. Forst. & G. Forst.

Synonyms: *Euodia longifolia* A. Rich.; *Euodia schullei* Warb. var. *simplicifolia* Guill.

Common name: Pungent shrub; bubu, karong on, sup, wareba (Papua New Guinea)

Habitat: Forests, cultivated

Distribution: Papua New Guinea to Samoa, Tonga, Niue, New Hebrides

Botanical observation: This handsome tree grows up to 15 tall. The plant is grown as ornamental in the Pacific. The leaves are opposite, unifoliolate or trifoliolate, and exstipulate. The petiole is 0.5–6.5 cm long. The folioles are discretly dotted, elliptical, linear, or obovate, 0.3–8 × 4–15 cm, wavy, attenuate at base, and obtuse to acuminate at apex. The racemes are terminal and up to about 30 cm long. The calyx is minute and develops 4 triangular lobes. The corolla is minute, whitish, and develops 4 triangular and oil gland-dotted lobes. A disc is present. The 4 stamens are about 6 mm long. The gynoecium includes 4 carpels merged at base. The fruits consist of 1–4 follicles which are about 4 mm long, each sheltering a reddish brown and somewhat glossy seed.

Medicinal uses: Boils (Solomons); fever (Pacific Islands; Papua New Guinea); toothache (Pacific Islands); cold, diarrhea, pneumonia (Papua New Guinea)

Commentaries

i. The plant has apparently not been examined for its antimicrobial constituents. Essential oil of the plant evoked some levels of antibacterial effects (Maharaj et al., 2016).

ii. In Papua New Guinea., *Euodia elleryana* F. Muell is used for fever (local name: *Kurih*) and in China *Euodia lepta* (Spreng.) Merr (local name: *Bei la gong dian*) is used for the treatment of flu and pneumonia.

iii. The indoloquinazoline-type alkaloid evodiamine isolated from the fruits of *Euodia rutaecarpa* Benth (synonym: *Tetradium ruticarpum* (A. Juss.) T.G. Hartley) inhibited the growth of *Klebsiella pneumoniae* (clinical; multidrug resistant isolate) with the MIC value of 128 µg/mL whereas cefotaxime, aztreonam, gentamicin, cefazolin, and cefutoxime demonstrated MIC values above 512 µg/mL (Wu et al., 2013). Evodiamine (Log D = 2.6 at pH 7.4; molecular mass = 303.3 g/mol) inhibited the growth of *Rhizoctonia solani* (Su et al., 2018).

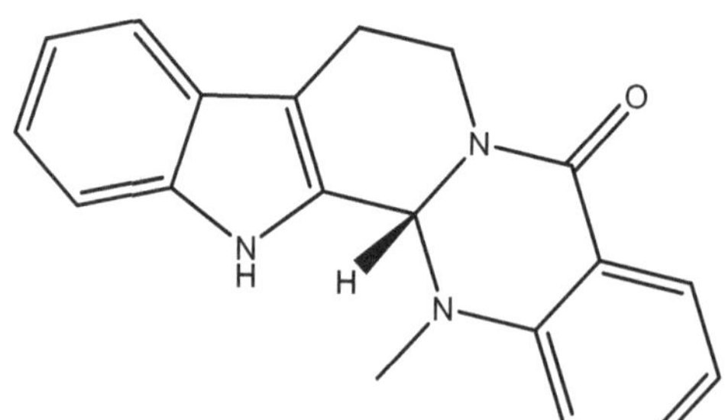

Evodiamine

The β-carboline alkaloid rhetsinine (also known as hydroxyevodiamine; Log D = 2.9 at pH 7.4; molecular mass = 319.3 g/mol) isolated from the fruits of *Euodia rutaecarpa* Benth inhibited the growth of *Xanthomonas oryxae* pv *oryzae and Xanthomonas oryxae* pv *oryzicola* with the EC_{50} values of 1 and 4.5 µg/mL, respectively (Su et al., 2018).

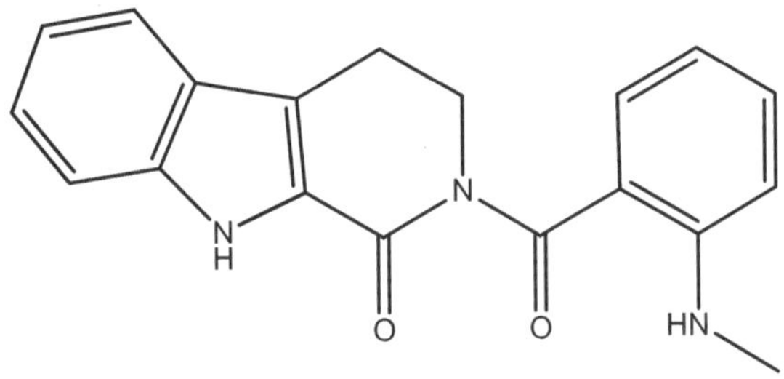

Rhetsinine

Dehydroevodiamine (molecular mass = 301.3 g/mol) from the fruits of *Euodia rutaecarpa* Benth inhibited *Xanthomonas oryxae* pv *oryzae* with an EC_{50} value of 1.4 µg/mL (Su et al., 2018).

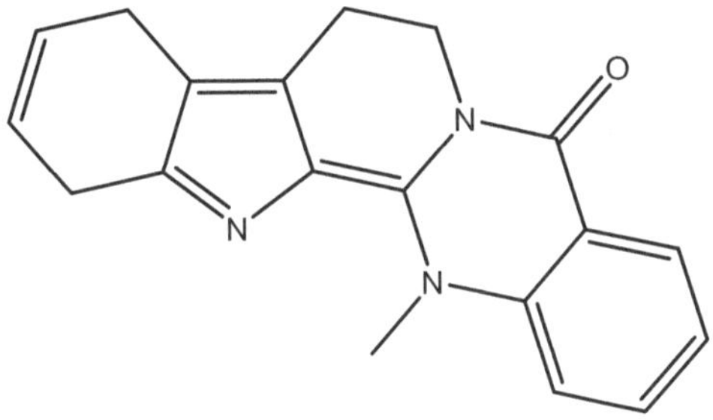

Dehydroevodiamine

Indoloquinazoline-type alkaloids are low molecular weight amphiphilic natural products which being planar intercalate in bacterial and fungal DNA and probably interfere with viral genome replication. Evodiamine inhibited plasmid DNA relaxation in *Escherichia coli* catalyzed by topoisomerase I at 5 µM (Wu et al., 2013). Such alkaloids might be antiviral. Aqueous extract of leaves at the concentration of 250 µg/mL inhibited Human immunodeficiency virus type-1 protease by 37.4% (Xu et al., 1996). Indoloquinazoline-type alkaloid and β-carboline alkaloid often bind to smooth muscle receptors and neuronal receptors (Chiou et al., 1996) and as such have potential as antibiotic/antifungal potentiators by inhibition of as bacterial and/or fungal efflux pumps.

iv. Members of the genus *Euodia* J.R. Forst. & G. Forst. produce antibacterial (Gram-positive) long chain quinolinones. 1-Methyl-2-nonyl-4(1H)-quinolone isolated from the fruits of *Euodia rutaecarpa* Benth (synonym: *Tetradium ruticarpum* (A. Juss.) T.G. Hartley) inhibited the growth of *Staphylococcus aureus* (ATCC 25393), *Staphylococcus epidermidis* (ATCC 12228), and *Bacillus subtilis* (ATCC 6633) with the MIC values of 32, 8, and 16 µg/mL, respectively (Xu et al., 2002). 1-Methyl-2-nonyl-4(1H)-quinolone (Log D = 6.4 at pH 7.4; molecular mass = 285.4 g/mol) isolated from the fruits from the fruits of *Euodia rutaecarpa* Benth inhibited the growth of *Staphylococcus aureus* (ATCC 25393), *Staphylococcus epidermidis* (ATCC 12228), and *Bacillus subtilis* (ATCC 6633) with the MIC values of 4, 4, and 8 µg/mL, respectively (Xu et al., 2002). Being lipophilic, this quinolone is probably antimycobacterial. 1-Methyl-2-[(Z)-5′-pentadecenyl]-4(1H)-quinolone isolated from the fruits of *Euodia rutaecarpa* Benth inhibited the growth of *Staphylococcus aureus* (ATCC 25393), *Staphylococcus epidermidis* (ATCC 12228), and *Bacillus subtilis* (ATCC 6633) with the MIC values of 16, 4, and 16 µg/mL, respectively (Xu et al., 2002).

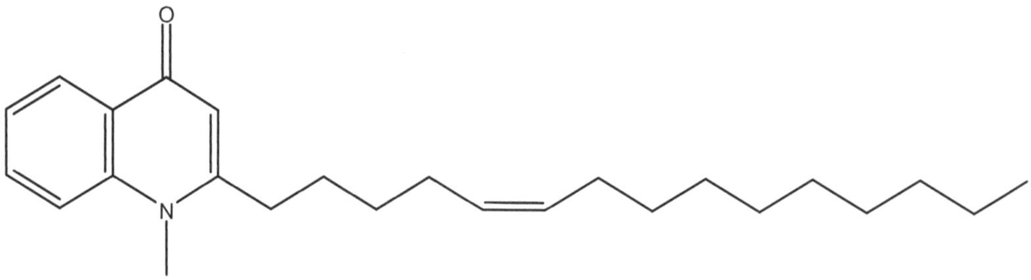

1-Methyl-2-[(*Z*)-5′-pentadecenyl]-4(1H)-quinolone

In a subsequent study, the lipophilic long chain quinolinone alkaloid evocarpine (Log D 5.9 = at pH 7.4; molecular mass = 339.5 g/mol) isolated from the fruits of *Euodia rutaecarpa* Benth inhibited the growth of methicillin-resistant *Staphylococcus aureus* (ATCC 33591) and *Staphylococcus aureus* (ATCC 25923) with the MIC values of 8 and 8 μg/mL, respectively, and was inactive against *Escherichia coli* (NCIMB 11190) (Pan et al., 2014). Evocarpine has a long lipophilic alkyl chain with probable affinity to mycolic acid in *Mycobactera* as well as a quinolinone moiety which probably targets DNA making it an interesting antimycobacterial agent. Evocarpine inhibited the growth of *Mycobacterium fortuitum* (ATCC 6841), *Mycobacterium smegmatis* (ATCC 19429), and *Mycobacterium phlei* (ATCC 19249) with the MIC values of 2, 2, and 2 μg/mL, respectively (Adams et al., 2005).

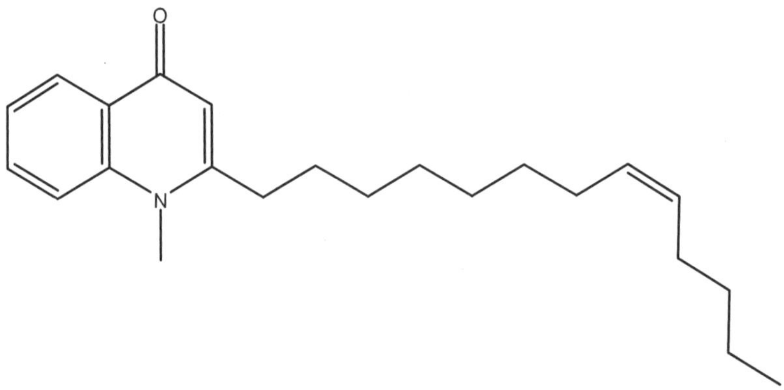

Evocarpine

The limonoids limonin and 6β-acetoxyl-5-epilimonin isolated from the fruits of *Euodia rutaecarpa* Benth at the concentration of 330 μg/mL inhibited the growth of *Escherichia coli* (CMCCB 44113) and *Staphylococcus aureus* (CMCCB 26003) (Liang et al., 2017).

v. *Euodia* J.R. Forst. & G. Forst. belongs to the Subfamily Rutioideae and the tribe Zanthoxyleae as well as the genus *Lunasia* Blanco. *Lunasia amara* Blanco produces the quinoline alkaloids 4-methoxy-2-phenylquinoline and 4-methoxy-2-(3′,4′-methylenedioxy) phenyl quinoline which inhibited *Mycobacterium tuberculosis* H37Rv with the MIC values of 16 μg/mL (rifampicin MIC = 0.1 μg/mL). 7,8-Methylenedioxydictamine from this plant inhibited *Mycobacterium tuberculosis* H37Rv with the MIC values of 33 μg/mL (Aguinaldo et al., 2007).

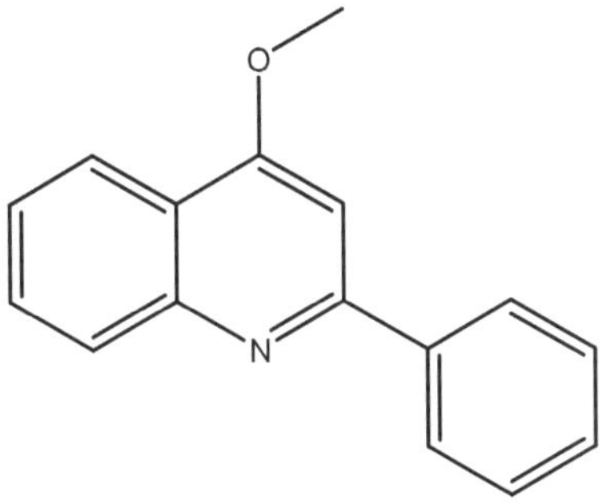

4-Methoxy-2- phenylquinoline

REFERENCES

Adams, M., Wube, A.A., Bucar, F., Bauer, R., Kunert, O. and Haslinger, E., 2005. Quinolone alkaloids from *Evodia rutaecarpa*: a potent new group of antimycobacterial compounds. *International Journal of Antimicrobial Agents*, *26*(3), pp. 262–264.

Aguinaldo, A.M., Dalangin-Mallari, V.M., Macabeo, A.P.G., Byrne, L.T., Abe, F., Yamauchi, T. and Franzblau, S.G., 2007. Quinoline alkaloids from *Lunasia amara* inhibit *Mycobacterium tuberculosis* H37Rv *in vitro*. *International Journal of Antimicrobial Agents*, *29*(6), pp. 744–746.

Chiou, W.F., Liao, J.F., Shum, A.Y.C. and Chen, C.F., 1996. Mechanisms of vasorelaxant effect of dehydroevodiamine: a bioactive isoquinazolinocarboline alkaloid of plant origin. *Journal of Cardiovascular Pharmacology*, *27*(6), pp. 845–853.

Liang, X., Li, B., Wu, F., Li, T., Wang, Y., Ma, Q. and Liang, S., 2017. Bitterness and antibacterial activities of constituents from *Evodia rutaecarpa*. *BMC Complementary and Alternative Medicine*, *17*(1), p. 180.

Maharaj, P.P., Devi, R. and Prasad, S., 2016. Antimicrobial effect of essential oils of some Fijian medicinal plant leaves on pathogenic bacteria. *The South Pacific Journal of Natural and Applied Sciences*, *34*(2), pp. 35–39.

Pan, X., Bligh, S.A. and Smith, E., 2014. Quinolone alkaloids from Fructus Euodiae Show activity against Methicillin-resistant *Staphylococcus aureus*. *Phytotherapy Research*, *28*(2), pp. 305–307.

Su, X.L., Xu, S., Shan, Y., Yin, M., Chen, Y., Feng, X. and Wang, Q.Z., 2018. Three new quinazolines from *Evodia rutaecarpa* and their biological activity. *Fitoterapia*, *127*, pp. 186–192.

Wang, X.X., Zan, K., Shi, S.P., Zeng, K.W., Jiang, Y., Guan, Y., Xiao, C.L., Gao, H.Y., Wu, L.J. and Tu, P.F., 2013. Quinolone alkaloids with antibacterial and cytotoxic activities from the fruits of *Evodia rutaecarpa*. *Fitoterapia*, *89*, pp. 1–7.

Wu, J.Y., Chang, M.C., Chen, C.S., Lin, H.C., Tsai, H.P., Yang, C.C., Yang, C.H. and Lin, C.M., 2013. Topoisomerase I inhibitor evodiamine acts as an antibacterial agent against drug-resistant *Klebsiella pneumoniae*. *Planta Medica*, *79*(1), pp. 27–29.

Xu, H.X., Wan, M., Loh, B.N., Kon, O.L., Chow, P.W. and Sim, K.Y., 1996. Screening of traditional medicines for their inhibitory activity against HIV-1 Protease. *Phytotherapy Research*, *10*(3), pp. 207–210.

8.5.4.7 *Feronia elephantum* Corrêa

Synonyms: *Feronia limonia* (L.) Swingle, *Limonia acidissima* L.; *Schinus limonia* L.

Common names: Elephant apple, wood apple; kathbel (Bangladesh); kramsang (Cambodia); dadhithya, kaith, kapittha, nela veraga, veela marum (India); mafit (Laos); kauista (Indonesia); belinggai (Malaysia); thibin (Myanmar); pamunoan (the Philippines); câń thăng (Vietnam)

Habitat: Cultivated

Distribution: Tropical Asia

Botanical observation: This handsome tree grows to about 9 m tall and is sacred in India. The bark exudes a gum upon incision. The stems are spiny at node, the spines woody, axillary, and about 2 cm long. The leaves are imparipinnate, spiral, and exstipulate. The petiole and rachis are winged and fleshy. The blade presents 2–4 opposite folioles plus a terminal one, which are 2.5–3.5 × 1.2–2.5 cm, glossy, fleshy, with 3–5 pairs of secondary nerves, aromatic, spathulate to obovate, glossy, dark green, about 5 mm long and emarginated at apex. The inflorescences are terminal or axillary panicles of minute reddish flowers. The calyx is minute and includes 6 lobes which are triangular. The corolla presents 6 petals which are elliptic and about 5 mm long. The androecium includes 12 reddish stamens. The ovary is minute. The berry is globose (somewhat like *Aegle marmelos* (L.) Corrêa) light-brown, rought, about 10 cm diameter, woody, and containing seeds in an edible pulp.

Medicinal uses: Diarrhea, dysentery, leucorrhea, pimples, tuberculosis (India); diarrhea, dysentery (India; Bangladesh)

Weak broad-spectrum antibacterial polar extract: Extract of fruits inhibited the growth of *Streptococcus pyogenes* (ATCC 10096), *Staphylococcus aureus* (ATCC 10390), *Proteus vulgaris* (ATCC 27968), *Escherichia coli* (ATCC 12632), and *Klebsiella pneumoniae* (ATCC 10031) with the MIC values of 960, 400, 1000, 440, and 940 μg/mL, respectively (Nithya & Saraswathi, 2010).

Moderate broad-spectrum antibacterial essential oil: Essential oil of leaves inhibited the growth of *Micrococcus luteus* and *Proteus mirabilis*, with the MIC values of 310 and 520 μg/mL, respectively (Joshi et al., 2011).

Moderate broad-spectrum antibacterial flavanols: 5,7,4′-Trihydroxy-6,8-di-(3-methylbut-2-enyl)-flavanone isolated from the stembark inhibited the growth of *Staphylococcus aureus* (NCTC10788), *Escherichia coli* (NCTC 9001), *Enterobacter eloacae* (NCTC 9238), and *Klebsiella aerogenes* (Welcome Res. Lab. CM345) (Rahman & Gray, 2002). (−)-(2S)-5,3′-Dihydroxy-4′-methoxy-6″, 6″-dimethylchromeno-(7,8,2″,3″)-flavanone isolated from the stembark inhibited the growth of *Staphylococcus aureus* (NCTC 10788), *Escherichia coli* (NCTC9001), *Enterobacter eloacae* (NCTC 9238), and *Klebsiella aerogenes* (Welcome Res. Lab. CM345 (Rahman & Gray, 2002).

Moderate broad-spectrum antibacterial prenylated coumarin: Demethylsuberosin isolated from the stembark inhibited the growth of *Staphylococcus aureus* (NCTC10788), *Escherichia coli* (NCTC 9001), *Enterobacter eloacae* (NCTC 9238), and *Klebsiella aerogenes* (Welcome Res. Lab. CM345) (Rahman & Gray, 2002).

Moderate broad-spectrum antibacterial acridone: 1-Hydroxy-3-methoxy-*N*-methyl-acridan-9-one isolated from the stembark inhibited the growth of *Staphylococcus aureus* (NCTC 10788), *Escherichia coli* (NCTC 9001), *Enterobacter eloacae* (NCTC 9238), and *Klebsiella aerogenes* (Welcome Res. Lab. CM345) (Rahman & Gray, 2002).

Moderate broad-spectrum antibacterial sterol: (*E*)-5,8-diepoxy-24(ξ)-methylcholesta-6,22-dien-3-ol isolated from the stembark inhibited the growth of *Staphylococcus aureus* (NCTC10788), *Escherichia coli* (NCTC 9001), *Enterobacter eloacae* (NCTC 9238), and *Klebsiella aerogenes* (Welcome Res. Lab. CM345) with the MIC values of 117, 234, 234, and 234 μg/mL, respectively (Rahman & Gray, 2002).

Moderate broad-spectrum antibacterial benzofuran: 5-(3-Acetoxypropenyl)-2-(4-hydroxy-3-methoxyphenyl)-7-methoxy-2,3-dihydroxybenzofuran-3ylmethyl acetate isolated from the stembark inhibited the growth of *Staphylococcus aureus* (NCTC 10788), *Escherichia coli* (NCTC9001), *Enterobacter eloacae* (NCTC 9238), and *Klebsiella aerogenes* (Welcome Res. Lab. CM345) (Rahman & Gray, 2002).

Moderate broad-spectrum antifungal flavanol: 5,7,4′-Trihydroxy-6,8-di-(3-methylbut-2-enyl)-flavanone isolated from the stembark inhibited the growth of *Candida albicans* (IMI 149007) and *Aspergillus niger* (NCPF 3149) (Rahman & Gray, 2002).

Moderate broad-spectrum antifungal prenylated coumarin: Demethylsuberosin isolated from the stembark inhibited the growth of *Candida albicans* (IMI149007) and *Aspergillus niger* (NCPF3149) (Rahman & Gray, 2002).

Moderate broad-spectrum antifungal sterol: (*E*)-5,8-diepoxy-24(ξ)-methylcholesta-6,22-dien-3-ol isolated from the stembark inhibited the growth of *Candida albicans* (IMI 149007) and *Aspergillus niger* (NCPF 3149) with the MIC values of 117 and 234 μg/mL, respectively (Rahman & Gray, 2002).

Moderate broad-spectrum antifungal benzofuran: 5-(3-Acetoxypropenyl)-2-(4-hydroxy-3-methoxyphenyl)-7-methoxy-2,3-dihydroxybenzofuran-3yl methyl acetate isolated from the stembark inhibited the growth of *Candida albicans* (IMI 149007) and *Aspergillus niger* (NCPF3149) (Rahman & Gray, 2002).

Moderate antifungal (filamentous) essential oil: Essential oil of leaves inhibited the growth of *Penicillium chrysogenum* and *Aspergillus niger* with the MIC values of 200 and 260 µg/mL, respectively (Joshi et al., 2011).

Commentary: Why are flavanones often antiviral *in vitro*?

REFERENCES

Joshi, R.K., Badakar, V.M., Kholkute, S.D. and Khatib, N., 2011. Chemical composition and antimicrobial activity of the essential oil of the leaves of *Feronia elephantum* (Rutaceae) from North West Karnataka. *Natural Product Communications*, *6*(1), pp. 141–143.

Nithya, N. and Saraswathi, U., 2010. *In vitro* antioxidant and antibacterial efficacy of *Feronia elephantum* Correa fruit. *Indian Journal of Natural Products and Resources*, *1*(3), pp. 301–305.

Rahman, M.M. and Gray, A.I., 2002. Antimicrobial constituents from the stem bark of *Feronia limonia*. *Phytochemistry*, *59*(1), pp. 73–77.

8.5.4.8 *Glycosmis pentaphylla* (Retz.) DC.

Synonyms: *Glycosmis arborea* (Roxb.) DC.; *Glycosmis chylocarpa* Wight & Arn.; *Glycosmis quinquefolia* Griff.; *Limonia arborea* Roxb.; *Limonia pentaphylla* Retz.; *Myxospermum chylocarpum* (Wight & Arn.) M. Roem.

Common names: Orange berry; ashaora, atissorah, athishadla, bujuraful, datmajan; tatiang (Bangladesh); ban nimbu, konda giluguru, panal (India); khoei tai, som chuen (Thailand)

Habitat: Forests, roadsides, around villages

Distribution: Pakistan, India, Nepal, Bhutan, Sri Lanka, Myanmar, Bangladesh, Cambodia, las, Vietnam, China, Thailand, Malaysia, Indonesia, the Philippines

Botanical observation: This tree grows to about 5 m tall. The leaves are imparipinnate, spiral, and exstipulate. The blade includes 2–4 folioles which are sessile, sub-opposite plus a terminal one. The folioles are dull or glossy, wavy, crenate, darkish green, 10–25 × 3–7 cm, glossy, somewhat assymetrical, oblong lanceolate, acute at apex, and marked with about 8–12 pairs of secondary nerves. The panicles are axillary or terminal. The calyx develops 5 ovate and minute lobes. The 5 petals are pure white, about 5 mm long, and elliptic. A disc is present. The 10 stamens have white and fleshy filaments. The ovary is minute globose and develops a short style. The berries are globose to somewhat oblate, white to light pink and somewhat transluscent-like, and about 8 mm in diameter.

Medicinal uses: Cough, boils, skin infections (Bangladesh); leucorrhoea (India); fever, jaundice, toothbrush (Bangladesh; India); burns, herpes (Thailand)

Antibacterial halo developed by carbazole-type alkaloid: Glycozolidol (200 µg/mL/well) inhibited the growth of *Staphylococcus aureus*, *Bacillus firmis*, *Sarcina lutea*, *Agrobacterium tumefaciens*, and *Proteus vulgaris* (Bhattacharyya et al., 1985).

Moderate antibacterial (Gram-negative) quinolinone: 3-(3′, 3′-Dimethylallyl)-4,8-dimethoxy-*N*-methylquinolin-2-one, the furoquinoline skimmianine and 1-hydroxy-3,4-dimethoxy-10-methylacridan-9-one inhibited the growth of *Escherichia coli* and *Salmonella typhimurium* (TISTR 292) (Sripisut et al., 2012).

Moderate antibacterial (Gram-negative) furoquinoline: Skimmianine and 1-hydroxy-3,4-dimethoxy-10-methylacridan-9-one inhibited the growth of *Escherichia coli* and *Salmonella typhimurium* (TISTR 292) (Sripisut et al., 2012).

Moderate antibacterial (Gram-negative) acridone: 1-Hydroxy-3,4-dimethoxy-10-methylacridan-9-one inhibited the growth of *Escherichia coli* and *Salmonella typhimurium* (TISTR 292) (Sripisut et al., 2012).

Commentaries

i. Members of the genus *Glycosmis* Corrêa upon microbial infection produce carbazole-type alkaloid phytoalexins of which glycozolidol (Pacher et al., 2001). Once could suggest that isolating antimicrobial principles from wounded or infected plants might be a mean to find antimicrobial leads.

ii. (*E*)-3-(3-hydroxymethyl-2-butenyl)-7-(3-methyl-2-butenyl)-1H-indole, glybomine B, and glycoborinine isolated from the stems and leaves of *Glycosmis montana* Pierre inhibited the replication of the Human immunodeficiency virus type-1 (IIIB strain) in C8166 cells with IC_{50} values of 1.1, 9.7, and 4.4 μg/mL, respectiverly and selectivity indexes of 11.6, 1.4, and 4.6 (Wang et al., 2005).

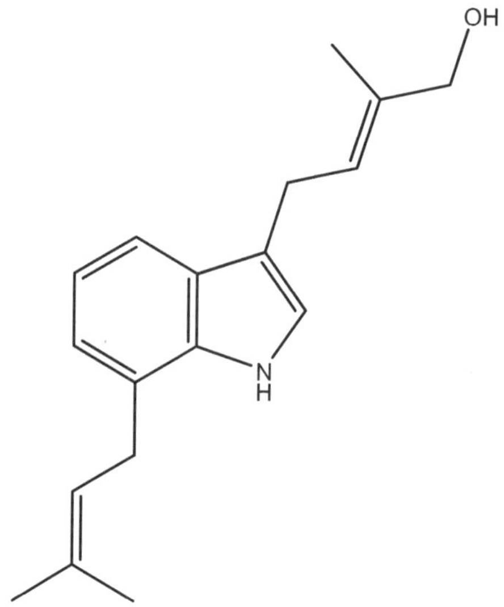

(*E*)-3-(3-hydroxymethyl-2-butenyl)-7-(3-methyl-2-butenyl)-1H-indole

iii. Members of the genus *Glycosmis* Corrêa produce unusual series of sulfur-containing propanamides of which (+)-*S*-deoxydihydroglyparvin isolated from the stems and leaves of *Glycosmis parva* Craib which inhibited the replication of the Human immunodeficiency virus type-1 (KOS) and Human immunodeficiency virus type-2 (Baylor186) in Vero cells with the EC_{50} values of 12.5 and 18.7 μg/mL (CC_{50}: 37.5 μg/mL) (Chansriniyom et al., 2009).

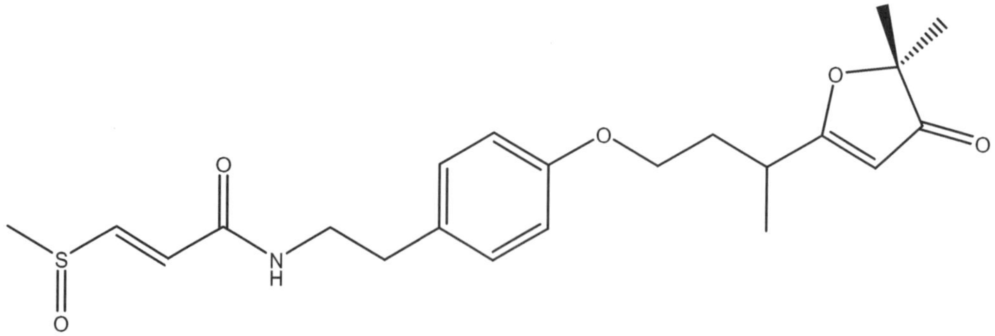

iv. *Glycosmis arborea* (Roxb.) A. DC. is used in Bangladesh (local name: *Bonjamir*) to treat flu, cough, and fever.

REFERENCES

Bhattacharyya, P., Chakrabartty, P.K. and Chowdhury, B.K., 1985. Glycozolidol, an antibacterial carbazole alkaloid from *Glycosmis pentaphylla*. *Phytochemistry*, *24*(4), pp. 882–883.

Chansriniyom, C., Ruangrungsi, N., Lipipun, V., Kumamoto, T. and Ishikawa, T., 2009. Isolation of acridone alkaloids and N-[(4-monoterpenyloxy) phenylethyl]-substituted sulfur-containing propanamide derivatives from *Glycosmis parva* and their anti-herpes simplex virus activity. *Chemical and Pharmaceutical Bulletin*, *57*(11), pp. 1246–1250.

Pacher, T., Bacher, M., Hofer, O. and Greger, H., 2001. Stress induced carbazole phytoalexins in Glycosmis species. *Phytochemistry*, *58*(1), pp. 129–135.

Sripisut, T., Ritthiwigrom, T., Promgool, T., Yossathera, K., Deachathai, S., Phakhodee, W., Cheenpracha, S. and Laphookhieo, S., 2012. Glycopentaphyllone: the first isolation of hydroperoxyquinolone from the fruits of *Glycosmis pentaphylla*. *Phytochemistry Letters*, *5*(2), pp. 379–381.

Wang, J., Zheng, Y., Efferth, T., Wang, R., Shen, Y. and Hao, X., 2005. Indole and carbazole alkaloids from *Glycosmis montana* with weak anti-HIV and cytotoxic activities. *Phytochemistry*, *66*(6), pp. 697–701.

8.5.4.9 *Melicope bonwickii* (F. Muell.) T.G. Hartley

Synonym: *Euodia bonwickii* F. Muell.

Common names: Abal, awal (Indonesia); kamal (the Philippines)

Habitat: Forests

Distribution: Indonesia, the Philippines, Papua New Guinea, Australia

Botanical observation: This handsome tree grows up to 40 m tall. The leaves are trifoliolate, opposite, and exstipulate. The folioles are obovate, somewhat asymmetrical, coriaceous, marked with 11–15 pairs of secondary nerves, 10–30 × 5–15 cm, glossy, tapered at base, and acuminate at apex. The cymes are cauliflorous, densely flowered, and up to about 10 cm long. The calyx is 4-lobed. The corolla includes 4 minute white oblong petals. A disc is present. The androecium presents 4 stamens longer than the petals. The follicles are obovoid, 4–6 mm long, woody, rough and open to disclose black and glossy seeds.

Medicinal use: Dysentery (Indonesia)

Commentaries

i. The genus *Melicope* J.R. Forst. & G. Forst. and the genus *Euodia* J.R. Forst. & G. Forst. are very similar and are examples of taxa whose division (based on minute difference) uncessesarily complicate things. Also, some taxononists tend to keep modifying the name of plants. Some advocate the use of genetic material to identify plants and agreeing to such idea would limit the world of botanists to a few labs equipped with expensive machines. Botany is all about the sensible and simple conscious observation of plants with common sense and logic.

ii. Members of the genus *Melicope* J.R. Forst. & G. Forst. are used for the treatment of microbial infections in Asia and the Pacific. *Melicope denhamii* (Seem.) T.G. Hartley is used in the Solomon Islands to treat boils. Essential oil of leaves of *Melicope denhamii* (Seem.) T.G. Hartley at 1:1 dilution inhibited the growth of *Bacillus subtilis* (MTCC No. 441) and *Escherichia coli* (MTCC No. 443) with inhibition zone diameters of 21 and 10 mm, respectively (George et al., 2015).

iii. In Taiwan, *Melicope lunu-ankenda* (Gaertn.) T.G. Hartley is used to treat colds and in India it is used against fever (local name: *Kambili*). In Papua New Guinea *Melicope triphylla* (Lam.) Merr. (local names: *Kupun*, *Wajo*) affords a treatment for diarrhoea, and *Melicope xanthoxyloides* (F. Muell.). T.G. Hartley is used (local name: *Gandembo*) to treat cough.

The flavone tamarixetin 3-robinobioside (molecular mass = 624.5 g/mol) isolated from the leaves of *Melicope pteleifolia* (Champ. ex Benth.) T.G. Hartley inhibited the neuraminidase activity of H1N1, H9N2, H1N1 (wild type), and H274Y with IC_{50} values of 24.9, 23.1, 26.6, and 40.6 µM, respectively (Nguyen et al., 2016). Tamarixetin 3-robinobioside inhibited the cytopathic effects of H1N1 (A/PR/8/34 virus) in Madin-Darby canine kidney cells at a concentration of 60 µM (Nguyen, et al., 2016). This plant is used in China for the treatment of rheumatic arthritis (local name: *Hā diao liān*) and this adds more interest for its possible development as an anti-influenza phytomedicaton (if not toxic) since inflammation is part of the pathophysiology of flu. From this plant, a cyclopeptide antibiotic is produced. Although antimicrobial cyclopeptides are not covered in the present work, it must be grasped that cyclopeptides from medicinal plants are often produced by symbiotic bacteria or fungi, extracted by biochemical techniques, and of interest for the discovery of antimicrobial drugs. Melicopteline C from *Melicope pteleifolia* (Champ. ex Benth.) T.G. Hartley inhibited the Influenza A virus (H1N1) with an EC_{50} value of 2.5 µM (Lee et al., 2020).

iv. Antibacterial and antifungal acetophenones are produced by plants in the genus *Melicope* J.R. Forst. & G. Forst.: xanthoxylin (Log D = 2.2 at pH 7.4; molecular mass = 196.2 g/mol) inhibited the growth of *Candida albicans* and *Penicillium expansum* with the minimum inhibitory masses of 25 and 15 µg/TLC respectively (Simonsen et al., 2004).

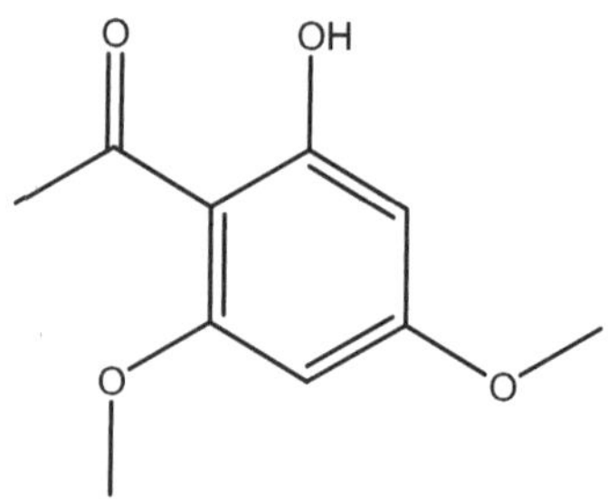

Xanthoxylin

Meliviticine A isolated from *Melicope viticina* (Wall. ex Kurtz) T.G. Hartley (local name in China: *Dan ye mi zhu yu*) inhibited the growth of *Micrococcus lysideikticus*, *Bacillus subtilis*, *Bacterium paratyphosum* B, methicillin-resistant *Staphylococcus aureus*, *Salmonella typhi*, and *Pseudomonas aeruginosa* with the MIC values of 50, 50, 25, 50, 50, and 50 µg/mL, respectively (Li et al., 2019). Meliviticine A inhibited the growth of *Verticillum dahlia*, *Alternaria alternata*, *Rhizoctonia solani*, *Scleritina scelrotiorum*, *Phytophthora parasitica*, and *Gibberella saubinetii* with the MIC values of 50, 25, 100, 100, 25, and 50 µg/mL, respectively (Li et al., 2019). Respective (+) and (–) isomers of meliviticine A had the same antibacterial and antifungal activities implying that the bacterial and fungal targets are multiple and not via a specific enzyme inhibition or an enzyme with broad substrates. Natural products from plants can differ from antibiotics, antifungals, and antivirals by having not only multiple targets but also having mode of actions depending on their concentrations. This is sensible as these products are naturally synthetized by plant to control the growth of microbes growing on them as well as inhibiting their efflux pumps and warding off other life forms threatening their survival. Antibiotics produced by bacteria and fungi are specifically designed for microbial homeostasis.

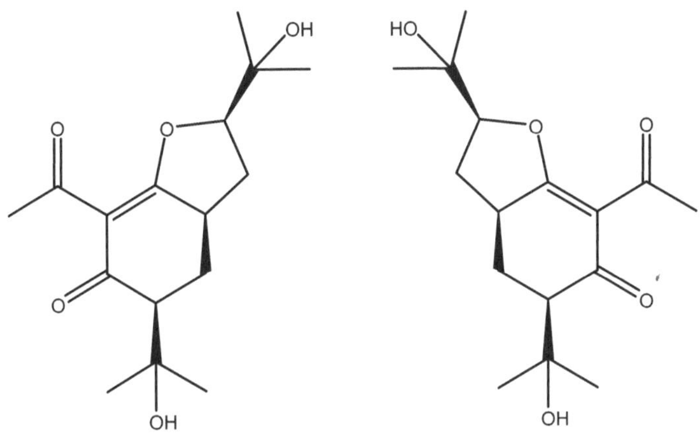

(+/-) Meliviticine A

The phloroglucinols sessiliflorene and sessiliflorol A isolated from *Melicope sessiliflora* (Merr.) T.G. Hartley inhibited the replication of the Herpes simplex virus type-1 with the IC_{50} values of 0.3 and 3 μM, respectively and the Herpes simplex virus type-2 with the IC_{50} values of 22.3 and 10.4 μM, respectively (Tada et al., 1990).

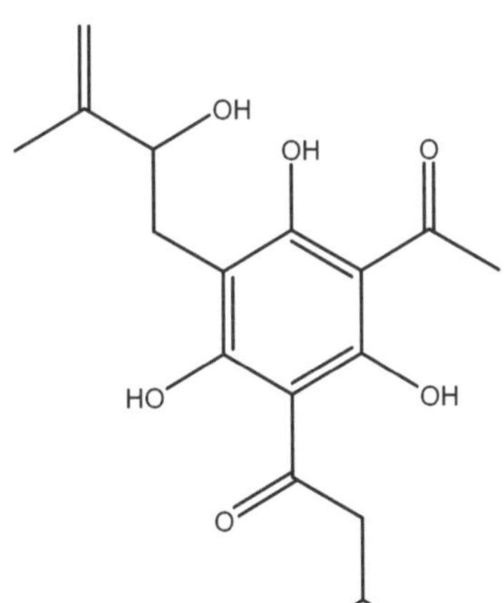

Sessiliflorene

v. From the fruits of *Melicope latifolia* (DC.) T.G. Hartley, the coumarins *O*-methyloktadrenolon, alloxanthoxyletin, and the quinoline alkaloid *N*-methylflindersine (Log D = 2.9 at pH 7.4; and molecular mass = 241.2 g/mol) inhibited the replication of Hepatitis C virus in Huh7it-1 cells with IC_{50} values of 10.9, 21.7, and 3.8 μg/mL, respectively and selectivity indexes of 5.8, above 24.3, and 25.5 respectively (Widyawaruyanti et at., 2021).

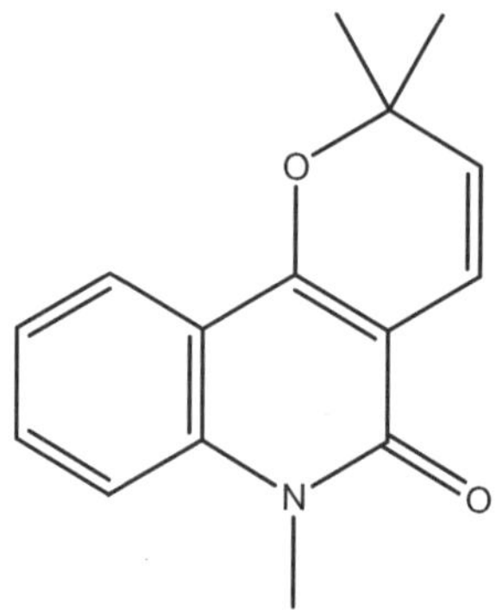

N-methylflindersine

REFERENCES

George, S., Nair, S.A., Venkataraman, R. and Baby, S., 2015. Chemical composition, antibacterial and anticancer activities of volatile oil of *Melicope denhamii* leaves. *Natural Product Research*, *29*(20), pp. 1959–1962.

Lee, B.W., Quy Ha, T.K., Park, E.J., Cho, H.M., Ryu, B., Doan, T.P., Lee, H.J. and Oh, W.K., 2020. Melicopteline A–E, unusual cyclopeptide alkaloids with antiviral activity against Influenza A virus from *Melicope pteleifolia*. *The Journal of Organic Chemistry*, *86*, pp. 1437–1447.

Li, W., Rao, L., Liu, Y., He, Q., Fan, Y., You, Y.X., Su, Y., Hu, F., Xu, Y.K., Lin, B. and Yin, S., 2019. (±)-Melivitícines A and B: rearranged prenylated acetophenone derivatives from *Melicope viticina* and their antimicrobial activity. *Bioorganic Chemistry*, *90*, p. 103099.

Nguyen, N.H., Quy Ha, T.K., Choi, S., Eum, S., Lee, C.H., Chinh, V.T. and Oh, W.K., 2016. Chemical constituents from *Melicope pteleifolia* leaves. *Phytochemistry*, *130*, pp. 291–300.

Simonsen, H.T., Adsersen, A., Bremner, P., Heinrich, M., Wagner Smitt, U. and Jaroszewski, J.W., 2004. Antifungal constituents of *Melicope borbonica*. *Phytotherapy Research: An International Journal Devoted to Pharmacological and Toxicological Evaluation of Natural Product Derivatives*, *18*(7), pp. 542–545.

Tada, M., Takakuwa, T., Nagai, M. and Yoshii, T., 1990. Antiviral and antimicrobial activity of 2, 4-diacylphloroglucinols, 2-acylcyclohexane-1, 3-diones and 2-carboxamidocyclo-hexane-1, 3-diones. *Agricultural and Biological Chemistry*, *54*(11), pp. 3061–3063.

Widyawaruyanti, A., Tanjung, M., Permanasari, A.A., Saputri, R., Tumewu, L., Adianti, M., Aoki-Utsubo, C., Hotta, H., Hafid, A.F. and Wahyuni, T.S., 2021. Alkaloid and benzopyran compounds of *Melicope latifolia* fruit exhibit anti-hepatitis C virus activities. *BMC Complementary Medicine and Therapies*, *21*(1), pp. 1–9.

8.5.4.10 *Murraya paniculata* (L.) Jack

Synonyms: *Murraya exotica* L.

Common names: China box, orange jessamine; kamini (Bangladesh); shan fan (China); konji, madhukamini, maramulla (India); kemuning (Indonesia; Malaysia); makay, yuzana (Myanmar); bajardanta (Nepal); sika (Papua New Guinea); kamuning (the Philippines); kaeo (Thailand); cut gie, nget koi (Vietnam)

Habitat: Gardens, parks

Distribution: India to the Pacific

Botanical description: This handsome tree grows up to 5 m tall. The leaves are pinnate, alternate, and exstipulate. The blade is up to 11 cm long and consists of 2–7 pairs of folioles which are sessile, alternate, spathulate, asymmetrical, 2–9 × 1.5–6 cm, dark green, somewhat coriaceous, obscurely oil gland-dotted, and glossy. The inflorescences are arranged in terminal or axillary panicles of showy, pure white and heavenly fragrant flowers. The 5 sepals are minute and triangular. The 5 petals are about 2 cm long, elliptic to spathulate, oil gland-dotted, and pure white. The androecium comprises 10 stamens with slightly flattened filaments around a well-developed nectary disc.

The style is slender and the stigma globular. The drupes are about 1 cm long, marked with oil-glands, green at first then reddish.

Medicinal uses: Cough (Papua New Guinea); cuts, wounds (Bangladesh); dysentery (China, India, Nepal, the Philippines); venereal diseases, toothache (Nepal); diarrhea, leprosy (India); fever (India, Nepal).

Anticandidal halo developed by mid-polar extract: Chloroform extract of leaves (1 mg/6 mm disc) inhibited the growth of *Candida albicans* with the inhibition zone diameter of 19.7 mm (Phongpaichit et al., 2005).

Antifungal (filamentous) essential oil: Essential oil of leaves (300 µL on agar) inhibited by 91.2% the mycelial growth of *Sclerotinia sclerotiorum* (BRM 29673) (Silva et al., 2019).

Moderate broad-spectrum antibacterial essential oil: Essential oil of ripe fruits inhibited the growth of *Streptococcus salivarius* (ATCC 25975), *Streptococcus sobrinus* (ATCC 33478), *Streptococcus mutans* (ATCC 25175), *Streptococcus mitis* (ATCC 49456), and *Streptococcus sanguinis* (ATCC 10556) with the MIC values of 250, 400, 250, 100, and 200 µg/mL, respectively (Silva et al., 2020).

Moderate antimycobacterial essential oil: Essential oil of ripe fruits inhibited the growth of *Mycobacterium tuberculosis* (H37Rv) (ATCC 27294), *Mycobacterium kansasii* (ATCC 12478), and *Mycobacterium avium* (ATCC 25291) with the MIC values of 500, 250, and 2000 µg/mL, respectively (Silva et al., 2020).

Commentaries

i. The plant yields prenylated coumarins as well as methoxylated flavones and quinoline alkaloids including edulitine (Imai et al., 1989). The antimicrobial mode of action of quinolines involves at least in part intercalation with DNA (Hanawa et al., 2004).

ii. In the Subfamily Aurantioideae, tribe Clauseneae *Micromelum integerrimum* Wight. & Arn. is used for the treatment of tuberculosis in India as well as *Micromelum pubescens* Bl. in Indonesia and Malaysia. *Micromelum hirsutum* Merr. brings to being micromolide (also known as (–)-Z-9-octadecene-4-olide), and the carbazole-type alkaloids lansine and 3-formyl-6-methoxycarbazole inhibited the growth of *Mycobacterium tuberculosis* (H37R*v*) with the MIC values of 1.5, 14.3, and 15.6 µg/mL, respectively (Ma et al., 2005). Micromolide is produced by some insects as antimicrobial agents to protect the growth of larvae (Herzner et al., 2013). Preclinical trials of micromolide (or hemisynthetic derivatives) as antimycobacterial agent are warranted.

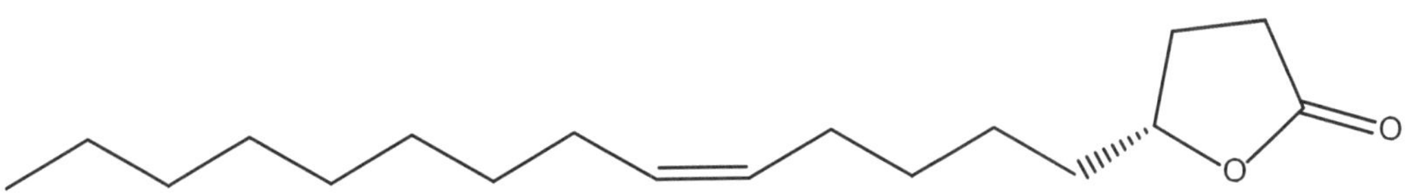

Micromolide

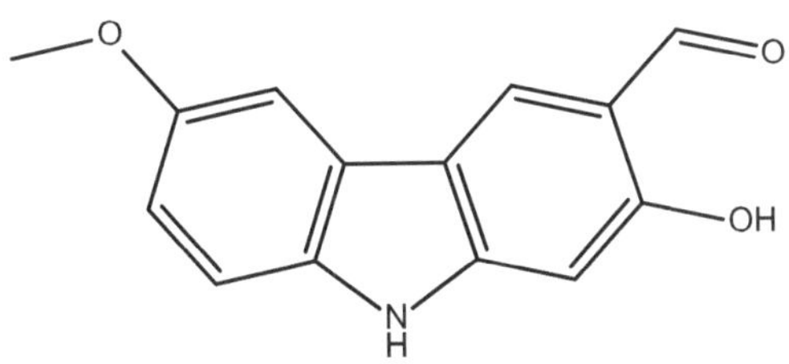

Lansine

Micromelum falcatum (Lour.) Tanaka is used to treat abscesses in China (local name: *Dan gan*).

The coumarins 7-methoxy-8-(4′-methyl-3′-furanyl)coumarin, micromarin B, and isomicromelin isolated from the fruits of *Micromelum falcatum* (Lour.) Tanaka inhibited *Pythium insidiosum* with inhibition zones of 15.5, 19.2, and 21 mm, respectively (20 μL of a 20 mg/100 μL/disc) (Suthiwong et al., 2014).

iii. The leaves of *Murraya koenigii* (L.) Spreng. are used to flavor the delicious Indian curry, and the plant affords in Bangladesh (local name: *Kari patta*) a medicine for fever, diarrhea, skin infection. In India, this aromatic treelet affords a treatment for cholera and dysentery (local name: *Kari vepa*). Essential oil of leaves of *Murraya koenigii* (L.) Spreng. inhibited the growth of *Listeria innocua* with the MIC value of 600 μg/mL (Erkan et al., 2012). The carbazole-type alkaloid murrayamine inhibited the growth of *Staphylococcus aureus* (ATCC 29213) and *Bacillus cereus* (IIIM 25) with the IC_{50} values of 11.7 and 23.2 μM (Nalli et al., 2016). Girinimbine (Log D = 4.9 at pH 7.4; molecular mass = 263.3 g/mol) isolated from *Murraya paniculata* (L.) Jack. inhibited *Bacillus cereus* (IIIM 25) with the IC_{50} values of 3.4 μM (Nalli et al., 2016). Girinimbine inhibited the growth of *Staphylococcus aureus, Bacillus subtilis, Escherichia coli, Proteus vulgaris, Aspergillus niger*, and *Candida albicans* (Joshi et al., 2018).

Koenimbine isolated from *Murraya koenigii* (L.) Spreng. inhibited *Staphylococcus aureus* (ATCC 29213) and *Bacillus cereus* (IIIM 25) with IC_{50} values of 17 and 22.5 μM (Nalli et al., 2016).

The carbazole-type alkaloids koenine, koenigine, and mahanine, isolated from the leaves of *Murraya koenigii* (L.) Spreng. inhibited the growth of *Streptococcus aureus* and *Klebsiella pneumoniae* with the MIC_{90} values ranging from 3.12 to 12.5 μg/mL (Kanamycin MIC_{90} 12.5 μg/mL). Koenigine was active against variety of *Candida* spp. with MIC_{90} ranging from 12.5–100 μg/mL (Joshi et al., 2018). 3-Formyl-1-1methoxycarbazole isolated from the stem bark of *Murraya koenigii* (L.) Spreng. inhibited the growth of *Staphylococcus aureus, Bacillus subtilis, Escherichia coli, Proteus vulgaris, Aspergillus niger*, and *Candida albicans* with the MIC values of 325, 50, 50, 50, 25, and 100 μg/mL, respectively (Joshi et al., 2018). 3,3′-[Oxybis(methylene)]bis(9-methoxy-9*H*-carbazole isolated from the stem bark of *Murraya koenigii* (L.) Spreng., inhibited the growth of *Staphylococcus aureus, Escherichia coli, Proteus vulgaris, Aspergillus niger*, and *Candida albicans* with the MIC values of 50, 25, 6.2, 25, and 25 μg/mL, respectively (Rahman & Gray, 2005).

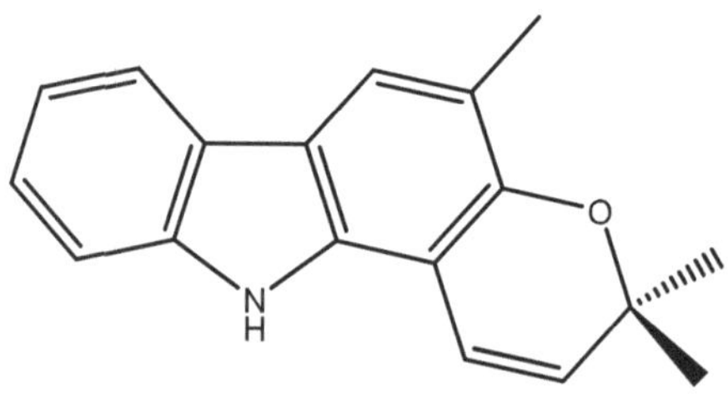

Girinimbine

REFERENCES

Erkan, N., Tao, Z., Rupasinghe, H.P., Uysal, B. and Oksal, B.S., 2012. Antibacterial activities of essential oils extracted from leaves of *Murraya koenigii* by solvent-free microwave extraction and hydro-distillation. *Natural Product Communications*, *7*(1), pp. 121–124.

Hanawa, F., Fokialakis, N., Skaltsounis, A.L., 2004. Photo-activated DNA binding and antimicrobial activities of furoquinoline and pyranoquinolone alkaloids from rutaceae. *Planta Medica*, *70*, pp. 531–535.

Herzner, G., Schlecht, A., Dollhofer, V., Parzefall, C., Harrar, K., Kreuzer, A., Pilsl, L. and Ruther, J., 2013. Larvae of the parasitoid wasp *Ampulex compressa* sanitize their host, the American cockroach, with a blend of antimicrobials. *Proceedings of the National Academy of Sciences*, *110*(4), pp. 1369–1374.

Imai, F., Itoh, K., Kishibuchi, N., Kinoshita, T. and Sankawa, U., 1989. Constituents of the root bark of *Murraya paniculata* collected in Indonesia. *Chemical and Pharmaceutical Bulletin*, *37*(1), pp. 119–123.
Joshi, T., Jain, T., Mahar, R., Singh, S.K., Srivastava, P., Shukla, S.K., Mishra, D.K., Bhatta, R.S., Banerjee, D. and Kanojiya, S., 2018. Pyranocarbazoles from *Murraya koenigii* (L.) Spreng. as antimicrobial agents. *Natural Product Research*, *32*(4), pp. 430–434.
Ma, C., Case, R.J., Wang, Y., Zhang, H.J., Tan, G.T., Van Hung, N., Cuong, N.M., Franzblau, S.G., Soejarto, D.D., Fong, H.H. and Pauli, G.F., 2005. Anti-tuberculosis constituents from the stem bark of *Micromelum hirsutum*. *Planta Medica*, *71*(3), p. 261.
Nalli, Y., Khajuria, V., Gupta, S., Arora, P., Riyaz-Ul-Hassan, S., Ahmed, Z. and Ali, A., 2016. Four new carbazole alkaloids from *Murraya koenigii* that display anti-inflammatory and anti-microbial activities. *Organic & Biomolecular Chemistry*, *14*(12), pp. 3322–3332.
Nutan, M.T.H., Hasnat, A. and Rashid, M.A., 1998. Antibacterial and cytotoxic activities of *Murraya koenigii*. *Fitoterapia*, *69*, pp. 173–175.
Phongpaichit, S., Subhadhirasakul, S. and Wattanapiromsakul, C., 2005. Antifungal activities of extracts from Thai medicinal plants against opportunistic fungal pathogens associated with AIDS patients. *Mycoses*, *48*(5), pp. 333–338.
Rahman, M.M. and Gray, A.I., 2005. A benzoisofuranone derivative and carbazole alkaloids from *Murraya koenigii* and their antimicrobial activity. *Phytochemistry*, *66*(13), pp. 1601–1606.
Silva, F.F.A.D., Alves, C.C.F., Oliveira Filho, J.G.D., Vieira, T.M., Crotti, A.E.M. and Miranda, M.L.D., 2019. Chemical constituents of essential oil from *Murraya paniculata* leaves and its application to *in vitro* biological control of the fungus *Sclerotinia sclerotiorum*. *Food Science and Technology*, *39*, pp. 413–417.
Silva, F.F.A.D., Fernandes, C.C., Santiago, M.B., Martins, C.H.G., Vieira, T.M., Crotti, A.E.M. and Miranda, M.L.D., 2020. Chemical composition and *in vitro* antibacterial activity of essential oils from *Murraya paniculata* (L.) Jack (Rutaceae) ripe and unripe fruits against bacterial genera Mycobacterium and Streptococcus. *Brazilian Journal of Pharmaceutical Sciences*, *56*. doi:10.1590/s2175-97902019000418371.
Suthiwong, J., Sriphana, U., Thongsri, Y., Promsuwan, P., Prariyachatigul, C. and Yenjai, C., 2014. Coumarinoids from the fruits of *Micromelum falcatum*. *Fitoterapia*, *94*, pp. 134–141.

8.5.4.11 *Ruta angustifolia* (L.) Pers.

Synonym: *Ruta chalepensis* var. *angustifolia* (Pers.) Willk. & Lange

Common names: Garuda (Malaysia), godong minggu (Indonesia); cu li huong (Vietnam)

Habitat: Cultivated

Distribution: Tropical Asia

Botanical observation: This fragrant herb native to Southern Europe grows up to 1 m tall. The stem is terete and glabrous. The leaves are sessile, minute, oblong, dotted with oil glands, and are about 1 cm × 0.5 cm. The cymes are axillary. The 4 triangular sepals are 2 mm long. The 4 petals are obovate, of a peculiar yellow, ciliate, and up to 1 cm long. A disc is present. The 8 stamens are as long as the petals. The ovary is conspicuous, lobed, somewhat conical, and oil gland-dotted. The capsules are oil-gland dotted and 4-lobed.

Medicinal use: Jaundice (Indonesia)

Strong antiviral (enveloped monopartite single-stranded (+) RNA) furanocoumarin: Chalepin isolated from the leaves inhibited the replication of the Hepatitis C virus (J6/JFH1-P47 strain) with the IC_{50} values of 1.7 μg/mL and the selectivity index of 8.2 via a postentry mechanism and decreased levels of virus RNA replication and viral protein synthesis (Wahyuni et al., 2014).

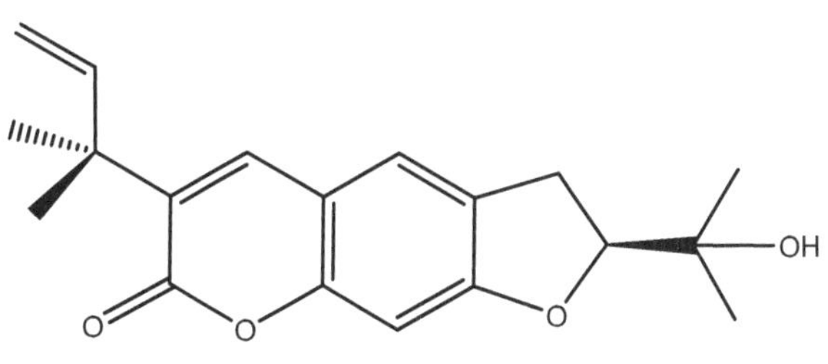

Chalepin

Strong antiviral (enveloped monopartite single-stranded (+) RNA) furanoquinoline: γ-Fagarine and kokusaginine isolated from the leaves inhibited the replication of the Hepatitis C virus (J6/JFH1-P47 strain) with the IC_{50} values of 20.4 and 6.4 μg/mL, respectively, and selectivity indexes above 1.5 and 4.7, respectively (Wahyuni et al., 2014).

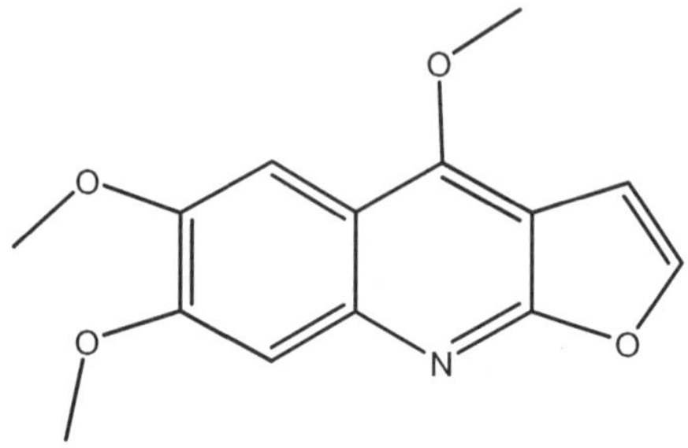

Kokusaginine

Strong antiviral (enveloped monopartite single-stranded (+) RNA) acridone: Arborinine isolated from the leaves inhibited the replication of the Hepatitis C virus (J6/JFH1-P47 strain) with the IC_{50} value of 6.4 μg/mL and a selectivity index above 1.5 (Wahyuni et al., 2014).

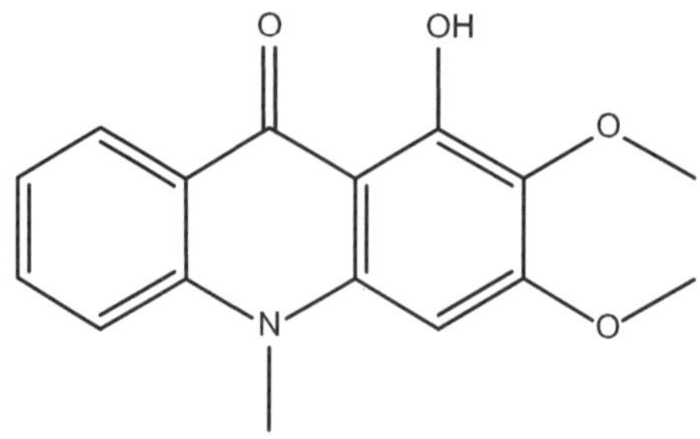

Arborinine

Strong antiviral (enveloped monopartite single-stranded (+) RNA) quinolinone: Pseudane IX isolated from the leaves inhibited the replication of the Hepatitis C virus (J6/JFH1-P47 strain) with the IC_{50} value of 1.4 μg/mL and a selectivity index of 18.6 via a post entry mechanism and decreased levels of virus RNA replication and viral protein synthesis (Wahyuni et al., 2014).

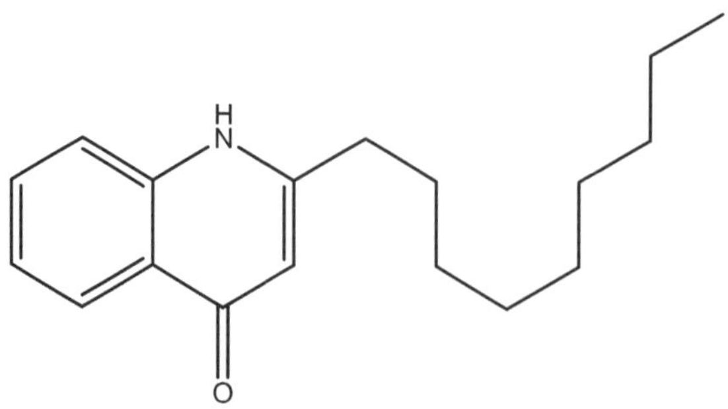

Pseudane IX

Ruta angustifolia Pers.

Commentaries

i Here are examples of natural products with different chemical frameworks from a same plant with activity against the same microbe. This principles might have activity against Coronaviruses including the Severe acute respiratory syndrome-associated coronavirus.

ii Methanol extract of aerial parts of *Ruta chalepensis* L. (6 mm disc impregnated with 5 mg/mL solution) inhibited the growth of *Escherichia coli*, *Enterococcus faecalis*, *Klebsiella pneumoniae*, *Klebsiella oxytocoa*, *Proteus mirabilis*, *Staphylococcus aureus*, and *Pseudomonas aeruginosa* with inhibition zone diameter of 23, 17, 13, 18, 20, 18, and 15 mm, respectively (Priya et al., 2009). Methanol extract of leaves of *Ruta chalepensis* L. inhibited the growth of *Aspergillus flavus*, *Aspergillus niger*, and *Candida albicans* with inhibition zone diameters of 18, 16, and 14 mm, respectively (15 μL/6 mm disc) (Aouadhi et al., 2013). Essential oil from the aerial parts of *Ruta chalepensis* L. inhibited the growth of *Candida albicans* (ATCC 10231) and *Trichophyton rubrum* (SNB-TR1) with the MIC values of 512 and 512 μg/mL, respectively (Khoury et al., 2014). Essential oil of *Ruta chapelensis* inhibited the growth of *Streptococcus mutans* with the MIC of 2.5 μg/mL (Miller et al., 2015).

REFERENCES

Aouadhi, C., Ghazghazi, H., Hamrouni, S., Hasnaoui, B. and Maaroufi, A., 2013. *In vitro* antifungal activity of the essential oil and the methanolic extract of *Ruta chalepensis*. *Archives de l'Institut Pasteur de Tunis*, *90*(1–4), p. 39.

Khoury, M., Stien, D., Ouaini, N., Eparvier, V., Arnold Apostolides, N. and El Beyrouthy, M., 2014. Chemical composition and antimicrobial activity of the essential oil of *Ruta chalepensis* L. growing wild in Lebanon. *Chemistry & Biodiversity*, *11*(12), pp. 1990–1997.

Miller, A.B., Cates, R.G., Lawrence, M., Soria, J.A.F., Espinoza, L.V., Martinez, J.V. and Arbizú, D.A., 2015. The antibacterial and antifungal activity of essential oils extracted from Guatemalan medicinal plants. *Pharmaceutical Biology*, *53*(4), pp. 548–554.

Priya, P.S., Sasikumar, J.M. and Gowsigan, G., 2009. Antibacterial activity of methanol extract of *Ruta chalapensis* (L), *Quercus infectoria* (Oliver) and *Canthium parviflorum* (Lam). *Ancient Science of Life*, *29*(2), p. 28.

Wahyuni, T.S., Widyawaruyanti, A., Lusida, M.I., Fuad, A., Fuchino, H., Kawahara, N., Hayashi, Y., Aoki, C. and Hotta, H., 2014. Inhibition of hepatitis C virus replication by chalepin and pseudane IX isolated from *Ruta angustifolia* leaves. *Fitoterapia*, *99*, pp. 276–283.

8.5.4.12 *Toddalia asiatica* (L.) Lam.

Synonyms: *Paullinia asiatica* L.; *Toddalia aculeata* Pers.; *Toddalia floribunda* Wall.; *Toddalia tonkinensis* Guillaumin

Common names: Lopez root; kanchana, kaka toddali, mirapagandra (India); duri keng-keng (Indonesia); dahanbaz-dahan (Iran); pem paks (Laos); akar kucing (Malaysia); kaboat (the Philippines); cam núi (Vietnam)

Habitat: Forests, seashores

Distribution: India, Sri Lanka, Southeast Asia, China

Botanical examination: This robust climber grows up to a length of 20 m. The main stem can reach 10–15 cm in diameter. The stems are spiny. The leaves are spiral, 3-foliolate, and exstipulate. The petioles are 2.5–4 cm long. The folioles are glossy, very dark green above, 3–10 × 1–4 cm, oil gland-dotted and show several pairs of secondary nerves as well as an intramarginal nerve. The racemes are axillary and terminal, up to about 20 cm long, and minute. The calyx is tubular and develops 4–5 lobes. The 4–5 petals are white cream, elliptic, and about 4 mm long. A disc is present. The 4–5 stamens are about 4 mm long. The ovary is ovoid. The berries are up to 1 cm long.

Medicinal uses: Jaundice (Iran); fever (India, Iran)

Broad-spectrum antibacterial polar extract: Ethanol extract (500 μg/disc) inhibited the growth of multidrug resistance *Escherichia coli*, *Klebsiella pneumonia*, *Pseudomonas aeruginosa*, *Proteus mirabilis*, *Enterobacter aerogenes*, *Enterococcus faecalis*, *Vibrio* cholerae, *Staphylococcus aureus*, *Staphylococcus epidermidis*, *Acetobacter* sp., and *Bacillus subtilis* with the inhibition zone diameters of 12, 9, 9, 12, 9, 10, 11, 13, 13, 13, and 13 mm, respectively (Venkatadri et al., 2015).

Strong broad-spectrum antibacterial hydrophilic benzo[c]phenanthridine alkaloids: Avicine (molecular mass = 367.7 g/mol) inhibited the growth of *Staphylococcus epidermidis* (ATCC 12228), *Staphylococcus aureus* (ATCC 6538), *Streptococcus pyogenes* (ATCC 19615), *Bacillus subtilis* (ATCC 6633), *Klebsiella pneumoniae* (ATCC 13883), *Escherichia coli* (ATCC 25922) (Tavares et al., 2014). Chelerythrine (molecular mass = 348.1 g/mol) inhibited the growth of *Staphylococcus aureus, Escherichia coli,* and *Aeromonas hydrophila* with the MIC/MBC values of 12.5/25, 25/50, and 50/>200 µg/mL, respectively (Miao et al., 2011).

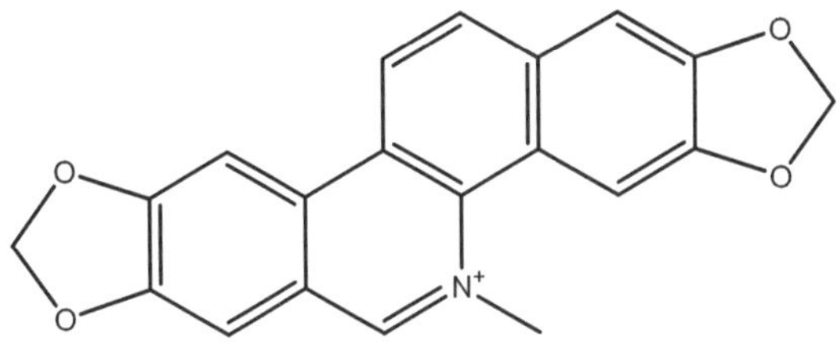

Avicine

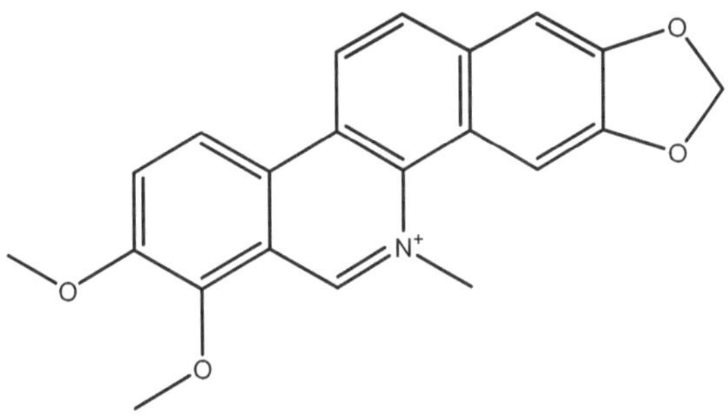

Chelerythrine

Strong antibacterial (Gram-positive) amphiphilic prenylated coumarin: Ulopterol (also know as peucedanol; Log D = 1.2 at pH 7.4; molecular mass: 278.3 g/mol) inhibited the growth of *Pseudomonas aeruginosa* (MTCC 741), *Shigella flexneri* (MTCC 1457), *Enterobacter aerogenes* (MTCC 111), *Staphylococcus aureus* (MTCC 96), methicillin-resistant *Staphylococcus aureus, Klebsiella pneumoniae* (ESBL-3967), *Staphylococcus epidermidis* (MTCC 3615), and *Echerischia coli* (ESBL-3984) with the MIC values ranging from 62.5 to 250 µg/mL, respectively (Raj et al., 2012).

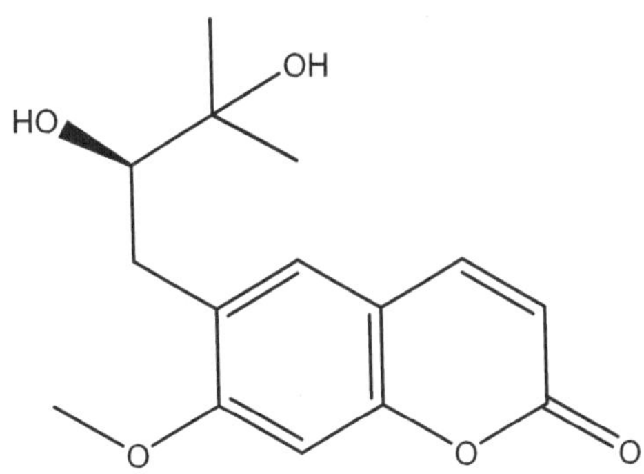

Ulopterol

Moderate broad-spectrum antibacterial benzo[c]phenanthridine alkaloids: 11-Demethylrhoifoline B isolated from the roots inhibited the growth of *Staphylococcus aureus, Staphylococcus epidermidis, Escherichia coli, Enterococcus cloacae, Klebsiella pneumoniae, Pseudomonas aeruginosa*, and *Shigella dysenteriae* with the MIC values of 500, 400, 400, 400, 300, 300, and 400 μg/mL, respectively. 8-Acetylnorchelerythrine isolated from the roots inhibited the growth of *Staphylococcus aureus, Staphylococcus epidermidis, Escherichia coli, Enterococcus cloacae, Klebsiella pneumoniae, Pseudomonas aeruginosa*, and *Shigella dysenteriae*. Rhoifoline B isolated from the roots inhibited the growth of *Staphylococcus aureus, Staphylococcus epidermidis, Escherichia coli, Enterococcus cloacae, Klebsiella pneumoniae, Pseudomonas aeruginosa*, and *Shigella dysenteriae* (Hu et al., 2014).

Strong antimycobacterial benzophenanthridine alkaloid: Norchelerythrine from the stem bark inhibited the growth of *Mycobacterium tuberculosis* with the MIC value of 25 μg/mL (Phatchana & Yenjai, 2014).

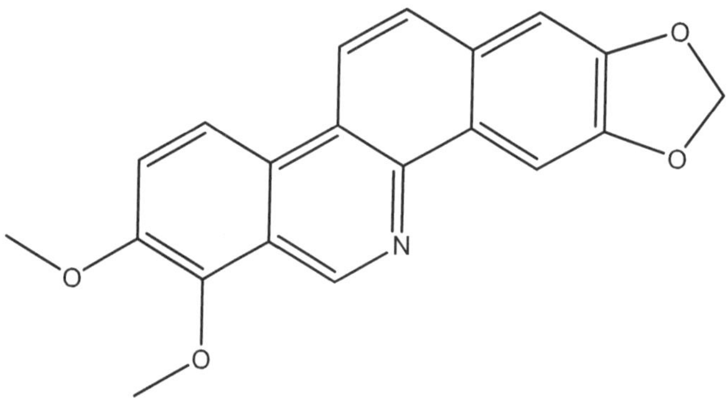

Norchelerythrine

Moderate antimycobacterial phenolic: Nelumol A from the stem bark inhibited the growth of *Mycobacterium tuberculosis* with the MIC value of 50 μg/mL (Phatchana & Yenjai., 2014).

Moderate antimycobacterial prenylated coumarin: 8-Geranyloxy-5,7-dimethyloxycoumarin from the stem bark inhibited the growth of *Mycobacterium tuberculosis* with the MIC value of 50 μg/mL (Phatchana & Yenjai, 2014).

Strong broad-spectrum antifungal hydrophilic benzo[c]phenanthridine alkaloids: Avicine (molecular mass = 367.7 g/mol) inhibited the growth of *Candida albicans* (ATCC 10231), *Saccharomyces cerevisae* (ATCC 2601), and *Cryptococcus neoformans* (ATCC 28952) with the MIC/MBC values of 6.2/25, 12.5/25, and 12.5/25 μg/mL, respectively (Tavares et al., 2014). Chelerythrine (molecular mass = 348.1 g/mol) inhibited the growth of *Ustilaginoidea virens* and *Cochliobolus miyabeanus* with the EC_{50} values of of 6.5 and 5.6 μg/mL, respectively.

Strong antifungal (filamentous) amphiphilic prenylated coumarin: Ulopterol (also know as peucedanol; Log D = 1.2 at pH 7.4; molecular mass: 278.3 g/mol) inhibited the growth of *Candida albicans* (MTCC 227), *Trichophyton rubrum* (57/01), *Trichophyton mentagrophytes* (66/01), and *Aspergillus flavus* with the MIC valiues of 250, 250, 250, and 15.6 μg/mL, respectively (Raj et al., 2012).

Moderate anticandidal benzo[c]phenanthridine alkaloids: 11-Demethylrhoifoline B, 8-acetylnorchelerythrine, and rhoifoline B isolated from the roots inhibited the growth of *Candida albicans* with the MIC values of 300, 90, and 400 μg/mL, respectively (Hu et al., 2014).

Strong antiviral (enveloped segmented linear single stranded (-) RNA) coumarins: (±)-Spirotriscoumarin A and (±)-spirotriscoumarin B inhibited the replication of the Influenza virus A (H3N2) with the IC_{50} values of 3.1 and 2.8 μM, respectively (Tang et al., 2016).

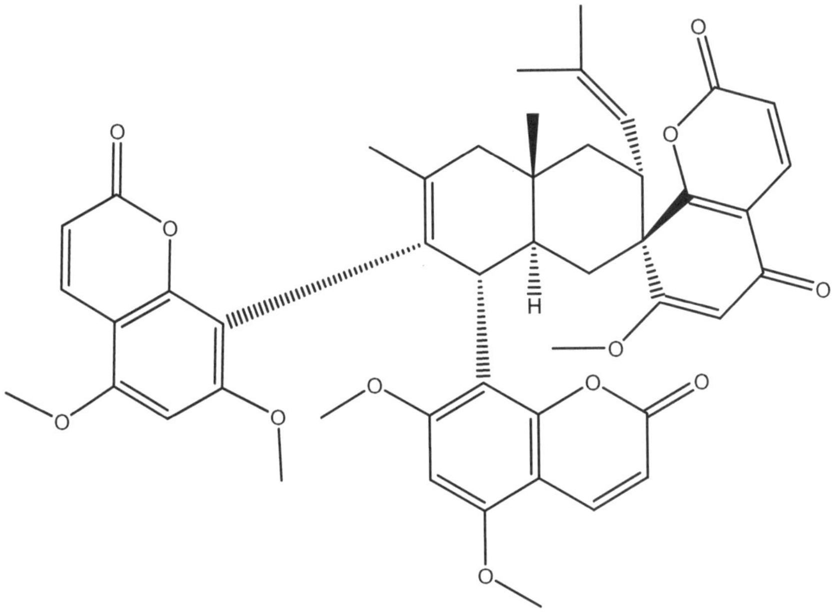

(-)-Spirotriscoumarin B

Commentaries

i. The plant brings to being series of coumarins including isopimpinellin as well at the benzo[c]phenanthridine alkaloids avicine and chelerythrine and the quinoline alkaloids *N*-methylflindersine and skimmianine (Ishii et al., 1991).

ii. Natural products with chiral centers often exists as enantiomers in plants. Mixtures of enantiomers can have higher antimicrobial activities than R or S enantiomers themselves. Why? The pure enantiomers, (+) spirotriscoumarin A, (–)-spirotriscoumarin A, (+) spirotriscoumarin B, and (–)-spirotriscoumarin B, exhibited 3- to 6-fold weaker activity against influenza virus A (IC_{50}: 9.8, 11.1, 17.4, and 8.6 µM, respectively) than their corresponding racemic mixtures. Further synergistic effects between (+) spirotriscoumarin A and (–) spirotriscoumarin A and between (+) spirotriscoumarin B and (–)-spirotriscoumarin B against Influenza virus A may exist (Tang et al., 2016). Antimicrobial activity in plants is the result of the combined action and interactions of multiple natural products including their own enantiomers.

iii. In the Subfamily Toddalioideae and tribe Toddalieae, *Phellodendron amurense* Rupr. is used in China for fever and aqueous extract of bark of this handsome tree at the concentration of 250 µg/mL inhibited Human immunodeficiency virus type-1 protease by 100% (Xu et al., 1996).

iv. In the Subfamily Toddalioideae and tribe Toddalieae, *Skimmia laureola* (DC.) Decne. is used in China (local name: *Yue gui yin yu*) for the treatment of chicken pox and measles. Essential oil of this beautiful bush inhibited the growth of methicillin-resistant *Staphylococcus aureus* and *Staphylococcus epidermidis* with the MIC values 32 and 64 µg/mL, respectively. The oil inhibited the growth of *Aspergillus niger* and *Penicillium chrysogenum* with the MIC value in the range 32–16 µg/ mL (Shah et al., 2013).

v. Consider that chelerythrine binds to single-stranded RNA (Basu & Kumar, 2017) and as such could be of value against RNA viruses.

REFERENCES

Basu, P. and Kumar, G.S., 2017. Small molecule–RNA recognition: binding of the benzophenanthridine alkaloids sanguinarine and chelerythrine to single stranded polyribonucleotides. *Journal of Photochemistry and Photobiology B: Biology, 174*, pp. 173–181.

Hu, J., Shi, X., Chen, J., Mao, X., Zhu, L., Yu, L. and Shi, J., 2014. Alkaloids from *Toddalia asiatica* and their cytotoxic, antimicrobial and antifungal activities. *Food Chemistry, 148*, pp. 437–444.

Ishii, H., Kobayashi, J., Ishikawa, M., Haginiwa, J. and Ishikawa, T., 1991. Studies on the chemical constituents of Rutaceae++ plants. LXVI. The chemical constituents of *Toddalia asiatica* (L.) Lam. (*T. aculeata* Pers.).(1). Chemical constituents of the root bark. *Yakugaku zasshi: Journal of the Pharmaceutical Society of Japan, 111*(7), pp. 365–375.

Miao, F., Yang, X.J., Zhou, L., Hu, H.J., Zheng, F., Ding, X.D., Sun, D.M., Zhou, C.D. and Sun, W., 2011. Structural modification of sanguinarine and chelerythrine and their antibacterial activity. *Natural Product Research, 25*(9), pp. 863–875.

Phatchana, R. and Yenjai, C., 2014. Cytotoxic coumarins from *Toddalia asiatica. Planta Medica, 80*(8/9), pp. 719–722.

Raj, M.K., Balachandran, C., Duraipandiyan, V., Agastian, P. and Ignacimuthu, S., 2012. Antimicrobial activity of Ulopterol isolated from *Toddalia asiatica* (L.) Lam.: a traditional medicinal plant. *Journal of Ethnopharmacology, 140*(1), pp. 161–165.

Shah, W.A., Dar, M.Y., Zagar, M.I., Agnihotri, V.K., Qurishi, M.A. and Singh, B., 2013. Chemical composition and antimicrobial activity of the leaf essential oil of *Skimmia laureola* growing wild in Jammu and Kashmir, India. *Natural Product Research, 27*(11), pp. 1023–1027.

Tang, Z.H., Liu, Y.B., Ma, S.G., Li, L., Li, Y., Jiang, J.D., Qu, J. and Yu, S.S., 2016. Antiviral Spirotriscoumarins A and B: two pairs of oligomeric coumarin enantiomers with a spirodienone–sesquiterpene skeleton from *Toddalia asiatica. Organic Letters, 18*(19), pp. 5146–5149.

Tavares, L.D.C., Zanon, G., Weber, A.D., Neto, A.T., Mostardeiro, C.P., Da Cruz, I.B., Oliveira, R.M., Ilha, V., Dalcol, I.I. and Morel, A.F., 2014. Structure-activity relationship of benzophenanthridine alkaloids from *Zanthoxylum rhoifolium* having antimicrobial activity. *PloS One, 9*(5), p. e97000.

Venkatadri, B., Arunagirinathan, N., Rameshkumar, M.R., Ramesh, L., Dhanasezhian, A. and Agastian, P., 2015. *In vitro* antibacterial activity of aqueous and ethanol extracts of *Aristolochia indica* and *Toddalia asiatica* against multidrug-resistant bacteria. *Indian Journal of Pharmaceutical Sciences, 77*(6), p. 788.

Xu, H.X., Wan, M., Loh, B.N., Kon, O.L., Chow, P.W. and Sim, K.Y., 1996. Screening of traditional medicines for their inhibitory activity against HIV-1 Protease. *Phytotherapy Research, 10*(3), pp. 207–210.

8.5.4.13 *Zanthoxylum piperitum* (L.) DC.

Common names: Japanese pepper tree, red pepper, Sichuan pepper; shu chiao, ch'uan chiao, nan chiao, hwa-tseaou (China); hajikami, sanshou (Japan); chopinamu (Korea)

Habitat: Shrublands

Distribution: China, Japan, Korea

Botanical observation: This thorny shrub grows up to 3–4 m high. The leaves are compound, spiral, and exstipulate. The blade consists of 6–7 pairs of folioles which are thick and glossy, about 3–6 cm long, elliptic-ovate, resinous, fragrant, and crenate. The rachis is thorny. The follicles are 4 mm in diameter, bright red becoming purple when ripe and split into two to expose a single black, glossy seed which is pungent and tastes like lemon or ginger.

Medicinal uses: Dysentery (China); cat bites (Korea)

Strong broad-spetrum antibacterial essential oil: Essential oil inhibited the growth of *Bacillus cereus, Shigella choleraesuis, Staphylococcus aureus, Escherichia coli* (O157:H7), and *Vibrio parahaemolyticus* with the MIC values of 1.2, 2.5, 2.5, 5, and 1.2 μg/mL, respectively (Choi et al., 2008).

Moderate antibacterial (Gram-positive) condensed tannins: Proanthocyanidins from the follicles inhibited the growth of methicillin-resistant *Staphylococcus aureus* (OM 481), methicillin-resistant *Staphylococcus aureus* (OM 505), methicillin-resistant *Staphylococcus aureus* (OM 584), and methicillin-resistant *Staphylococcus aureus* (OM 623), and *Staphylococcus aureus* (209P) (Kusuda et al. 2006).

Strong antifungal (filamentous) essential oil: Essential oil of seeds inhibited the growth of *Aspergillus flavus* with the MIC value of 4.5 µL/mL (Phuangsri et al., 2017).

Strong antiviral (non-enveloped monopartite linear single-stranded (+) RNA) polar extract: Methanol extract of leaves inhibited the replication of the Human rhinovirus 2, Human rhinovirus 3, Coxsackie A16, Coxsackie B3, Coxsackie B4, and Enterovirus 71 virus with the IC_{50} values of 59.4, 39.9, 45.8, 68.5, 93.5, and 4.4 µg/mL, respectively (Choi, 2016).

Strong antiviral (enveloped monopartite linear single-stranded (+) RNA) condensed tannins: Proanthocyanidins isolated from the follicles inhibited the replication of the Severe acute respiratory syndrome-associated coronavirus in Vero cells with the IC_{50} value of 5.5 µg/mL· via inhibition of protein synthesis (Maeda et al., 2011).

Antiviral (enveloped segmented linear single-stranded (−) RNA) flavonol glycosides: Quercetin 3-*O*-β-D-galactopyranoside, quercetin 3-O-α-L-rhamnopyranoside, and kaempferol 3-O-α-L-rhamnopyranoside isolated from the leaves of inhibited at 250 µg/mL the formation of plaques in Madin-Darby canine kidney cells by influenza (A/NWS/33 (H1N1) (Ha et al., 2014). 3-*O*-β-D-galactopyranoside, quercetin 3-*O*-α-L-rhamnopyranoside, and kaempferol 3-*O*-α-L-rhamnopyranoside inhibited neuraminidase of Influenza (A/NWS/33 (H1N1)) with IC_{50} values of 434, 211, and 273 µg/mL, respectively (Ha et al., 2014).

Commentaries

i. *Zanthoxylum piperitum* (L.) DC. is used in Nepal to prevent COVID-19 and being edible (?), the fruits could be developed as functional food to prevent or manage COVID-19 and other coming pandemic coronaviruses. The concept of antiviral functional food is captivating and a number of edible medicinal plants and fungi are known to boost immunity such as *Rosmarinus officinalis* L. or *Ganoderma lucidum Curtis* P. Karst.

ii. In India, *Zanthoxylum budrunga* (Roxb.) Roxb. Wall. ex DC. (local name: *Ashvaghra*) is used to treat diarrhoea, bronchitis, and skin infection. *Zanthoxylum oxyphyllum* Edgew. is used for fever, bronchitis, and toothache, and *Zanthoxylum alatum* Roxb. is used (local name: *Bokey timbur*) to treat cholera.

iii. *Zanthoxylum armatum* DC (Indian thorny ash) is used in India (local names: *Nyang*, *Muthrubi*, *Timur*) for dysentery, fever, and toothache and in Nepal (local name: *Timur*) for diarrhea. The flavonol ombuin (100 µg/4 mm disc) isolated from the fruits of this thorny tree inhibited the growth of *Staphylococcus aureus*, *Escherichia coli*, *Streptococcus mutans*, *Pseudomonas aeruginosa*, *Salmonella typhimurium*, *Salmonella typhi*, *Enterococcus faecalis*, *Bacillus subtilis*, *Streptococcus pyogenes*, and *Staphylococcus epidermidis* (Nooreen et al., 2017).

iv. *Zanthoxylum acanthopodium* DC. (Japanese pepper) used for fever in Myanmar (local names: *Chy-inbawngla*, *Sen ho hling*) for cough, fever, flu, dysentery, and toothache, in China for fever and dysentery, and in India (local name: *Mukthrubi*) is used to treat fever, cough, and bronchitis.

v. Aqueous extract of fruit peels of *Zanthoxylum bungeanum* Maxim. is used in China (local name: *Hu jiao*) to treat diarrhea and at the concentration of 250 µg/mL inhibited Human immunodeficiency virus type-1 protease by 42.5% (Xu et al., 1996).

vi. *Zanthoxylum rhetsa* DC. is used in Iran (local name*: Dahan bastah-baz*) to treat cholera, cystitis, and gonorrhea.

vii. *Zanthoxylum schinifolium* Siebold & Zucc. is a spice in Korea (local name: *Sanchonamu*) and produces the prenylated coumarin collinin (Log D = 4.9 at pH 7.4; molecular mass = 328.1 g/mol) which inhibited the growth of both *Mycobacterium tuberculosis* (H37Rv) and extensively drug-resistant *Mycobacterium tuberculosis* with the MIC_{50} values ranging from 3.1 to 6.2 µg/mL (Kim et al., 2018).

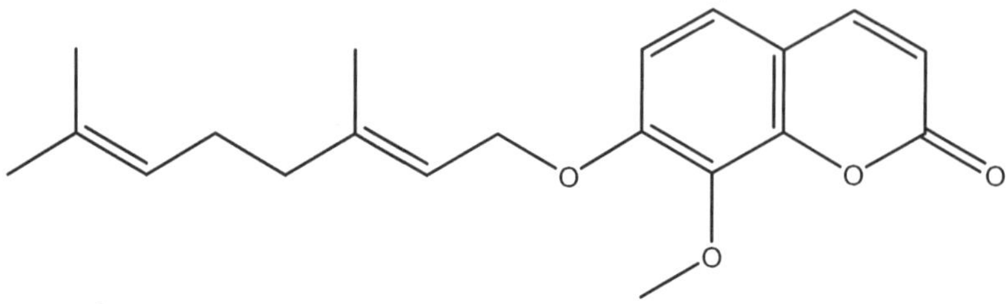

Collinin

Aqueous extract of fruit peels of *Zanthoxylum schinifolium* Siebold & Zucc. at the concentration of 250 μg/mL inhibited Human immunodeficiency virus type-1 protease by 15.8% (Xu et al., 1996).

viii. Wutaipyranol A, 8-methoxywutaipyranol A, and demethoxywutaiensal isolated from the root wood of *Zanthoxylum wutaiense* I.S. Chen inhibited the growth of *Mycobacterium tuberculosis* (H37Rv) with the MIC values of 52.4, 55.6, and 45.8 μg/mL, respectively (Huang et al., 2011). Methyl 7-methoxyanodendroate, 7-methoxywutaifuranal, wutaiensal, dictamnine, and γ-fagarine isolated from the root wood of *Zanthoxylum wutaiense* I.S. Chen exhibited antitubercular activity against *Mycobacterium tuberculosis* (H37Rv) with the MIC values of 35, 35, 30, 30, and 30 μg/mL, respectively (Huang et al., 2008).

ix. *Zanthoxylum limonella* Alston is used in Thailand for fever. Essential oil of fruits of this plant (15 μL/disc) inhibited the growth of *Salmonella typhi*, *Streptococcus enteriditis*, *Escherichia coli*, Clostridium perfringens, and *Campylobacter jejuni* with inhibition zone diameters of 20.5, 16.3, 13.5, 27, and 18 mm, respectively (Wannissorn et al., 2005).

x. 6-Acetonyldihydronitidine, decarine, dihydroavicine, haplopine, (+) hinokinin, isopimpinellin, 4-methyl-1-methyl-2-quinolone, *O*-methylcedrelopsin, *O*-methyltembamide, oxyavicine, and (+) tembamide isolated from the root bark of *Zanthoxyllum ailanthoides* Siebold & Zucc. inhibited the replication of the Human immunodeficiency virus type-1 (IIIB strain) replication in H9 cells with EC_{50} values of <0.1, <0.1, <0.1, 0.6, <0.1, 0.6, 0.6, 0.5, <0.1, <0.1, and <0.1 μg/mL, respectively (Cheng et al., 2005).

Nitidine from a member of the genus *Zanthoxylum* L. evoked 50% inhibition of growth of *Micrococcus luteus*, *Staphylococcus aureus*, *Mycobacterium smegmatis*, and *Mycobacterium avium* at a concentration of 200, 300, 1000, and 3000 μg/mL, respectively (Gu et al., 2004). Nitidine (molecular mass = 348.1 g/mol) evoked 50% inhibition of growth of *Candida albicans*, *Cryptococcus neoformans*, *Saccharomyces cerevisiae*, and *Aspergillus niger* at a concentration of 20, 80, 300, and 3000 μg/mL, respectively (Gu et al., 2004).

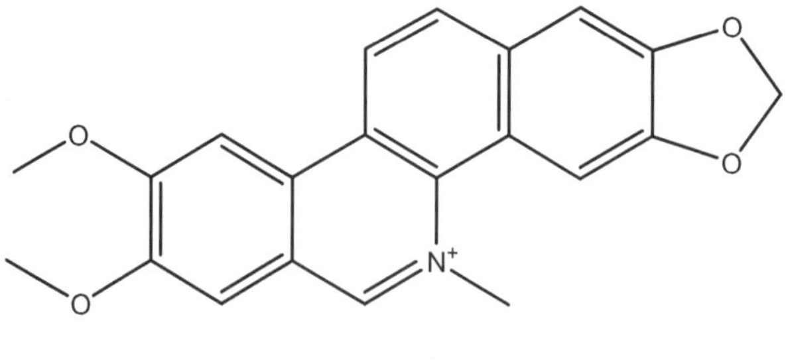

Nitidine

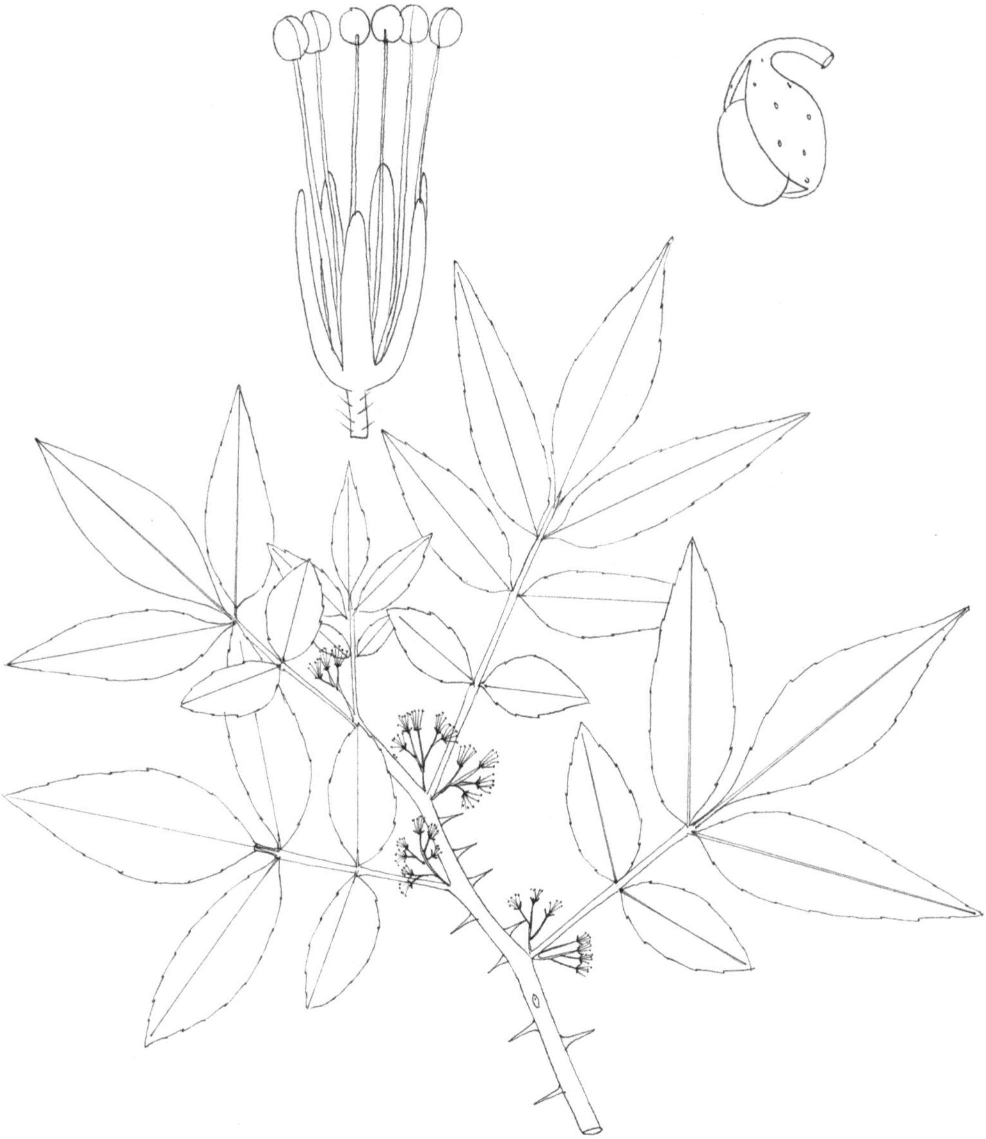

Zanthoxylum acanthopodium DC.

REFERENCES

Cheng, M.J., Lee, K.H., Tsai, I.L. and Chen, I.S., 2005. Two new sesquiterpenoids and anti-HIV principles from the root bark of *Zanthoxylum ailanthoides*. *Bioorganic & Medicinal Chemistry*, *13*(21), pp. 5915–5920.

Choi, H.J., 2016. Evaluation of antiviral activity of zanthoxylum species against picornaviruses. *Osong Public Health and Research Perspectives*, *7*(6), pp. 400–403.

Choi, S.I., Chang, K.M., Lee, Y.S. and Kim, G.H., 2008. Antibacterial activity of essential oils from *Zanthoxylum piperitum* AP DC. and Zanthoxylum schinifolium. *Food Science and Biotechnology*, *17*(1), pp. 195–198.

Gu, J.Q., Graf, T.N., Lee, D., Chai, H.B., Mi, Q., Kardono, L.B., Setyowati, F.M., Ismail, R., Riswan, S., Farnsworth, N.R. and Cordell, G.A., 2004. Cytotoxic and antimicrobial constituents of the bark of diospyros ma ritima collected in two geographical locations in Indonesia. *Journal of Natural Products*, *67*(7), pp. 1156–1161.

Huang, H.Y., Ishikawa, T., Peng, C.F., Chen, S. and Chen, I.S., 2011. Secondary metabolites from the root wood of *Zanthoxylum wutaiense* and their antitubercular activity. *Chemistry & Biodiversity*, *8*(5), pp. 880–886.

Huang, H.Y., Ishikawa, T., Peng, C.F., Tsai, I.L. and Chen, I.S., 2008. Constituents of the root wood of *Zanthoxylum wutaiense* with antitubercular activity. *Journal of Natural Products*, *71*(7), pp. 1146–1151.

Kim, S., Seo, H., Al Mahmud, H., Islam, M.I., Lee, B.E., Cho, M.L. and Song, H.Y., 2018. *In vitro* activity of collinin isolated from the leaves of *Zanthoxylum schinifolium* against multidrug-and extensively drug-resistant *Mycobacterium tuberculosis*. *Phytomedicine*, *46*, pp. 104–110.

Kusuda, M., Inada, K., Ogawa, T.O., Yoshida, T., Shiota, S., Tsuchiya, T. and Hatano, T., 2006. Polyphenolic constituent structures of *Zanthoxylum piperitum* fruit and the antibacterial effects of its polymeric procyanidin on methicillin-resistant *Staphylococcus aureus*. *Bioscience, Biotechnology, and Biochemistry*, *70*(6), pp. 1423–1431.

Maeda, T., Yoshinaka, Y., Yonemoto, Y., Higuchi, H., Kitabayashi, T. and Hattori, K., 2011. Anti SARS-CoV activity of extracts from Japanese pepper (*Zanthoxylum piperitum* (L.) DC. f. inerme Makino). *Horticultural Research (Japan)*, *10*(2), pp. 267–272.

Nooreen, Z., Singh, S., Singh, D.K., Tandon, S., Ahmad, A. and Luqman, S., 2017. Characterization and evaluation of bioactive polyphenolic constituents from *Zanthoxylum armatum* DC., a traditionally used plant. *Biomedicine & Pharmacotherapy*, *89*, pp. 366–375.

Phuangsri, C., Nuntawong, N. and Niamsup, H., 2017. Antifungal activity of the essential oil extracted from *Zanthoxylum piperitum* seeds against *Aspergillus flavus*. *Chiang Mai Journal of Science*, *44*(2), pp. 584–594.

Xu, H.X., Wan, M., Loh, B.N., Kon, O.L., Chow, P.W. and Sim, K.Y., 1996. Screening of traditional medicines for their inhibitory activity against HIV-1 Protease. *Phytotherapy Research*, *10*(3), pp. 207–210.

8.5.5 FAMILY SAPINDACEAE A.L DE JUSSIEU (1789)

The family Sapindaceae consists of 130 genera and 1500 species of mostly trees and shrubs. The leaves are simple or compound, alternate, exstipulate, or stipulate. The infloresснces are panicles, cymes, or fascicles of minute flowers. The perianth comprises 3–5 sepals and 3–5 petals. The androecium is made of 5–10 stamens. The ovary is 1–6-locular, each locule sheltering 1–2 ovules growing from axile placentas. The fruits are nuts, samaras, berries, or capsules. Plants in this family produce triterpene saponins.

8.5.5.1 *Acer truncatum* Bunge

Synonyms: *Acer lobulatum* Nakai

Common names: Shantung maple; yuan bao feng (China)

Habitat: Forests, cultivated

Distribution: China, Korea

Botanical observation: This handsome tree grows up to a height of 10 m. In Autumn, its crown displays a mind-blowing reddish-orange color. The leaves are simple, spiral, and exstipulate. The petiole is slender and up to 9 cm long. The blade is pale green, edible (Mongolia), palmately 5–7 lobed (the lobes triangular), glossy above, 8–12 × 5–10 cm, truncate at base, and acuminate or caudate at apex. The inflorescences are axillary corymbs. The 5 sepals are oblong, about 4–5 mm long, and round to acute at apex. The 5 petals are oblong-obovate, 5–7 mm long, light yellow (at first glance somewhat Brassicaceous). The 8 stamens are minute and around a minute lobed disc. The ovary is compressed. The samara presents flat, thick, about 1 cm long nutlets, as well as a pair of wings which are greenish white, and about 2 cm long.

Medicinal uses: Apparently none for microbial infections

Antibacterial halo (Gram-positive) halo developed by hydrophilic aryltetralin lignan glycoside: Schizandriside (Log D = 0.4 at pH 7.4; molecular mass = 492.5 g/mol) (2 μg/6 mm disc) isolated from the leaves inhibited the growth of *Staphylococcus aureus* with an inhibition zone of 11 mm (Dong et al., 2006).

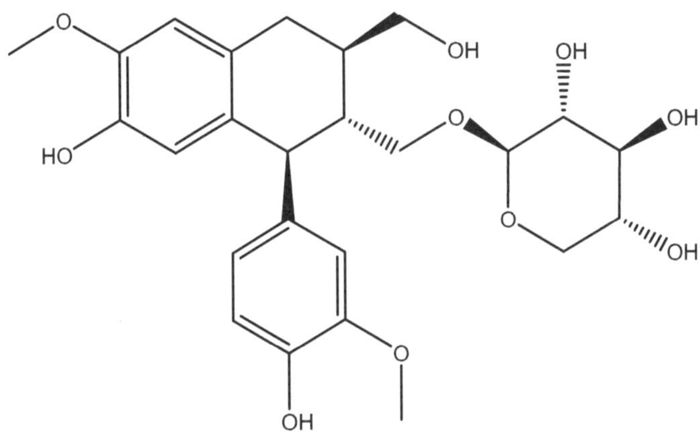

Schizandriside

Moderate antibacterial (Gram-positive) mid-polar extract: Ethylacetate extract of leaves inhibited the growth of *Staphylococcus aureus* (ATCC 25923), *Staphylococcus epidermidis*, *Escherichia coli* (ATCC 25922), and *Pseudomonas aeruginosa* (ATCC 27853) with the MIC values of 200, 100, 200, and 200 μg/mL, respectively, and inhibited β-oxoacyl-acyl carrier protein reductase of *Escherichia coli* with the IC_{50} value of 2.7 μg/mL (Zhang et al., 2008).

Weak antibacterial (Gram-positive) gallotannin: 1,2,3,4,6-Penta-*O*-galloyl-β-D-glucose isolated from the leaves inhibited the growth of *Staphylococcus aureus* (ATCC 25923), *Staphylococcus epidermidis*, *Escherichia coli* (ATCC 25922), and *Pseudomonas aeruginosa* (ATCC 27853) with the MIC values of 200, 60, 200, and 100 μg/mL, respectively (Zhang et al., 2008) and inhibited β-oxoacyl-acyl carrier protein reductase of *Escherichia coli* with the IC_{50} value of 0.9 μg/mL (Zhang et al., 2008). Tannic acid from the leaves inhibited the growth of *Staphylococcus aureus*, *Staphylococcus epidermidis* (ATCC 12228), *Shigella flexneri*, and *Shigella dysenteriae* and inhibited β-Ketoacyl-acyl carrier protein reductase with IC_{50} value of 0.7 μM (Wu et al., 2010).

Commentary: β-ketoacyl-acyl carrier protein reductase is involved in the synthesis of fatty acids in bacteria (Gerusv, 2010) and potentially an interesting antibacterial target. Is schizandriside antiviral?

REFERENCES

Dong, L.P., Ni, W., Dong, J.Y., Li, J.Z., Chen, C.X. and Liu, H.Y., 2006. A new neolignan glycoside from the leaves of *Acer truncatum*. *Molecules*, *11*(12), pp. 1009–1014.

Gerusz, V., 2010. Recent advances in the inhibition of bacterial fatty acid biosynthesis. *Annual Reports in Medicinal Chemistry*, *45*, pp. 295–311.

Wu, D., Wu, X.D., You, X.F., Ma, X.F. and Tian, W.X., 2010. Inhibitory effects on bacterial growth and b-ketoacyl-ACP reductase by different species of maple leaf extracts and tannic acid. *Phytotherapy Research*, *24*(S1), pp. S35–S41.

Zhang, F., Luo, S.Y., Ye, Y.B., Zhao, W.H., Sun, X.G., Wang, Z.Q., Li, R., Sun, Y.H., Tian, W.X. and Zhang, Y.X., 2008. The antibacterial efficacy of an aceraceous plant [Shantung maple (*Acer truncatum* Bunge)] may be related to inhibition of bacterial β-oxoacyl-acyl carrier protein reductase (FabG). *Biotechnology and Applied Biochemistry*, *51*(2), pp. 73–78.

8.5.5.2 *Allophylus cobbe* (L.) Forsyth f.

Synonym: *Rhus cobbe* L.

Common names: Tit-berry; mukkannan perukilam, patta nai, triputa (India); sicancang (Indonesia); kulimpapa burung (Malaysia); zaung gale (Myanmar); haim (Papua New Guinea); balic (the Philippines); to sai (Thailand)

Habitat: Mangroves, seashores, muddy coasts

Distribution: Tropical Asia

Botanical observation: This shrub grows up to a height of 20 m. The stems are hairy. The leaves are trifoliolate, spiral, and exstipulate. The petiole is up to about 10 cm long. The folioles are laxly and irregularly serrate, 8–12 × 5–7 cm, cuneate at base, oblong-elliptic, and acute at apex. The spikes are axillary and about 15 cm long. The 4 sepals and the 4 petals are minute. The androecium presents 8 stamens. The ovary is hairy and the stigma bifid. The drupe is edible, about 5 mm across, juicy, bright orange to red and glossy.

Medicinal uses: Diarrhea (India); trush (Malaysia); fever, wounds (Thailand)

Strong broad-spectrum antibacterial halo developed by non-polar extract: Petroleum ether extract inhibited the growth of *Bacillus subtilis*, *Bacillus megaterium*, *Bacillus cereus*, *Shigella sonnei*, and *Salmonella paratyphi* with the MIC values of 62.5, 62.5, 125, 31.2, and 31.2 µg/mL, respectively (Islam et al., 2012).

Strong broad-spectrum antifungal halo developed by non-polar extract: Petroleum ether extract inhibited the growth of *Blastomyces dermatitidis*, *Candida albicans*, and *Cryptococcus neoformans* with MIC values of 125, 62.5, and 31.2 µg/mL, respectively (Islam et al., 2012).

Commentary: Not much is apparently known about the antimicrobial principles of this plant. Extract of *Allophyllus serratus* inhibited the replication of the Newscastle disease virus (Babbar et al., 1982).

REFERENCES

Babbar, O.P., Joshi, M.N. and Madan, A.R., 1982. Evaluation of plants for antiviral activity. *Indian Journal of Medical Research*, *76*(S), pp. 54–65.

Islam, M.T., Noor, M.A., Karon, B. and de Freitas, R.M., 2012. *In vitro* antimicrobial and brine shrimp lethality of Allophylus cobbe L. *Ayu*, *33*(2), p. 299.

8.5.5.3 *Cardiospermum halicacabum* L.

Common names: Balloon vine, heart pea; bhado; ketafoksa maal-karanga (Bangladesh); phong thuyen kat (Cambodia, Laos, Vietnam), gan da, tian deng long (China); karavi, latafatki, lataphatki, Mudakkathan, pokapiya (India), ketipes (Indonesia), peri belun (Malaysia), kala-myetsi, malalai (Myanmar), bangcolon (the Philippines); penelvel (Sri Lanka); phok om, poo om (Thailand)

Synonyms: *Cardiospermum corindum* L., *Cardiospermum microcarpum* Kunth, *Cardiospermum molle* Kunth

Habitat: Wastelands, roadsides, vacant plots, open forest

Distribution: Tropical Asia and Pacific

Cardiospermum halicacabum L.

Botanical observation: This slender climber native to tropical America grows up to about 3 m long. The stem produces tendrils which are 7–10 cm long. The leaves are alternate, alternate, and compound. The petiole is 1.5–3 cm long. The folioles are 3-partite and pinnately lobed. The inflorescence is axillary on a 7–10 cm long axe which is hairy, with a pair of tendrils. The 4 sepals are broadly ovate, about 2 mm long, and marked with red patches. The 4 petals are orbicular, the 2 upper emarginated at apex, about 2 mm long, and whitish-yellow. The 8 stamens are about 5 mm long. The ovary is angled with a 3-lobed stigma. The capsules are pyriform, ballon to lantern-like, trigonous, papery, pendulous, and includes 3 locules. The seeds are globose, 4 mm in diameter, dull-black, and smooth side.

Medicinal uses: Chicken pox, small pox (Bangladesh), fever (Myanmar, Thailand), sores (India), wounds (Thailand)

Moderate broad-spectrum antibacterial polar extract: Ethanol extract of leaves inhibited the growth of *Escherichia coli* (MTCC 739), *Pseudomonas aeruginosa* (MTCC 2453), *Staphylococcus aureus*, *Shigella* sp., *Pseudomonas aeruginosa* (MTCC 1934), *Aeromonas hygrophila*, *Enterococcus faecalis*, and *Escherichia coli* (MTCC 227) with MIC values ranging from 90 to 125 µg/mL (Jeyadevi et al., 2013).

Moderate antifungal (filamentous) polar extract: Ethanol extract of leaves inhibited the growth of *Candida albicans* (MTCC 227) with MIC values of 190 µg/mL (Jeyadevi et al., 2013).

Antifungal halo developed by polar extract: Methanol extract of seeds inhibited the growth of *Trichophyton rubrum* (500 µg/disc) with an inhibition zone diameter equivalent to that of amphotericin B (12 µg/disc) (Gaziano et al., 2018).

Antifungal (filamentous) halo developed by flavonol: Luteolin isolated from of seeds inhibited the growth of *Trichophyton rubrum* (500 µg/disc) with an inhibition zone diameter equivalent to that of amphotericin B (3 µg/disc) (Gaziano et al., 2018).

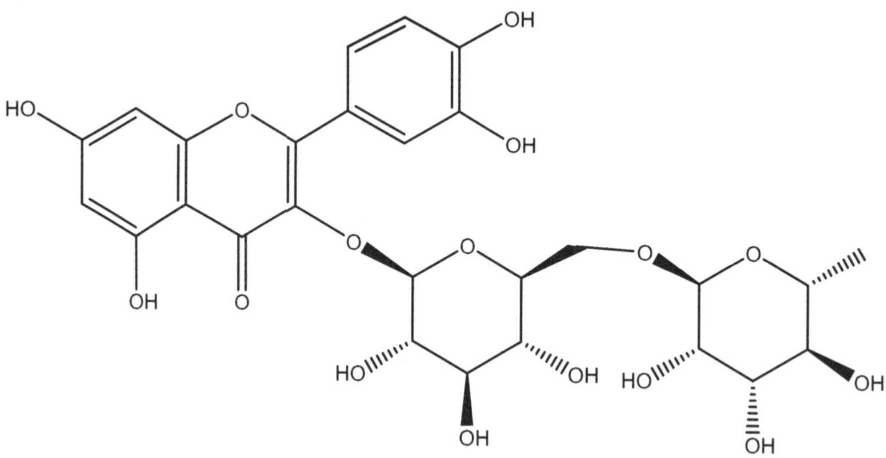

Rutin

Antifungal (filamentous) halo developed by hydrophilic flavonol glycoside: Rutin (Log D = – 1.7 at pH 7.4; molecular mass = 610.5 g/mol) isolated from of seeds inhibited the growth of *Trichophyton rubrum* (500 µg/disc) with an inhibition zone diameter equivalent to that of amphotericin B (1.5 µg/disc) (Gaziano et al., 2018).

Moderate antiviral (monopartite linear dimeric single stranded (+) RNA) polar extract: Ethanol extract of aerial parts inhibited the replication of the Human immunodeficiency virus type-1 in

MT-4 cells with the EC_{50} value of 202.2 μg/mL and the selectivity index of 4.9 (Premanathan et al., 2000).

Antiviral (enveloped monopartite double-stranded DNA) polar extract: Methanol extract inhibited the secretion of the Hepatitis B virus surface-antigen (Murugan et al., 2011).

Viral enzyme inhibition by polar extract: Methanol extract inhibited the Human immunodeficiency virus reverse transcriptase (Murugan et al., 2011).

Commentaries: Antibacterial activity of aqueous extract of this climber against *Staphylococcus aureus* was observed by Osborn (1943).

REFERENCES

Gaziano, R., Campione, E., Iacovelli, F., Marino, D., Pica, F., Di Francesco, P., Aquaro, S., Menichini, F., Falconi, M. and Bianchi, L., 2018. Antifungal activity of *Cardiospermum halicacabum* L. (Sapindaceae) against *Trichophyton rubrum* occurs through molecular interaction with fungal Hsp90. *Drug Design, Development and Therapy*, *12*, p. 2185.

Murugan, K., Prabu, R.V., Sangeetha, S. and Al-Sohaibani, S., 2011. Antiviral activity of *Cardiospermum halicacabum* L. extract against coinfecting agents HIV and HBV. *Journal of Herbs, Spices & Medicinal Plants*, *17*(4), pp. 403–418.

Osborn, E.M., 1943. On the occurrence of antibacterial substances in green plants. *British Journal of Experimental Pathology*, *24*(6), p. 227.

Premanathan, M., Rajendran, S., Ramanathan, T. and Kathiresan, K., 2000. A survey of some Indian medicinal plants for anti-human immunodeficiency virus (HIV) activity. *Indian Journal of Medical Research*, *112*, p. 73.

8.5.5.4 *Dimocarpus longan* Lour.

Synonyms: *Euphoria longan* (Lour.) Steud., *Euphoria longana* Lam., *Nephelium longana* Camb., *Nephelium longan* (Lour.) Hook

Common name: Eye ball tree, longan; mien (Cambodia); long yan (China); poovuttu mararn (India); medaru (Indonesia); nam nhai (Laos); mata kucing, mato dot tungau (Malaysia); ga-naing-gyo, kyet mouk, (Myanmar); lamyai pa (Thailand); lengkeng (Malaysia); nhan (Vietnam)

Habitat: Cultivated

Distribution: India, Bangladesh, South China, Southeast Asia

Botanical observation: This graceful tree grows to 40 m tall. The bole is straight with a smooth and skin-like bark. The stem is terete, lenticelled, grooved, and hairy by apex. The leaves are compound, spiral and exstipulate. The rachis grows up to 20 cm long and bears 2–6 pairs of folioles which are alternate. The petiolule is 0.5–3 cm long. The foliole is elliptical, 10–20 × 3.5–5 cm, glossy, leathery, ovate oblong, marked with 9–15 pairs of secondary nerves, hairy below, wedge-shaped at base, and blunt to acuminate at apex. The panicles are terminal. The calyx is minute comprises 5 lobes which are about 3 mm long. The 5 petals are minute and hairy. The 8 stamens are inserted on a disc. The drupe is globose, warty, 1–3 cm in diameter and contains a globose, beautiful, dark brown, and glossy seed embedded in a translucent and delicious aril, the whole looking like eye globes of cats whence the Malay name "*Mata kucing*".

Medicinal uses: Leucorrhoea, syphilis (Cambodia, Laos, Vietnam)

Broad-spectrum antibacterial halo developed by mid-polar extract: Chloroform extract of stems and leaves (500 μg/6 mm disc) inhibited the growth of *Bacillus subtilis*, *Bacillus cereus*, *Staphylococcus aureus*, *Escherichia coli*, *Salmonella typhi*, *Shigella dysenteriae*, *Vibrio mimicus*, and *Vibrio parahemolyticus* with inhibition zone diameters of 20, 12, 13, 13, 18, 14, 18, and 13 mm, respectively (Ripa et al., 2010).

Strong antifungal (yeast) phenolic: Ellagic acid isolated from the seeds inhibited the growth of *Candida albicans* (ATCC 90028) and *Cryptococcus neoformans* (clinical isolate) with the MIC values of 62.5 and 15.6 μg/mL, respectively (Rangkadilok et al., 2012).

Dimocarpus longan Lour.

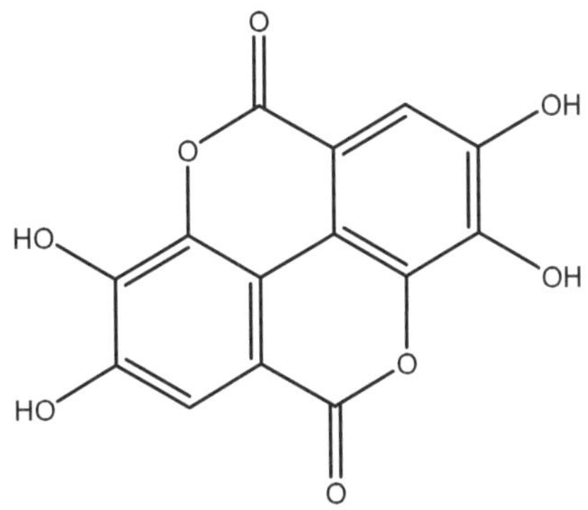

Ellagic acid

Strong antiviral (enveloped monopartite single-stranded (+) RNA) polar extract: Methanol extract of leaves inhibited the replication of the Hepatitis C virus (JFH1 strain of genotype 2a) in Huh7it-1 cells with an EC_{50} value of 19.4 µg/mL and the selectivity index of 35.1. The extract at 50 µg/mL mixed with Hepatitis C virus 2 hours before inoculation evoked 99.7% inhibition in virus indicating virucidal effect (Apriyanto et al., 2016). The extract also evoked a marginal inhibition of Hepatitis C virus entry in host cell and viral RNA replication (Apriyanto et al., 2016). The extract also inhibited the replication of the Dengue virus type 2 (DS-18/09 strain) with the EC_{50} value of 40.9 µg/mL in Vero cells (Apriyanto et al., 2016).

Commentary: The fruit contains gallic acid, corilagin, and ellagic acid. One could examine the probable presence of anti-coronaviruses and other antiviral agents in this tree.

REFERENCES

Apriyanto, D.R., Aoki, C., Hartati, S., Hanafi, M., Kardono, L.B.S., Arsianti, A., Louisa, M., Sudiro, T.M., Dewi, B.E., Sudarmono, P. and Soebandrio, A., 2016. Anti-hepatitis C virus activity of a crude extract from longan (*Dimocarpus longan* Lour.) leaves. *Japanese Journal of Infectious Diseases*, *69*(3), pp. 213–220.

Ripa, F.A., Haque, M. and Bulbul, I.J., 2010. *In vitro* antibacterial, cytotoxic and antioxidant activities of plant *Nephelium longan*. *Pakistan Journal of Biological Sciences*, *13*(1), p. 22.

Rangkadilok, N., Tongchusak, S., Boonhok, R., Chaiyaroj, S.C., Junyaprasert, V.B., Buajeeb, W., Akanimanee, J., Raksasuk, T., Suddhasthira, T. and Satayavivad, J., 2012. *In vitro* antifungal activities of longan (*Dimocarpus longan* Lour.) seed extract. *Fitoterapia*, *83*(3), pp. 545–553.

8.5.5.5 *Dodonaea viscosa* Jacq.

Common names: Hopseed bush; che sang zi, ken di (China); santha (India); hmaing (Myanmar); ghawraskay, sanatha (Pakistan); kelne (Papua New Guinea); alipata (the Philippines); goichap (Yap)

Synonyms: *Dodonaea burmanniana* DC.; *Dodonaea dioica* Roxb. ex DC.; *Ptelea viscosa* L.

Habitat: Forests margins, savannahs, beaches

Distribution: Tropica and sub-tropical Asia and Pacific

Botanical observation: This shrub grows to a height of 3 m. The stems are winged and exsude a resin. The leaves are simple, spiral, subsessile, and exstipulate. The blade is linear, oblong, or spathulate, 5–12 × 0.5–4 cm, sticky, and acute at apex. The panicles are terminal or axillary. The 4 sepals are lanceolate, about 3 mm long, and obtuse at apex. Corolla: none. A disc is present. The androecium includes 7 or 8 stamens. The ovary is trigonal and develops a 3 mm style and a trilobed stigma. The capsules are obcordate, dehiscent, 2- or 3-winged, reddish, and about 2 cm long, each wing sheltering 1 or 2 seeds.

Medicinal uses: Burn, colds, caries (Pakistan); sorethroat (India); ulcers (China), fever (India; the Philippines; Taiwan, Palau); ulcers (the Phiippines); wounds (India; Pakistan; Yap); boils, sores, dysentery (Papua New Guinea).

Broad-spectrum antibacterial halo developed by polar extract: Ethanol extract of aerial parts (3.2 mg/mL extract, 70 μL/well) inhibited the growth of *Staphylococcus aureus*, *Micrococcus luteus*, *Bacillus subtilis*, *Escherichia coli*, and *Pseudomonas aeruginosa* (Khurram et al., 2009).

Moderate broad-spectrum antibacterial halo developed by clerodane-type diterpenes: 6β-Hydroxy-15,16-epoxy-5β,8β,9β,10α-cleroda-3,13(16) and 14-trien-18-oic acid isolated from the aerial parts inhibited the growth of *Staphylococcus aureus* (NCIMB 6571) and *Escherichia coli* (NCIMB 8797) (Khurram et al., 2015).

Moderate antibacterial (Gram-positive) polar extract: Methanol extract inhibited the growth of *Streptococcus mutans* with the MIC/MBC values of 700/3100 μg/mL (Naidoo et al., 2012).

REFERENCES

Khurram, M., Khan, M.A., Hameed, A., Abbas, N., Qayum, A. and Inayat, H., 2009. Antibacterial activities of *Dodonaea viscosa* using contact bioautography technique. *Molecules*, *14*(3), pp. 1332–1341.

Khurram, M., Lawton, L.A., Edwards, C., Iriti, M., Hameed, A., Khan, M.A. and Khan, F.A., 2015. Rapid bioassay-guided isolation of antibacterial Clerodane type Diterpenoid from *Dodonaea viscosa* (L.) Jaeq. *International Journal of Molecular Sciences*, *16*(9), pp. 20290–20307.

Naidoo, R., Patel, M., Gulube, Z. and Fenyvesi, I., 2012. Inhibitory activity of *Dodonaea viscosa* var. angustifolia extract against *Streptococcus mutans* and its biofilm. *Journal of Ethnopharmacology*, *144*(1), pp. 171–174.

8.5.5.6 *Lepisanthes rubiginosa* (Roxb.) Leenh.

Synonym: *Erioglossum rubiginosum* (Roxb.) Bl.; *Sapindus rubiginosus* Roxb.

Common names: Aul, chagalnadi, kalayo (Bangladesh); chi cai (China); ma huat (Thailand)

Habitat: Open forests, thickets, beach

Distribution: India, Cambodia, Laos, Vietnam, Thailand, Malaysia, Malaysia, the Philippines, Papua New Guinea, Northern Australia

Botanical observation: This shrub grows to a height of about 3 m tall. The stems are hairy. The leaves are compounds, spiral, and exstipulate. The blade is up to about 50 cm long, presents 2–8 pairs of folioles larger toward the apex of the leaf, dark green above, glossy, elliptic to ovate, hairy below, marked with about 10 pairs of secondary nerves, cuneate to rounded at base, and and obtuse or rounded at apex. The racemes are terminal and bear numerous flowers which are about 5 mm across. The 4 sepals are about 2.5 mm long. The 4 petals are obovate and are about 5 mm long. The androecium presents 8 stamens. The drupes are yellow to red, magnificent, glossy, edible (Thailand) about 1.5 cm long, and Araceous-like.

Medicinal uses: Fever, skin diseases (Bangladesh)

Antibacterial halo developed by polar extracts: Methanol extract of bark inhibited the growth of *Staphylococcus aureus* (ATCC 29740), *Salmonella typhi* (ATCC14028), and *Shigella dysenteriae* (ATCC 9027) (Barua et al., 2013). Methanol extract of bark of *Lepisanthes rubiginosa* (Roxb.) Leenh. inhibited the growth of *Staphylococcus aureus* (ATCC 29740) and *Salmonella* sp. (ATCC 14028) (Rana et al., 2014).

Commentary: The plant yields oleanane-type triterpene saponins (hederagenin aglycone) as well as the linear sesquiterpene glycoside rubiginoside (Adesanya et al., 1999). *Sapindus mukorossi* Gaertn. is used as a source of soap in Bangladesh (local name = *Ritha*). From the fruits of this tree, the oleanane-type triterpene saponins oleanolic acid. 3-O-α-L-arabinopyranosyl-(1 → 3)-α-L-rhamnopyranosyl-(1 → 2)-α-L-arabinopyranoside and α-hederin inhibited the growth of *Trichophyton rubrum* (ATCC 28188) with the MIC^{80} values of 8 and 16 μg/mL (Li et al., 2016; Hu et al., 2018). Saponins bind to and form complexes with ergosterol in the cell membrane of fungi, leading to the formation of transmembrane pores and consequent loss of membrane integrity and loss of cytoplasmic contents (Mert-Türk, 2006). On account of their amphiphilic nature, saponins are often responsible for the mucolytic and detergent activity of medicinal plants: *Sapindus laurifolius* Valh is used in India (local name = *Arishtaka*) as soap.

REFERENCES

Adesanya, S.A., Martin, M.T., Hill, B., Dumontet, V., Van Tri, M., Sévenet, T. and Païs, M., 1999. Rubiginoside, a farnesyl glycoside from *Lepisanthes rubiginosa*. *Phytochemistry*, *51*(8), pp. 1039–1041.

Barua, S., Rana, S.M., Billah, M., Naim, Z. and Sarwar, G., 2013. Pharmacological, phytochemical and physicochemical properties of methanol extracts of *Erioglossum rubiginosum* barks. *Journal of Health Sciences*, *3*(11), pp. 51–62.

Hu, Q., Chen, Y.Y., Jiao, Q.Y., Khan, A., Li, F., Han, D.F., Cao, G.D. and Lou, H.X., 2018. Triterpenoid saponins from the pulp of Sapindus mukorossi and their antifungal activities. *Phytochemistry*, *147*, pp. 1–8.

Mert-Türk, F., 2006. Saponins versus plant fungal pathogens. *Journal of Cellular and Molecular Medicine*, *5*, pp. 13–17.

Rana, S.M., Billah, M.M., Hossain, M.S., Saifuddin, A.K.M., Islam, S.A., Banik, S., Naim, Z. and Raju, G.S., 2014. Susceptibility of microorganism to selected medicinal plants in Bangladesh. *Asian Pacific Journal of Tropical Biomedicine*, *4*(11), pp. 911–917.

8.5.5.7 *Harpullia petiolaris* Radlk.

Habitat: Forests

Distribution: Malaysia, Indonesia, Papua New Guinea

Botanical observation: This tree grows to a height of up 20 m tall. The bark is grey and fissured and the wood is yellowish. The stems are hairy. The leaves are compound, spiral, and exstipulate. The rachis is 10–30 cm long. The blade presents 4–7 pairs of folioles. The folioles are sessile or not, 10–40 × 3.5–15 cm, hairy below, subcordate to rounded at base. The cymes are cauliflorous and up to 50 cm long. The 5 sepals are elliptic and about 5 mm long. The 5 petals are white, oblanceolate, about 1 cm long. A disc is present. The androecium includes 5 stamens. The ovary is bilobed and develops a 6 mm long style. The capsules are ellipsoid, red, about 3 cm across, coriaceous, and shelter a few black seeds in a yellow aril.

Medicinal use: Skin diseases (Papua New Guinea)

Broad-spectrum antibacterial halo developed by mid-polar extract: Dichloromethane extract of stem bark (4 mg/disc) inhibited the growth of *Bacillus cereus*, *Bacillus subtilis*, *Staphylococcus aureus*, *Escherichia coli*, *Klebsiella pneumoniae*, *Pseudomonas aeruginosa*, and *Salmonella typhi* (Khan & Omoloso, 2002). This extract inhibited the growth of *Neisseria gonorrhoae* with inhibition zone diameter of 16 mm (Khan & Omoloso, 2002).

Broad-spectrum antifungal halo developed by mid-polar extract: Dichloromethane extract of stem bark (4 mg/disc) inhibited the growth of *Aspergillus niger*, *Candida albicans*, *Trichophyton mentagrophytes*, *Trichophyton rubrum*, and *Trichophyton tonsurum* with the inhibition zone diameters of 10, 12, 12, 10, and 12 mm, respectively (Khan & Omoloso, 2002).

Commentary: Antimicrobial principles in *Harpullia petiolaris* Radlk. are apparently unknown. Members of the genus *Harpullia* Roxb. produce oleanane-type triterpenoid saponins (Voutquenne et al., 2002) as well as flavonol glycosides and phenolic glycosides. The phenolic glycoside harpulliaside A isolated from the leaves of *Harpullia pendula* Planch. ex F. Muell. inhibited the growth of *Staphylococcus aureus*, *Bacillus subtilis*, and *Vibrio parahemolyticus* (Abdelkader et al., 2016). Ethanol extract of leaves of *Harpullia arborea* (Blanco) Radlk. inhibited the replication of the Hepatitis C virus (JFH1a strain) and Hepatitis C virus (J6/JFH1 strain) in Huh7 cells with the IC_{50} values of 17.5 and 12.4 µg/mL, respectively (Tumewu et al., 2014).

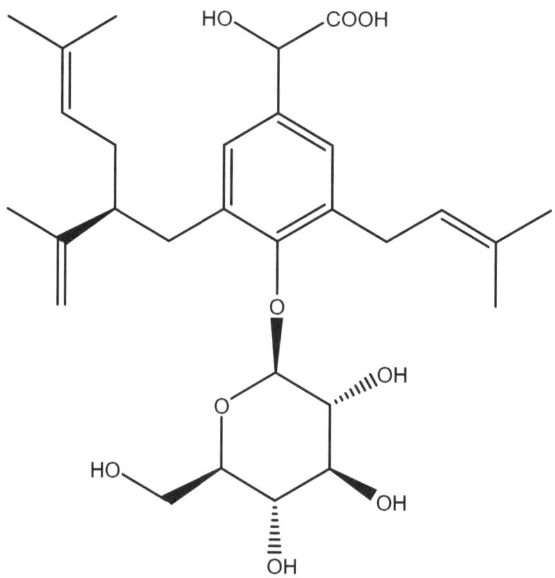

Harpulliaside A

REFERENCES

Abdelkader, M.S.A., Rateb, M.E., Mohamed, G.A. and Jaspars, M., 2016. Harpulliasides A and B: two new benzeneacetic acid derivatives from *Harpullia pendula*. *Phytochemistry Letters*, *15*, pp. 131–135.

Khan, M.R. and Omoloso, A.D., 2002. Antibacterial, antifungal activity of *Harpullia petiolaris*. *Fitoterapia*, *73*(4), pp. 331–335.

Tumewu, L., Apryani, E., Santi, M.R., Wahyuni, T.S., Permanasari, A.A., Adianti, M., Aoki, C., Widyawaruyanti, A., Hafid, A.F., Lusida, M.I. and Hotta, H., 2014. Anti Hepatitis C virus activity screening on *Harpullia arborea* extracts and isolated compound. *Proceedings ICPPS* 2014. pp. 165–167.

Voutquenne, L., Kokougan, C., Lavaud, C., Pouny, I. and Litaudon, M., 2002. Triterpenoid saponins and acylated prosapogenins from *Harpullia austro-caledonica*. *Phytochemistry*, *59*(8), pp. 825–832.

8.5.5.8 *Nephelium mutabile* Bl.

Synonyms: *Nephelium chryseum* auct. non Bl., *Nephelium ramboutan-ake* (Labill.) Leenh

Common names: Rambutan king; meritam, pulasan (Malaysia), kapoelasan (Indonesia), bulala (the Philippines)

Habitat: Cultivated

Distribution: Southeast Asia

Botanical observation: This tree can reach 15 m tall. The stem is hairy at apex. The leaves are compound, spiral, and exstipulate. The rachis is 15–30 cm and presents 2–5 pairs of subopposite folioles. The folioles are dark green above, lanceolate, 6–15 × 2–5 cm, hairy below, and marked with 6–8 pairs of secondary nerves. The panicle is terminal or axillary and present minute yellowish flowers. The calyx comprises 5 hairy lobes. The androecium comprises 5–8 stamens. The capsule is ovoid, dark purplish, up to about 6 cm long, covered with oblong tubercles. The seed is ovoid, oblong, about 2.5 cm long and embedded in a translucent and edible aril.

Medicinal use: Fever (Malaysia)

Commentaries

i. The plant has apparently not been much examined for antimicrobial properties.
ii. *Nephelium lappaceum* L. is used in the Philippines (local name: *Rambutan*) is used to treat dysentery and fever. The monoterpene lactone lappaceolide A (30 µg/disc) isolated from the seeds of *Nephelium lappaceum* L. inhibited the growth of *Escherichia coli* (UPCC 1195), *Pseudomonas aeruginosa* (UPCC 1244), *Staphylococcus aureus* (UPCC 1143), *Candida albicans* (UPCC 2168), *Trichophyton mentagrophytes* (UPCC 4193), and *Aspergillus niger* (UPCC 4219) with inhibition zone diameters of 11, 12, 11, 13, 14, and 12 mm, respectively (Ragasa et al., 2005).

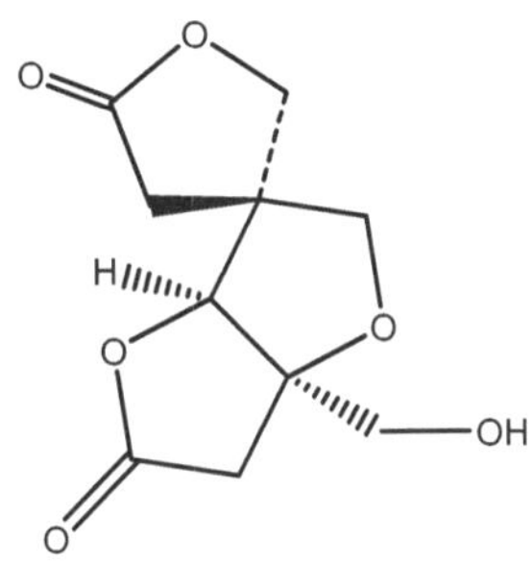

lappaceolide A

The monoterpene lactone lappaceolide B (30 µg/disc) isolated from the seeds of *Nephelium lappaceum* L. inhibited the growth of *Escherichia coli* UPCC 1195, *Pseudomonas aeruginosa* UPCC 1244, *Staphylococcus aureus* UPCC 1143, *Candida albicans* UPCC 2168, *Trichophyton mentagrophytes* UPCC 4193, and *Aspergillus niger* UPCC 4219 with inhibition zone diameters of 11, 12, 11, 13, 13, and 11 mm, respectively (Ragasa et al., 2005).

REFERENCE

Ragasa, C.Y., De Luna, R.D., Cruz, W.C. and Rideout, J.A., 2005. Monoterpene lactones from the seeds of *Nephelium lappaceum*. *Journal of Natural Products*, *68*(9), pp. 1394–1396.

8.5.5.9 *Schleichera trijuga* Willd.

Common names: Ceylon oak, gum-lac, lac tree, Sri Lankan oak; kusum (Bangladesh); puvu, pumarum, vanamra (India), kasambi (Indonesia); kon (Sri Lanka)

Synonyms: *Pistacia oleosa* Lour., *Schleichera oleosa* (Lour.) Oken

Habitat: Forests

Distribution: India, Sri Lanka, Nepal, Myanmar, Cambodia, Vietnam, Laos, Thailand, Malaysia, Indonesia, Papua New Guinea, Australia

Botanical observation: This handsome tree which grows to 20 m tall. The bark is pale brown, scaly and thin. The stem is terete, lenticelled and glabrous. The wood is hard and durable. The leaves are pinnate, alternate and exstipulate. The leave can reach 40 cm long. The rachis bears 2–4 pairs of folioles, which are 4–20 × 1–4 cm, and elliptic with 10–12 pairs of conspicuous secondary

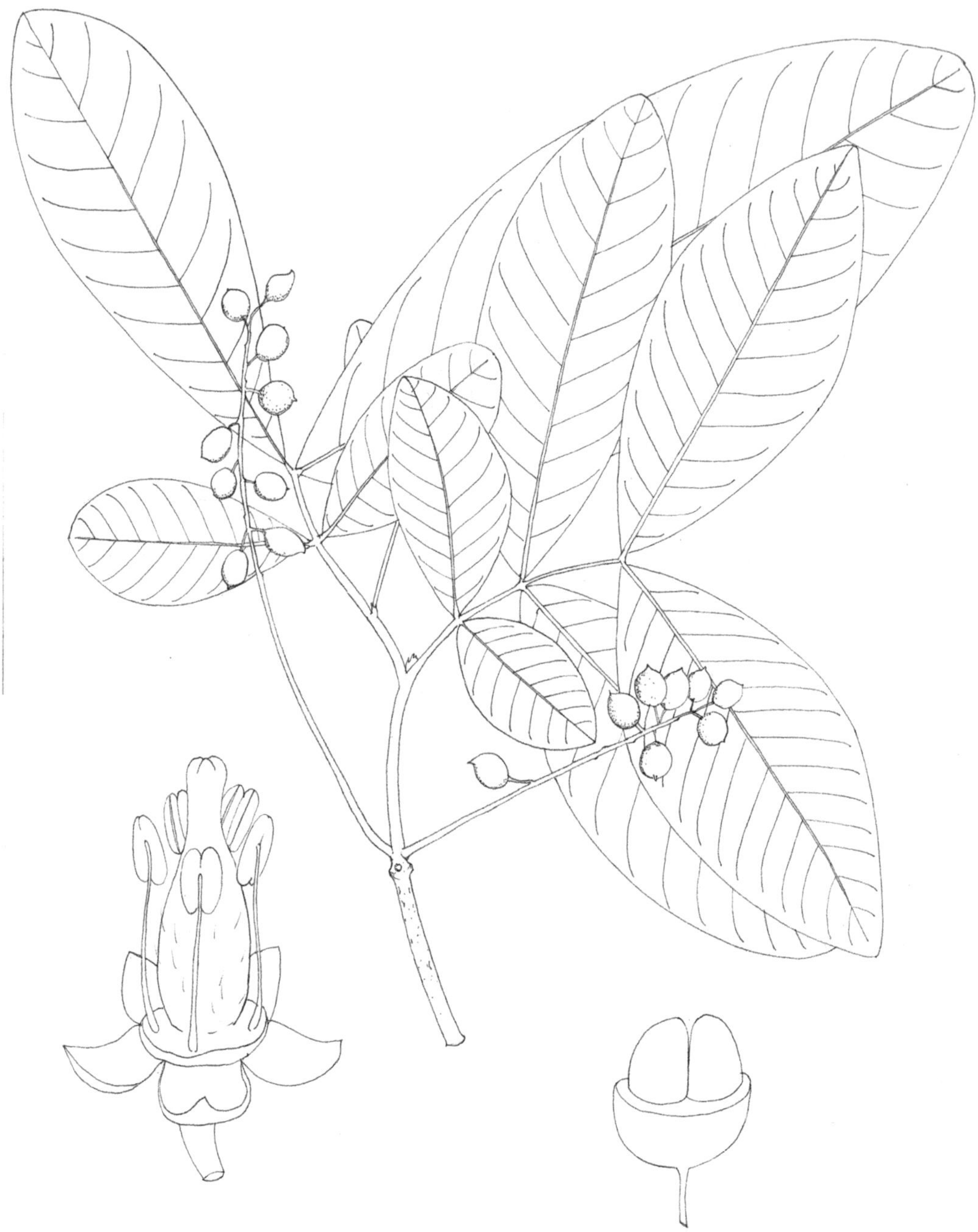

Schleichera trijuga willd.

nerves. The raceme is 7.5–12.5 cm long and opposite the leaves. The calyx presents 5 sepals which are triangular. The 5 stamens are inserted on a disc around a bottle-shaped ovary. The berries are ovoid, beaked, 2.5–3.8 cm long and contain 1–2 seeds which are 1.5 cm long, smooth, brown, and embedded in a waxy, glossy, and orangish aril.

Medicinal uses: Skin diseases and sores (Bangladesh; India); boils (India, Myanmar); wounds (Indonesia)

Moderate broad-spectrum antibacterial taraxastane-type triterpene: Taraxerone isolated from the bark inhibited the growth of *Escherichia coli* and *Staphylococcus aureus* with the MIC values of 100 μg/mL, respectively (Ghosh et al., 2011). Tricadenic acid isolated from bark inhibited the growth of *Escherichia coli* and *Staphylococcus aureus* (Ghosh et al., 2011).

Strong broad-spectrum antifungal lipophilic taraxastane-type triterpene: Taraxerone (Log D = 9.8 at pH 7.4; molecular mass = 424.7 g/mol) inhibited the growth of *Colletotrichum gloeosporioides, Fusarium equisiti, Curvularia eragrostidis, Alternaria alternata*, and *Colletotrichum camelliae* with the MIC values of <5, 20, 40, 10, and <5 μg/mL, respectively (Ghosh et al., 2011). Tricadenic acid inhibited *Colletotrichum gloeosporioides, Fusarium equisiti, Curvularia eragrostidis, Alternaria alternata*, and *Colletotrichum camelliae* with the MIC values of 4.8, 19.5, 40, 19.5, and 39 μg/mL, respectively (Ghosh et al., 2011).

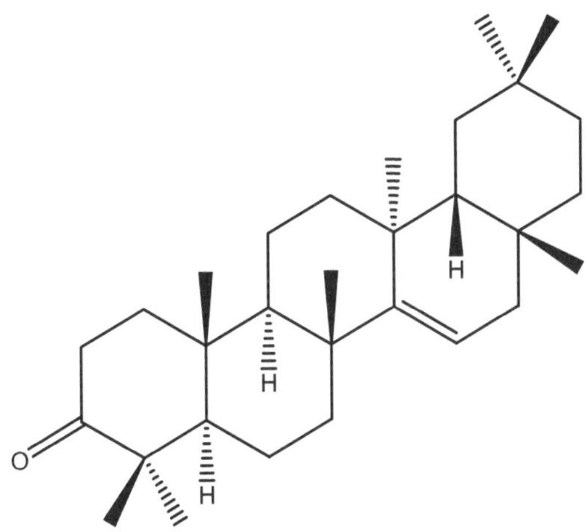

Taraxerone

Strong antiviral (enveloped monopartite linear double-stranded DNA) polar extract: Aqueous extract of fruits at the concentration of 100 μg/mL inhibited the replication of the Herpes simplex type-1 (strain 7401H) in Vero cells by about 65% (Nawawi et al., 1999).

Antiviral (enveloped monopartite linear double-stranded DNA) taraxasterane-type triterpene: Taraxasterone inhibited the replication of the Herpes simplex virus type-1 and type-2 (Kuljanabhagavad et al., 2009).

Commentaries

i. The plant synthetises series of cholestane-type sterols (Pettit et al., 2000).
ii. In the family Sapindaceae, *Holigarna arnottiana* Hook.f is used in India (local name = *Cheru*) to heal wounds and *Pentaspadon velutinus* Hook is used in Malaysia (local name = *Pelong beludu*) for ringworms.
iii. Consider that proanthocyanidins in this family are antiviral: *Litchi sinensis* Sonn. is used for the treatment of smallpox in China (local name = *Li zhi*) and in the Philippines as

Litchi sinensis Sonner

postpartum protective remedy (local name = *Alupag*). Epicatechin-(2β → *O* → 7,4β → 6)-epicatechin-(2β → O → 7,4β → 6)-epicatechin isolated from the seeds of *Litchi chinensis* Sonn. inhibited Coxsackie virus B3 (CVB3) (Nancy strain) with an IC_{50} value of 35.2 µg/mL and selectivity index of 3.2 (virazole IC_{50} value of 69.2 µg/mL and selectivity index of 6.2). Aesculitannin A isolated from the seeds inhibited Herpes simplex virus 1 with an IC_{50} value of 27.1 µg/mL and selectivity index of 2 (Xu et al., 2010).

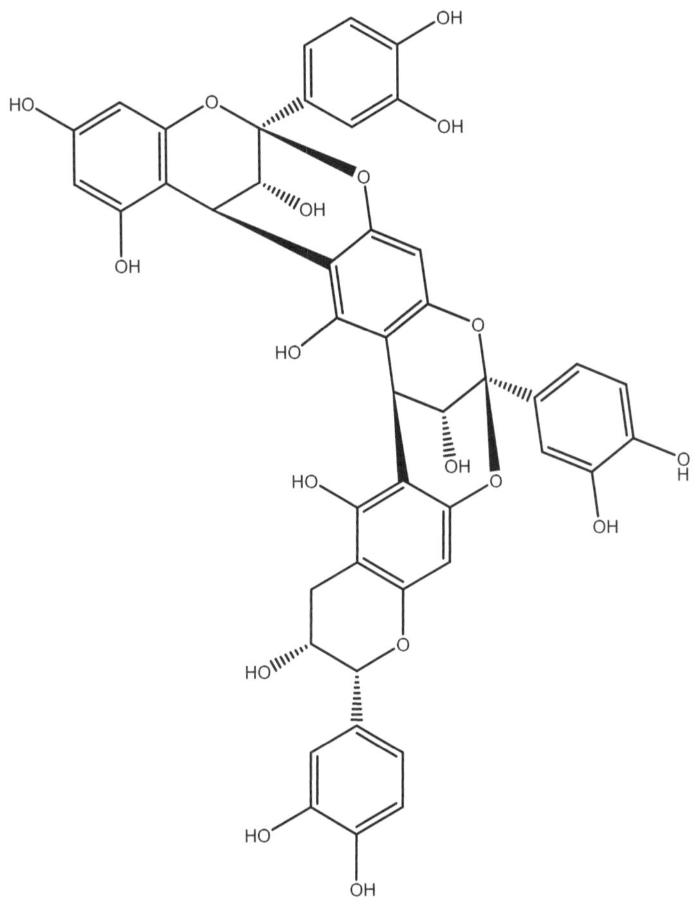

Epicatechin-(2β→*O*→7,4β→6)-epicatechin-(2β→O→7,4β→6)-epicatechin

Pometia pinnata J.R. Forst. & G. Forst. is used in Papua New Guinea (local name = *Taun*) for the treatment of dysentery and fever. Ethanol extract of leaves of this tree inhibited the Human immunodeficiency virus type-1 integrase with the IC_{50} value of 8.8 µg/mL (Suedee et al., 2013). From this extract proanthocyanidin A2 inhibited the Human immunodeficiency virus type-1 integrase with the IC_{50} value of 30.1 µg/mL (Suedee et al., 2013).

REFERENCES

Ghosh, P., Chakraborty, P., Mandal, A., Rasul, M.G., Chakraborty, M. and Saha, A., 2011. Triterpenoids from *Schleichera oleosa* of Darjeeling foothills and their antimicrobial activity. *Indian Journal of Pharmaceutical Sciences*, *73*(2), p. 231.

Kuljanabhagavad, T., Suttisri, R., Pengsuparp, T. and Ruangrungsi, N., 2009. Chemical structure and antiviral activity of aerial part from *Laggera pterodonta*. *Journal of Health Research*, *23*(4), pp. 175–177.

Nawawi, A.A., Nakamura, N., Hattori, M., Kurokawa, M. and Shiraki, K., 1999. Inhibitory effects of Indonesian medicinal plants on the infection of herpes simplex virus type 1. *Phytotherapy Research: An International Journal Devoted to Pharmacological and Toxicological Evaluation of Natural Product Derivatives*, *13*(1), pp. 37–41.

Pettit, G.R., Numata, A., Cragg, G.M., Herald, D.L., Takada, T., Iwamoto, C., Riesen, R., Schmidt, J.M., Doubek, D.L. and Goswami, A., 2000. Isolation and structures of schleicherastatins 1–7 and schleicheols 1 and 2 from the teak forest medicinal tree *Schleichera oleosa*. *Journal of Natural Products*, *63*(1), pp. 72–78.

Suedee, A., Tewtrakul, S. and Panichayupakaranant, P., 2013. Anti-HIV-1 integrase compound from *Pometia pinnata* leaves. *Pharmaceutical Biology*, *51*(10), pp. 1256–1261.

Xu, X., Xie, H., Wang, Y. and Wei, X., 2010. A-type proanthocyanidins from lychee seeds and their antioxidant and antiviral activities. *Journal of Agricultural and Food Chemistry*, *58*(22), pp. 11667–11672.

8.5.6 FAMILY SIMAROUBACEAE A. P. DE CANDOLLE (1811)

The family Simaroubaceae consists of about 20 genera and 150 species of handsome trees or shrubs. The bark is often bitter and somewhat resinous. The leaves are simple or compound, alternate or spiral and stipulate or exstipulate. The stipules are ephemeral and minute or leafy. The folioles are not uncommonly somewhat assymetrical. Inflorescences are axillary racemes, panicles or cymes of tiny flowers. The calyx develops 3–5 lobes. The corolla includes 3–6 petals. A disc is present. The androecium includes 3–6 petals inserted at the base of the disc. The gynoecium includes 2–5 free or fused carpels with free or connate styles, each carpel sheltering 1–2 ovules growing from axil placentas. The fruits are drupes, capsules or samaras. Plants in this family generate quassinoids and dammarane-type triterpenes as well as β-carboline alkaloids of antimicrobial interest.

8.5.6.1 *Ailanthus excelsus* Roxb.

Common names: Indian tree-of-heaven, Persian sumach; arala, aralu, arru, kanjal, mahanimba, puutivrksha (India)

Habitat: Arid fields, open lands, roadside, cultivated

Distribution: Pakistan, India, China

Botanical observation: This tree grows up to about 25 m tall. The bark is smooth, light grey, aromatic, bitter, and exudes a reddish gum after incision. The leaves are pinnate, spiral, and exstipulate. The leaves can reach almost 1 m long. The rachis is reddish and bears 8 to about 14 pairs of folioles. The folioles are subopposite, 8.5–15 × 4–6 cm, dark green, glossy, mostly ovate, lobed, the lobes with glands below, and assymetrical. The petiolules are up to 3 cm long. The inflorescences are lax panicles of minute flowers. The calyx is 5-lobed. The 5 petals are white and oblong. A lobed disc is present. The androecium consists of 10 stamens. The samara is triangular to lanceolate, twisted at base, looks like somewhat an eye, up to 4.5 cm long, nerved, red, coriaceous, and 1-seeded.

Medicinal uses: Bronchitis, diarrhea, dysentery, fever, sores (India)

Antibacterial (Gram-negative) halo developed by mid-polar extract: Ethylacetate extract of stem bark (3 mg/disc) inhibited the growth of *Pseudomonas aeruginosa* (MTCC 424) and *Pseudomonas cichorii* (PC 1) (Shrimali et al., 2001).

Moderate broad-spectrum antibacterial polae extract: Acetone extract inhibited the growth of *Bacillus cereus*, *Enterococcus faecalis*, *Pseudomonas aeruginosa*, *Salmonella typhimurium*, and *Staphylococcus aureus* with the MIC values of 600, 600, 300, 200, and 200 μg/mL, respectively (Poljuha et al., 2017).

Strong broad-spectrum antibacterial lipophilic dammarane-type triterpene: Ocotillone (also known as (24R)-25-hydroxy-20,24-epoxydammaran-3-one; Log D = 6.9 at pH 7.4; molecular mass = 458.7 g/mol) isolated from the stem bark inhibited the growth of *Bacillus subtilis*, *Staphylococcus epidermidis*, *Pseudomonas aeruginosa*, and *Salmonella typhimurium* with the MIC values of 25, 50, 12.5, and 12.5 μM, respectively (Lee et al., 2002).

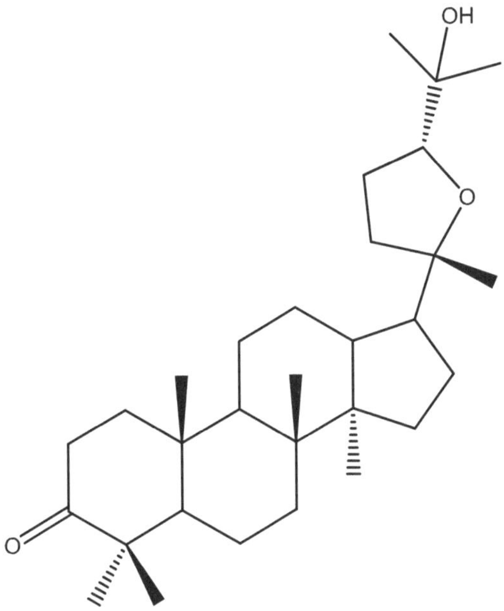

Ocotillone

Broad-spectrum antifungal halo developed by dammarane-type triterpenes: Ocotillone, ailexcelone, and ailexcelone (100 µg/mL/7 mm well) isolated from the heartwood inhibited the growth *Aspergillus niger* with the inhibition zone diameters of 11, 12, and 10 mm, respectively (Srinivas et al., 2006). Ocotillone and ailexcelone (100 µg/mL/7 mm well) inhibited the growth of *Rhizopus oryzae* with the inhibition zone diameters of 11 and 12 mm, respectively (Srinivas et al., 2006). Ailexcelone and ailexcelone (100 µg/mL/7 mm well) inhibited the growth of *Candida albicans* with the inhibition zone doiameters of 14 and 11 mm, respectively (Srinivas et al., 2006).

Broad-spectrum antifungal malabarican-type triterpene: Malabaricol inhibited the growth of *Aspergillus niger* and *Candida albicans* (Srinivas et al., 2006). Epoxymalabaricol inhibited the growth of *Rhizopus oryzae* and *Candida albicans* (Srinivas et al., 2006).

Moderate antiviral (monopartite linear dimeric single stranded (+) RNA) dammarane-type triterpene: Ocotillone inhibited the replication of the Human immunodeficiency virus in $CD4^+$ cells by 70.7% at the concentration of 100 µg/mL (Chang et al., 2003).

Moderate antiviral (non-enveloped monopartite linear single-stranded (+) RNA) decanortriterpene-type quassinoid glycosides: Chuglycosides J and K exhibited inhibitory effects against the Tobacco mosaic virus replication with the IC_{50} values of 56.2 and 137.7 µM, respectively (Tan et al., 2020).

Commentaries

i. The plant produces flavonoid glycosides inhibiting angiotensin converting enzyme of which luteolin-7-*O*-β-glucopyranoside (Loizzo et al., 2007) as well as quassinoids (Tamura et al., 2003).
ii. Dammarane-type triterpenes from medicinal plants in general occur in resins and often strongly inhibit the replication on the Herpes simplex type 1 and 2 like notably the lipophilic dammaradienol (Log D = 10.6 at pH 7.4; molecular mass = 426.7 g/mol) with IC_{50} values of 2.5 and 3 µg/mL, respectively (Poehland et al., 1987).

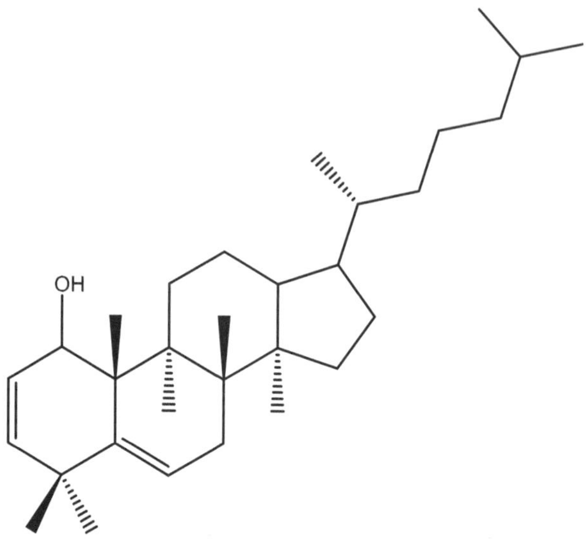

Dammaradienol

iii. The plant abounds with quassinoids of which glaucarubine, ailanthinone, glaucarubinone, and glaucarubolone (Joshi et al., 2003) as well as the β-carboline canthin-6-one (Cho et al., 2018).
iv. In India, *Ailanthus malabarica* (local name: *Perumaram*) is used for bronchitis, fever, typhoid, and dysentery.

REFERENCES

Chang, Y.S., Moon, Y.H. and Woo, E.R., 2003. Virus-cell fusion inhibitory compounds from *Ailanthus altissima* Swingle. *Korean Journal of Pharmacognosy*, *34*(1), pp. 28–32.

Cho, S.K., Jeong, M., Jang, D.S. and Choi, J.H., 2018. Anti-inflammatory effects of Canthin-6-one alkaloids from *Ailanthus altissima*. *Planta Medica*, *50*(8), pp. 527–535.

Joshi, B.C., Pandey, A., Sharma, R.P. and Khare, A., 2003. Quassinoids from *Ailanthus excelsa*. *Phytochemistry*, *62*(4), pp. 579–584.

Lee, D.G., Chang, Y.S., Park, Y.K., Hahm, K.S. and Woo, E.R., 2002. Antimicrobial effects of ocotillone isolated from stem bark of *Ailanthus altisshima*. *Journal of Microbiology and Biotechnology*, *12*(5), pp. 854–857.

Loizzo, M.R., Said, A., Tundis, R., Rashed, K., Statti, G.A., Hufner, A. and Menichini, F., 2007. Inhibition of angiotensin converting enzyme (ACE) by flavonoids isolated from *Ailanthus excelsa* (Roxb) (Simaroubaceae). *Phytotherapy research: An International Journal Devoted to Pharmacological and Toxicological Evaluation of Natural Product Derivatives*, *21*(1), pp. 32–36.

Poehland, B.L., Carte, B.K., Francis, T.A., Hyland, L.J., Allaudeen, H.S. and Troupe, N., 1987. *In vitro* antiviral activity of dammar resin triterpenoids. *Journal of Natural Products*, *50*(4), pp. 706–713.

Poljuha, D., Sladonja, B., Šola, I., Dudaš, S., Bilić, J., Rusak, G., Motlhatlego, K.E. and Eloff, J.N., 2017. Phenolic composition of leaf extracts of *Ailanthus altissima* (Simaroubaceae) with antibacterial and antifungal activity equivalent to standard antibiotics. *Natural Product Communications*, *12*(10), p. 1934578X1701201021.

Shrimali, M., Jain, D.C., Darokar, M.P. and Sharma, R.P., 2001. Antibacterial activity of *Ailanthus excelsa* (Roxb). *Phytotherapy Research*, *15*(2), pp. 165–166.

Srinivas, P.V., Rao, R.R. and Rao, J.M., 2006. Two new tetracyclic triterpenes from the heartwood of *Ailanthus excelsa* RoXB. *Chemistry & Biodiversity*, *3*(8), pp. 930–934.

Tamura, S., Fukamiya, N., Okano, M., Koyama, J., Koike, K., Tokuda, H., Aoi, W., Takayasu, J., Kuchide, M. and Nishino, H., 2003. Three new quassinoids, ailantinol E, F, and G, from *Ailanthus altissima*. *Chemical and Pharmaceutical Bulletin*, *51*(4), pp. 385–389.

Tan, Q.W., Ni, J.C., Shi, J.T., Zhu, J.X. and Chen, Q.J., 2020. Two novel quassinoid glycosides with antiviral activity from the Samara of *Ailanthus altissima*. *Molecules*, *25*(23), p. 5679.

8.5.6.2 *Brucea javanica* (L.) Merr.

Synonyms: *Brucea sumatrana* Roxb.; *Gonus amarissimus* Lour.

Common names: Java Brucea; damli thnang (Cambodia); hiang ying, kushan shu, ya dan zi (China); dahun belur, kuwalot (Indonesia); bi khon, kom roi (Laos); ladah pahit (Malaysia); bogo bogo, magkapayos (the Philippines); ratchadat (Thailand); cay suot (Vietnam)

Habitat: Forest, cultivated

Distribution: From India to Papua New Guinea.

Botanical description: This tree grows to a height of 10 m. The leaves are compound, 20–50 cm long, exstipulates, and spiral. The blade comprises 3–15 pairs of folioles. The folioles are oblong, lanceolate or ovate, asymmetrical, 1.5–5 × 3.5–11 cm, very bitter, serrate, and hairy. The petiolules are up to 9 mm long. The panicles are cauliflorous, hairy, axillary, and bear minute flowers. The calyx comprises 4 sepals united at base. The 4 petals are greenish. A thick and 4-lobed disc is present. The androecium consists of 4 stamens. The gynoecium consists of 4 carpels which are free, ellipsoid, glossy, each developing a linear stigma. The drupes are black, glossy, ovoid, and about 5 mm long.

Medicinal uses: Dysentery, furncles, sores (China); fever (Laos); ringworm (Indonesia)

Strong antibacterial (Gram-positive) amphiphilic β-carboline alkaloid: Canthin-6-one (Log D = 2.9 at pH 7.4; molecular mass = 220.2 g/mol) inhibited the growth of *Staphylococcus aureus* (1199B), *Staphylococcus aureus* (XU212), and *Staphylococcus aureus* (ENRSA-15) with the MIC values of 8, 8, and 32 μg/mL, respectively (O'Donnell & Gibbons, 2007).

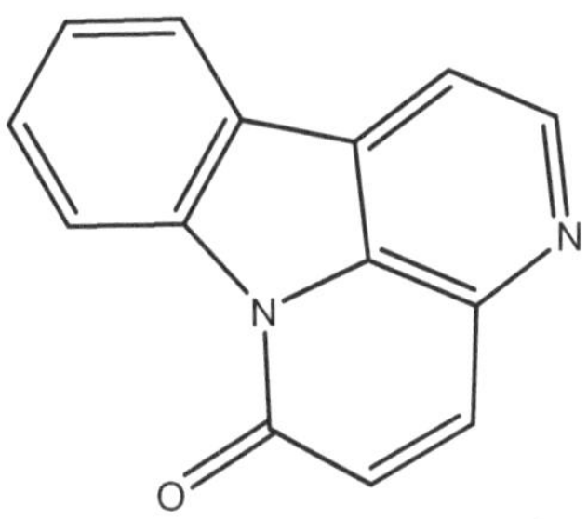

Canthin-6-one

Strong antimycobacterial β-carboline alkaloid: Canthin-6-one inhibited the growth of *Mycobacterium fortuitum* (ATCC 6841), *Mycobacterium smegmatis* (ATCC 14468), *Mycobacterium smegmatis* (mc²22700), *Mycobacterium phlei* (ATCC 111758), and *Mycobacterium abcessus* (ATCC 19977) with the MIC values of 16, 8, 8, 8, and 16 μg/mL, respectively (O'Donnell & Gibbons, 2007).

Broad-spectrum antibacterial polar extract: Methanol extract of seeds inhibited the growth of *Staphylococcus aureus* (ATCC 25923), *Staphylococcus aureus* (COWAN 1), *Staphylococcus epidermidis* (ATCC 12228), *Chromobacterium violaceum* (ATCC 12472), *Escherichia coli* (ATCC 25922), *Escherichia coli* (ATCC 35218), *Pseudomonas aeruginisa* (Paol), and *Streptococcus mutans* (ATCC 14721) (Simamora et al., 2019).

Anticandidal halo developed by polar extract: Aqueous extract of seeds (100 mg/mL/6 mm disc) inhibited the growth of *Candida albicans* (ATCC 14053), *Candida dubliniensis* (ATCC MYA-2975), *Candida glabrata* (ATCC 90030), *Candida krusei* (ATCC14243), *Candida lusitaniae* (ATCC 64125), *Candida parapsilosis* (ATCC 22019), and *Candida tropicalis* (ATCC 13803) with the inhibition zone diameters of 13.8, 22.7, 15.6, 12, 15.1, 15, and 11.3 mm, respectively (Harun & Razak, 2013).

Strong antiviral (non-enveloped monopartite linear single-stranded (+) RNA) decanortriterpene-type quassinoids: Brucatol and bruceine B inhibited the replication with the Tobacco mosaic virus (Yan et al., 2010).

Viral enzyme inhibition by polar extract: Aqueous extract of seeds at the concentration of 250 µg/mL inhibited Human immunodeficiency virus type-1 protease by 61.29% (Xu et al., 1996).

Commentaries: The plant brings to being series of β-carboline alkaloids of which canthin-6-one, 5-methoxycanthin-6-one, and 1-hydroxy-11-methoxycanthin-6-one (Harris et al., 1985) as well as quassinoids (Yan et al., 2012). Consider that quassinoids are often strongly active against the phyto-pathogenic Tobaccos mosaic virus and are probably used as plant defense (phytoalexin). Canthin-6-one (O'Donnell & Gibbons, 2007) is planar and intercalates with bacterial DNA (Cao et al., 2007). One could examine the antiviral properties of canthin-6-one and derivates against coronaviruses.

REFERENCES

Cao, R., Peng, W., Wang, Z. and Xu, A., 2007. β-Carboline alkaloids: biochemical and pharmacological functions. *Current Medicinal Chemistry*, *14*(4), pp. 479–500.

Harris, A., Anderson, L.A., Phillipson, J.D. and Brown, R.T., 1985. Canthin-6-one alkaloids from *Brucea antidysenterica* root bark. *Planta Medica*, *51*(2), pp. 151–153.

Harun, W.H.A.W. and Razak, F.A., 2013. Antifungal susceptibility and growth inhibitory response of oral Candida species to *Brucea javanica* Linn. extract. *BMC complementary and Alternative Medicine*, *13*(1), pp. 1–8.

O'Donnell, G. and Gibbons, S., 2007. Antibacterial activity of two canthin-6-one alkaloids from *Allium neapolitanum*. *Phytotherapy Research: An International Journal Devoted to Pharmacological and Toxicological Evaluation of Natural Product Derivatives*, *21*(7), pp. 653–657.

Simamora, A., Timotius, K.H. and Santoso, A.W., 2019. Antidiabetic, antibacterial and antioxidant activities of different extracts from *Brucea javanica* (L.) merr seeds. *Pharmacognosy Journal*, *11*(3), pp. 479–485.

Xu, H.X., Wan, M., Loh, B.N., Kon, O.L., Chow, P.W. and Sim, K.Y., 1996. Screening of Traditional medicines for their inhibitory activity against HIV-1 Protease. *Phytotherapy Research*, *10*(3), pp. 207–210.

Yan, X.H., Chen, J., Di, Y.T., Fang, X., Dong, J.H., Sang, P., Wang, Y.H., He, H.P., Zhang, Z.K. and Hao, X.J., 2010. Anti-tobacco mosaic virus (TMV) Quassinoids from Brucea javanica (L.) Merr. *Journal of Agricultural and Food Chemistry*, *58*(3), pp. 1572–1577.

8.5.6.3 *Eurycoma longifolia* Jack

Common names: Bittu bark; antung sa (Cambodia); tongkat Ali, bidara pahit (Malaysia); ba binh, hau phat (Vietnam).

Habitat: Forests

Distribution: Myanmar, Cambodia, Laos, Vietnam, Thailand, Malaysia, Indonesia, the Philippines

Botanical observation: This mysterious tree grows to a height of 10 m from a characteristically elongated and more or less straight taperoot. The wood is whitish and very bitter. The leaves are compound, spiral, exstipulate, gathered at apex of the bole, and up to about 1 m long. The rachis presents up to about 20 pairs of folioles which are sessile, 5–20 × 1.5–5 cm, lanceolate, of a peculiar green, and somewhat coriaceous. The base of the folioles is asymmetric. The panicles are axillary, tomentose, and bear minute flowers. The calyx is 5–6-lobed. The 5–6 petals are hairy, oblong, burgundy, and about 5 mm long. A disc is present. The androecium comprises 5–6 stamens. The 5–6 carpels are free with peltate stigma. The drupes are ellipsoid and about 1.5 cm long.

Medicinal use: Dysentery (Myanmar, Vietnam)

Broad-spectrum antibacterial polar extract: Methanol extract of stems (2.5 mg/mL/6 mm disc) inhibited the growth of *Staphylococcus aureus*, *Micrococcus luteus*, *Enterococcus faecalis*, *Staphylococcus aureus*, and *Proteus vulgaris* with the inhibition zone diameters of 10, 11, 11, 14, and 12 mm, respectively (Farouk & Benafri, 2007).

Strong antiviral (enveloped monopartite linear single-stranded (+) RNA) extract: Aqueous extract inhibited the replication of the Dengue virus serotype-1, serotype-2, serotype-3, and serotype-4 in C6/36 cells with the IC_{50} values of 33.8, 33.5, 58.3 and 119 µg/mL, respectively (George et al. 2019).

In vivo antiviral polar extract: Aqueous extract given orally (100 mg/Kg twice a day for 6 days) to AG129 mouse afforded some levels of protections against the Dengue virus type-2 (George et al., 2019).

Commentaries

i. The plant assembles series of quassinoids of which longilactone, 11-dehydroklaineanone, and eurycomanone as well as series of β-carboline indole alkaloids including canthin-6-one (Mitsunaga et al., 1994).

ii. In Laos, *Eurycoma hernandiana* (local name: *Hak ien don*) is traditionally used to treat fever.

iii. *Eurycoma longifolia* Jack has been traditionally used by Malays as tonic and the plant was taken one or twice a month in the form of a decoction. It is now sold as male aphrodisiac in the form of capsules or mixed with drinks and taken daily posing the possibility of chronic toxicity. Nevertheless, this plant could be of value for COVID-19. Traditional use of plants depends on healers (for instance "*bomohs*" in Malaysia, or "*herbalistos*" in the Philippines, or "*hakeems*" in Pakistan) who know (from oral or written tradition or information given from the unseen) what plant to use, when, and for how long. Traditional healers very seldom recommend daily intake of medicinal plants on a long term as compared with Western medicine's health supplements. In the Philippines for instance, Cebuanos living in mountain will consult herbalisto when one is ill and the herbalisto will recommend (often based on what is called "supernatural" guidance) some leaves and roots to be taken for a short period of time and it often save lifes. Thus, it is not possible to develop herbal products without a deep understanding of the medicinal tradition of the plant. This tradition is not taught in most so-called schools of pharmacy which under financial pressure of accreditation and university ranking, have put aside their own traditional cultural medicinal wealth (except India and China) to please accreditation boards which according to some serve the sole interest of the corporate world (Osippian, 2008; Schwartz, 2017). Such schools become unable to produce pharmacists and researchers able to use the wealth of medicinal plants and developing them for the well being of their respective countries (Abdel-Azim, 2011). The COVID-19 chaos and the other waves of microbial pandemics are a wake-up call that schools of pharmacy must train over a period of at least 5years their students with serious and solid teaching in medicinal plants (pharmacognosy) because forming pharmacy graduates skilled in pharmacognosy will guarantee protection to the community against pandemics, the manufacture of efficient local herbal products, creation of jobs in rural areas, and the formation of skilled students that will be able to enroll for PhD locally and become themselves skilled academics. The Western world has lost its traditional medicine practice. In France, for instance, fully trained herbalists were allowed to have shops and prescribe herbs of excellent quality and this was abrogated in 1941. Herbs are now in the West purchased online exposing the population to fraud and poisoning. In China, where traditional medicine practice has been preserved to some extent, COVID-19 appears to be controlled evidencing the fact that depriving pharmacy students from serious medicinal plant training (pharmacognosy) is synonym of depriving the population from simple but efficient herbals that could save their lifes. The only beneficiaries from the removal of medicinal plant teaching from pharmacy syllabus is the pharmaceutical industry and its share-holders. The COVID-19 tragedy also warns us that academics should be given the time to sciences and let administrators do the administration. Promotions should be given based on research outputs and not following any corporate criteria such as communication skills, popularity with students, indoor awards, and other unintelligent criteria. Further, the corporate model followed in universities is synonym with the implementation of some sort of culture of repression, mobbying, and lack of intellectual freedom which are detrimental to the development of science (Chomsky, 1999).

iv. Extract of the plant being active against Dengue virus which is a (+)-RNA virus could have activity against COVID-19 and other upcoming pandemic (+)-RNA viruses.
v. The structure of canthin-6 – one has similarity with the structure of serotonin and thus might inhibit bacteria and/or fungal efflux pumps.

REFERENCES

Abdel-Azim, N.S., Shams, K.A., Shahat, A.A., El Missiry, M.M., Ismail, S.I. and Hammouda, F.M., 2011. Egyptian herbal drug industry: challenges and future prospects. *Research Journal of Medicinal Plant*, *5*(2), pp. 136–44.

Chomsky, N., 1999. *Profit over people: Neoliberalism and global order.* Seven Stories Press, USA.

Farouk, A.E. and Benafri, A., 2007. Antibacterial activity of *Eurycoma longifolia* Jack. *Saudi Medical Journal*, *28*(9), pp. 1422–1424.

George, A., Zandi, K., Biggins, J., Chinnappan, S., Hassandarvish, P. and Yusof, A., 2019. Antiviral activity of a standardized root water extract of Eurycoma longifolia (Physta®) against dengue virus. *Tropical Biomedicine*, *36*(2), pp. 412–421.

Mitsunaga, K., Koike, K., Tanaka, T., Ohkawa, Y., Kobayashi, Y., Sawaguchi, T. and Ohmoto, T., 1994. Canthin-6-one alkaloids from Eurycoma longifolia. *Phytochemistry*, *35*(3), pp. 799–802.

Osipian, A.L., 2008. Corruption in higher education: does it differ across the nations and why?. *Research in Comparative and International Education*, *3*(4), pp. 345–365.

Schwartz, M.A., 2017. From the ordinary to corruption in higher education. *Journal of Management Inquiry*, *26*(3), pp. 270–279.

8.5.6.4 *Picrasma javanica* Bl.

Common names: Quassia wood; bonpashala (India); nan pawkyaw, taung-kamaka (Myanmar); tuba ulet (Indonesia); balimbing (Malaysia); nalis (the Philippines); yee nam bai lek yee num (Thailand); khor diejp (Vietnam)

Habitat: Forests

Distribution India, Myanmar, Cambodia, Laos, Vietnam, Thailand, Malaysia, Indonesia, the Philippines, Papua New Guinea, Solomon Islands

Botanical description: This handsome tree grows up to a height of 25 m. The bark is dark, smooth, bitter, and the inner bark yellowish with brownish sapwood. The leaves are compound, spiral, stipulate, and fragrant. The stipules are orbicular, leaf-like, and veined. The petiole grows up to about 10 cm long. The blade comprises 5–7 membranaceous, wavy, folioles which are oblong to elliptic, 4–20 × 1–10 cm, wavy, marked with about 5 pairs of secondary nerves, tapering and somewhat asymmetrical at base, and acuminate at apex. The cymes are axillary, showy, up to about 25 cm long, and present numerous tiny flowers. The 4 sepals are triangular. The corolla presents 4 whitish petals. A 4-lobed disc is present. The corolla presents 4 petals which are up to about 5 mm long. A 4-lobed disc is present. The androecium comprises 4 stamens. The drupes are green, red or blue, glossy, up to about 1 cm across and globose to subglobose.

Medicinal uses: Fever (Myanmar, Thailand, Indonesia); sores (Indonesia)

Broad-spectrum antibacterial non-polar extract: Petroleum ether of stem bark (3 mg/disc) inhibited the growth of *Bacillus cereus*, *Bacillus coagulans*, *Bacillus megaterium*, *Bacillus subtilis*, *Micrococcus luteus*, *Micrococcus roseus*, *Staphylococcus albus*, *Staphylococcus aureus*, *Staphylococcus epidermidis*, *Streptococcus faecalis*, *Streptococcus pneumoniae*, *Escherichia coli*, *Klebsiella pneumoniae*, *Proteus vulgaris*, *Pseudomonas aeruginosa*, *Salmonella typhimurium*, and *Serratia marscencens* (Khan et al., 2001).

Strong antibacterial decanortriterpene-type quassinoids: (16*R*)-methoxyjavanicin B, (16*S*)-methoxyjavanicin B, javanicin B, and picrajavanicin B inhibited the growth of *Bacillus subtilis* with the MIC values of 1.6, 1.6, 6.3, and 3.1 μM, respectively (Wong et al., 2019).

Commentaries

i. The plant produces series of quassinoids and β-carboline alkaloids (Wong et al., 2019) with possible antimicrobial properties.

ii. In India, *Picrasma quassioides* Bennett. (local name: *Nimatotaa*) is traditionally used as anthelminthic and as such could have antibiotic potentiating effects (See earlier).
iii. What is the antibacterial mode of action of quassinoids? In Eukaryotic cells quassinoids inhibit protein synthesis and are inactive towards protein synthesis in *Escherichia coli* (probably because of intrisinc resistance such as permeability or efflus pumps) (Liao et al., 2016).
iv. The plant has apparently not been studied for possible antiviral properties. Note that *Picrasma quassioides* (D. Don) Benn. generates antiviral quassinoids (Chen et al., 2009).

REFERENCES

Chen, J., Yan, X.H., Dong, J.H., Sang, P., Fang, X., Di, Y.T., Zhang, Z.K. and Hao, X.J., 2009. Tobacco mosaic virus (TMV) inhibitors from *Picrasma quassioides* Benn. *Journal of Agricultural and Food Chemistry*, *57*(15), pp. 6590–6595.

Khan, M.R., Kihara, M. and Omoloso, A.D., 2001. Antibacterial activity of *Picrasma javanica*. *Fitoterapia*, *72*(4), pp. 406–408.

Liaoo, L.L., Kupchan, S.M. and Horwitz, S.B., 1976. Mode of action of the antitumor compound bruceantin, an inhibitor of protein synthesis. *Molecular Pharmacology*, *12*(1), pp. 167–176.

Wong, C.P., Nugroho, A.E., Awouafack, M.D., Win, Y.Y., Win, N.N., Ngwe, H., Morita, H. and Morita, H., 2019. Two new quassinoids and other constituents from *Picrasma javanica* wood, and their biological activities. *Journal of Natural Medicines*, *73*(3), pp. 589–596.

8.5.6.5 *Quassia indica* (Gaertn.) Noot.

Synonyms: *Samadera indica* Gaertn.

Common names: Bitter wood, niepa bark tree; karinghota, nibam (India); kayu pahit (Indonesia); kacang kacang (Malaysia); le-seik-shin (Myanmar); age, daraput (the Philippines); samadara gas (Sri Lanka)

Habitat: Watery spots in forests, mangroves

Botanical observation: This handsome tree grows up to 20 m tall. The leaves are simple, alternate, and exstipulate. The petiole is 1–2.5 cm long. The blade is broadly lanceolate to elliptic, dark green, smooth, and glossy above, about 10–18 × 4–12 cm, and with a somewhat translucent margin. The umbels are axillary or terminal on about 20 cm stalks. The calyx is about 2 mm long, hairy, and develops 4 lobes. The 4 petals are somewhat yellowish, glossy, obovate, recurved, and up to 3 cm long. A disc is present. The 8 stamens are spreading, the filaments with the same color as the petals and pure white anthers. The fruits are glossy, transported by the seas, somewhat reddish, smooth, somewhat vaguely pod or blade-shaped, up to about 9 cm long, and contain a single seed.

Medicinal uses: Cholera (the Philippines); fever (India, Indonesia; Myanmar; the Philippines; Solomon Islands)

Broad-spectrum antibacterial halo developed by polar extract: Methanol extract of leaves (1 mg/mL/5 mm well) inhibited the growth of *Escherichia coli* and *Staphylococcus aureus* with the inhibition zone diameters of 22 and 24 mm, respectively (Anusha & Sudha Bai, 2017).

Broad-spectrum antifungal halo developed by polar extract: Extract of leaves (1 mg/mL/5 mm well) inhibited the growth of *Aspergillus niger* and *Candida albicans* with the inhibition zone diameters of 12 and 10 mm, respectively (Anusha & Sudha Bai, 2017).

Commentaries

i. The plant has apparently not been studied for antiviral properties.The quassinoid simalikalactone D (Log D = 1 at pH 7.4; molecular mass = 478.5 g/mol) isolated from a member of the genus *Quassia* L. inhibited the replication of the Herpes simplex, Semliki forest, Coxsackie, and Vesicular stomatitis *viruses* (Apers et al., 2002). Thus, quassinoids could be explored for their activity against coronaviruses and other forthcoming pandemic viruses.
ii. In the closely related family Surianaceae Arnott (1834) *Suriana maritima* L. is to treat dysentery by the Islanders of the India ocean.

REFERENCES

Anusha, P. and Sudha Bai, R., 2017. Phytochemical profile and antimicrobial potential of methanolic extracts of bark and leaf of *Quassia indica* (Gaertn.) Nooteb. *The Journal of Phytopharmacology*, *6*(5), pp. 269–276.

Apers, S., Cimanga, K., Berghe, D.V., Van Meenen, E., Longanga, A.O., Foriers, A., Vlietinck, A. and Pieters, L., 2002. Antiviral activity of simalikalactone D, a quassinoid from *Quassia africana*. *Planta Medica*, *68*(1), pp. 20–24.

8.6 Order Santalales R. Br. ex Bercht. & J. Presl (1820)

8.6.1 FAMILY LORANTHACEAE A. L. DE JUSSIEU (1808)

The family Loranthaceae consists of about 70 genera and 700 species of parasitic plants growing on trees. The leaves are simple, mostly opposite or whorled, sometimes reduced to scales, and without stipules. The blade is often characteristically somewhat spongy to coriaceous and glaucous and often without distinct secondary nerves. The flowers are actinomorphic, often somewhat elongated, and brightly colored. The calyx is adnate to the ovary and annular or cupular. The 4–6 petals are free or united into a tube, and often split down one side. The stamens are as numerous as the petals and inserted at base of the corolla. The anthers are 2-locular. The gynoecium consists of 3–4 carpels forming a compound, unilocular and inferior ovary containing 4–12 ovules attached to a free and basal central columnar placenta. The fruits are berries or drupes.

8.6.1.1 *Dendrophthoe pentandra* (L.) Miq.

Synonyms: *Loranthus pentandrus* L.

Common names: Malayan mistletoe; wu rui ji sheng (China); benalu, kemlandean (Indonesia); dedalu api jantan (Malaysia); bogto (the Philippines); ka fak (Thailand); dây chùm gớ'i, tang kis sin (Vietnam)

Habitat: Forests

Distribution: India, Myanmar, Laos, Cambodia, Thailand, Malaysia, Indonesia, the Philippines

Botanical observation: This epiphytic shrub grows up to 2 m tall and emits somewhat an aura of dryness. The leaves are simple, opposite, and exstipulate. The stems are smooth, glossy, and at apex hairy. The petiole is 0.5–2 cm long. The blade is lanceolate to elliptic, attenuate to cuneate at base, somewhat wavy, 5–13 × 2.5–8 cm, coriaceous, marked with 2–4 pairs of secondary nerves, and acute or rounded at apex. The racemes are axillary, few-flowered, and hairy. The calyx is cupular, hairy, about 3 mm long and 5-lobed. The corolla is tubular, up to 2 cm long, inflated at base, and develops 5 narrowly lanceolate lobes which are about 1 cm long and reflexed. The androecium includes 5 stamens. The berries are red, ovoid, up to 1 cm long.

Medicinal use: Cough (Vietnam); postpartum remedy, wounds (Malaysia); sores, ulcers (India; Malaysia)

Broad-spectrum antibacterial halo developed by flavonol glycoside: Quercitrin isolated from the leaves (100 μg/mL/6 mm disc) inhibited the growth of *Escherichia coli*, *Salmonella typhi*, *Staphylococcus aureus*, and *Pseudomonas aeruginosa* with the inhibition zone diameters of 7.1, 7.1, 9.3, and 7.6 mm, respectively (Hardiyanti et al., 2019).

Commentaries

i. The plant abounds with phenolics of which methyl gallate 3-*O*-β-D-glucopyranoside, methyl gallate 3-*O*-(6′-*O*-galloyl)-β-D glucopyranoside, (+)-catechin, procyanidin B-1 and procyanidin B-3 (Sahakitpichan et al., 2017) which probably work synergistically as astringents.
ii. *Dendrophthoe falcata* L.f. is used as postpartum protective remedy and to heal wounds in India (Local name: *Bandaaka, Ithil*) and Bangladesh (local name: *Sat-jat parsallo*).

DOI: 10.1201/9781003180371-3

In Malaysia, *Dendrophthoe incarnata* (Jack) Miq. (local name: *Dedalu api gajah*) is used to treat ringworm and *Loranthus cochinchinensis* Lour. is used to treat diarrhea (Local name: *Dedalu api*).

iii. Aqueous extract of *Loranthus parasiticus* (L.) Merr. (synonym: *Taxillus parasiticus* (L.) S.T. Chiu) at the concentration of 250 µg/mL inhibited Human immunodeficiency virus type-1 protease by 59.8% (Xu et al., 1996).

REFERENCES

Hardiyanti, R., Marpaung, L., Adnyana, I.K. and Simanjuntak, P., 2019. Isolation of quercitrin from *Dendrophthoe pentandra* (L.) Miq leaves and it's antioxidant and antibacterial activities. *Rasayan Journal of Chemistry*, *12*(4), pp. 1822–1827.

Sahakitpichan, P., Disadee, W., Buntawong, R., Chimnoi, N., Ruchirawat, S. and Kanchanapoom, T., 2017. A furan-2-carbonyl C-glucoside and an alkyl glucoside from the parasitic plant, *Dendrophthoe pentandra*. *Phytochemistry Letters*, *21*, pp. 90–93.

Xu, H.X., Wan, M., Loh, B.N., Kon, O.L., Chow, P.W. and Sim, K.Y., 1996. Screening of traditional medicines for their inhibitory activity against HIV-1 Protease. *Phytotherapy Research*, *10*(3), pp. 207–210.

8.6.1.2 *Scurrula atropurpurea* (Blume) Danser

Synonyms: *Cichlanthus philippensis* (Cham. & Schltdl.) Tiegh.; *Loranthus atropurpureus* Blume; *Loranthus philippensis* Cham. & Schltdl.; *Scurrula philippensis* (Cham. & Schltdl.) G. Don

Common names: Benalu (Indonesia); muc vij tias (Vietnam)

Habitat: Forests

Distribution: Cambodia, Laos, Vietnam, Thailand, Indonesia, the Philippines

Botanical observation: This epiphytic shrub grows to 1 m tall. The stems are hairy. The leaves are simple, opposite, and exstipulate. The petiole is 0.7–1.2 cm long and hairy. The blade is ovate or oblong, wavy, 5–10 × 2–6 cm, hairy below, smooth and glossy, marked with lateral 4 or 5 pairs of secondary nerves, cuneate to obtuse at base, and acute or obtuse at apex. The racemes are axillary, solitary or fascicled. The calyx is about 2 mm long and somewhat 4-denticulate. The corolla is red, about 2.5 cm long, tubular, somewhat curved, and develops 4 lobes which are narrowly lanceolate, 6–8 mm long, and reflexed. The androecium includes 4 stamens. The ovary develops a capitate stigma. The berries are greenish, pyriform, about 8 mm long, and hairy.

Medicinal use: Mouthwash (the Philippines)

Broad-spectrum antibacterial polar extract: Methanol extract of leaves (400 µg/disc) inhibited the growth of *Bacillus subtilis*, *Escherichia coli*, and *Staphylococcus aureus* with the inhibition zone diameters of 11.3, 13.6, and 14.9 mm, respectively (Elfita et al., 2020).

Strong antiviral (non-enveloped linear monopartite single-stranded (+) RNA) polar extract: Methanol extract inhibited the replication of the Poliovirus in Vero cells with the EC_{50} value of 62 µg/mL and a selectivity index above 8 (Lohézic-Le Dévéhat et al., 2012).

Commentaries

i. Members of the family Loranthaceae (and often in this order Santalales) produce polyacetylenic fatty acids which are often antibacterial and or antifungal. Octadeca-8,10,12-triynoic acid is produced by *Scurrula atropurpurea* (Blume) Danser (Ohashi et al., 2003). Further, epiphytic Loranthaceae contain secondary metabolites from their host tree (caffeine for instance if it grows on *Camellia sinensis* L.) (Ohashi et al., 2003).

ii. *Scurrula ferruginea* (Jack) Danser is used in Malaysia (local name: *Dedalu api merah*) to heal wounds and as postpartum protective remedy. Like most Loranthaceae it abounds with phenolics including flavonols and is thus astringent. Consider that that medicinal plants used to heal wounds and/or treat diarrhea often accumulates phenolics (including tannins).

iii. *Scurrula gracilifolia* Danser is used in Cambodia, Laos, and Vietnam to treat furuncles.

REFERENCES

Elfita, E., Muharni, M., Mardiyanto, M., Fitrya, F., Fera, F. and Widjajanti, H., 2020. Antibacterial activity of traditional medicine *Scurrula atropurpurea* (BL) DANS and their endophytic fungi. In Wahyuningsih, T.D., et al. (eds) *Key Engineering Materials* (Vol. 840, pp. 205–213). Trans Tech Publications Ltd, Stafa-Zurich.

Lohézic-Le Dévéhat, F., Bakhtiar, A., Bezivin, C., Amoros, M. and Boustie, J., 2002. Antiviral and cytotoxic activities of some Indonesian plants. *Fitoterapia*, *73*(5), pp. 400–405.

Ohashi, K., Winarno, H., Mukai, M., Inoue, M., Prana, M.S., Simanjuntak, P. and Shibuya, H., 2003. Indonesian medicinal plants. XXV. Cancer cell invasion inhibitory effects of chemical constituents in the parasitic plant *Scurrula atropurpurea* (Loranthaceae). *Chemical and Pharmaceutical Bulletin*, *51*(3), pp. 343–345.

8.6.1.3 *Taxillus umbellifer* (Schult.) Danser

Synonyms: *Loranthus umbellifer* Schult. f.; *Scurrula umbellifer* (Schult. f.) G. Don

Common names: An hua dun guo ji sheng (China); hagero (Nepal)

Habitat: Forests, mountain slopes.

Distribution: India, Myanmar, Bhutan, China;

Botanical observation: This epiphytic shrub grows up to 2 m tall. The stems are hairy at apex. The leaves are simple, opposite or subopposite, and exstipulate. The petiole is 0.8–1.2 cm long. The blade is oblong-ovate or elliptic, 10–12 × 5–6 cm, coriaceous, hairy below, marked with 5–7 pairs of secondary nerves, cuneate at base, and shortly acuminate or obtuse at apex. The umbel-like fascicles are axillary and hairy. The calyx is tubular, 4–5 mm long, and develops 4 lobes. The corolla is red, curved, hairy, and develops 4 lobes which are 8 mm long and reflexed. The ovary develops a capitate stigma. The berries are subglobose.

Medicinal uses: Skin infection (Nepal)

Commentaries

(i) The plant has apparently not been examined for antimicrobial activities. Plants in the genus *Taxillus* Tiegh. accumulate flavonoids of which the flavone glycosides avicularin, hyperin, quercitrin, and taxillusin as well as (+)-catechin and the condensed tannins procyanidins B-1 and B-3 (Su et al, 2017). Polar extracts (aqueous, methanol, ethanol, acetone) of plants accumulating phenolics are always broadly antimicrobial in *vitro*. Extract of *Taxillus chinensis* (DC.) Danser inhibited the Human immunodeficiency type-1 protease (Dũng et al., 2015).

(ii) Hexane extract of stems and leaves of *Taxillus chinensis* (DC.) Danser (local name: *Sāng jì sheng*) (China) inhibited the replication of the Severe acute respiratory syndrome-acquired coronavirus in Vero cells with the IC_{50} value of 5.3 µg/mL and a selectivity index above 92.8 (Wen et al., 2011). The active principle here is not known. Hexane is a non-polar solvent and thus there is (are) lipophilic antisevere acute respiratory syndrome-acquired coronavirus principles in this plant.

(iii) Kaempferol-3,7-bisrhamnoside and (3*S*)-3-hydroxy-1,7-bis(4-hydroxy-phenyl)-6*E*-hepten-5-one isolated from *Taxillus sutchuenensis* (Lecomte) Danser inhibited the enzymatic activity of the Hepatitis C virus NS3 protease with the IC_{50} values of 19.4 and 28.7 µM, respectively (Yang et al., 2017).

(iv) *Taxillus kaempferi* (DC) Danser is used in China (local name: *Xiao ye dun guo ji sheng*) as an antiseptic. In India, *Taxillus tomentosus* (B. Heyne ex Roth) Tiegh. (local name: *Tellanaguis*) is used to treat leprosy.

REFERENCES

Dũng, N.Q., Châu, L.T.K., Loan, N.T.H., Phương, N.T., Thương, P.T., Nghĩa, P.T. and Thuân, B.P., 2015. Hoạt tính ức chế Pepsin và Protease HIV-1 của các cao chiết và hoạt chất Acid maslinic từ dược liệu. *VNU Journal of Science: Natural Sciences and Technology*, *31*(2), pp. 18–27.

SU, W.B., Wang, H., LI, H.Y., Pei, H.H., Zhu, X.K. and Dong, L.U., 2017. Content analysis of avicularin, quercitrin and quercetin in Herba Taxilli from different host plants. *Journal of International Pharmaceutical Research*, *44*(7), pp. 738–742.

Yang, L., Lin, J., Zhou, B., Liu, Y. and Zhu, B., 2017. Activity of compounds from *Taxillus sutchuenensis* as inhibitors of HCV NS3 serine protease. *Natural Product Research*, *31*(4), pp. 487–491.

Wen, C.C., Shyur, L.F., Jan, J.T., Liang, P.H., Kuo, C.J., Arulselvan, P., Wu, J.B., Kuo, S.C. and Yang, N.S., 2011. Traditional Chinese medicine herbal extracts of *Cibotium barometz*, *Gentiana scabra*, *Dioscorea batatas*, *Cassia tora*, and *Taxillus chinensis* inhibit SARS-CoV replication. *Journal of Traditional and Complementary Medicine*, *1*(1), pp. 41–50.

8.6.2 FAMILY VISCACEAE BATSCH (1802)

The family Viscaceae consists of about 7 genera and 400 species of shrubs or herbs growing on trees. The stems are characteristically articulated with nodes. The leaves are simple, opposite, and often minute and scaly, and exstipulate. The inflorescences are spikes, cymes or solitary and are terminal or axillary. The perianth presents 3–4-lobes. The androecium includes 3–4 stamens. The gynoecium consists of a 1-loculed ovary and a short style. The drupes contain a sticky gel.

8.6.2.1 *Viscum articulatum* Burm. F.

Synonyms: *Viscum pedunculatum* Barlow

Common name: Liu guo hu ji sheng (China)

Habitat: Forests

Distribution: India, Bangladesh, Myanmar, Cambodia, Laos, Vietnam, Thailand, China, Malaysia, Indonesia, the Philippines, Australia

Botanical observation: This epiphytic and pendulous herb grows to a length of about 50 cm. The stems grow dichotomously, are flattened, somewhat fleshy of a peculiar light green, and longitudinally 3-ridged. The leaves are reduced to minute pairs of scales. The cymes are axillary, minute, and few flowered. The perianth develops 4 broadly lanceolate lobes. The berries are whitish or greenish white, globose, smooth, glossy, somewhat pearl-like and translucent, about 5 mm across, and apiculate.

Medicinal uses: Bronchitis (Cambodia, Laos, Vietnam); tuberculosis (China); dysentery, jaundice, wounds (India)

Moderate broad-spectrum antibacterial polar extract: Methanol extract inhibited the growth of *Staphylococcus aureus* and *Escherichia coli* with the MIC/MBC values of 728/1456 and 920/1840 μg/mL (Ray et al., 2014).

Strong antiviral (monopartite linear dimeric single stranded (+) RNA) flavanol glycoside: Homoeriodictyol-7-*O*-β-D-glucopyranoside-4′-*O*-β-D-apiofuranoside inhibited the replication of the Human immunodeficiency virus with the EC_{50} value of 18 μg/mL and a selectivity index above 11 (Li et al., 2008).

Commentary: The plant abounds with flavonoid glycosides as well as triterpenes of which oleanolic acid (Patel et al., 2018).

REFERENCES

Li, Y., Zhao, Y.L., Huang, N., Zheng, Y.T., Yang, Y.P. and Li, X.L., 2008. Two new phenolic glycosides from *Viscum articulatum*. *Molecules*, *13*(10), pp. 2500–2508.

Patel, B.P. and Singh, P.K., 2018. *Viscum articulatum* Burm. f.: a review on its phytochemistry, pharmacology and traditional uses. *Journal of Pharmacy and Pharmacology*, *70*(2), pp. 159–177.

Ray, S., Tripathi, S., Mondal, A.K. and Verma, N.K., 2014. Antimicrobial activities of leaf extract of some hemi parasitic Taxa against some common human pathogen: first time report in the World. *Bulletin of Environment, Pharmacology and Life Sciences*, *3*, pp. 21–28.

8.6.3 FAMILY OLACACEAE A. L. DE JUSSIEU EX R. BROWN (1818)

The family Olacaceae consists of about 25 genera and 250 species of trees, shrubs, or climbers. The leaves are alternate or opposite and exstipulate. The inflorescences are axillary or terminal cymes or racemes. The flowers are actinomorphic. The calyx is tubular and truncated or develops 4–6 lobes. The corolla consists of 3–6 petals. A disc is present. The androecium comprises 3–15 stamens. The ovary is made of 2–5 locules, each locule sheltering a single pendulous ovule. The style ends up into a 2–5-lobed stigma. The fruits are drupes or capsules. So far, medicinal plants in this family are known to produce antimicrobial alkyl sulfurs and polyacetylenic fatty acids.

8.6.3.1 *Scorodocarpus borneensis* Becc.

Synonym: *Ximenia borneensis* Baill.

Common names: Wood garlic; bawang hutan, pokok kulim (Malaysia)

Habitat: Forests

Distribution: Thailand, Malaysia, Indonesia

Botanical description: This majestic tree grows to a height of 25 m. The area in which grows this tree smells garlic. The bark is brown and scaly, and the inner bark is pinkish – red, juicy, and with strong garlic smell. The leaves are simple, spiral, and without stipules. The petiole is 1.5–1.8 cm long, keeled at apex and woody. The blade is glossy, greyish – green, coriaceous, obovate, 17–24 × 6–9.5 cm, acuminate at apex, and marked with 3–6 pairs of secondary nerves sunken above and a few tertiary nerves below. The racemes are axillary. The flowers are greenish-white and 6 mm long. The drupes are green, about 5 cm accross, and contain a seed enclosed in a shell with somewhat longitudinal fissures. The seeds have a very pungent and turn purplish within at germination.

Medicinal use: Ringworm (Malaysia)

Moderate broad-spectrum antibacterial alkyl sulfur: 2,4,6-Trithiaheptane-5–thione 2,2 dioxide isolated from the fruits inhibited the growth of *Bacillus subtilis* (PCI 219) and *Escherichia coli* (NIHJ) with the MIC values of 50 and 50 µg/mL, respectively (Lim et al., 1998).

Moderate broad spectrum antifungal alkyl sulfur: 2,4,6-Trithiaheptane-5–thione 2,2 dioxide isolated from the fruits inhibited the growth of *Candida albicans* (KF1), *Saccharomyces cerevisiae* (ATCC 9763), *Mucor racemosus* (IFO 4581), and *Aspergillus niger* (KF 105) with the MIC values of 50, 50, 50, 25, and 50 µg/mL, respectively (Lim et al., 1998).

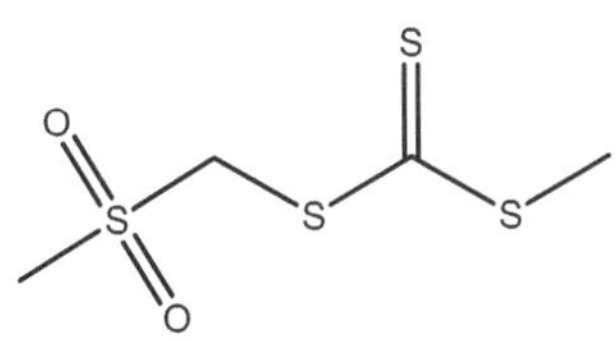

2,4,6-Trithiaheptane-5–thione 2,2 dioxide

Commentaries

i. The garlic smell is due to series of aliphatic sulfur and sulfoxide compounds including 2,4,5,7-tetrathiaoctane 4,4-dioxide, 5-thioxo-2,4,6-trithiaheptane 2,2-dioxide, and O-ethyl S-methylthiomethyl thiosulfite (Lim et al., 1998).
ii. Petroleum ether extract of the fresh seeds inhibited the growth of a broad spectrum of fungi *in vitro* and was antibacterial *in vivo* (Wiart, 2001).
iii. This tree is on its way to extinction owed to increasing logging and burning for palm oil businesses. The oil of its fruits treats ringworm infection and could be used as

ingredient for antifungal creams. Note that strains of multidrug-resistant pathogenic and phytopathogenic fungi are on our way (Kainz et al., 2020) and when these will impact us, *Scorodocarpus borneensis* Becc. and many other life-saving plants will be extinct as well as their antifungal products. There is the hope that palm oil production could be controlled and recent decision of the European Union goes in this direction.

iv. Malaferin B, malaferin C, garcimangosone D, (-) epicatechin-3-*O*-benzoate, and procyanidin B5 isolated from *Malania oleifera* Chun & S.K. Lee inhibited Human immunodeficiency virus type-1 (IIIB strain) with EC_{50} values of 18, 15.8, 18, 17, and 31.4 µg/mL, respectively (Wu et al., 2012). The selectivity index for garcimangosone D was 11 (Wu et al., 2012).

v. *Ximenia americana* L. is used in India (local name: *Chiru-illantai*) for fever, jaundice, diarrhea, and venereal diseases. Methanol extract of bark of *Ximenia americana* L. (100 mg/mL; 0.1 mL/well) inhibited the growth of *Bacillus subtilis, Staphylococcus aureus, Escherichia coli*, and *Pseudomonas aeruginosa* with inhibition zone diameters of 23, 30, 19, and 22 mm, respectively. Methanol extract of bark (100 mg/mL; 0.1 mL/well) inhibited *Candida albicans, Saccharomyces cerevisiae*, and *Aspergillus niger* with inhibition zone diameters of 12, 26, and 14 mm, respectively (Omer & Elnima, 2003). Ethanol extract of roots of *Ximenia americana* L. inhibited *Staphylococcus aureus* (ATCC 29213) with the MIC of 188 µg/mL and inhibited the growth of 7 clinical trains of methicillin-resistant *Staphylococcus aureus* with minimum inhibiting concentrations ranging from 188 to >1500 µg/mL (Kone et al., 2004). This extract inhibited the growth of *Enterococcus faecalis* (ATCC 29212) with the MIC of 188 µg/mL and inhibited the growth of 7 clinical strains of aminoside resistant with the MIC ranging from 94 to 375 µg/mL (Kone et al., 2004). The extract inhibited *Streptococcus pyogenes* with the minimum inhibiting concentration of 94 µg/mL (Kone et al., 2004).

vi. *Ochanostachys amentacea*. Mast. is used as a postpartum protective remedy in Malaysia. The polyacelylene fatty acid minquartynoic acid (Log D = 13.4 at pH 7.4; molecular mass = 284.3 g/mol) isolated from *Onachanostachys amentacea* Mast. at a concentration of 2–5 µg/mL inhibited human lymphoblastoid cell killing by the Human immunodeficiency virus type-1 (Rashid et al., 2001).

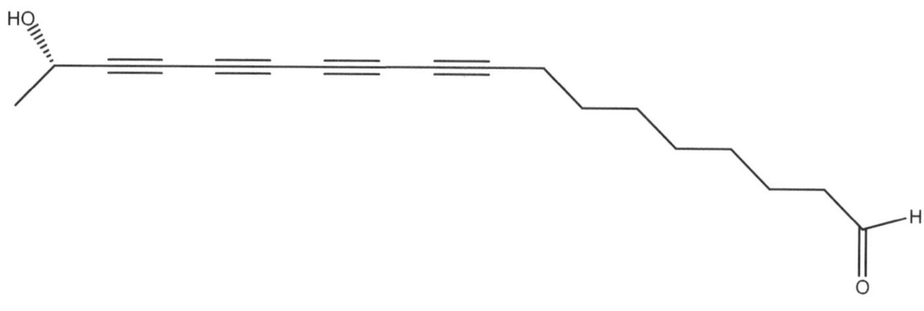

Minquartynoic acid

vii. *Olax scandens* Roxb. is used in India (local name: *Kaattu pavalam*) and Myanmar (local name: *Dheniaani*) for fever and in Thailand (local name: *Joh to*) for oral abscesses. Hexane extract of leaves of this shrub (5 mg/disc) inhibited the growth of *Bacillus subtilis, Enterococcus faecalis, Pseudomonas aeruginosa*, and *Klebsiella pneumoniae* with the inhibition zone diameters of 8, 13, 12, and 9 mm, respectively (Duraipandiyan et al., 2006), on probable account of polyacetylenic fatty acids because they are common in the family Olacaceae and liposoluble. Polyacetylenic fatty acids are strongly antibacterial, antimycobacterial, antifungal, and antiviral (Negri, 2015). The mechanism of action of these phytoalexins (De Wit & Kodde, 1981) is not yet fully understood but being lipophilic, they probably accumulate in and destabilize the hydrophobic internal layer of cytoplasmic membranes of bacteria of fungi (Kim et al., 1988). These are however cytotoxic for mammalian cells and of no clinical systemic antimicrobial usefulness.

viii. In the family Opiliaceae Valeton (1886), *Champereia griffithiana* Planch. ex Kurz is used in Malaysia to Ethylacetate. Ethylacetate extract of stems of *Lepionurus sylvestris* Bl. inhibited the growth of *Streptococcus mitis* (ATCC 49456T) with the MIC/MBC values of 19.5/19.5 μg/mL (Meerungrueang & Panichayupakaranant, 2014).

REFERENCES

De Wit, P.J.G.M. and Kodde, E.M.G.M., 1981. Induction of polyacetylenic phytoalexins in *Lycopersicon esculentum* after inoculation with *Cladosporium fulvum* (syn. Fulvia fulva). *Physiological Plant Pathology*, *18*(2), pp. 143–148.

Duraipandiyan, V., Ayyanar, M. and Ignacimuthu, S., 2006. Antimicrobial activity of some ethnomedicinal plants used by Paliyar tribe from Tamil Nadu, India. *BMC Complementary and Alternative Medicine*, *6*(1), pp. 1–7.

Kainz, K., Bauer, M.A., Madeo, F. and Carmona-Gutierrez, D., 2020. Fungal infections in humans: the silent crisis. *Microbial Cell*, *7*(6), p. 143.

Kim, H., Lee, Y.H. and Kim, S.I., 1988. A possible mechanism of polyacetylene: membrane cytotoxicity. *Toxicological Research*, *4*(2), pp. 95–105.

Kone, W.M., Atindehou, K.K., Terreaux, C., Hostettmann, K., Traore, D. and Dosso, M., 2004. Traditional medicine in North Côte-d'Ivoire: screening of 50 medicinal plants for antibacterial activity. *Journal of Ethnopharmacology*, *93*(1), pp. 43–49.

Lim, H., Kubota, K., Kobayashi, A. and Sugawara, F., 1998. Sulfur-containing compounds from *Scorodocarpus borneensis* and their antimicrobial activity. *Phytochemistry*, *48*(5), pp. 787–790.

Meerungrueang, W. and Panichayupakaranant, P., 2014. Antimicrobial activities of some Thai traditional medical longevity formulations from plants and antibacterial compounds from *Ficus foveolata*. *Pharmaceutical Biology*, *52*(9), pp. 1104–1109.

Negri, R., 2015. Polyacetylenes from terrestrial plants and fungi: recent phytochemical and biological advances. *Fitoterapia*, *106*, pp. 92–109.

Omer, M.E.F.A. and Elnima, E.I., 2003. Antimicrobial activity of *Ximenia americana*. *Fitoterapia*, *74*(1–2), pp. 122–126.

Rashid, M.A., Gustafson, K.R., Cardellina, J.H. and Boyd, M.R., 2001. Absolute stereochemistry and anti-HIV activity of minquartynoic acid, a polyacetylene from *Ochanostachys amentacea* 1a. *Natural Product Letters*, *15*(1), pp. 21–26.

Wiart, C., 2001. *Antibicrobial and cytotoxic compounds of Scorodocarpus borneensis (Olacaceae) and Glycosmis calcicola (Rutaceae)* (Doctoral dissertation, Universiti Putra Malaysia).

Wu, X.D., Cheng, J.T., He, J., Zhang, X.J., Dong, L.B., Gong, X., Song, L.D., Zheng, Y.T., Peng, L.Y. and Zhao, Q.S., 2012. Benzophenone glycosides and epicatechin derivatives from *Malania oleifera*. *Fitoterapia*, *83*(6), pp. 1068–1071.

8.6.4 FAMILY SANTALACEAE R. BROWN 1810

The family Santalaceae consists of about 35 genera and 450 species of herbs, shrubs, or trees which are often parasitic. The leaves are simple, alternate or opposite, and exstipulate. The inflorescences are racemes, spikes, or heads. The perianth is tubular and 4–5-lobed. A disc is present. The androecium includes 4–5 stamens. The gynoecium comprises a unilocular ovary with 1–3 pendulous ovules from a basal placenta and a short style. The fruit is a nut or a drupe.

8.6.4.1 *Dendrotrophe umbellata* (Blume) Miq.

Synonyms: *Henslowia umbellata* (Blume) Blume; *Viscum umbellatum* Blume

Common names: Selam phao (Cambodia); dalu dalu (Malaysia); ti mia muea yang (Thailand); hen tasn (Vietnam)

Habitat: Forests

Distribution: Cambodia, Laos, Vietnam, Thailand, Malaysia, Indonesia

Botanical observation: This parasitic shrub grows up to about 5 m long on trees. The leaves are simple, spiral, and exstipulate. The petiole is 5 mm long. The blade is obovate, 2.5–7 × 2–4 cm, cuneate at base, rounded at apex, coriaceous, dark green and glossy above, marked with 3–5 pairs of longitudinal secondary nerves (with a *Smilax* or *Viscum* look). The inflorescences are umbel-like. The perianth is minute and 5-lobed. The androecium includes 5 stamens inserted at base of the perianth lobes. The ovary develops a lobed stigma. The drupe is subglobose, blackish when ripe, glossy, and up to 1 cm long.

Medicinal use: Fever (Malaysia)

Broad-spectrum antibacterial halo developed by polar extract: Methanol extract of roots inhibited the growth of *Bacillus subtilis* (NRRL 1459IB) and *Pseudomonas aeruginosa* (UJ 60690) (Ali et al., 1995).

Broad-spectrum antifungal halo developed by polar extract: Methanol extract of roots inhibited the growth of *Aspergillus ochraceous* (NRRL398), *Aspergillus niger* (ATCC 11391), *Cunninghamella elegans* (ATCC 9244), *Candida albicans* (ATCC 10231), and *Candida intermedia* (ATCC 5159) (Ali et al., 1995).

Commentaries: In Malaysia, *Dendrotrophe varians* (Blume) Miq.is used for fever and *Dendrophthoe grandifrons* (King) Danser is used to treat ringworms. Plants in the genus *Dendrotrophe* Miq. produce phenolic glycosides (Hou et al., 2013) with probable antimicrobial properties.

REFERENCES

Ali, A.M., El-Sharhawy, S.H., Hamid, J.A., Ismail, N.H. and Lajis, N.H., 1995. Antimicrobial activity of selected Malaysian plants. *Pertanika Journal of Tropical Agricultural Science, 18*, pp. 57–62.

Hou, X., Sun, M., Gao, H., Cui, Y. and Xiao, K., 2013. Chemical constituents from the stems of *Dendrotrophe frutescens. Biochemical Systematics and Ecology, 51*, pp. 156–159.

8.6.4.2 *Santalum album* L.

Common names: Indian sandalwood; chandan; sada chandan, shet chandan (Bangladesh); bai tan xiang (China); shandanum (India); cendana (Indonesia; Malaysia); san ta ku (Myanmar); sandal-i-safid (Iran); nanttha hpyu (Myanmar); chantana (Thailand)

Habitat: Forests, cultivated

Distribution: Tropical Asia

Botanical observation: This tree probably native to Indonesia grows up to about 6 m tall. The leaves are simple, decussate, and exstipulate. The wood is aromatic and used as Holy incense by Bhuddists, Hindus, and Muslims. The petiole is slender and grows up to 2.5 cm long. The blade is lanceolate, obovate, or elliptic, 3.5–6 × 1.5–2.5 cm, coriaceous, fleshy, light green, obtuse at base and apex, and with inconspicuous secondary nerves. The inflorescences are axillary or terminal panicles of minute flowers. The perianth is tubular, develops 4 trianglular and recurved lobes, about 5 mm long and burgundy to red. The androecium comprises 4 stamens. The ovary is minute and develops a slender style which slightly protrudes out of the corolla and a 4-lobed stigma. The fruits are ovoid, about 1 cm long, and smooth.

Medicinal uses: Burn (India); cholera, tuberculosis (Bangladesh); lung diseases (China); urinary tract infection (India; Iran); fever (India; Myanmar); gonorrhea (Bangladesh; India; Myanmar)

Strong broad-spectrum antibacterial mid-polar extract: Dichloromethane/methanol extract inhibited the growth of *Escherichia coli* (2931), *Klebsiella aerogenes* (2098), *Arthrobacter nicotianae* (2460), *Bacillus subtilis* (2063), and *Staphylococcus aureus* (5021) with the MIC_{90} value of 10, 10, 10, 10, and 0.6 µg/mL, respectively (Misra & Dey, 2012).

Strong broad-spectrum antibacterial essential oil: Essential oil inhibited the growth of *Staphylococcus aureus*, *Escherichia coli*, *Pseudomonas aeruginosa*, and *Klebsiella pneumonia* with the MIC values of 6, 60, 60, and 600 ppm, respectively (Jirovetz et al., 2006).

Strong antifungal (filamentous) amphiphilic santalane-type sesquiterpenes: α-Santalol (Log D = 4.6 at pH 7.4; molecular mass = 220.3 g/mol) and β-santalol isolated from the wood inhibited the growth of *Trichophyton rubrum* with the MIC values of 12.5 and 25 µg/mL, respectively (Kim et al., 2017).

Strong antiviral (enveloped monopartite linear double-stranded DNA) essential oil: Essential oil inhibited the replication of the Herpes simplex virus type-2 (strain HG52) in RC-37 cells with the IC_{50} values of 0.0005% and a selectivity index of 3 (Koch et al., 2008).

Strong antiviral (enveloped segmented linear single stranded (−) RNA) amphiphilic santalane-type sesquiterpene: β-Santalol of (Log D = 4.5 at pH 7.4; molecular mass = 220.3 g/mol) at the concentration of 100 µg/mL inhibited the replication of the influenza A/HK (H3N2) virus in in Madin-Darby canine kidney cells by 86% (Paulpandi et al., 2012).

Commentaries

i. The plant is used for the treatment of COVID-19 in China (Wang et al., 2020) and the active principle is apparently yet unknown. The plant yields santalane-type sesquiterpenes which could be examined for their anticoronaviral activities.
ii. Essential oil extracted from the plant is used in perfumery on account of santalol which is very weakly antibacterial (Jirovetz et al., 2006). Santalol is neuroactive (Okugawa et al., 1995) and as such is probably able to inhibit bacterial and or fungal efflux pumps.
iii. In the "*Tractato contra la peste*", our peer the Italian physician Giovanni Manardi (1462–1536) recommended wearing bags of aromatic plants including sandalwood as a mean of protection against plagues (Vincentini, 2020).

REFERENCES

Jirovetz, L., Buchbauer, G., Denkova, Z., Stoyanova, A., Murgov, I., Gearon, V., Birkbeck, S., Schmidt, E. and Geissler, M., 2006. Comparative study on the antimicrobial activities of different sandalwood essential oils of various origin. *Flavour and Fragrance Journal*, *21*(3), pp. 465–468.

Kim, T.H., Hatano, T., Okamoto, K., Yoshida, T., Kanzaki, H., Arita, M. and Ito, H., 2017. Antifungal and ichthyotoxic sesquiterpenoids from Santalum album heartwood. *Molecules*, *22*(7), p. 1139.

Koch, C., Reichling, J., Schneele, J. and Schnitzler, P., 2008. Inhibitory effect of essential oils against herpes simplex virus type 2. *Phytomedicine*, *15*(1–2), pp. 71–78.

Misra, B.B. and Dey, S., 2012. Comparative phytochemical analysis and antibacterial efficacy of *in vitro* and *in vivo* extracts from East I ndian sandalwood tree (*Santalum album* L.). *Letters in Applied Microbiology*, *55*(6), pp. 476–486.

Okugawa, H., Ueda, R., Matsumoto, K., Kawanishi, K. and Kato, A., 1995. Effect of α-santalol and β-santalol from sandalwood on the central nervous system in mice. *Phytomedicine*, *2*(2), pp.119–126.

Paulpandi, M., Kannan, S., Thangam, R., Kaveri, K., Gunasekaran, P. and Rejeeth, C., 2012. In vitro anti-viral effect of β-santalol against influenza viral replication. *Phytomedicine*, *19*(3–4), pp. 231–235.

Vicentini, C.B., Manfredini, S., Mares, D., Bonacci, T., Scapoli, C., Chicca, M. and Pezzi, M., 2020. Empirical "integrated disease management" in Ferrara during the Italian plague (1629–1631). *Parasitology international*, *75*, p. 102046.

Wang, S.X., Wang, Y., Lu, Y.B., Li, J.Y., Song, Y.J., Nyamgerelt, M. and Wang, X.X., 2020. Diagnosis and treatment of novel coronavirus pneumonia based on the theory of traditional Chinese medicine. *Journal of Integrative Medicine*, *18*(4), pp. 275–283.

8.6.4.3 *Thesium chinense* Turcz.

Synonym: *Thesium rugulosum* Bunge ex A. DC.

Common name: Bai rui cao (China)

Habitat: Wet grasslands

Distribution: China, Korea, Mongolia, Japan

Botanical observation: This gracile and diffuse herb grows up to about 50 cm tall. The leaves are linear, somewhat vaguely rosemary-like, 1.5–3.5 × 0.5–1.5 mm, and acute or acuminate at apex. The racemes are many flowered. The perianth is tubular, whitish, about 3 mm long, 5-lobed, the lobes are triangular. The 5 stamens grow from the perianth tube. The ovary develops a short style and a somewhat capitate stigma. The capsules are minute, veined, greenish, and marked at apex with upper part of the perianth.

Medicinal uses: Abscesses (Korea); bronchitis, cough, mastitis, pneumonia (China)

Strong broad-spectrum antibacterial amphiphilic acetylenic fatty acid: Exocarpic acid (also known as (13E)-13-Octadecene-9,11-diynoic acid; Log D = 3.4 at pH 7.4; molecular mass= 274.3 g/mol) inhibited the growth of oral pathogenic bacterial strains, *Porphyromonas gingivalis*, *Fusobacterium nucleatum*, and *Streptococcus mutans* with the MIC values of 0.8, 3.4, and 13.7 µg/mL, respectively (Liu et al., 2018).

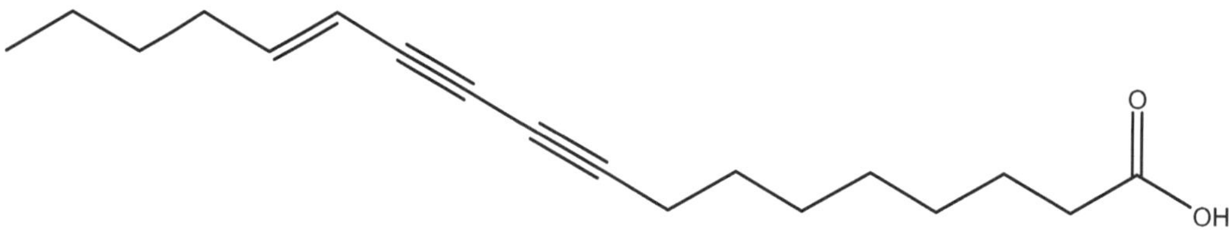

Exocarpic acid

8.7 Order Caryophyllales Juss. ex Bercht. & J. Presl (1820)

8.7.1 FAMILY AMARANTHACEAE A. L. DE JUSSIEU (1789)

The family Amaranthaceae includes about 70 genera and 2400 species of herbs. The leaves are simple, spiral or opposite, and exstipulate. The inflorescences are terminal or axillary spikes, heads, racemes, or cymes. The flowers are minute. The perianth comprises 3–5 veined, membranaceous or scarious tepals. The androecium comprises 3–5 stamens united at base. The ovary is unilocular with filiform stigma and contains 1-many ovules. The fruits are tiny capsules or utricles reniform or subglobose seeds.

8.7.1.1 *Achyranthes aspera* L.

Synonyms: *Achyranthes argentea* Lam.; *Achyranthes aspera* var. *indica* L.; *Achyranthes aspera* var. *obtusifolia* Suess.; *Achyranthes indica* (L.) Mill.; *Achyranthes obtusifolia* Lam.; *Achyranthes robusta* C.H. Wright; *Achyranthes sicula* Roth; *Cadelaria indica* (L.) Raf.; *Cadelaria sicula* Raf.; *Centrostachys aspera* (L.) Standl.; *Centrostachys indica* (L.) Standl.; *Cyathula geniculata* Lour.; *Stachyarpagophora aspersa* (L.) M. Gómez

Common names: Prickly chaff-flower; kakra lata, miming khachi, upolengra (Bangladesh); apang (Bangladesh; India); ciao kou cao, tǔ niú xī (China); apamarga, apang, chirchiti, kukkurudhanthi (India); jarong (Indonesia); murasaki-no-kozuchi (Japan); khoy ngou (Laos); ara songsang (Malaysia); kyet-mauk-pyan (Myanmar); ultekuro (Nepal); azgikai, puthkanda, (Pakistan); saramat (the Philippines); khuai nguu (Thailand)

Habitat: Wastelands, roadsides

Distribution: Tropical Asia

Botanical observation: This herb grows to 1.2 m tall. The stems are articulated, slender, smooth, and somewhat angled. The leaves are simple, opposite, and exstipulate. The petiole is 0.5–1.5 cm long. The blade is dull green, elliptic to spathulate, 1.5–10 × 1–6 cm, with about 4–8 pairs of secondary nerves, wavy, and obtuse at apex. The spikes are slender and can reach up to 30 cm long. The perianth consists of 5 tepals which are lanceolate and minute. The androecium includes 5 stamens. The gynoecium is minute and oblong. The utricles are ovoid, pendulous, and contain numerous tiny seeds.

Medicinal uses: Bronchitis, cough, sores, tuberculosis, wounds (India); dysentery, laryngitis (China; India); boil, fever, gonorrhea (Bangladesh; India); cuts, boils diarrhea, dysentery, tonsillitis, toothbrush (Nepal); toothache (Pakistan); jaundice (Bangladesh); pneumonia (Pakistan; India).

Moderate broad-spectrum antibacterial polar extract: Aqueous acetone extract of leaves inhibited the growth of *Escherichia coli*, *Klebsiella pneumoniae*, *Bacillus subtilis*, and *Staphylococcus aureus* with the MIC values of 390, 90, 90, and 390 µg/mL, respectively (Ndhlala et al., 2015).

Moderate antibacterial (Gram-negative) mid-polar extract: Ethylacetate extract of leaves inhibited the growth of *Salmonella typhi* with the MIC value of 290 µg/mL (Ndhlala et al., 2015).

Moderate anticandidal polar extract: Aqueous acetone extract of leaves inhibited the growth of *Candida albicans* with an MIC value of 390 µg/mL (Ndhlala et al., 2015).

Moderate antifungal (filamentous) mid-polar extract: Chloroform extract of leaves inhibited the growth of *Microsporum canis* with an MIC value of 250 µg/mL (Ndhlala et al., 2015).

DOI: 10.1201/9781003180371-4

Strong antifungal (filamentous) mid-polar extract: Ethylacetate extract of leaves inhibited the growth of *Trichophyton mentagrophytes*, *Trichophyton rubrum*, *Microsporum gypseum*, *Microsporum canis*, *Epidermophyton floccossum* with the MIC/MFC values of 31.2/62.5, 7.8/15.6, 15.6/31.2, 15.6/31.2, and 31.2/62.5 µg/mL, respectively (Kalaivanan et al., 2013).

Antiviral (enveloped monopartite linear double-stranded DNA) lipophilic triterpene: Oleanolic acid (Log D = 5.9 at pH 7.4; molecular mass=456.7 g/mol) from the roots inhibited the formation of plaques by the Herpes simplex virus type-1 and Herpes simplex virus type-2 by with EC_{50} of 6.8 and 7.8 µg/mL, respectively (Mukherjee et al., 2013). At 13.8 µg/mL, this pentacyclic triterpene inhibited the replication of Herpes simplex virus type-1 by 100% (Mukherjee et al., 2013). In Vero cells, oleanolic acid had a CC_{50} value of 98 µg/mL (Mukherjee et al., 2013).

Commentaries

i. The plant yields pentacyclic triterpenoid saponins (Michl et al., 2000) as well as phenolics (Narayan & Kumar, 2014) which probably account for the medicinal properties. Work is needed to elucidate the chemical structures of the polar and mid-polar antimicrobial compounds produced by this common weed.
ii. Methanol extract of roots of *Achyranthes japonica* Nakai at the concentration of 100 µg/mL inhibited RNA-dependent DNA polymerase and ribonuclease H activities of Human immunodeficiency virus type-1 reverse transcriptase by 5.3% and 29.7%, respectively (Min et al., 2001).
iii. *Achyranthes bidentata* Bl. is used in Nepal (local name: *Datiwan*) for throat, boils and toothache and in India, it is used for cough (local name: *Shveta apaamaarga*).
iv. In Bangladesh, pieces of the plant are worn as necklace for jaundice suggesting the presence of hepatoprotective or anti-Hepatitis B virus volatile principles. This plant is one of the most widely used medicinal plants in Asia.

REFERENCES

Kalaivanan, C., Chandrasekaran, M. and Venkatesalu, V., 2013. Screening of selected medicinal plants for *in vitro* antidermatophytic activity. *Journal de mycologie medicale*, *23*(4), pp. 247–254.

Michl, G., Abebe, D., Bucar, F., Debella, A., Kunert, O., Schmid, M.G., Mulatu, E. and Haslinger, E., 2000. New triterpenoid saponins from *Achyrantes aspera* Linn. *Helvetica Chimica Acta*, *83*(2), pp. 359–363.

Min, B.S., Kim, Y.H., Tomiyama, M., Nakamura, N., Miyashiro, H., Otake, T. and Hattori, M., 2001. Inhibitory effects of Korean plants on HIV-1 activities. *Phytotherapy Research*, *15*(6), pp. 481–486.

/Mukherjee, H., Ojha, D., Bag, P., Chandel, H.S., Bhattacharyya, S., Chatterjee, T.K., Mukherjee, P.K., Chakraborti, S. and Chattopadhyay, D., 2013. Anti-herpes virus activities of *Achyranthes aspera*: an Indian ethnomedicine, and its triterpene acid. *Microbiological Research*, *168*(4), pp. 238–244.

Narayan, C. and Kumar, A., 2014. Antineoplastic and immunomodulatory effect of polyphenolic components of *Achyranthes aspera* (PCA) extract on urethane induced lung cancer *in vivo*. *Molecular Biology Reports*, *41*(1), pp. 179–191.

Ndhlala, A.R., Ghebrehiwot, H.M., Ncube, B., Aremu, A.O., Gruz, J., Šubrtová, M., Doležal, K., du Plooy, C.P., Abdelgadir, H.A. and Van Staden, J., 2015. Antimicrobial, anthelmintic activities and characterisation of functional phenolic acids of *Achyranthes aspera* Linn.: a medicinal plant used for the treatment of wounds and ringworm in East Africa. *Frontiers in Pharmacology*, *6*, p. 274.

8.7.1.2 *Achyranthes ferruginea* Roxb.

Synonyms: *Psilotrichum ferrugineum* (Roxb.) Moq.

Common name: Lin di xian (China)

Habitat: Wet lands

Distribution: India, Nepal, Philippines, Cambodia, Laos, Myanmar, Thailand, Vietnam, Malaysia, Nepal, the Philippines

Botanical observation: This herb grows up to 60 cm tall. The stems are articulated, slender, purplish, smooth, hairy at nodes, and somewhat angled. The leaves are simple, opposite, and exstipulate. The petiole is up to 1.2 cm long. The blade is elliptic, lanceolate, to obovate, 1.5–7.5×0.5–1.5 cm, and acute at apex. The spikes are terminal and about 3cm long. The 5 tepals are veined, lanceolate and minute. The androecium includes 5 stamens. The gynoecium is minute and bottled-shaped. The utricles are ellipsoid and contain numerous tiny seeds.

Medicinal uses: Boils (Bangladesh)

Broad-spectrum antibacterial halo developed by mid-polar extract: Chloroform extract (200 μg/disc) inhibited the growth of *Bacillus subtilis, Sarcina lutea, Staphylococcus aureus, Bacillus megaterium, Streptococcus β-haemolyticus, Shigella dysenteriae, Shigella boydii, Shigella sonnei, Shigella flexneri, Escherichia coli, Shigella typhii, Klebsiella* sp., and *Pseudomonas aeruginosa.*

Broad-spectrum antibacterial halo developed by amide alkaloid: N-trans-feruloyl-4-methyldopamine (200 μg/disc) inhibited the growth of *Bacillus subtilis, Sarcina lutea, Staphylococcus aureus, Bacillus magaterium, Streptococcus β-haemolyticus, Shigella dysenteriae, Shigella boydii, Shigella sonnei, Shigella flexneri, Escherichia coli, Salmonella typhi, Klebsiella* sp., and *Pseudomonas aeruginosa* with inhibition zone diameters of 10, 9, 11, 6, 10, 14, 14, 14, 14, 12, 7, 10, and 6 mm, respectively (Rahman et al., 2007).

Antifungal (filamentous) halo developed by amide alkaloid: N-trans-feruloyl-4-methyldopamine from the plant at (200 μg/disc) inhibited the growth of *Aspergillus flavus* and *Rhizopus oryzae* with inhibition zone diameters of 12 and 14 mm, respectively (Rahman et al., 2007).

REFERENCE

Rahman, M.M., Alam, A.K., Sadik, G., Islam, M.R., Khondkar, P., Hossain, M.A. and Rashid, M.A., 2007. Antimicrobial and cytotoxic activities of *Achyranthes ferruginea. Fitoterapia, 78*(3), pp. 260–262.

8.7.1.3 *Aerva javanica* (Burm. f.) Juss.

Synonyms: *Achyranthes javanica* Pers.; *Aerva persica* (Burm. f.) Merr.; *Aerva tomentosa* Forssk.; *Aerva wallichii* Moq.; *Celosia lanata* L.; *Iresine javanica* Burm. f.; *Iresine persica* Burm. f.

Common names: Desert cotton, Javanese wool plant, kapok bush; mao vic loong (China); bur (India); rumput upas-upasan (Indonesia); on-hnye (Myanmar); bhoo, patharphori (Pakistan); taba-ahas (the Philippines)

Habitat: Dry and sun exposed soils, deserts

Distribution: Tropical and sub-tropical Asia

Botanical observation: This herb grows about 1 m tall. The stems are diffuse, woody, and woolly. The leaves are simple, alternate, sessile, woolly, and exstipulate. The blade is elliptic, to spathulate, 2.5–6.5×0.3–1.5 cm, and with about 3–4 pairs of secondary nerves. The spikes are sessile, whitish, woolly, cylindrical and up to about 10 cm long. The perianth consists of 5 tepals. The androecium includes 5 stamens. The gynoecium is minute and develops a forked stigma. The utricles are minute.

Medicinal uses: Wounds, ringworm (Pakistan)

Broad-spectrum antibacterial halo developed by polar extract: Methanol extract (10 mg/mL, 200 μL/well) inhibited the growth of *Escherichia coli, Pseudomonas aeruginosa, Klebsiella pneumoniae*, methicillin-resistant *Staphylococcus aureus, Staphylococcus aureus, Staphylococcus epidermidis*, and *Salmonella typhi* with inhibition zone diameter of 18, 14, 10, 14, 18, 8, and 6 mm, respectively (Mufti et al., 2012).

Strong antiviral (monopartite linear dimeric single-stranded (+) RNA) amphiphilic lupane-type triterpene: Betulinic acid (Log D = 5.1 at pH 7.4; molecular mass=456.7 g/mol) inhibited Human immunodeficiency virus type-1 syncitium formation with the IC_{50} of 27.2 µg/mL and a selectivity index of 2.6 (Reutrakul et al., 2010).

Viral enzyme inhibition by lupane-type triterpene: Betulinic acid inhibited Human immunodeficiency virus type-1 reverse transcriptase activity with IC_{50} of 41.7 µg/mL (Reutrakul et al., 2010).

Commentaries

i. This plant produces series of flavonoid glycosides (Mussadiq et al., 2013) as well as ecdysteroids (Saleem et al., 2013), and betulinic acid (Baltina et al., 2003).
ii. Triterpenes are often able to inhibit the replication of the Human immunodeficiency virus *in vitro.*

REFERENCES

Baltina, L.A., Flekhter, O.B., Nigmatullina, L.R., Boreko, E.I., Pavlova, N.I., Nikolaeva, S.N., Savinova, O.V. and Tolstikov, G.A., 2003. Lupane triterpenes and derivatives with antiviral activity. *Bioorganic & Medicinal Chemistry Letters*, *13*(20), pp. 3549–3552.

Mufti, F.U.D., Ullah, H., Bangash, A., Khan, N., Hussain, S., Ullah, F., Jamil, M. and Jabeen, M., 2012. Antimicrobial activities of *Aerva javanica* and *Paeonia emodi* plants. *Pakistan Journal of Pharmaceutical Sciences*, *25*(3), pp. 565–569.

Mussadiq, S., Riaz, N., Saleem, M., Ashraf, M., Ismail, T. and Jabbar, A., 2013. New acylated flavonoid glycosides from flowers of *Aerva javanica*. *Journal of Asian Natural Products Research*, *15*(7), pp. 708–716.

Reutrakul, V., Anantachoke, N., Pohmakotr, M., Jaipetch, T., Yoosook, C., Kasisit, J., Napaswa, C., Panthong, A., Santisuk, T., Prabpai, S. and Kongsaeree, P., 2010. Anti-HIV-1 and anti-inflammatory lupanes from the leaves, twigs, and resin of *Garcinia hanburyi*. *Planta Medica*, *76*(4), pp. 368–371.

Saleem, M., Musaddiq, S., Riaz, N., Zubair, M., Ashraf, M., Nasar, R. and Jabbar, A., 2013. Ecdysteroids from the flowers of *Aerva javanica*. *Steroids*, *78*(11), pp. 1098–1102.

8.7.1.4 *Aerva lanata* (L.) A. L. Juss. ex Schultes

Synonyms: *Achyranthes lanata* L.; *Achyranthes lanata* Roxb.; *Aerva elegans* Moq.; *Illecebrum lanatum* (L.) L.

Common names: Mountain knot-grass; chaya, dhuri bui, paashaanabheda, sirupeelai (India); on-hnye (Myanmar)

Habitat: Sun exposed dry soils, deserts

Distribution: Tropical and sub-tropical Asia

Botanical observation: This prostate herb grows up to 2 m long. The stems are terete and woolly. The leaves are simple, alternate, and exstipulate. The petiole can reach 2 cm long. The blade is orbicular, spathulate, or elliptic, cuneate at base, wavy, rounded at apex, somewhat dark green above, woolly below, marked with 3–4 pairs of secondary nerves, and 1–5×0.5–3.5 cm. The spikes are axillary, woolly, whitish and up to 1.5 cm long. The perianth is made of 5 tepals. The androecium includes 5 stamens. The ovary develops 2 stigmas. The utricles are minute and contain numerous seeds.

Medicinal uses: Acnea (Myanmar); boils, cough, diarrhea, hepatitis (India); skin infection, cough (Bangladesh); cough, jaundice, wounds (Pakistan)

Broad spectrum antibacterial halo developed by a mid-polar extract: Ethylacetate extract (200 µg/disc) inhibited the growth of *Bacillus subtilis, Bacillus cereus, Staphylococcus aureus, Escherichia coli, Shigella dysenteriae, Shigella sonnei, Shigella flexneri*, and *Shigella boydii* with zone of inhibition of 12, 9, 13, 18, 12, 10, 7, and 9 mm respectively (Chowdhury et al., 2002). In a subsequent study ethanol extract of aerial parts (30 µg/disc, 6 mm disc) inhibited the growth of *Escherichia coli*, *Pseudomonas areruginosa*, *Salmonella typhi*, *Serratia marcescens*, and *Staphylococcus aureus* (Mickymaray et al., 2016).

Commentaries

i. The plant generates the antimicrobial β-carboline alkaloid canthin-6-one (O'Donnell & Gibbons, 2007) as well as pimmarane-type diterpenes (Bharitkar et al., 2015)
ii. *Aerva persica* Merr. is used to heal wounds in Iran (local name: *Gormenaku*).

REFERENCES

Bharitkar, Y.P., Hazra, A., Apoorva Poduri, N.S., Ash, A., Maulik, P.R. and Mondal, N.B., 2015. Isolation, structural elucidation and cytotoxicity evaluation of a new pentahydroxy-pimarane diterpenoid along with other chemical constituents from *Aerva lanata*. *Natural Product Research*, *29*(3), pp. 253–261.

Cao, R., Peng, W., Wang, Z. and Xu, A., 2007. β-Carboline alkaloids: biochemical and pharmacological functions. *Current Medicinal Chemistry*, *14*(4), pp. 479–500.

Chowdhury, D., Sayeed, A., Islam, A., Bhuiyan, M.S.A. and Khan, G.A.M., 2002. Antimicrobial activity and cytotoxicity of *Aerva lanata*. *Fitoterapia*, *73*(1), pp. 92–94.

Mickymaray, S., Al Aboody, M.S., Rath, P.K., Annamalai, P. and Nooruddin, T., 2016. Screening and antibacterial efficacy of selected Indian medicinal plants. *Asian Pacific Journal of Tropical Biomedicine*, *6*(3), pp. 185–191.

O'Donnell, G. and Gibbons, S., 2007. Antibacterial activity of two canthin-6-one alkaloids from *Allium neapolitanum*. *Phytotherapy Research: An International Journal Devoted to Pharmacological and Toxicological Evaluation of Natural Product Derivatives*, *21*(7), pp. 653–657.

8.7.1.5 *Aerva sanguinolenta* (L.) Blume

Synonyms: *Achyranthes sanguinolenta* L., *Achyranthes scandens* Roxb., *Aerva scandens* (Roxb.) Moq.; *Aerva timorensis* Moq.; *Aerva velutina* Moq.

Common name: Kapok bush; chaya, raktahata (Bangladesh); bai hua xian (China); astamabayda, bhadra, nuriya, (India); daun erpa, sambang solok (Indonesia)

Habitat: Sun exposed wastelands, crops, roadsides

Distribution: Pakistan, India, Nepal, Sikkim, Bangladesh, Bhutan, Cambodia, Laos, Myanmar, Vietnam, China, Thailand, Malaysia, Indonesia, Philippines, Taiwan

Botanical observation: This herb grows up to 1 m long. The stems are terete and somewhat hairy. The leaves are simple, exstipulate, alternate or sub-opposite. The petiole is about 2 cm long. The blade is elliptic, spathulate, or lanceolate, cuneate at base, acuminate at apex, hairy, and 1.5–10×0.8–6 cm. The spikes are axillary or terminal, sessile, bullet-shaped, woolly, whitish, and about 1.5 cm long. The 5 tepals are minute, lanceolate and membranaceous. The androecium comprises 5 minutes stamens fused at base. The gynoecium is bottle shaped. The utricles contain numerous reniform, black, and glossy seeds.

Medicinal uses: Boils (India); wounds (Bangladesh); vaginal infection (Indonesia)

Moderate antibacterial (Gram-positive) polar extract: Ethanol extract of leaves inhibited the growth of *Staphylococcus aureus* (ATCC 6538P) with the MIC value of 512 μg/mL (Romulo et al., 2018). Ethanol extract of leaves inhibited the growth of *Enterococcus faecalis* (ATCC 29212) with the MIC value of 512 μg/mL (Romulo et al., 2018).

Moderate anticandidal polar extract: Ethanol extract of leaves inhibited the growth of *Candida albicans* (ATCC 10231) with the MIC value of 512 μg/mL (Romulo et al., 2018).

Strong antibacterial (Gram-positive) amphiphilic meroterpene: Bakuchiol (Log D = 5.6 at pH 7.4; molecular mass=256.3 g/mol) isolated from the leaves inhibited the growth of *Streptococcus mutans* (MTCC 497), *Actinomyces viscosus* (ATCC 15987), and *Streptococcus sanguinis* (ATCC 10556) with the MIC values of 0.9, 0.9, and 0.9 μg/mL, respectively (Rao et al., 2012).

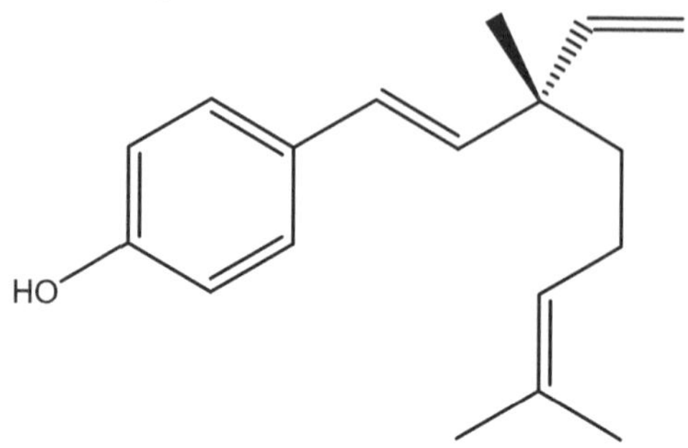

Bakuchiol

Moderate antifungal (yeast) meroterpene: Bakuchiol isolated from the leaves inhibited the growth of *Malassezia furfur* (MTCC-1374) with the MIC values of 250 μg/mL (Rao et al., 2012).

Commentaries: In Afghanistan, *Amaranthus retroflexus* L. (local name: *Taj khorosak*) is used for the treatment of tonsillitis. In Pakistan, *Amaranthus viridis* L. (local names: *Gunhar, mariro*) is used for diarrhea and in Thailand (local name: *Phak khom*) it is used for fever. In China, *Amaranthus lividus* L is used for diarrhea and mastititis (local name*: Gan jiang mi*). In India, *Amaranthus tricolor* L. (local name: *Maarisha-rakta*) is used for diarrhea. A member of the genus *Amaranthus* L. is used for the treatment of COVID-19 in China (Yang et al., 2020).

REFERENCES

Rao, G.V., Kavitha, K., Gopalakrishnan, M. and Mukhopadhyay, T., 2012. Isolation and characterization of a potent antimicrobial compound from *Aerva sanguinolenta* Blume: An alternative source of bakuchiol. *Journal of Pharmacy Research*, *5*(1), pp. 174–176.

Romulo, A., Zuhud, E.A., Rondevaldova, J. and Kokoska, L., 2018. Screening of *in vitro* antimicrobial activity of plants used in traditional Indonesian medicine. *Pharmaceutical Biology*, *56*(1), pp. 287–293.

Yang, Q.X., Zhao, T.H., Sun, C.Z., Wu, L.M., Dai, Q., Wang, S.D. and Tian, H., 2020. New thinking in the treatment of 2019 novel coronavirus pneumonia. *Complementary Therapies in Clinical Practice*, *39*, p. 101131.

8.7.1.6 *Amaranthus blitum* L.

Synonyms: *Albersia blitum* (L.) Kunth; *Amaranthus ascendens* Loisel.; *Amaranthus lividus* L.; *Euxolus ascendens* (Loisel.) H. Hara; *Euxolus blitum* Gren.

Common names: Slender amaranth; gobura nutya (Bangladesh); ao tou xian (China); jangli choulai (India)

Habitat: Wastelands, roadside, garden

Distribution: India, Bangladesh, Sikkim, Nepal, Laos, Vietnam, Malaysia, Thailand, China, Japan

Botanical observation: This erect herb grows up to 30 cm tall. The stem is terete and glabrous. The leaves are simple, spiral, and exstipulate. The petiole is 1–3.5 cm long. The blade is broadly lanceolate to rhombic, edible, dull light green, somewhat fleshy, 1.5–4.5 × 1–3 cm, cuneate at base, and notched at apex. The spikes are axillary or terminal, and about 3–8.5 cm long. The corolla comprises 5 tepals. The androecium includes 5 stamens. The gynoecium develops 2–3 stigmas. The utricules are ovoid and contain circular seeds.

Medicinal use: Ringworm (Bangladesh)

Antibacterial (Gram-positive) halo developed by a polar extract: Methanol extract of the whole plant (1 mg/6 mm disc) inhibited the growth of *Bacillus cereus* with inhibition zone diameter of 9 mm (Wiart et al., 2004).

REFERENCE

Wiart, C., Mogana, S., Khalifah, S., Mahan, M., Ismail, S., Buckle, M., Narayana, A.K. and Sulaiman, M., 2004. Antimicrobial screening of plants used for traditional medicine in the state of Perak, Peninsular Malaysia. *Fitoterapia*, *75*(1), pp. 68–73.

8.7.1.7 *Amaranthus gangeticus* L.

Synonyms: *Amaranthus tricolor* L.

Common name: Chinese spinach, tampala, Joseph's coat; garden amaranth; lal shak (Bangladesh); xian (China); tanduliya, appulakittantu (India)

Habitat: Sun exposed and moist lands

Distribution: Tropical and Subtropical Asia

Botanical observation: This erect herb grows to 1.5 m tall. The stem is terete, fleshy, purplish, and striated. The leaves are simple, spiral, and exstipulate. The petiole is 1.5–5 cm long. The blade is green and purple, fleshy, edible, 5–10 × 2–6.5 cm, broadly lanceolate to rhombic, wavy, tapering at base, and acute at apex. The spikes are axillary or terminal, elongated, and up to 15cm long. The androecium includes 3 stamens. The ovary develops 3 stigmas. The utricles containing tiny seeds.

Medicinal uses: Diarrhea (Bangladesh; India); leucorrhea (Bangladesh)

Broad-spectrum antibacterial halo developed by n-alkanes: Long chain n-alkanes fraction (100 μg/agar well) inhibited the growth of *Staphylococcus albus*, *Staphylococcus aureus*, *Streptococcus viridans*, *Escherichia coli*, *Pseudomonas pyocyanea*, and *Klebsiella* sp. (Sharma, 1993).

Broad-spectrum antibacterial halo developed by phytosterol: β-Sitosterol (100 μg/agar well) inhibited the growth of *Staphylococcus albus*, *Staphylococcus aureus*, *Streptococcus viridans*, *Escherichia coli*, *Pseudomonas pyocyanea*, and *Klebsiella* sp. (Sharma 1993). In a subsequent study β-sitosterol (1 mg/6 mm disc) inhibited the growth of *Staphylococcus aureus* (ATCC 25923), *Escherichia coli* (ATCC 5922), *Klebsiella pneumoniae* (ATCC 10031) and *Bacillus thuringiensis* (Singh & Singh, 2003).

Antifungal (filamentous) halo developed by phytosterol: β-Sitosterol inhibited the growth (1 mg/6 mm disc) of *Aspergillus niger*, *Aspergillus flavus*, *Rhizoctonia phaseoli*, and *Penicillium chrysogenum* (Singh & Singh, 2003).

REFERENCES

Sharma, R.K., 1993. Phytosterols: wide-spectrum antibacterial agents. *Bioorganic Chemistry*, *21*(1), pp. 49–60.
Singh, B. and Singh, S., 2003. Antimicrobial activity of terpenoids from *Trichodesma amplexicaule* Roth. *Phytotherapy Research*, *17*(7), pp. 814–816.

8.7.1.8 *Amaranthus spinosus* L.

Synonyms: *Amaranthus caracasanus* Kunth; *Amaranthus diacanthus* Raf.; *Amaranthus spinosus* fo. *Inermis* Lauterb. & K. Schum.; *Amaranthus spinosus* var. *basiscissus* Thell.; *Amaranthus spinosus* var. *circumscissus* Thell.; *Amaranthus spinosus* var. *indehiscens* Thell.; *Amaranthus spinosus* var. *purpurascens* Moq.; *Amaranthus spinosus* var. *pygmaeus* Hassk.; *Amaranthus spinosus* var. *rubricaulis* Hassk.; *Amaranthus spinosus* var. *viridicaulis* Hassk.; *Galliaria spinosa* (L.) Nieuwl.

Common names: Prickly amaranth, spiny amaranth, pigweed; kanta maris, khuira kanta (Bangladesh); ci xian, cì xiàn cài (China); cherucheera, chaulai, kaata khutara, tanduliuyah (India); hen nu new, hin-nu-new-subauk (Myanmar); phak khom nam (Thailand)

Habitat: Roadsides, waste places

Distribution: Tropical Asia

Botanical observation: This erect herb grows up to 1 m tall. The stem is terete, fleshy, and subglabrous. The leaves are simple, spiral, and exstipulate. The petiole is 1–8 cm long with a pair of spines at base. The blade is lanceolate, tapering at base along the petiole, acute at apex, wavy and presents 3–10 pairs of secondary nerves. The spikes are axillary or terminal and about 8–15 cm long. The 5 tepals are linear, minute, and green. Five stamens are present. The gynoecium is ovoid and develops 3 stigmas. The utricles contain subglobose seeds.

Medicinal uses: Blenorrhea, gonorrhea, boils, jaundice (Bangladesh); boils, fever, bronchitis, diarrhea, dysentery, leprosy, gonorrhea, ringworm (India); fever (Cambodia); wounds (Indonesia); bronchitis, boils (Malaysia); aphthous ulcers (Thailand); dysentery, sore throat (China); gonorrhea (Myanmar)

Strong broad-spectrum antibacterial fatty acid: (14*E*, 18*E*, 22*E*, 26*E*)-methyl nonacosa-14,18,22,26 tetranoate acid inhibited the growth of *Staphylococcus aureus* (ML-59), *Bacillus licheniformis*, *Staphylococcus aureus* (29737), *Escherichia coli* (K-12 ROW), *Shigella sonnei*, *Vibrio cholera*, *Vibrio algynolyteus*, *Shigella boydii* and *Salmonella typhimurium* (NCTC 74) with the MIC values ranging from 25 to 100 µg/mL (Mondal and Maity, 2016). This fatty acid evoked membrane insults in *Staphylococcus aureus* (ML-59) and *Escherichia coli* (K-12 ROW) (Mondal & Maity, 2016).

Strong antifungal (filamentous) non-polar extract: Hexane extract inhibited the growth of *Fusarium proliferatum* (MRC 6908) with an MIC value of 40 µg/mL (Mondal & Maity, 2010) Hexane extract inhibited the growth *of Fusarium verticilloides* (MRC 8559) with an MIC value of 20 µg/mL (Thembo et al., 2010).

Commentary: *Alternanthera paronychioides* L. Hill. is used to treat burns in India (local name *Julsachi*).

REFERENCES

Mondal, A. and Maity, T.K., 2016. Antibacterial activity of a novel fatty acid (14*E*, 18*E*, 22*E*, 26*E*)-methyl nonacosa-14, 18, 22, 26 tetraenoate isolated from *Amaranthus spinosus*. *Pharmaceutical Biology*, *54*(10), pp. 2364–2367.

Thembo, K.M., Vismer, H.F., Nyazema, N.Z., Gelderblom, W.C.A. and Katerere, D.R., 2010. Antifungal activity of four weedy plant extracts against selected mycotoxigenic fungi. *Journal of Applied Microbiology*, *109*(4), pp. 1479–1486.

8.7.1.9 *Alternanthera philoxeroides* (Mart.) Griseb.

Synonyms: *Achyranthes philoxeroides* (Mart.) Standl.; *Bucholzia philoxeroides* Mart.; *Telanthera philoxeroides* (Mart.) Moq.

Common names: Alligator-weed, xi han lian zi cao (China); ponne (India); mokunuwann (Sri Lanka)

Habitat: Riverbanks, lakes, moist wastelands

Distribution: India, Bangladesh, Thailand, Malaysia, Indonesia

Botanical observation: This semi-aquatic erect herb grows up to 1 m tall and has somewhat at first glance and from some distance resemblance with *Eclipta alba* (L.) Hassk. in the family Asteraceae. The stem is terete, purplish, fleshy, hairy, and rooting at nodes. The leaves are edible, simple, sessile, opposite, exstipulate, and sheathing at base. The blade is elliptic to spathulate, somewhat fleshy, dark green, glossy, with about 8–11 discrete pairs of secondary nerves, attenuate at base, acute at apex, and 2.5–5 × 0.7–2 cm. The inflorescence is a peduncled and axillary globose head which is about 1.5 cm in diameter. The 5 tepals are white, oblong, and about 5 mm long. The androecium comprises 5 stamens fused at base. The ovary develops a globose stigma. The fruits are tiny utricles.

Medicinal use: Fever (Malaysia)

Broad-spectrum strong antibacterial polar extract: Methanol extract of leaves inhibited the growth of *Staphylococcus aureus* (MTCC 2940), *Bacillus subtilis* (MTCC 441), *Escherichia coli* (MTCC 739), and *Pseudomonas aeruginosa* (MTCC 2453) with the MIC values of 35.2, 45.2, 45, and 40 µg/mL, respectively (Rawani et al., 2011).

Strong antiviral (enveloped monopartite linear, single-stranded (+) RNA) non-polar extract: Petroleum ether extract inhibited the replication of the Dengue virus with an IC_{50} value of 47.4 µg/mL (Jiang et al., 2005).

Moderate antiviral (enveloped monopartite double-stranded DNA) C-glycosyl flavonol glycoside: Luteolin-6-*C*-β-*D*-boivinopyranoside inhibited the secretion of the Hepatitis B virus surface antigen and Hepatitis B virus e-antigen in HepG2.2.15 cells with IC_{50} of 11.3 and 39.7 µM, respectively (Li et al., 2016).

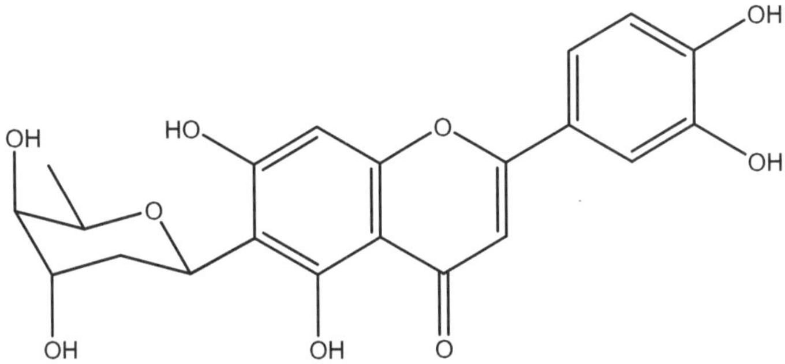

Luteolin-6-*C*-β-*D*-boivinopyranoside

Antiviral (enveloped monopartite linear double-stranded DNA) oleanane-type triterpene saponin: Chikusetsusaponin IVa inhibited the replication of the Herpes simplex virus type-1, Herpes simplex virus type-2, and Human cytomegalovirus (Rattanathongkom et al., 2009). For Herpes simplex virus type-2 Chikusetsusaponin IVa inactivated virus particles and inhibited the release of viruses from host cells (Rattanathongkom et al., 2009).

Moderate antiviral (enveloped monopartite linear single-stranded (−) RNA) triterpene saponins: Chikusetsusaponin IVa inhibited the replication of the Measles virus and Mumps virus (Rattanathongkom et al., 2009).

Commentaries

i. The plant contains quercetin and β-carboline (Zhang et al., 2018).
ii. *Alternanthera bettzickiana* (Regel) G. Nicholson is used in Thailand (local name: *Phak pet daeng*) for fever. *Alternanthera paronychioides* A. St.-Hil. is used in India (local name: *Sanchi*) for the treatment of fever.

REFERENCES

Jiang, W.L., Luo, X.L. and Kuang, S.J., 2005. Effects of *Alternanthera philoxeroides* Griseb against dengue virus *in vitro*. *Di 1 jun yi da xue xue bao= Academic Journal of the First Medical College of PLA*, *25*(4), pp. 454–456.

Li, B., Guo, Q.L., Tian, Y., Liu, S.J., Wang, Q., Chen, L. and Dong, J.X., 2016. New anti-HBV C-boivinopyranosyl flavones from *Alternanthera philoxeroides*. *Molecules*, *21*(3), p. 336.

Rattanathongkom, A., Lee, J.B., Hayashi, K., Sripanidkulchai, B.O., Kanchanapoom, T. and Hayashi, T., 2009. Evaluation of chikusetsusaponin IVa isolated from *Alternanthera philoxeroides* for its potency against viral replication. *Planta Medica*, *75*(8), pp. 829–835.

Rawani, A., Pal, S. and Chandra, G., 2011. Evaluation of antimicrobial properties of four plant extracts against human pathogens. *Asian Pacific Journal of Tropical Biomedicine*, *1*(1), pp. S71–S75.

Zhang, X., Li, P., Guo, S., Wang, S. and Liu, D., 2018. Quantitation of β-carboline and quercetin in alligator weed (*Alternanthera philoxeroides* (Mart.) Griseb.) by LC-MS/MS and evaluation of cardioprotective effects of the methanol extracts. *Drug Discoveries & Therapeutics*, *12*(6), pp. 341–346.

8.7.1.10 *Alternanthera pungens* Kunth

Synonyms: *Achyranthes leiantha* (Seub.) Standl.; *Achyranthes lorentzii* (Uline) Standl.; *Achyranthes mucronata* Lam.; *Achyranthes radicans* B. Heyne ex Roth; *Achyranthes repens* L.; *Alternanthera achyrantha* (L.) R. Br.; *Alternanthera achyrantha* (L.) R. Br. ex Sweet; *Alternanthera achyrantha* var. *leiantha* Seub.; *Alternanthera echinata* Sm.; *Alternanthera lorentzii* Uline; *Alternanthera pungens* fo. *Pauciflora* Suess.; *Alternanthera pungens* var. *leiantha* Suess.; *Alternanthera repens* (L.) J. F. Gmel.; *Alternanthera repens* (L.) Kuntze; *Alternanthera repens* (L.) Link; *Celosia echinata* Willd. ex Roem. & Schult.; *Desmochaeta sordida* Bunbury; *Guilleminea procumbens* Rojas Acosta; *Illecebrum achyranthum* L.; *Illecebrum pungens* (Kunth) Spreng.; *Pityranthus crassifolius* Mart.; *Pupalia sordida* Moq.; *Telanthera pungens* (Kunth) Moq.

Common names: Creeping chaffweed, khaki weed; keshtto lata (Bangladesh); ottara mul (India)

Habitat: Wastelands

Distribution: Tropical and Subtropical Asia

Botanical observation: This prostrate herb grows up to 50 cm long. The stem is terete, purplish, and hairy. The leaves are simple, exstipulate, and arranged in clusters. The blade is obovate to spathulate, fleshy, glossy, 1.5–4.5 × 0.3–2.5 cm, rounded to acute at the apex and hairy. The flowers are arranged in axillary sessile heads which are globose and about 1.5 cm long. The 5 petals are minute. The androecium includes 5 stamens and the gynoecium is minute. The utricle is orbicular and contains tiny discoid, brown, and glossy seeds.

Medicinal use: Gonorrhea (Bangladesh)

Broad-spectrum antibacterial halo developed by polar extract: Aqueous acetone extract (1 mg/6 mm disc) inhibited the growth of *Enterococcus faecalis* (10907 CIP), *Shigella dysenteriae* (5451 CIP), *Proteus mirabilis* (ATCC 35659), *Salmonella enteritica* (CIP 105150), *Enterococcus aerogenes* (CIP 104725), *Bacillus subtilis* (ATCC 21332), *Listeria monocytogenes* (CRBIB 13134), and *Staphylococcus aureus* (ATCC 9144) with inhibition zone diameter of 10, 8, 9, 10, 8, 10, 8, and 8 mm, respectively (Zongo et al., 2011).

Clinical antiviral (enveloped monopartite linear dimeric single stranded (+)RNA) polar extract: AIDS patients taking infusion of the plant for a year every 2 days had increased T lymphocytes count (Djohan et al., 2009).

REFERENCES

Djohan, Y.F., Camara, C., Mondé, A.A., Koffi, G., Niamké, G., Déré, L., Tiahou, G., Djessou, P. and Sess, D., 2009, September. Interest of antioxidants in the care of the patients infected by the HIV: the experience of long term administration of *Alternanthera pungens* herb tea. *Annales de biologie clinique*, *67*(5), pp. 563–568.

Zongo, C., Savadogo, A., Somda, K.M., Koudou, J. and Traore, A.S., 2011. *In vitro* evaluation of the antimicrobial and antioxidant properties of extracts from whole plant of *Alternanthera pungens* HB & K. and leaves of *Combretum sericeum* G. Don. *International Journal of Phytomedicine*, *3*(2), p. 9.

8.7.1.11 *Alternanthera sessilis* (L.) DC.

Synonyms: *Achyranthes ficoidea* var. *sessilis* (L.) Pers.; *Achyranthes sessilis* (L.) Desf. ex Steud.; *Achyranthes triandra* Roxb.; *Alternanthera achyranth* Forssk.; *Alternanthera achyranthoides* Forssk.; *Alternanthera denticulata* R. Br.; *Alternanthera ficoidea* (L.) P. Beauv.; *Alternanthera nodiflora* R. Br.; *Alternanthera polygonoides* (L.) R. Br.; *Alternanthera sessilis* (L.) R. Br.; *Alternanthera tenella* Moq.; *Alternanthera triandra* Lam.; *Gomphrena polygonoides* L.; *Gomphrena sessilis* L.; *Illecebrum indicum* Houtt.; *Illecebrum sessile* (L.) L.; *Paronychia sessilis* (L.) Desf.

Common names: Sessile joyweed; chenchi shak, sarhanchi (Bangladesh); lian zi cao (China); garundi, matsyaakshi, phakchet, ponnankanni (India); bhiringi jhar (Nepal); kachari (Pakistan); phak pet khao (Thailand)

Habitat: Wet wastelands, roadsides, gardens, swamps

Distribution: Tropical and Subtropical Asia

Botanical observation: This prostate herb grows to 30cm long. The stem is terete, hairy rooting at nodes, and somewhat purplish. The leaves are simple, exstipulate, and opposite. The petiole is absent or about 5mm long. The blade is elliptic, tapering at base, acute at apex, elliptic, 1–8×0.2–2.5cm, and subglabrous. The spikes are axillary, sessile, ovoid, and about 5mm in diameter. The perianth includes 5 lanceolate, 1-nerved, and white tepals. The androecium consists of 5 stamens. The gynoecium is bottle-shaped. The utricles contains minute discoid and brown glossy seeds

Medicinal uses: Burns, bronchitis, diarrhea, jaundice, leprosy (India); abscesses, infections (Bangladesh); venereal disease, wounds (Nepal); aphthous ulcers (Thailand); tuberculosis (China); fever (India; Nepal; Thailand); boils (Bangladesh; India)

Broad-spectrum antibacterial halo developed by polar extract: Methanol extract (500 µg/disc) inhibited the growth of *Escherichia coli*, *Pseudomonas aeruginosa*, *Salmonella typhi*, *Salmonella paratyphi*, *Shigella boydii*, *Vibrio parahemolyticus*, *Bacillus subtilis*, *Bacillus megaterium*, *Staphylococcus aureus*, and *Sarcina lutea* with the inhibition zones diameters of 16, 6, 20, 21, 6, 6. 7, 6, 6, and 13mm, respectively (Ullah et al., 2013).

Strong antiviral (enveloped monopartite linear double standed DNA) polar activity: Ethanol extract inhibited the growth of Herpes simplex virus type-1 with IC_{50} value of 10 µg/mL and was not toxic for host cells (Hamidi et al., 1996).

Commentaries

i. From the leaves, a mixture of diastereomers of an ionone displayed some levels of activity against *Pseudomonas aeruginosa* and *Trichophyton mentagrophytes* (Rasaga et al., 2002).
ii. The plants contain oleanane-type triterpene saponins (Penders et al., 1992).

REFERENCES

Hamidi, J.A., Ismaili, N.H., Ahmadi, F.B. and Lajisi, N.H., 1996. Antiviral and cytotoxic activities of some plants used in Malaysian indigenous medicine. *Pertanika Journal of Tropical Agricultural Science*, *19*(2/3), pp. 129–136.

Penders, A., Delaude, C. and Kapundu, M., 1992. Principal structure of Saponin from *Alternanthera sessilis*. *Bulletin des Societes Chimiques Belges (Belgium)*, *101*(3), pp. 227–232.

Ragasa, C.Y., Tremor, N. and Rideout, J.A., 2002. Ionone derivatives from *Alternanthera sessilis*. *Journal of Asian Natural Products Research*, *4*(2), pp. 109–115.

Ullah, M.O., Haque, M., Urmi, K.F., Zulfiker, A.H.M., Anita, E.S., Begum, M. and Hamid, K., 2013. Antibacterial activity and brine shrimp lethality bioassay of methanolic extracts of fourteen different edible vegetables from Bangladesh. *Asian Pacific Journal of Tropical Biomedicine*, *3*(1), pp. 1–7.

8.7.1.12 *Celosia argentea* L.

Synonyms: *Amaranthus purpureus* Nieuwl.; *Celosia argentea* var. *cristata* (L.) Kuntze; *Celosia coccinea* L.; *Celosia cristata* L.; *Celosia debilis* S. Moore; *Celosia huttonii* Mast.;*Celosia japonica* Houtt.; *Celosia margaritacea* L.; *Celosia marilandica* Retz.; *Celosia pallida* Salisb.; *Celosia pyramidalis* Burm. f.; *Celosia swinhoei* Hemsl.

Common names: Celosia, cockscomb; ranga chuma, swetmurge (Bangladesh); gāng bā lān, niú mǔ wo, qing xiang, ts'ingsiang (China); choal, kukarsira, vitunna, safed murga, shitivaaraka, thathania, (India); bayam ekor belanda, bunga merah (Indonesia); gul-i-halva (Iran); nogeito (Japan); bayam merah (Malaysia); kyet-mauk (Myanmar); mao gat rang (Vietnam)

Habitat: Grassy roadsides, wastelands

Distribution: Pakistan, India, Nepal, Bhutan, Cambodia, Vietnam, Laos, Thailand, Malaysia, Indonesia, China, Korea, Japan, the Philippines

Botanical observation: This erect herb grows up to 1 cm tall. The stem is terete. The leaves are simple, exstipulate, and alternate. The petiole is absent or about 1.5 cm long. The blade is oblong-lanceolate, attenuate at base, edible (India, China), acute at apex, elliptic, 5–8 × 1–3 cm, and subglabrous. The spikes are terminal, cylindrical to fusiform, and up about 10 cm long. The 5 petals are whitish-pink and 1 cm long. The 5 stamens are fused at base and with purplish anthers. The gynaecium is bottle-shaped with a slender and purplish style. The utricles are ovoid and contain tiny reniform seeds.

Medicinal use: Boils, dysentery, sores, stomatitis (India); eye infection (China; India); fever, wounds (Myanmar); diarrhea (India; Iraq); cough (Indonesia; Iraq)

Broad-spectrum antibacterial halo developed by polar extract: Methanol extract of bark (1 mg/6 mm disc) inhibited the growth of *Bacillus cereus*, *Pseudomonas aeruginosa*, *Staphylococcus aureus* with inhibition zone diameter of 8, 15, and 9 mm, respectively (Wiart et al., 2004).

Strong broad-spectrum antibacterial polar extract: Ethanol extract inhibited the growth of *Staphylococcus* sp., *Klebsiella* sp., *Escherichia coli*, *Streptococcus* sp., *Shigella* sp., *Pseudomonas* sp., *Salmonella* sp., and *Vibrio* sp. with the MIC of 0.06, 0.07, 0.06, 0.07, 0.03, 0.06, 0.1, and 0.07 mg/mL, respectively (Gnanamani et al., 2003).

Anticandidal halo developed by polar extract: Methanol extract of bark (1 mg/6 mm disc) inhibited the growth of *Candida albicans* with inhibition zone diameter of 11 mm (Wiart et al., 2004).

Commentary: The plant produces oleanane-type triterpene saponins (Pang et al., 2014) which may participate in the antibacterial and antifungal properties.

REFERENCES

Gnanamani, A., Priya, K.S., Radhakrishnan, N. and Babu, M., 2003. Antibacterial activity of two plant extracts on eight burn pathogens. *Journal of Ethnopharmacology*, *86*(1), pp. 59–61.

Pang, X., Yan, H.X., Wang, Z.F., Fan, M.X., Zhao, Y., Fu, X.T., Xiong, C.Q., Zhang, J., Ma, B.P. and Guo, H.Z., 2014. New oleanane-type triterpenoid saponins isolated from the seeds of *Celosia argentea*. *Journal of Asian Natural Products Research*, *16*(3), pp. 240–247.

Wiart, C., Mogana, S., Khalifah, S., Mahan, M., Ismail, S., Buckle, M., Narayana, A.K. and Sulaiman, M., 2004. Antimicrobial screening of plants used for traditional medicine in the state of Perak, Peninsular Malaysia. *Fitoterapia*, *75*(1), pp. 68–73.

8.7.1.13 *Gomphrena celosioides* Mart

Common names: Water globehead; trochiek toun say (Cambodia); yin hua xian (China); lalchita, pedda goranta (India); botonsilyong-gapang (the Philippines); kunyee (Thailand)

Habitat: Roadsides, wastelands

Distribution: Tropical and Subtropical Asia

Botanical observation: This prostrate herb grows up to 30 cm long. The stem is terete, striated, and hairy. The leaves are simple, exstipulate, and opposite. The petiole is up to about 1.5 cm long. The blade is broadly lanceolate, acute at apex, 1.5–4.5 × 0.5–1.3 cm, and somewhat hairy. The spikes are terminal, ovoid to globose, white, and about 7 cm long. The 5 tepals are up to 5 mm long. The androecium consists of 5 stamens fused into a tube. The gynoecium is minute, obovoid, and develops 2 stigmas. The utricles are subglobose and contain tiny ovoid and glossy seeds.

Medicinal use: Postpartum protective remedy (Thailand)

Broad-spectrum antibacterial halo developed by polar extract: Ethanol extract (10 mg/mL, 20 mL/disc) inhibited the growth of *Staphylococcus aureus* (ATCC 12598) and *Salmonella typhi* (ATCC 19430) with inhibition zone diameters of 7.6 and 7.3 mm, respectively (de Moura et al., 2004).

Broad-spectrum antibacterial halo developed by mid-polar extract: Ethylacetate extract (12.5 mg/mL/7 mm well) inhibited the growth of *Staphylococcus aureus* (NCTC 6571), *Escherichia coli* (NCTC 9001), *Pseudomonas aeruginosa* (NCTC 6750), *Bacillus subtilis*, and *Salmonella typhi* (ATCC 14028) with inhibition zone diameters of 13, 12, 12, 13, and 12 mm respectively (Dosumu et al., 2010).

Broad-spectrum antibacterial halo developed by simple phenolic: Vanillic acid (10 mg/mL, 20 mL/disc) inhibited the growth of *Staphylococcus aureus* (ATCC 12598) with the inhibition zone diameter of 7.3 mm and *Salmonella typhi* (ATCC 19430) with the inhibition zone diameter of 7.3 mm (de Moura et al., 2004).

Broad-spectrum antibacterial halo developed by phytosterol glycoside: Stigmast-6-en-3-*O*-β-(D-glucopiranoside) (10 mg/mL, 20 mL/disc) inhibited the growth of *Staphylococcus aureus* (ATCC 12598) with the inhibition zone diameter of 7.6 mm and *Salmonella typhi* (ATCC 19430) with the inhibition zone diameter of 7.3 mm (de Moura et al., 2004).

Broad-spectrum antifungal halo developed by mid-polar extract: Ethylacetate extract (12.5 mg/mL/7 mm well) inhibited the growth of *Candida albicans* (NCTC 7534), *Aspergillus niger*, and *Trichophyton* sp. (clinical isolates) with inhibition zone diameters of 20, 17, and 14 mm respectively (Dosumu et al., 2010).

Commentaries

i. When reporting antibacterial or antifungal activities using paper disc or agar well indicating a concentration of extract is not recommended. What is useful is to indicate the exact weight of extract or compound loaded per disc or well.
ii. From members of the genus *Gomphrena* L., isorhamnetin 3-*O*-robinobioside inhibited *Mycobacterium phlei* with the MIC value of 50 μg/mL (Pomilio et al., 1992). Kaempferol inhibited *Mycobacterium phlei* with the MIC value of 50 μg/mL (Pomilio et al., 1992). Galangin inhibited *Mycobacterium phlei* with the MIC value of 50 μg/mL (Pomilio et al., 1992).
iii. Kaempferol inhibited *Staphylococcus aureus, Bacillus cereus, Staphylococcus epidermidis, Escherichia coli, Salmonella typhimurium, Shigella flexneri*, and *Pseudomonas aeruginosa* with the MIC/MBC of 2.5/5, 2.5/2.5, 1.2/1.2, 2.5/5, 2.5/5, 1.2/5, and 1.2/5 mg/mL, respectively (Fu et al., 2016).
iv. Flavonol glycosides are often antimycobacterial *in vitro*.

REFERENCES

Dosumu, O.O., Idowu, P.A., Onocha, P.A. and Ekundayo, O., 2010. Isolation of 3-(4-hydroxyphenyl) methylpropenoate and bioactivity evaluation of *Gomphrena celosioides* extracts. *EXCLI Journal*, *9*, p. 173.

de Moura, R.M.X., Pereira, P.S., Januario, A.H., de Castro França, S. and Dias, D.A., 2004. Antimicrobial screening and quantitative determination of benzoic acid derivative of Gomphrena celosioides by TLC-densitometry. *Chemical and Pharmaceutical Bulletin*, *52*(11), pp. 1342–1344.

Fu, L., Lu, W. and Zhou, X., 2016. Phenolic compounds and *in vitro* antibacterial and antioxidant activities of three tropic fruits: persimmon, guava, and sweetsop. *BioMed Research International*, *2016*. doi:10.1155/2016/4287461.

Pomilio, A.B., Buschi, C.A., Tomes, C.N. and Viale, A.A., 1992. Antimicrobial constituents of *Gomphrena martiana* and *Gomphrena boliviana*. *Journal of Ethnopharmacology*, *36*(2), pp. 155–161.

8.7.2 FAMILY AIZOACEAE MARTINOV (1820)

The family Aizoaceae includes 130 genera and 2500 species of elegant, often small, fleshy herbs often growing in sandy and sunny places. The leaves are simple, alternate, or opposite, and exstipulate. The inflorescences are terminal or axillary cymes. The flowers are minute. The perianth is tubular and develops 5–8 tepals. The androecium comprises at least 3 free stamens. The gynoecium is made of at least 2 carpels fused into a multilocular ovary each locule sheltering 1-many ovules growing from axil or parietal placentas. The capsules are minute and circumscissile.

8.7.2.1 *Sesuvium portulacastrum* (L.) L.

Synonyms: *Halimus portulacastrum* (L.) Kuntze; *Pharnaceum rubens* Adamson; *Portulaca portulacastrum* L.; *Portulaca venezuelensis* Urb.; *Sesuvium acutifolium* Miq.; *Sesuvium ortegae* Spreng.; *Sesuvium pedunculatum* Pers.; *Sesuvium revolutifolium* Ortega; *Sesuvium sessile* Larrañaga; *Sesuvium sessile* Pers.; *Sesuvium sessiliflorum* Dombey ex Rohrb.; *Trianthema americana* Gillies ex Arn.

Common names: Sea purslane; jadu palang (Bangladesh); hai ma chi (China); dhapa, nttnia (India); bilang bilang (the Philippines)

Habitat: Seashores

Distribution: Tropical Asia

Botanical observation: This prostrate halophytic herb grows up to 50 cm long. The stems are terete, glabrous, rooting at nodes, and somewhat purplish. The leaves are simple, sessile, opposite, and exstipulate. The blade is fleshy, glossy, oblong to lanceolate or elliptical, and 1.5–5 cm × 0.2–1 cm. The flowers are axillary and solitary. The 5 tepals are triangular and pink and give to the perianth a star-shaped appearance. The androecium includes 15 stamens. The ovary is ovate and develops 3–5 stigmas. The capsule is ovate, up to about 1 cm long, and contains numerous tiny glossy and black seeds.

Medicinal uses: Burns, gonorrhea, wounds (India)

Moderate broad-spectrum fatty acid fraction: Fatty acids fraction of leaves (containing mainly palmitic acid, oleic acid, linoleic acid and linolenic acid) inhibited the growth of *Bacillus subtilis* (NCIM 2063), *Bacillus pumilus*, (NCIM 2327), *Micrococcus luteus* (NCIM 2376), *Staphylococcus aureus* (NCIM 2901), methicillin-resistant *Staphylococcus aureus* (NCTC 6571), *Pseudomonas aeruginosa* (NCIM 5031), *Klebsiella pneumoniae* (NCIM 2957), *Escherichia coli* (NCIM 2256) (Chandrasekaran et al., 2011).

Very weak broad-spectrum antifungal fatty acids fraction: Fatty acid fraction of leaves (containing mainly palmitic acid, oleic acid, linoleic acid and linolenic acid) inhibited the growth of *Candida albicans*, *Candida krusei*, *Candida tropicalis*, *Candida parapsilosis*, *Aspergillus niger*, and *Aspergillus fumigatus* with the MIC/MBC values of 0.5/1, 2/4, 2/4, 2/4, 8/16, and 4/8, mg/mL, respectively (Chandrasekaran et al., 2011).

Broad-spectrum antibacterial halo developed by essential oil: Essential oil of leaves (25 μL/4mm well) inhibited the growth of *Acinetobacter calcoaceticus* (NCIB 8250), *Bacillus subtilis* (NCIB 3610), *Citrobacter freundii* (NCIB 11490), *Clostridium perfringens* (ATCC 36200), *Clostridium sporogenes* (NCIB 10696), *Escherichia coli* (NCIB 8879), *Klebsiella pneumoniae* (NCIB 4184), *Proteus vulgaris* (NCIB 4175), *Pseudomonas aeruginosa* (NCIB 950), *Salmonella typhi* (ATCC 6539), *Staphylococcus aureus* (NCIB 6751), and *Yersinia enterocolitica* (NCIB 10460), with the inhibition zone diameters of 14.5, 12, 11, 15, 14.9, 19.2, 5.3, 12, 8.5, 13.2, 9.5, and 9.7 mm, respectively (Magwa et al., 2006).

Broad-spectrum antifungal halo developed by essential oil: Essential oil of leaves (10 μg/mL) inhibited the growth of *Candida albicans* (ATCC 10231), *Aspergillus niger* (ATCC 10535), *Aspergillus flavus* (ATCC 10124), and *Penicillium notatum* (ATCC 9178) by 72.2, 66.8, 64.1, and 56.9%, respectively (Magwa et al., 2006).

Antiviral (enveloped monopartite double-stranded DNA) polar extract: Ethanol extract of leaves inhibited the binding to the Hepatitis B virus to host cells (Premanathan et al., 1999).

REFERENCES

Chandrasekaran, M., Senthilkumar, A. and Venkatesalu, V., 2011. Antibacterial and antifungal efficacy of fatty acid methyl esters from the leaves of *Sesuvium portulacastrum* L. *European Review for Medical and Pharmacological Sciences*, *15*(7), pp. 775–780.

Magwa, M.L., Gundidza, M., Gweru, N. and Humphrey, G., 2006. Chemical composition and biological activities of essential oil from the leaves of *Sesuvium portulacastrum*. *Journal of Ethnopharmacology*, *103*(1), pp. 85–89.

Premanathan, M., Kathiresan, K. and Nakashima, H., 1999. Mangrove halophytes: a source of antiviral substances. *South Pacific Study*, *19*(1–2), pp. 49–57.

8.7.2.3 *Trianthema portulacastrum* L.

Synonyms: *Portulacastrum monogynum* (L.) Medik.; *Trianthema flexuosa* Schumach. & Thonn.; *Trianthema littoralis* Cordem.; *Trianthema monanthogyna* L.; *Trianthema monogyna* L.; *Trianthema procumbens* Mill.

Common names: Black pig weed, desert horse purslane; pti thmar (Cambodia); jia hai ma ch (China); vizakh; (Iran); biskhapra, mookaratai, nadaraku, sabuni, saaranatthi, shvetapunarnava, varshaabhu, (India); ghra, itsit, pandraoosh (Pakistan); toston (the Philippines); sam bien (Vietnam); ayam, pahk bia hin (Thailand)

Habitat: Open sandy soil, fields

Distribution: Tropical Asia

Botanical observation: This halophyte herb grows up to 35 cm long. The stems are fleshy, somewhat purplish, and hairy. The leaves are simple, alternate, and exstipulate. The petiole is up to 1 cm long. The blade is spathulate, 1.5–2.6 cm × 0.6–2.5 cm, cuneate at base, acute or bilobed at apex, dull green, and with about 4 pairs of secondary nerves. The flowers are graceful, axillary, and solitary. The 5 tepals are 3 mm long, light pink, and oblong. The androecium is made of 10–25 stamens. The ovary is ovate and develops a single stigma. The capsules are bilobed and contain numerous tiny glossy and black seeds.

Medicinal uses: Bronchitis, jaundice, eye infection, diarrhea (India); hepatitis, urinary tract infection (Pakistan).

Weak broad-spectrum polar extract: Methanol extract of leaves inhibited the growth of *Staphylococcus aureus*, *Proteus vulgaris*, *Pseudomonas aeruginosa*, *Shigella flexneri*, *Salmonella typhi*, and *Escherichia coli* (Kavitha et al., 2014).

Moderate antifungal tetraterpene: Trianthenol inhibited at a concentration of 400 µg/mL the growth of *Trichophyton scheonleinii*, *Pseudallescheria boydii*, *Trichophyton simii*, *Fusarium solani var. lycopersici*, *Macrophomia phaseolina*, and *Rhizoctomia solani* by 50.7, 45.5, 48.9, 58.3, 40, and 69.5%, respectively (Nawaz et al., 2001).

Commentary: The plant probably synthetises anti- Hepatitis B principles because extract of *Sesuvium portulacastrum* (L.) L. has antiviral property against the Hepatitis B virus (see earlier) and *Trianthema portulacastrum* L. and the following plant (*Zaleya pentandra* (L.) C. Jeffrey) are used for jaundice.

REFERENCES

Kavitha, D., Parvatham, R. and Padma, P.R., 2014. Assessment of *Trianthema portulacastrum* for its antimicrobial potential and investigation of their phytochemicals using HPTLC, GC-MS, and IR. *International Journal of Pharmacy and Pharmaceutical Sciences*, *6*(1), p. 675.

Nawaz, H.R., Malik, A. and Ali, M.S., 2001. Trianthenol: an antifungal tetraterpenoid from *Trianthema portulacastrum* (Aizoaceae). *Phytochemistry*, *56*(1), pp. 99–102.

8.7.2.4 *Zaleya pentandra* (L.) C. Jeffrey

Synonyms: *Zaleya govindia* (G. Don) Nair; *Trianthema govindia* F. Ham.; *Trianthema pentandra* L.

Common names: African purslane; vellai sharanai (India); acho waho, itsit, slathi (Pakistan)

Habitat: Coastal sandy shores, Deserts

Distribution: Iran, Pakistan, India

Botanical observation: This halophyte herb grows up to 40 cm long. The stems are terete, somewhat fleshy, and glabrous. The leaves are simple, opposite, and exstipulate. The petiole is up to 1.8 cm long and sheathing at base. The blade is spathulate, fleshy, papillose, 1.5–4 cm × 0.6–2.5 cm, cuneate at base, round or notched at apex, with a conspicuous sunken midrib and no apparent secondary nerves. The flowers are minute and packed in dense axillary clusters. The perianth is 5-lobed, the lobes are lanceolate, purplish, and up to 4 mm long. The androecium includes 5 stamens. The ovary is somewhat cylindrical and develops 2 styles. The capsules are dehiscent, about 4 mm long and contain tiny and black seeds.

Medicinal uses: Flu, jaundice, hepatitis (Pakistan)

Broad-spectrum antibacterial polar extract: Methanol extract of whole plant (3 mg/8 mm well) inhibited the growth of *Salmonella typhi* and *Staphylococcus aureus* with inhibition zone diameters of 13 and 17.5 mm, respectively (Afzal et al., 2016).

Antifungal (filamentous) mid-polar extract: Dichloromethane extract of at a concentration of 250 µL/mL inhibited the growth of *Aspergillus niger* and *Fusarium solani* by 89 and 49%, respectively (Afzal et al., 2015).

Commentary: The medicinal uses suggest some antiviral properties and some research could be done in this direction. Note that an ethanol extract of *Galenia africana* L. inhibited *Mycobacterium smegmatis* and *Mycobacterium tuberculosis* with the MIC values of 0.7 and 1.2 mg/mL respectively (Mativandlela et al., 2008). From this extract, 5,7,2′-trihydroxyflavone (Log D = 1.3 at pH 7.4; molecular mass = 270.2 g/mol) inhibited *Mycobacterium smegmatis* and *Mycobacterium tuberculosis* (Mativandlela et al., 2008). Other antimycobacterial principles may await discovery from this family.

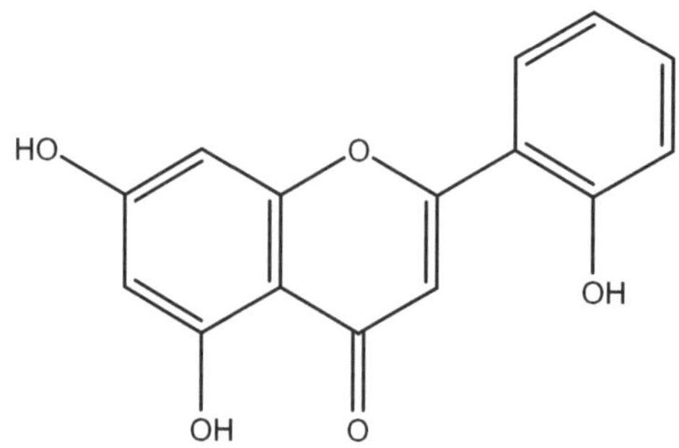

5,7,2'-Trihydroxyflavone

REFERENCES

Afzal, S., Chaudhary, B.A., Ahmad, A. and Afzali, K., 2015. Prelimiary phytochemical analysis and antifungal activities of crude extracts of *Zaleya pentandra* and Corchorus depressus Linn. *Acta Poloniae Pharmaceutica*, *72*, pp. 329–334.

Afzal, S., Chaudhry, B.A., Saeed, J., Afzal, K., Ahmed, B. and Qadir, M.I., 2016. Antibacterial and antioxidant activity of methanolic extract of *Zaleya pentandra*. *Acta Poloniae Pharmaceutica*, *73*(1), pp. 147–151.

Mativandlela, S.P.N., Meyer, J.J.M., Hussein, A.A., Houghton, P.J., Hamilton, C.J. and Lall, N., 2008. Activity against *Mycobacterium smegmatis* and *Mycobacterium tuberculosis* by extract of South African medicinal plants. *Phytotherapy Research: An International Journal Devoted to Pharmacological and Toxicological Evaluation of Natural Product Derivatives*, *22*(6), pp. 841–845.

8.7.3 FAMILY BASELLACEAE RAFINESQUE (1837)

The family Basellaceae includes 4 genera and 20 species of fleshy climbers growing from tuberous rhizomes. The leaves are simple, alternate, and exstipulate. The inflorescences are spikes, panicles, or racemes. The flowers are minute. The perianth includes 5 fleshy and glossy tepals. The androecium comprises 5 free stamens. The gynoecium consists of 3 carpels fused into an unilocular ovary containing a single basal ovule and produces 3 styles. The fruits are indehiscent capsules surrounded by persistent tepals.

8.7.3.1 *Anredera cordifolia* (Ten.) Steenis

Synonyms: *Boussingaultia cordifolia* Ten.; *Anredera baselloides* (Kunth) Baill.; *Boussingaultia baselloides* Kunth; *Boussingaultia cordata* Sprenger; *Boussingaultia cordifolia* (Moq.) Volkens; *Boussingaultia cordifolia* Ten.; *Boussingaultia gracilis* Miers

Common names: Madeira-vine; phak pang, luo kui shu (China); binahong (Indonesia); dear chun (Thailand); mồng tơi củ (Vietnam)

Habitat: Fences, thickets, gardens, wastelands

Distribution: Tropical and Subtropical Asia

Botanical observation: This ornamental yet poisonous climber native to South America grows to about 6 m long from a stout rhizome. The stems are terete and develop aerial tubers which have a repulsive aura. The leaves are simple, spiral, and exstipulate. The petiole is up to 3 cm long. The blade is ovate, fleshy, glossy, wavy, 1–11 × 0.5–8 cm, more less cordate base, and acute at apex. The spikes are axillary, showy, somewhat tail-like, whitish, and up to 25 cm long. The 5 tepals are pure white-to-yellowish, 5 mm long, and ovate. The androecium includes 5 white to yellowish, stamens. The gynoecium develops 3 styles. The capsules are subglobose, 3-lobed, wrinkled, and shelter a few black and glossy seeds.

Medicinal use: Wounds (Indonesia)

Strong antibacterial (Gram-positive) essential oil: Essential oil from leaves inhibited the growth of *Bacillus cereus* (ATCC 1177), *Bacillus subtilis* (ATCC 6633), and *Staphylococcus epidermidis* (ATCC 12228) with the MIC values of 100, 50, and 25 µg/mL, respectively (Souza et al., 2014). MBC for *Staphylococcus epidermidis* (ATCC 12228) was 50 µg/mL (Souza et al., 2014).

Moderate broad-spectrum antibacterial non polar extract: Hexane extract of leaves inhibited the growth of *Staphylococcus aureus*, methicillin-sensitive *Staphylococcus aureus*, methicillin-resistant *Staphylococcus aureus*, *Bacillus subtilis*, *Bacillus cereus*, and extended spectrum β-lactamase *Escherichia coli* with the MIC/MBC values of 256/>4096, 2048/4096, 1024/2048, 512/4096, 512/1024, and 1024/2048 µg/mL, respectively (Leliqia et al., 2017).

Moderate antimycobacterial non polar extract: Hexane extract of leaves inhibited the growth of *Mycobacterium tuberculosis* (H37/Rv), streptomycin-rifampicin-resistant *Mycobacterium tuberculosis*, and isoniazid-ethambutol-resistant *Mycobacterium tuberculosis* with the MIC values of 500, 500, and 250 µg/mL, respectively (Pitaloka & Sukandar, 2018).

Antimycobacterial non-polar antibiotic potentiator: Hexane extract of leaves increased the sensitivity of *Mycobacterium tuberculosis* (H37/Rv) to streptomycin (Pitaloka & Sukandar, 2018). This extract increased the sensitivity of streptomycin-rifampicin-Resistant *Mycobacterium tuberculosis* to izoniasid and isoniazid-ethambutol-resistant *Mycobacterium tuberculosis* to rifampicin (Pitaloka & Sukandar, 2018).

Strong antiviral (enveloped linear monopartite double-stranded DNA) polar extract: Aqueous extract inhibited Adenovirus-3 and -8 with EC_{50} of 44.1 and 44.7 µg/mL, respectively, (Chiang et al 2003).

Commentaries

i. The non-polar antibacterial and antimycobacterial constituents of this plant are apparently yet to be identified.
ii. The strong antiviral property of aqueous extract might involve saponins (Astuti et al., 2011) or polysaccharides (Zhang et al., 2017). Polysaccharides are primary metabolites (Cooper & Nicola, 2014) and therefore not discussed in the present work.
iii. The leaves have similarity with the leaves of *Basella alba* L. posing the risk of poisoning. Consider that good botanical training prevents poisoning.
iv. *Anredera scandens* (L.) Moq. is used for boils in the Philippines (local name: *Malabato*).

REFERENCES

Astuti, S.M., Sakinah, M.A., Andayani, R.B. and Risch, A., 2011. Determination of saponin compound from *Anredera cordifolia* (Ten) Steenis plant (binahong) to potential treatment for several diseases. *Journal of Agricultural Science*, *3*(4), p. 224.

Chiang, L.C., Cheng, H.Y., Liu, M.C., Chiang, W. and Lin, C.C., 2003. *In vitro* anti-herpes simplex viruses and anti-adenoviruses activity of twelve traditionally used medicinal plants in Taiwan. *Biological and Pharmaceutical Bulletin*, *26*(11), pp. 1600–1604.

Cooper, R. and Nicola, G., 2014. *Natural Products Chemistry: Sources, Separations and Structures*. CRC Press, Boca Raton, FL.

Leliqia, N.P.E., Sukandar, E.Y. and Fidrianny, I., 2017. Antibacterial activities of *Anredera Cordifolia* (Ten.) V. Steenis leaves extracts and fractions. *Asian Journal of Pharmaceutical and Clinical Research*, *10*(12), pp. 175–178.

Pitaloka, D.A.E. and Sukandar, E.Y., 2018. Synergistic study on n-hexane extract of *Anredera cordifolia* (Ten.) v Steenis (binahong) leaves combined with antituberculosis drugs against drug-sensitive and drug-resistant of *Mycobacterium tuberculosis*. *Journal of Applied Pharmaceutical Science*, *8*(5), pp. 134–138.

Souza, L.F., de Barros, I.B.I., Mancini, E., Martino, L.D., Scandolera, E. and Feo, V.D., 2014. Chemical composition and biological activities of the essential oil from *Anredera cordifolia* grown in Brazil. *Natural Product Communications*, *9*(7), pp. 1003–1006.

Zhang, Z.P., Shen, C.C., Gao, F.L., Wei, H., Ren, D.F. and Lu, J., 2017. Isolation, purification and structural characterization of two novel water-soluble polysaccharides from *Anredera cordifolia*. *Molecules*, *22*(8), p. 1276.

8.7.3.2 *Basella alba* L.

Synonyms: *Basella cordifolia* Lam.; *Basella nigra* Lour.; *Basella rubra* L.; *Gandola alba* Rumph. ex L.; *Gandola rubra* Rumph. ex L.

Common names: Ceylon spinach, Malabar nightshade, Malabar spinach; pui shak (Bangladesh); chrâlong (Cambodia); luo kui (China); basala, puin sak, upodika (India); genjerot (Indonesia); pang (Laos); gondola (Malaysia); kin peint (Myanmar); paro sag, poi sag (Nepal); poi (Pakistan); alugbati, libdto (the Philippines); phak plang (Thailand); mồng tơi, (Vietnam)

Habitat: Cultivated, villages, forests borders

Distribution: Tropical Asia

Botanical observation: This climber grows to about 2 m long and has an aura of positive life force. The stems are terete, purplish, glossy, and fleshy. The leaves are simple, spiral, and exstipulate. The petiole is up to 3 cm long. The blade is ovate, fleshy, glossy, edible, 2.5–13 × 5–8 cm, more or less cordate base, and acuminate at apex. The spikes are axillary or terminal and up to 20 cm long. The perianth is tubular, purplish-white, 3–4 mm long, and develops 5 fleshy and ovate lobes. Five stamens are present. The gynoecium includes 3 styles. The capsules are enclosed in tepals, red to dark red or black, glossy, and about 5 mm across.

Medicinal uses: Acnea, balanitis, burn (India); dysentery (Thailand); boils (India; the Philippines); diarrhea, fever, sores, urinary tract infections (Myanmar); tuberculosis (India)

Strong broad-spectrum antibacterial polar extract: Methanol extract inhibited the growth of *Pseudomonas aeruginosa* (MTCC 1688), *Staphylococcus aureus* (MTCC 737), *Bacillus subtilis* (MTCC 441), *Micrococcus luteus* (MTCC 106), *Escherichia coli* (MTCC 443) with the MIC values of 3.1, 3.1, 3.1, 3.1, and 6.2 µg/mL, respectively (Sushila et al., 2010).

Commentary: The plant produces the betacyanin pigments gomphrenin, betanin, and isobetanin (Kumar et al., 2015) which probably account, at least in part, for the antibacterial properties (Gengatharan et al., 2015).

REFERENCES

Gengatharan, A., Dykes, G.A. and Choo, W.S., 2015. Betalains: natural plant pigments with potential application in functional foods. *LWT-Food Science and Technology*, *64*(2), pp. 645–649.

Kumar, S.S., Manoj, P., Shetty, N.P., Prakash, M. and Giridhar, P., 2015. Characterization of major betalain pigments-gomphrenin, betanin and isobetanin from *Basella rubra* L. fruit and evaluation of efficacy as a natural colourant in product (ice cream) development. *Journal of Food Science and Technology*, *52*(8), pp. 4994–5002.

Sushila, R., Deepti, A., Permender, R., Madhavi, T., Dharmender, R. and Rathee, D., 2010. Cytotoxic and antibacterial activity of *Basella alba* whole plant: a relatively unexplored plant. *Pharmacologyonline*, *3*, pp. 651–658.

8.7.4 FAMILY ANCISTROCLADACEAE PLANCH. EX WALP. (1851)

The family Ancistrocladaceae consists of the single genus *Ancistrocladus* Wall.

8.7.4 *1 Ancistrocladus tectorius* (Lour.) Merr.

Synonyms: *Ancistrocladus carallioides* Craib; *Ancistrocladus cochinchinensis* Gagnep.; *Ancistrocladus hainanensis* Hayata; *Ancistrocladus harmandii* Gagnep.; *Ancistrocladus pinangianus* Wall. ex Planch.

Common names: Gou zhi teng (China); akar bebulus, akar julong hitam, trung bulus (Indonesia); khu hang kouy (Laos); akar julong hitam, lindah sapi (Malaysia); khon maa (Thailand); trung quân (Vietnam)

Habitat: Forests

Distribution: India, Andaman and Nicobar Islands, Myanmar, Cambodia, Laos, Thailand, Vietnam, Malaysia, Bangladesh, Indonesia

Botanical observation: This discrete woody climber grows to about 20 m long. The stems are terete and present woody hooks at nodes. The leaves are simple, spiral, sessile, and stipulate. The blade is 7–30 × 2–10 cm, clasping the stem, elliptic to obovate, somewhat coriaceous, smooth, glossy, tapering at base, marked with about 7–12 pairs of unconspicuous secondary nerves, and acute at apex. The panicles are terminal and lax. The 5 sepals are about 5 mm long and elliptic. The 5 petals are broadly elliptic and pinkish to purplish or burgundy and about 5 mm long. The androecium includes 10 stamens of irregular length. The gynoecium includes 3 styles. The nuts are about 1.5 cm long and present 5 spathulate wings of irregular length, up to about 6 cm long, and light brown.

Medicinal uses: Diarrhea, dysentery (Malaysia; Thailand)

Antibacterial (Gram-positive) halo developed by polar extract: Methanol extract of whole plant (1 mg/6 mm disc) inhibited the growth of *Bacillus cereus*, *Bacillus subtilis*, and *Staphylococcus aureus* with the inhibition zone diameters of 13, 22, and 12 mm, respectively (Wiart, 2004).

Anticandidal halo developed by polar extract: Methanol extract of whole plant (1 mg/6 mm disc) inhibited the growth of *Candida albicans* with inhibition zone diameter of 10 mm (Wiart, 2004).

Commentaries

i. Plants in the genus *Ancistrocladus* Wall. yield unique types of naphthylisoquinoline alkaloids such as korundamine A (Log D = 4.5 at pH 7.4; molecular mass = 768.9 g/mol) which inhibited the cytopathic effect of the Human immunodeficiency virus with the EC_{50} value of 2 µM (Hallock et al., 1998). This alkaloid inhibited resistant strains of Human immunodeficiency virus with the EC_{50} values of 8, 10, and 6 µM for the CEM-SS/OC100, MT2/A17 and MT2/G9106 host cell/virus, respectively (Hallock et al., 1998).

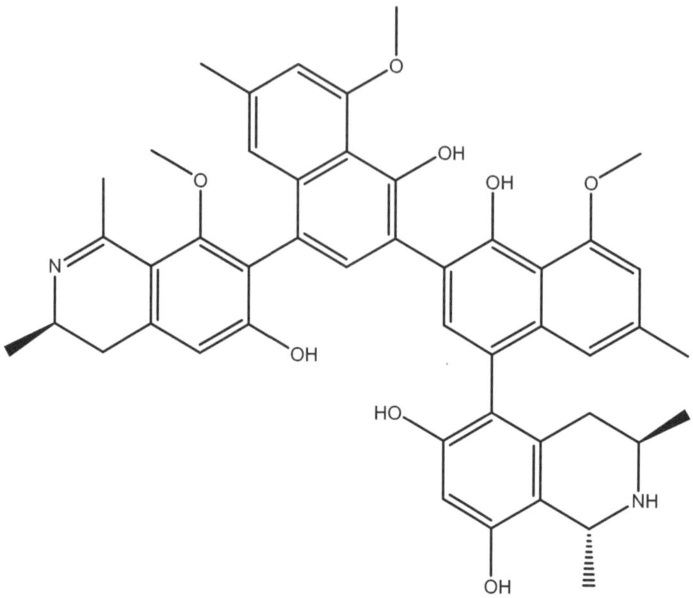

Korundamine A

The naphthylisoquinoline alkaloids michellamine A_2, A_3, and A_4 isolated from the root bark of a member of the genus *Ancistrocladus* Wall. inhibited Human immunodeficiency virus type-1 (strain IIIB) with an IC_{50} of 29.6, 15.2, and 20.4 µM, respectively in A3.01 T lymphoblast cells (Bringmann et al., 2016).

ii. In the Andamans, *Ancistrocladus attenuatus* Dyer (local name: *Khoon bel*) is used to treat jaundice. Medicinal plants used to treat jaundice often yield anti-Human immunodeficiency virus principles. Jaundice can be a symptom of viral hepatitis which is caused by Hepatitis A virus (single-stranded (+) RNA virus), Hepatitis B virus (double-stranded DNA virus), and Hepatitis C viruses (single-stranded (+) RNA virus). The Human immunodeficiency virus is a double-stranded (+) RNA virus.

REFERENCES

Bringmann, G., Steinert, C., Feineis, D., Mudogo, V., Betzin, J. and Scheller, C., 2016. HIV-inhibitory michellamine-type dimeric naphthylisoquinoline alkaloids from the Central African liana *Ancistrocladus congolensis*. *Phytochemistry*, *128*, pp. 71–81.

Hallock, Y.F., Cardellina, J.H., Schäffer, M., Bringmann, G., François, G. and Boyd, M.R., 1998. Korundamine A, a novel HIV-inhibitory and antimalarial "hybrid" naphthylisoquinoline alkaloid heterodimer from *Ancistrocladus korupensis*. *Bioorganic & Medicinal Chemistry Letters*, *8*(13), pp. 1729–1734.

Wiart, C., Mogana, S., Khalifah, S., Mahan, M., Ismail, S., Buckle, M., Narayana, A.K. and Sulaiman, M., 2004. Antimicrobial screening of plants used for traditional medicine in the state of Perak, Peninsular Malaysia. *Fitoterapia*, *75*(1), pp. 68–73.

8.7.5 FAMILY CACTACEAE A.L DE JUSSIEU (1789)

The family Cactaceae consists of about 100 genera and 1500 species of spiny and succulent herbs or treelets. The leaves are simple, alternate, exstipulate, or very much reduced. The flowers are often solitary and showy. The perianth includes numerous tepals. The androecium comprises numerous free stamens. The gynoecium consists of 3-many carpels fused into a unilocular ovary containing numerous ovules growing on parietal placentas. The fruits are berries. Members of this family produce notably isoquinoline alkaloids and betalains.

8.7.5.1 *Opuntia ficus-indica* (L.) Mill.

Synonyms: *Cactus chinensis* Roxb.; *Cactus ficus-indica* L.; *Cactus opuntia* L.; *Opuntia arcei* Cárdenas; *Opuntia chinensis* (Roxb.) K. Koch; *Opuntia compressa* J.F. Macbr.; *Opuntia cordobensis* Speg.; *Opuntia ficus-barbarica* A. Berger; *Opuntia megacantha* Salm-Dyck; *Opuntia paraguayensis* K. Schum.; *Opuntia tuna-blanca* Speg.; *Opuntia vulgaris* Mill.; *Platyopuntia cordobensis* (Speg.) F. Ritter

Common names: Indian fig, Barbary fig, prickly pear, nopal; chapati balli (India); li guo xian ren zhang (China)

Habitat: Sunny dry rocky lands

Distribution: Tropical Asia

Botanical observation: This succulent treelet grows up to about 3 m tall. The cladodes are dull green obovate, elliptic, or oblong, 25–55 × 7–19.5 cm, fleshy, thick, and somewhat hard. The areoles are elliptic, about 5 mm long, and develop 1–6 woody and very sharp spines which are up to 3 cm long. The flowers are solitary, sessile, and 5–8 cm in diameter. The perianth comprises numerous tepals which are yellowish red, ovate or obovate, oblong, emarginated or acute at apex, and about 3.5 cm long. The stamens are numerous, yellowish, and 8 mm long stamens. The gynoecium is about 1.5 cm long, bottle-shaped, and develops 7–10 stigmas. The berries are purplish, palatable, obovate, and up to 9 cm long.

Medicinal use: Diarrhea (the Philippines)

Antibacterial halo polar extract: Methanol extract of flowers (2mg/5mm well) inhibited the growth of *Escherichia coli* (ATCC 25922), *Staphylococcus aureus* (ATCC 25923), *Bacillus subtilis* (CTM 50077), and *Listeria monocytogenes* (ATCC 43251) with inhibition zone diameters of 10, 13, 12.5, and 25 mm, respectively (Ammar et al., 2015).

Moderate antibacterial (Gram-negative) extract: Methanol extract of cladodes inhibited *Campylobacter jejuni* (NCTC 5653), *Campylobacter jejuni*, and *Campylobacter coli* with MBC of 400, 300, and 400 µg/mL, respectively (Castillo et al., 2011).

Strong antiviral (enveloped monopartite linear double-stranded DNA) polar extract: Ethanol extract of stem inhibited the replication of the Herpes simplex type 2 with an IC_{50} of 63 µg/mL (Bouslama et al., 2011). From this extract, catabolites of chlorophyll pheophorbide a and pyropheophorbide a inhibited Herpes simplex type 2 replication with IC_{50} of 0.3 and 0.01 µg/mL, respectively (Bouslama et al., 2011).

Strong antiviral (non-enveloped monopartite linear single-stranded (+) RNA) chlorophyll catabollites: Pheophorbide a and pyropheophorbide a inhibited the Poliovirus type 1 replication with the IC_{50} values of 8.7 and 2.2 µg/mL, respectively (Bouslama et al., 2011).

Commentaries

i. Apparently, members of the family Cactaceae have not been much studied for their antibacterial properties (McCleary & Walkington, 1964).
ii. Numerous species in the genus *Opuntia* Mill are used as anti-infectious medicine in Asia. *Opuntia monacantha* Haw. is used in Pakistan (local name: *Chnutarthar*) for the treatment of fever, ulcers, leucoderma, and gonorrhea. In India, *Opuntia vulgaris* Mill. (local name: *Trappar sula*) is used to treat ear infections, *Opuntia cochinellifera* Mill. is used for burns and cough, *Opuntia dillenii* (Ker-Gawl.) Haw. (local names: *Naagaphani, Sappathi kalli*) is used for cough, gonorhoea, and wounds, *Opuntia vulgaris* Mill. is used for diarrhoea, and *Opuntia elatior* Mill. (local name: *Pah-nagpheni*) is applied to cuts.

REFERENCES

Ammar, I., Bardaa, S., Mzid, M., Sahnoun, Z., Rebaii, T., Attia, H. and Ennouri, M., 2015. Antioxidant, antibacterial and in vivo dermal wound healing effects of Opuntia flower extracts. *International Journal of Biological Macromolecules*, *81*, pp. 483–490.

Bouslama, L., Hayashi, K., Lee, J.B., Ghorbel, A. and Hayashi, T., 2011. Potent virucidal effect of pheophorbide a and pyropheophorbide a on enveloped viruses. *Journal of Natural Medicines*, *65*(1), pp. 229–233.

Castillo, S.L., Heredia, N., Contreras, J.F. and García, S., 2011. Extracts of edible and medicinal plants in inhibition of growth, adherence, and cytotoxin production of *Campylobacter jejuni* and *Campylobacter coli*. *Journal of Food Science*, *76*(6), pp. M421–M426.

McCleary, J.A. and Walkington, D.L., 1964. Antimicrobial activity of the Cactaceae. *Bulletin of the Torrey Botanical Club*, *91*(5), pp. 361–369.

8.7.6 FAMILY CARYOPHYLLACEAE A. L. DE JUSSIEU (1789)

The family Caryophyllaceae consists of about 100 genera and 2500 species of herbs and shrubs. The leaves are simple, opposite, decussate, exstipulate, or minutely stipulate. The interpetiolar node is swollen and this is a botanical characteristic of the family. The inflorescences are mainly cymes. The calyx comprises 5 sepals. The 5 petals comprise a claw, a limb and a scale. The androecium comprises 5–10 free stamens. The gynoecium is made of 2–5 carpels fused into a uni- or plurilocular ovary containing 1-many ovules attached to a central placenta. The capsules are dehiscent and filled with minute and reniform seeds. Plants in the family Caryophyllaceae accumulate triterpene saponins which are probably used as phytoanticipins: natural products inhibiting the growth of phytopathogenic microbes present prior to microbial infections.

8.7.6.1 *Dianthus chinensis* L.

Synonyms: *Dianthus amurensis* Jacq.; *Dianthus dentosus* Fisch. ex Rchb.; *Dianthus fischeri* Spreng.; *Dianthus morii* Nakai; *Dianthus sequieri* Chaix; *Dianthus subulifolius* Kitag.; *Dianthus versicolor* Fisch. ex Link

Common names: Chinese pink; shi zhu (China); pae raeng i kkot (Korea); sekitsuku (Japan)

Habitat: Sandy soils, Forests margins, steppes, seashores, mountains

Distribution: Kazakhstan, India, Kashmir, Nepal, China, Korea, Taiwan, Japan, Mongolia, Siberia

Botanical observation: This slender and graceful herb grows to about 50 cm tall. The leaves are simple, opposite, and exstipulate. The blade is linear-lanceolate, 0.2–0.4× 3–5 cm, and acuminate at apex. The inflorescence is solitary or a cyme. The calyx comprises 5 lobes which are lanceolate and about 5 mm long. The 5 petals are up to about 1.8 cm long, red, purple, pink, or white, somewhat spathulate, and toothed. The 10 stamens are as long as the petals claws and with blue anthers. The ovary is oblong and develops 2 linear stigmas. The capsules are elongated and 5-lobed and contain numerous tiny and black seeds.

Medicinal uses: Boils, urogenital infections (China)

Moderate antibacterial (Gram-positive) polar extract: Methanol extract inhibited the growth of *Propionibacterium acnes* with the MIC value of 400 μg/mL (Weaon et al., 2011).

Commentaries

i. In India, *Dianthus carophyllus* L. is used for viral infections and for diarrhea in Iraq. In Iran, *Dianthus persicus* Hausskn. is considered as "hot" as per the theory of humorism for the treatment of cough, gingivitis, and toothache. In Turkey, *Dianthus caryophyllus* L. (local name: *Dağ karanfili*) is used as antiseptic and *Gypsophila arrostii*

Guss. (local name: *Helvacı kökü*) is used for cough. Plants accumulating triterpene saponins are often used as expectorant since the amphiphilic nature of saponins dissolves mucus.

ii. The plant has apparently not been much studied for antimicrobial activities. It produces oleanane-type triterpene saponins (Hong-Yu et al., 1994) which probably account, at least partially, for its medicinal effects.

REFERENCES

Hong-Yu, L., Koike, K. and Ohmoto, T., 1994. Triterpenoid saponins from *Dianthus chinensis*. *Phytochemistry*, *35*(3), pp. 751–756.

Weon, J.B., Ahn, J.H. and Ma, C.J., 2011. Antibacterial activity of some medicinal plants against *Propionibacterium acnes*. *Korean Journal of Pharmacognosy*, *42*(1), pp. 98–101.

8.7.6.2 *Drymaria cordata* (L.) Willd. ex Schult.

Synonyms: *Drymaria adenophora* Urb.; *Drymaria diandra* (Sw.) Macfad.; *Drymaria diandra* Blume; *Drymaria procumbens* Rose; *Holosteum cordatum* L.; *Holosteum diandrum* Sw.; *Stellaria adenophora* (Urb.) León

Common names: Heart-leaf drymary, tropical chickweed; he lian dou cao (China); dehradun pithpapra, avijal, tandon-panbi (India); angling, jukut ibun (Indonesia); abhijalo, peperano (Nepal); lukumuaia (Papua New Guinea); bakalanga (the Philippines); yaa klet hoi (Thailand); tuft if (Vietnam)

Habitat: Wastelands, humid and shady spots, by streams, plantations, gardens, roadsides

Distribution: Tropical Asia

Botanical observation: This slender and creeping herb grows to about 90 cm long. The stems are rooting at node. The leaves are simple, opposite, and stipulate. The stipules are tiny and membranaceous. The petiole is up to about 7 mm long. The blade is broadly lanceolate to cordate, 0.2–3 × 1–3 cm, dull almost glaucous green, edible (Nepal), decurrent at base, marked with about 2 pairs of secondary nerves, and acuminate at apex. The cymes are terminal, lax, and few flowered. The calyx comprises 5 lobes which are lanceolate and about 3 mm long. The 5 petals are up to about 5 mm long, pure white, somewhat spathulate, and bifid. Five stamens are present. The ovary is ovoid and develops a trifid stigmas. The capsules are ovoid, elongated, 3-lobed, and contain numerous tiny, warty, and black seeds.

Medicinal uses: Boils, fever (Indonesia; Nepal), cough, cold, eye infection (Nepal); burn, diarrhea, dysentery, pneumonia, skin diseases, typhoid (India)

Broad-spectrum antibacterial halo developed by polar extract: Methanol extract (350 µg/mL/6 mm disc) inhibited the growth of *Staphylococcus aureus* (ATCC 29737), *Escherichia coli* (ATCC 10536), *Bacillus subtilis* (ATCC 6633), *Bacillus pumilus* (ATCC 14884), and *Pseudomonas aeruginosa* (ATCC 25614) with inhibition zone diameters of 20, 21, 15, 13, and 18 mm, respectively (Mukherjee et al., 1997).

Antimycobacterial halo developed by polar extract: Methanol extract inhibited the growth of *Mycobacterium phlei* (Taylor et al., 1995).

Antifungal (filamentous) halo developed by polar extract: Methanol extract inhibited the growth of *Microsporum gypseum* and *Trichophyton mentagrophytes* (Taylor et al., 1995).

Strong antiviral (monopartite linear dimeric single-stranded (+) RNA) amphiphilic β-carboline alkaloid: Drymaritin (Log D 2.8 at pH 7.4; molecular mass = 250.2 g/mol) inhibited the replication of the anti-Human immunodeficiency virus in H9 lymphocytes with an EC_{50} value of 0.6 µg/mL and a selectivity index of 20.6 (Hsieh et al., 2004).

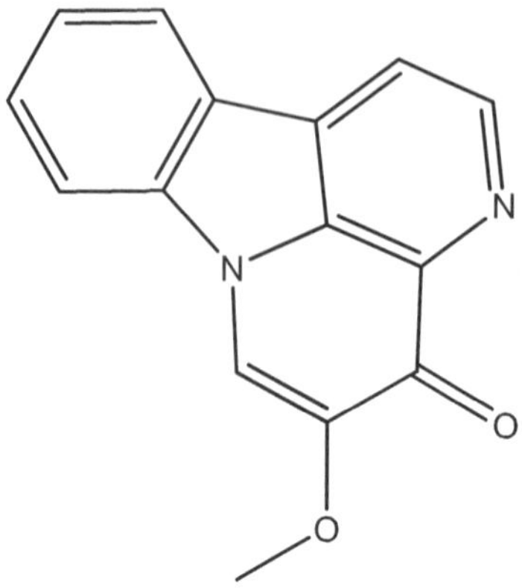

Drymaritin

Commentary: Drymaritin (Wetzel et al., 2009) could be a phytoalexin. Consider that amphiphilic β-carboline alkaloids are often antimycobacterial and antibacterial. The anti-Human immunodeficiency virus mechanism of drymaritn most probably involves interference with RNA replication. Drymaritin could be active against coronaviruses and other (+)-RNA viruses and be added to the list of natural products to be tested in emergency for the coming waves of pandemics.

REFERENCES

Hsieh, P.W., Chang, F.R., Lee, K.H., Hwang, T.L., Chang, S.M. and Wu, Y.C., 2004. A new anti-HIV alkaloid, drymaritin, and a new C-glycoside flavonoid, diandraflavone, from *Drymaria diandra*. *Journal of Natural Products*, *67*(7), pp. 1175–1177.

Mukherjee, P.K., Bhattacharya, S., Saha, K., Giri, S.N., Pal, M. and Saha, B.P., 1997. Antibacterial evaluation of *Drymaria cordata* Willd (Fam. Caryophyllaceae) extract. *Phytotherapy Research*, *11*(3), pp. 249–250.

Taylor, R.S., Manandhar, N.P. and Towers, G.H.N., 1995. Screening of selected medicinal plants of Nepal for antimicrobial activities. *Journal of Ethnopharmacology*, *46*(3), pp. 153–159.

Wetzel, I., Allmendinger, L. and Bracher, F., 2009. Revised structure of the alkaloid drymaritin. *Journal of Natural Products*, *72*(10), pp. 1908–1910.

8.7.6.3 *Gypsophila oldhamiana* Miq.

Common names: Manchurian baby's breath; chang rui shi tou hua (China)

Habitat: sandy shores, rocky soils

Distribution: China, Korea

Botanical observation: This magnificent herb grows to about 1 m tall and has an aura of dryness. The roots are stout and woody. The stems are puplish. The leaves are simple, opposite, sessile, and exstipulate. The blade is oblong, 4–8× 0.5–1.5 cm, somewhat fleshy, with 1–2 pairs of secondary nerves, and acute at apex. The cymes are lax, terminal, and corymbose. The calyx is tubular, 5-lobed, the lobes triangular and about 3 mm long. The 5 petals are 8 mm long, white or pink, spathulate, and emarginate. The androecium includes 10 stamens which protrude out of the corolla. The gynoecium is obovate and develops 2 linear stigmas. The capsules are ovoid and contain numerous tiny and brown seeds.

Medicinal use: Typhoid (China)

Commentaries

i. This plant has apparently not been studied for antimicrobial properties. It produces the β-carboline alkaloids oldhamiaines A and B (Zhang et al., 2015). β-Carboline alkaloids are planar molecules interacting with DNA and topoisomerase and inhibit Human immunodeficiency virus type-1, Herpes simplex virus, Gram-positive, Gram-negative bacteria, and fungi (Cao et al., 2007).

ii. Like many other plants in the Caryophyllaceae, the plant accumulates triterpene saponins, such as gypsogenin-type triterpene saponins (Bai et al., 2007). Note that the aglycone gypsogenin (Log D = 4.6 at pH 7.4; molecular mass=470.6 g/mol) inhibited the growth of *Bacillus cereus* (ATCC 11778), *Bacillus subtilis* (ATCC 6633) and *Bacillus thrungiensis* (Emirdağ-Öztürk et al., 2014).

iii. Saponins in Caryophyllaceae account for the use of these plants as soap, like for instance, *Gypsophila paniculata* L. in Iraq (local name: *El-sabuniyeh*) or for cough such as *Gypsophila arrostii* Guss. in Turkey (local name: *Helvacı kökü*).

iv. Note that medicinal plants for the treatment of typhoid are few and appear to be common in the Caryophyllaceae. One could examine the presence of anti-*Salmonella typhi* principles in this vast family.

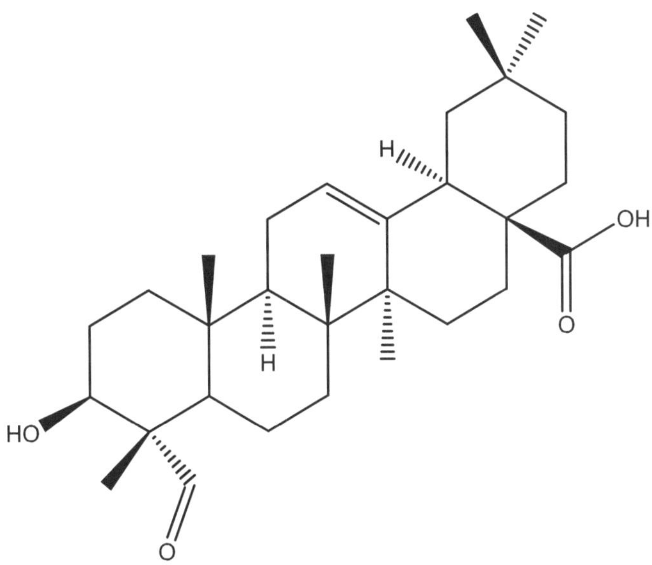

Gypsogenin

REFERENCES

Bai, H., Zhong, Y., Xie, Y.Y., Wang, Y.S., Liu, L., Zhou, L., Wang, J., Mu, Y.L. and Zuo, C.X., 2007. A major triterpenoid saponin from *Gypsophila oldhamiana*. *Chemistry & Biodiversity*, *4*(5), pp. 955–960.

Cao, R., Peng, W., Wang, Z. and Xu, A., 2007. β-Carboline alkaloids: biochemical and pharmacological functions. *Current Medicinal Chemistry*, *14*(4), pp. 479–500.

Emirdağ-Öztürk, S., Karayıldırım, T., Çapcı-Karagöz, A., Alankuş-Çalışkan, Ö., Özmen, A. and Poyrazoğlu-Çoban, E., 2014. Synthesis, antimicrobial and cytotoxic activities, and structure–activity relationships of gypsogenin derivatives against human cancer cells. *European Journal of Medicinal Chemistry*, *82*, pp. 565–573.

Zhang, Y., Wang, G., Lv, H., Luo, J. and Kong, L., 2015. Two new β-carboline alkaloids from the roots of *Gypsophila oldhamiana*. *Natural Product Research*, *29*(13), pp. 1207–1211.

8.7.6.4 *Lychnis coronaria* (L.) Desr.

Synonyms: *Agrostemma coronaria* L.; *Coronaria coriacea* (Moench) Schischk.; *Lychnis coriacea* Moench; *Silene coronaria* (L.) Clairv.

Common names: Rose campion, dusty-miller; jian chun luo (China); angaarda, shosh ghasa (India)

Habitat: Grassy lands, forests margin, cultivated

Distribution: Turkey, Iran, Turkestan, Pakistan, Turkey, India, Russia, China, Japan

Botanical observation: This magnificent herb grows to about 50 cm tall. The stems are somewhat glaucous and hairy. The leaves are simple, opposite, sessile, and exstipulate. The blade is spathulate, 2–9× 1.5–2.5 cm, hairy, somewhat fleshy, and acute at apex. The inflorescence is a lax panicle. The calyx is tubular, 5 lobed, ribbed, hairy, and up to about 2 cm long, the lobes linear. The 5 petals are up to about 8 mm long, bright pink, deltoid, up to about 2.5 cm long, and dentate at apex. The androecium includes 10 stamens. The gynoecium develops 5 stigmas. The capsules are oblong, 2 cm long, and contain numerous tiny reniform seeds.

Medicinal uses: Cough (India); wounds, cuts (Turkey)

Commentary: Apparently not investigated for its potential antimicrobial effects.

8.7.6.5 *Spergula arvensis* L.

Common names: Corn spurry; da zhua cao (China); gandel (India); hyau jhar (Nepal); jangli dhania (Pakistan)

Habitat: Fields, grasslands, riverbanks, moist and shady places

Distribution: Iran, Kazakhstan, Pakistan, India, Bhutan, China, Japan, The Philippines

Botanical observation: This slender and graceful herb grows up to about 50 cm tall. The leaves are simple, sessile, opposite, and stipulate. The stipules are tiny, triangular, and ephemeral. The blade is linear, 1–4×0.5–1 mm, somewhat hairy, and acute at apex. The cymes are lax, gracile, and terminal. The 5 sepals are triangular, 3 mm long, and somewhat hairy. The 5 petals are elliptic to ovate, pure white, and about as long as the sepals. Ten stamens are present. The ovary is ovoid and develops a 5-lobed stigma. The capsules are ovoid, about 3 mm long, and contain numerous tiny, edible, and kidney-shaped seeds.

Medicinal uses: Boils (Nepal); tuberculosis (India)

Commentary: Apparently not studied for antimicrobial properties. This plant was used for food in Prehistoric times (Behre, 2008).

REFERENCE

Behre, K.E., 2008. Collected seeds and fruits from herbs as prehistoric food. *Vegetation History and Archaeobotany*, *17*(1), pp. 65–73.

8.7.7 FAMILY CHENOPODIACEAE VENTENAT (1799)

The family Chenopodiaceae includes about 100 genera and 1500 species of herbs or shrubs often adapted to sandy and sun exposed areas. The leaves are simple, mostly reduced, alternate, and exstipulate. The inflorescences are terminal or axillary spikes. The flowers are minute. The perianth comprises 3–5 tepals. The androecium comprises 3–5 stamens united at base. The gynoecium consists of 2–5 carpels forming a unilocular ovary containing 1 ovule and developing a bifid stigma. The utricles are minute and contain numerous, reniform, or subglobose seeds.

8.7.7.1 *Arthrocnemum indicum* (Willd.) Moq.

Synonyms: *Arthrocnemum affine* Moss; *Arthrocnemum africanum* Moss; *Arthrocnemum ambiguum* (Michx.) Moq.; *Arthrocnemum capense* Moss; *Arthrocnemum decumbens* Toelken; *Arthrocnemum dunense* Moss; *Arthrocnemum hottentoticum* Moss; *Arthrocnemum littoreum* Moss; *Arthrocnemum mossianum* Toelken; *Arthrocnemum namaquense* Moss; *Arthrocnemum natalense* Moss; *Arthrocnemum perenne* Moss; *Arthrocnemum pillansii* Moss; *Arthrocnemum terminale* Toelken; *Arthrocnemum variiflorum* Moss; *Arthrocnemum xerophilum* Toelken; *Halosarcia indica* (Willd.) Moq; *Salicornia ambigua* Michx.; *Salicornia australasica* (Moq.) H. Eichler; *Salicornia brachiata* Miq.; *Salicornia fruticosa* Decne.; *Salicornia indica* Willd.; *Salicornia natalensis* Bunge ex Ung.-Sternb.; *Salicornia perennis* Mill.; *Salicornia radicans* Sm.

Common names: Crab grass, sea Asparagus; barillakoyalu, saakati pusalu, suvar, kemalahala (India); kotanai (Sri Lanka)

Habitat: Coastal clays, mangroves, salt marshes

Distribution: Tropical Asia

Botanical observation: This costal, halophytic, fleshy, edible, erect herb grows up to 25 cm tall. The stem is articulated. The internodes are smooth, about 1 cm long, and 6 mm in diameter. The leaves are abscent. The cymules are axillary and minute. The perianth is tubular and develops 3 lobes which persist in fruits. The androecium consists of 1 stamen. The gynoecium is minute and develops 2 stigmas. The utricles are triangular, about 7 mm in diameter, and contain tiny, smooth, and glossy seeds.

Medicinal use: Infection (Pakistan)

Weak antibacterial (Gram-positive) polar extract: Ethanol extract inhibited the growth of *Bacillus subtilis*, *Bacillus pumilus*, *Micrococcus luteus*, and *Staphylococcus aureus* with the MIC/MBC values of 1700/2000, 1700/3000, 1700/4000, and 1700/4000 µg/mL, respectively (Manikandan et al., 2009).

Moderate broad-spectrum non-polar extract: Fatty acid fraction (containing mainly palmitic acid, lauric acid, linoleic acid and linolenic acid) of aerial parts inhibited the growth of *Bacillus subtilis*, *Bacillus pumilus*, *Micrococcus luteus*, *Staphylococcus aureus*, *Pseudomonas aeruginosa*, *Klebsiella pneumonia*, *Escherichia coli*, and methicillin-resistant *Staphylococcus aureus* with the MIC/MBC values of 125/250, 250/500, 250/500, 125/250, 500/1000, 500/1000, 500/1000, and 125/250 µg/mL (Chandrasekaran et al., 2008).

Antifungal (filamentous) halo developed by polar extract: Methanol extract (500 µg/6 mm disc) inhibited the growth of *Aspergillus fresenii*, *Aspergillus japonicus*, and *Aspergillus niger* (Bhosale et al., 1999).

REFERENCES

Bhosale, S.H., Jagtap, T.G. and Naik, C.G., 1999. Antifungal activity of some marine organisms from India, against food spoilage Aspergillus strains. *Mycopathologia*, *147*(3), pp. 133–138.

Chandrasekaran, M., Kannathasan, K. and Venkatesalu, V., 2008. Antimicrobial activity of fatty acid methyl esters of some members of Chenopodiaceae. *Zeitschrift für Naturforschung C*, *63*(5–6), pp. 331–336.

Manikandan, T., Neelakandan, T. and Rani, G.U., 2009. Antibacterial activity of *Salicornia brachiata*, a halophyte. *Journal of Phytology*, *1*(6), pp. 441–443.

8.7.7.2 *Atriplex hortensis* Boiss

Synonym: *Atriplex nitens* Schkuhr

Common names: Orache, mountain spinach; yu qian bo cai (China); vasta haak (India); chandan-batva (Pakistan); pezik otu (Turkey)

Habitat: Wastelands, disturbed areas

Distribution: Temperate Asia

Botanical observation: This herb grows to 2 m tall. The stem is terete, stout, and somewhat purplish, angled and striated. The leaves are alternate and exstipulate. The petiole is up to 3 cm long. The blade is ovate-triangular, 5–25 ×3–18 cm, hastate, laxly toothed, dull green or purplish, somewhat fleshy, and acute at apex. The panicles are terminal or axillary, and present minute flowers. The perianth is comprises 5 oblong to elliptic lobes. Five stamens are present. The gynoecium is minute and develops a bifid stigma. The utricules are minute and contain tiny smooth and glossy seeds.

Medicinal uses: Jaundice (India)

Commentaries

i. *Atriplex hortensis* Boiss has apparently not been studied for its antimicrobial constituents. One could have some interest to look for anti Hepatitis B virus or anti-Human immunodeficiency virus principles. Note that a petroleum ether extract of a member of the genus *Atriplex* L. inhibited the growth of *Staphylococcus aureus* (CIP 4.83), *Pseudomonas aeruginosa* (CIP 82118), *Enterococcus hirae* (CIP 5855), *Escherichia coli* (CIP 53126), *Candida albicans* (IP 48.72), and *Aspergillus niger* (IP 1431.83) with the MIC values of 125, 62.5, 125, 125, 125, and 62.5 µg/mL, respectively (El Souda et al., 2015).

ii. The plant produces series of sulphated flavonoid glycosides of which kaempferol 3-*O*-sulphate-7-*O*-α-arabinopyranoside (Bylka et al., 2001). Flavonoid sulfates are found mainly in species occurring in coastal and swampy areas rich in mineral salts. Sulphated flavonoids are also isolated from plants occurring in arid habitats (Bylka et al., 2001).

iii. 7β,15α,16β-Trihydroxyolean-12-ene-28,30-dioic acid-3-*O*-*β*-D-xylopyranoside, rotundifolioside I, and corchorusin B from a member of the genus *Atriplex* L. inhibited the growth of *Escherichia coli* with the MIC values of 110, 66.2, and 114 µg/mL, respectively (Ali et al., 2015). 7β,15α,16β-Trihydroxyolean-12-ene-28,30-dioic acid-3-*O*-*β*-D-xylopyranoside, rotundifolioside I, and corchorusin B inhibited the growth of *Salmonella* sp. with the MIC values of 107, 111.7, and 102.5 µg/mL, respectively (Ali et al., 2015).

iv. The cholestane phytosteroids 20-hydroxyecdysone and septanoecdysone isolated from the aerial parts of *Atriplex portulacoides* C. Wright inhibited the growth of *Pseudomonas aeruginosa* with the MIC values of 60 and 100 µg/mL (Nejma et al., 2015). 20-Hydroxyecdysone and septanoecdysone inhibited the growth of *Escherichia coli* with the MIC values of 100 and 60 µg/mL (Nejma et al., 2015). 20-Hydroxyecdysone and septanoecdysone inhibited the growth of *Staphylococcus aureus* with the MIC values of 100 µg/mL (Nejma et al., 2015). Hydroxyecdysone and septanoecdysone inhibited the growth of *Enterococcus faecalis* (Nejma et al., 2015).

v. *Atriplex stocksii* Boiss is used for fever and jaundice in Pakistan.

REFERENCES

Ali, B., Tabassum, R., Riaz, N., Yaqoob, A., Khatoon, T., Tareen, R.B., Jabbar, A., Nasim, F.U.H. and Saleem, M., 2015. Bioactive triterpenoids from *Atriplex lasiantha*. *Journal of Asian Natural Products Research*, *17*(8), pp. 843–850.

Bylka, W., Stobiecki, M. and Frański, R., 2001. Sulphated flavonoid glycosides from leaves of *Atriplex hortensis*. *Acta Physiologiae Plantarum*, *23*(3), pp. 285–290.

El Souda, S.S.E.D., Matloub, A.A., Nepveu, F., Valentin, A. and Roques, C., 2015. Phenolic composition and prospective anti-infectious properties of *Atriplex lindleyi*. *Asian Pacific Journal of Tropical Disease*, *5*(10), pp. 786–791.

Nejma, A.B., Nguir, A., Jannet, H.B., Daïch, A., Othman, M. and Lawson, A.M., 2015. New septanoside and 20-hydroxyecdysone septanoside derivative from *Atriplex portulacoides* roots with preliminary biological activities. *Bioorganic & Medicinal Chemistry Letters*, *25*(8), pp. 1665–1670.

8.7.7.3 *Chenopodium album* L.

Synonyms: *Chenopodium browneanum* Roem. & Schult.; *Chenopodium centrorubrum* Nakai; *Chenopodium giganteum* D. Don; *Chenopodium iljinii* Golosk.; *Chenopodium probstii* Aellen; *Chenopodium striatiforme* Murr; *Chenopodium virgatum* Thunb.; *Chenopodium zobelli* A. Ludw. & Aellen

Common names: Lamb's quarters; betho chak (Bangladesh); li (China); papuruka, bathua, vastuccira (India); myu (Myanmar); bethe (Nepal); bathu, bathwa (Pakistan); unluca (Turkey)

Habitat: Roadside, garden, vacant lands, fields

Distribution: Subtropical and Temperate Asia

Botanical observation: This herb grows up to 1.5 m tall. The stem is terete, striated, zig-zag shaped, and somewhat fleshy. The leaves are spiral, simple, exstipulate, and edible. The petiole is up to about 6 cm long. The blade is ovate-triangular, 3–6×2.5–5 cm, hairy below, edible (Nepal), cuneate at base, toothed, and acute at apex. The inflorescences are axillary panicles of glomerules. The 5 tepals are ovate, hairy beneath, and menbranaceous. The androecium consists of 5 exerted stamens. The gynaecium is minute, flat, and develops 2 stigmas. The capsules are minute and open to expose a smooth, black, flattish, somewhat snail-like, and glossy seed.

Medicinal uses: Cuts, cough, diarrhea, fever, skin diseases, sore throat, wounds (India); eye infection, jaundice (Pakistan); sore throat (India; Pakistan); dysentery (Nepal); diarrhea (India; Myanmar; Nepal)

Broad-spectrum antibacterial halo developed by polar extract: Ethanol extract (1 mg/disc) inhibited the growth of *Escherichia coli*, *Salmonella typhimurium*, *Proteus vulgaris*, *Pseudomonas aeruginosa*, *Micrococcus luteus*, *Listeria monocytogenes*, *Bacillus cereus*, *Enterococcus faecalis*, and *Bacillus subtilis* with inhibition zone diameters of 12, 11, 12, 12, 9, 11, 12, 10, and 13 mm, respectively (Korcan et al., 2013).

Antimycobacterial polar extract: Aqueous extract of leaves inhibited the growth of *Mycobacterium tuberculosis* (Fitzpatrick, 1954).

Antifungal (filamentous) polar extract: Methanol extract of inflorescence inhibited the growth of *Macrophomina phaseolina* (Javaid et al., 2009). Methanol extract of roots inhibited the growth of *Sclerotium rolfsii* (Ali et al., 2017).

Commentaries

i. The plant yields oleanane-type triterpene saponins (Lavaud et al., 2000).
ii. In Pakistan, *Chenopodium murale* L. is used for fever (local name: *Bathu*).
iii. Saponin account for the anti-inflammatory (Just, 1998) and antipyretic (Adiukwu et al., 2012) properties of medicinal plants.

REFERENCES

Adiukwu, P.C., Amon, A., Nambatya, G., Adzu, B., Imanirampa, L., Twinomujuni, S., Twikirize, O., Amanya, M., Ezeonwumelu, J.O., Oloro, J. and Okoruwa, G.A., 2012. Acute toxicity, antipyretic and antinociceptive study of the crude saponin from an edible vegetable: Vernonia amygdalina leaf. *International Journal of Biological and Chemical Sciences*, *6*(3), pp. 1019–1028.

Ali, A., Javaid, A. and Shoaib, A., 2017. GC-MS analysis and antifungal activity of methanolic root extract of *Chenopodium album* against *Sclerotium rolfsii*. *Planta Daninha*, *35*, p. e017164713.

Fitzpatrick, F.K., 1954. Plant substances active against *Mycobacterium tuberculosis*. *Antibiotics & chemotherapy (Northfield, Ill.)*, *4*(5), p. 528.

Javaid, A. and Amin, M., 2009. Antifungal activity of methanol and n-hexane extracts of three Chenopodium species against *Macrophomina phaseolina*. *Natural Product Research*, *23*(12), pp. 1120–1127.

Just, M.J., Recio, M.C., Giner, R.M., Cuéllar, M.J., Máñez, S., Bilia, A.R. and Ríos, J.L., 1998. Anti-inflammatory activity of unusual lupane saponins from Bupleurum fruticescens. *Planta medica*, *64*(05), pp. 404–407.

Korcan, S.E., Aksoy, O., Erdoğmuş, S.F., Ciğerci, İ.H. and Konuk, M., 2013. Evaluation of antibacterial, antioxidant and DNA protective capacity of Chenopodium album's ethanolic leaf extract. *Chemosphere*, *90*(2), pp. 374–379.

Lavaud, C., Voutquenne, L., Bal, P., and Pouny, I., 2000. Saponins from *Chenopodium album*. *Fitoterapia*, *71*, pp. 338–340.

8.7.7.4 *Chenopodium botrys* L.

Synonyms: *Ambrina botrys* (L.) Moq.; *Atriplex botrys* (L.) Crantz; *Botrydium botrys* (L.) Small; *Dysphania botrys* (L.) Mosyakin & Clemants; *Neobotrydium botrys* (L.) Moldenke; *Roubieva botrys* (L.) Fuss; *Teloxys botrys* (L.) W.A. Weber; *Vulvaria botrys* (L.) Bubani

Common names: Jerusalem oak goosefoot; xiang li (China); ganda shorak (Afghanistan); hamaghee (Pakistan)

Habitat: Wastelands, roadsides, river beds

Distribution: Subtropical and Temperate Asia

Botanical observation: This herb grows up to 50 cm tall. The stem is terete, striated, zig-zag shaped, hairy, and somewhat fleshy. The leaves are fragrant, spiral, simple, and exstipulate. The petiole is up to about 1 cm long. The blade is oblong to lanceolate, 2–4 × 1–2 cm, hairy below, cuneate at base, lobed, and acute at apex. The cymes or glomerules are axillary and cauliflorous. The perianth comprises 5 ovate lobes which are hairy beneath and membranaceous at margin. The androecium consists of 1–3 stamens. The gynoecium is minute, flat, and develops 2 slender stigmas. The utricles are minutes, depressed, globose and contain tiny smooth, and flat glossy seeds.

Medicinal uses: Dysentery (Afghanistan); antiseptic, wounds (Pakistan)

Broad-spectrum antibacterial halo developed by mid-polar extract: Ethylacetate extract (1 mg/well) inhibited the growth of *Staphylococcus aureus*, *Klebsiella pneumoniae* and *Pseudomonas aeruginosa* with inhibition zone diameters of 14, 18.3, and 27.1 mm, respectively (Ullaha et al., 2017).

Strong antimycobacterial essential oil: Essential oil of aerial parts inhibited the growth of *Mycobacterium smegmatis* with an MIC value of 50 μg/mL (El-Sayed et al., 1989).

Commentary: The plant produces the flavones hispidulin, salvigenin, 5-methylsalvigenin, 3,7-methyleupatulin, and sinensetin (de Pascual-T et al., 1981) as well as eudesmane and guiaiane sesquiterpenes (de Pascual-T et al., 1980). Note that eudesmane sesquiterpenes often inhibit the replication of the Human immunodeficiency virus.

REFERENCES

de Pascual-T, J., Bellido, I.S. and González, M.S., 1980. Chenopodiaceae components: polyoxigenated sesquiterpenes from *Chenopodium botrys*. *Tetrahedron*, *36*(3), pp. 371–376.

de Pascual-T, J., Gonzalez, M.S., Vicente, S. and Bellido, I.S., 1981. Flavonoids from *Chenopodium botrys*. *Planta Medica*, *41*(4), p. 389.

El-Sayed, A.M., Al-Yahya, M.A. and Hassan, M.M., 1989. Chemical composition and antimicrobial activity of the essential oil of *Chenopodium botrys* growing in Saudi Arabia. *International Journal of Crude Drug Research*, *27*(4), pp. 185–188.

Ullaha, F., Iqbal, N., Ayaz, M., Sadiq, A., Ullah, I., Ahmad, S. and Imran, M., 2017. DPPH, ABTS free radical scavenging, antibacterial and phytochemical evaluation of crude methanolic extract and subsequent fractions of *Chenopodium botrys* aerial parts. *Pakistan Journal of Pharmaceutical Sciences*, *30*(3), pp. 761–766.

8.7.7.5 *Dysphania ambrosioides* (L.) Mosyakin & Clemants

Synonyms: *Ambrina ambrosioides* (L.) Spach; *Ambrina parvula* Phil.; *Ambrina spathulata* Moq.; *Atriplex ambrosioides* (L.) Crantz; *Blitum ambrosioides* (L.) Beck; *Chenopodium ambrosioides* L.; *Chenopodium integrifolium* Vorosch.; *Chenopodium spathulatum* (Moq.) Sieber ex Moq.; *Chenopodium suffruticosum* Willd.; *Teloxys ambrosioides* (L.) W.A. Weber

Common names: Mexican-tea, American wormseed; jing jie (China); baljawain, chulai, smrai (Pakistan); alpasotis (the Philippines); cay dau hoi (Vietnam)

Habitat: Wastelands, riverbanks

Distribution: China, Pakistan, India, Indonesia, Vietnam, the Philippines

Botanical observation: This herb grows up to 80cm tall. The stem is striated and hairy. The leaves are spiral and exstipulate. The petiole is up to 3cm long. The blade is oblong to lanceolate, dull light green, about 5 ×15cm, attenuate at base, incised, and acute or acuminate at apex. The spikes are axillary. The perianth develops 5 oblong lobes. Five stamens are present. The gynaecium is minute and develops 3–4 filiform stigmas. The utricles are minute and contain tiny, smooth, and glossy seeds.

Medicinal uses: Cough, dysentery (Pakistan, India), pulmonary obstruction (Pakistan)

Moderate broad-spectrum antibacterial essential oil: Essential oil of aerial parts inhibited the growth of *Staphylococcus aureus* (ATCC 25923) with the MIC/MBC values of 1950/7810 μg/mL (Boutkhil et al., 2009). Essential oil of leaves inhibited the growth of *Streptococcus mutans*, *Streptococcus mitis*, *Streptococcus sanguinis*, *Streptococcus sobrinus*, *Streptococcus salivarus*, *Lactobacillus casei*, and *Enterococcus faecalis* with the MIC values of 2000, 2000, 2000, 1000, 2000, 2000, and 1000 μg/mL, respectively (Soares et al., 2017). In a subsequent study, essential oil of leaves inhibited the growth of *Staphylococcus aureus* (ATCC 2592) and multidrug-resistant *Pseudomonas aeruginosa* (clinical strain) with the MIC value of 256 and 512μg/mL, respectively (Bezerra et al., 2019).

Strong broad-spectrum antibacterial mid-polar extract: Ethylacetate extract of leaves inhibited *Enterococcus faecalis*, *Staphylococcus aureus* and *Pseudomonas aeruginosa* with the MIC values of 4.2, 4.2, and 34.3 μg/mL, respectively (Jesus et al., 2017).

Moderate antimycobacterial mid-polar extract: Ethylacetate extract of leaves inhibited the growth of *Mycobacterium smegmatis*, *Mycobacterium avium*, and *Mycobacterium tuberculosis* with the MIC values of 312.5, >2500, and 1250 μg/mL, respectively (Jesus et al., 2017).

Antimycobacterial essential oil: Essential oil of flowers and fruits inhibited *Mycobacterium tuberculosis* (Opdyke, 1976).

Weak anticandidal essential oil: Essential oil of aerial parts inhibited the growth of *Candida albicans* (clinical) with the MIC value of 1950 μg/mL (Boutkhil et al., 2009). Essential oil of leaves (containing mainly α-terpinene, ascaridol, and acaridol epoxide) inhibited the growth *Candida albicans* (LM 77) and *Candida albicans* (A INCQS 400) with IC_{50} value of 19.3 and 25.2μg/mL, respectively (Bezerra et al., 2019).

Antifungal monoterpenes: The cyclic epoxide monoterpene ascaridole at ≤1mg inhibited the growth of *Sclerodum rolfsi* by more than 50% (Pare et al., 1993). α-Terpinene inhibited *Candida albicans* with the MIC value of 100ppm (Tampieri et al., 2005).

REFERENCES

Bezerra, J.W.A., Costa, A.R., de Freitas, M.A., Rodrigues, F.C., de Souza, M.A., da Silva, A.R.P., dos Santos, A.T.L., Linhares, K.V., Coutinho, H.D.M., de Lima Silva, J.R. and Morais-Braga, M.F.B., 2019. Chemical composition, antimicrobial, modulator and antioxidant activity of essential oil of *Dysphania ambrosioides* (L.) Mosyakin & Clemants. *Comparative Immunology, Microbiology and Infectious Diseases*, *65*, pp. 58–64.

Boutkhil, S., El Idrissi, M., Amechrouq, A., Chbicheb, A., Chakir, S. and El Badaoui, K., 2009. Chemical composition and antimicrobial activity of crude, aqueous, ethanol extracts and essential oils of *Dysphania ambrosioides* (L.) Mosyakin & Clemants. *Acta Botanica Gallica*, *156*(2), pp. 201–209.

Jesus, R.S., Piana, M., Freitas, R.B., Brum, T.F., Alves, C.F., Belke, B.V., Mossmann, N.J., Cruz, R.C., Santos, R.C., Dalmolin, T.V. and Bianchini, B.V., 2017. *In vitro* antimicrobial and antimycobacterial activity and HPLC–DAD screening of phenolics from *Chenopodium ambrosioides* L. *Brazilian Journal of Microbiology*, *49*(2), pp. 296–302.

Opdyke, D.L.J., 1976. Fragrance raw materials monographs. Chenopodium oil. *Food Chemistry and Toxicology*, *14*, pp. 713–715.

Pare, P.W., Zajicek, J., Ferracini, V.L. and Melo, I.S., 1993. Antifungal terpenoids from *Chenopodium ambrosioides. Biochemical Systematics and Ecology, 21*(6–7), pp. 649–653.

Soares, M.H., Dias, H.J., Vieira, T.M., de Souza, M.G., Cruz, A.F., Badoco, F.R., Nicolella, H.D., Cunha, W.R., Groppo, M., Martins, C.H. and Tavares, D.C., 2017. Chemical composition, antibacterial, schistosomicidal, and cytotoxic activities of the essential oil of *Dysphania ambrosioides* (L.) Mosyakin & Clemants (Chenopodiaceae). *Chemistry & Biodiversity, 14*(8), p. e1700149.

Tampieri, M.P., Galuppi, R., Macchioni, F., Carelle, M.S., Falcioni, L., Cioni, P.L. and Morelli, I., 2005. The inhibition of *Candida albicans* by selected essential oils and their major components. *Mycopathologia, 159*(3), pp. 339–345.

8.7.7.6 *Haloxylon salicornicum* (Moq.) Bunge ex Boiss.

Synonyms: *Caroxylon salicornicum* Moq.; *Hammada salicornica* (Moq.) Iljin

Common name: Lana (Pakistan)

Habitat: Desert, dunes, sandy plains

Distribution: Iran, Afghanistan, Pakistan, India

Botanical observation: This poisonous somewhat fleshy shrub grows from a deep tap root and up to about 1 m tall. The stems are smooth, articulated, and glabrous. The leaves are minute, sessile, triangular, scaly, and between stems internodes. The inflorescences are axillary and minute. The 5 tepals are light burgundy to white, broadly spathulate, membranaceous, and persistant in fruits. Five stamens are present. The 2 stigmas are hairy. The utricles are minute and membranaceous.

Medicinal uses: Internal ulcers (India); diarrhea, eye infection, wounds (Pakistan)

Moderate antimycobacterial lupane-type triterpenes: Lupeol and lupeol acetate, inhibited the growth of *Mycobacterium tuberculosis* (Bibi et al., 2010).

Moderate antimycobacterial oleanane-type triterpenes: β-Amyrin and β-amyrin acetate inhibited the growth of *Mycobacterium tuberculosis* with the MIC of 100 μg/mL (Bibi et al., 2010).

Moderate antimycobacterial ursane-type triterpene: Ursolic acid inhibited the growth of *Mycobacterium tuberculosis* with the MIC of 100 μg/mL (Bibi et al., 2010).

Moderate antimycobacterial cholestane-type phytosterol: 24-Ethyl cholesta-3,5-diene inhibited the growth of *Mycobacterium tuberculosis* with the MIC of 100 μg/mL (Bibi et al., 2010).

Moderate antimycobacterial stigmastane-type phytosterol: Stigmasterol inhibited the growth of *Mycobacterium tuberculosis* with the MIC of 100 μg/mL (Bibi et al., 2010).

Moderate antimycobacterial coumarin: 7,8-Dimethoxy-6-hydroxy-coumarin inhibited the growth of *Mycobacterium tuberculosis* with the MIC of 100 μg/mL (Bibi et al., 2010).

Strong antimycobacterial piperidine alkaloids: Haloxyline A and B inhibited the growth of *Mycobacterium tuberculosis* H37Rv with the MIC of 100 and 50 μg/mL, respectively (Bibi et al., 2010).

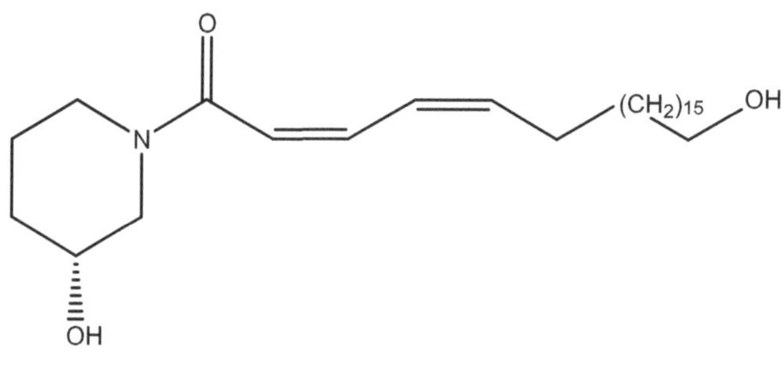

HaloxylineB

Moderate antimycobacterial phytosterol glycosides: 24 β-(24S)-ethyl cholesta-4,22 *E* diene 3-*O*-α-L-rhamnoside and stigmasterol 3-*O*-β-D-glucopyranoside inhibited the growth of *Mycobacterium tuberculosis* H37Rv with the MIC value of 50 µg/mL (Bibi et al., 2010).

Strong broad-spectrum antifungal piperidine alkaloids: Haloxiline A inhibited the growth of *Trichophyton longifusus*, *Candida albicans*, *Aspergillus flavus*, *Microsporum canis*, *Candida glabrata*, and *Fusarium solani* with the MIC values of 70, 65, 45, 70, 80, and 54 µg/mL, respectively (Bibi et al., 2010). Haloxiline B inhibited the growth of *Trichophyton longifusus*, *Candida albicans*, *Aspergillus flavus*, *Microsporum canis*, *Candida glabrata*, and *Fusarium solani* with the MIC values of 85, 80, 48, 75, 85, and 60 µg/mL, respectively (Ferheen et al., 2005).

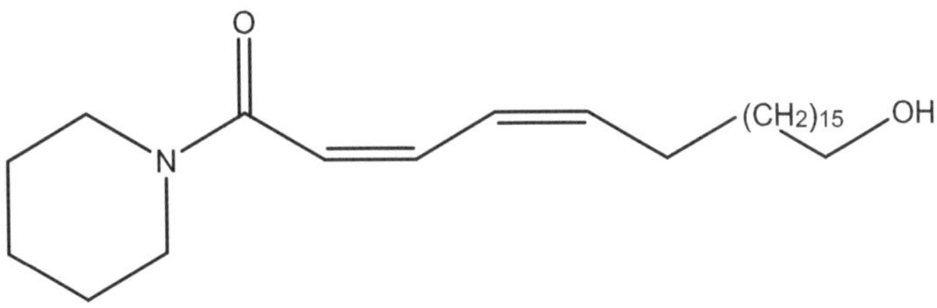

Haloxiline A

Commentary: Antifungal agents are often antimycobacterial and anticoronaviral. This is for instance the case for amphothericin (Mariotti et al., 2020; DeVald et al.,1984).

REFERENCES

Bibi, N., Tanoli, S.A.K., Farheen, S., Afza, N., Siddiqi, S., Zhang, Y., Kazmi, S.U. and Malik, A., 2010. *In vitro* antituberculosis activities of the constituents isolated from *Haloxylon salicornicum*. *Bioorganic & Medicinal Chemistry Letters*, *20*(14), pp. 4173–4176.

DeVald, B.L., Gerdes, J.C., Burks, J.S. and Kemp, M.C., 1984. Amphotericin inhibits coronavirus SD, SK and A59 growth. In *Molecular Biology and Pathogenesis of Coronaviruses* (pp. 151–152). Springer, Boston, MA.

Ferheen, S., Ahmed, E., Afza, N., Malik, A., Shah, M.R., Nawaz, S.A. and Choudhary, M.I., 2005. Haloxylines A and B, antifungal and cholinesterase inhibiting piperidine alkaloids from *Haloxylon salicornicum*. *Chemical and Pharmaceutical Bulletin*, *53*(5), pp. 570–572.

Mariotti, S., Teloni, R., de Turris, V., Pardini, M., Peruzzu, D., Fecchi, K., Nisini, R. and Gagliardi, M.C., 2020. Amphotericin B inhibits *Mycobacterium tuberculosis* infection of human alveolar type II epithelial A549 cells. *Antimicrobial Agents and Chemotherapy*, *64*(10), pp. e01164–20.

8.7.7.7 *Kochia scoparia* (L.) Schrad.

Synonyms: *Bassia scoparia* (L.) A.J. Scott; *Chenopodium scoparium* L.; *Kochia parodii* Aellen

Common names: Fireweed, summer cypress; ti fu (China)

Habitat: Dry and sandy open lands

Distribution: Temperate Asia

Botanical observation: This diffusely branched herb grows up to 1.5 m tall and turns beautifully reddish at some points of the year. The stems are terete and ribbed. The leaves are simple, spiral, exstipulate, and sessile. The blade is linear up to 5 cm long, somewhat hairy, attenuate at base, acuminate at apex, and with one pair of secondary nerves. The inflorescences are axillary and globose fascicles of minute flowers. The perianth is greenish, minute, membranaceous, and comprises 5 winged lobes. The androecium includes 5 stamens with slender filaments. The gynoecium is compressed and develops 2 slender purplish stigmas. The utricles are minute, globose, and contain tiny glossy and blackish seeds.

Medicinal use: Dysentery, urinary tract infection (China)

Strong broad-spectrum antibacterial essential oil: Essential oil (containing 22.9% of α-thujaplicin) of stems inhibited *Bacillus subtilis*, *Escherichia coli*, and *Staphylococcus aureus* with the MIC of 12.5, 12.5, and 25 μg/mL, respectively (El-Shamy et al., 2012).

Strong antibacterial (Gram-positive) non-polar extract: Hexane extract of fruits inhibited the growth of *Staphylococcus aureus* (ATCC 25923), methicillin-resistant *Staphylococcus aureus* (ATCC33591, expressing mecA gene, resistant to ampicillin and oxacillin), and methicillin-resistant *Staphylococcus aureus* (clinical strain, expressing mecA gene, resistant to oxacillin and oxacillin) (Joung et al., 2012). This extract increased the sensitivity of these bacterial strains to ampicillin (Joung et al., 2012).

Strong anticandidal essential oil: Essential oil (containing 22.9% of α-thujaplicin) aerial parts inhibited *Candida albicans* with the MIC value of 12.5 μg/mL (El-Shamy et al., 2012).

Commentaries

i. The seeds contain 20-hydroxyecdysone, momordin Ic, and oleanolic acid (Yoo et al., 2017). Oleanolic acid inhibited the growth of *Staphylococcus aureus* (ATCC 25923), *Staphylococcus aureus* (ATCC 33591), *Staphylococcus epidermidis* (ATCC12228), and *Enterococcus faecalis* (ATCC 29212) with the MIC_{50} of 31.2, 31.2, 62.5, and 7.8 μg/mL, respectively (Bonvicini et al., 2017).
ii. *Kochia prostrata* (L.) Schrad. is used to treat toothache in Pakistan (local name: *Us*).

REFERENCES

Al Bonvicini, F., Antognoni, F., Mandrone, M., Protti, M., Mercolini, L., Lianza, M., Gentilomi, G.A. and Poli, F., 2017. Phytochemical analysis and antibacterial activity towards methicillin-resistant *Staphylococcus aureus* of leaf extracts from *Argania spinosa* (L.) Skeels. *Plant Biosystems: An International Journal Dealing with all Aspects of Plant Biology, 151*(4), pp. 649–656.

El-Shamy, A.S.I., El-Beih, A.A. and Nassar, M.I., 2012. Composition and antimicrobial activity of essential oil of *Kochia scoparia* (L.) Schrad. *Journal of Essential Oil Bearing Plants, 15*(3), pp. 484–488.

Joung, D.K., Kim, Y.H., Yang, D.W., So, G.W., Lee, K.H., Kwon, D.Y., Choi, J.G., An, B.K., Ha, D.S. and Shin, D.W., 2012. Antibacterial and synergistic effects of *Kochia scoparia* extracts against methicillin-resistant *Staphylococcus aureus*. *African Journal of Microbiology Research, 6*(10), pp. 2455–2461.

Yoo, S.R., Jeong, S.J., Lee, N.R., Shin, H.K. and Seo, C.S., 2017. Quantification analysis and In vitro anti-inflammatory effects of 20-hydroxyecdysone, momordin ic, and oleanolic acid from the fructus of *Kochia scoparia*. *Pharmacognosy Magazine, 13*(51), p. 339.

8.7.7.8 *Suaeda maritima* (L.) Dumort.

Synonyms: *Atriplex maritima* (L.) Crantz; *Chenopodina australis* (R. Br.) Moq.; *Chenopodina maritima* (L.) Moq.; *Chenopodium australe* R. Br.; *Chenopodium maritimum* L.; *Dondia maritima* (L.) Druce; *Kochia sedoides* (L.) Schrad.; *Lerchea maritima* (L.) Kuntze; *Salsola indica* Willd.; *Salsola maritima* (L.) Poir.; *Salsola maritima* (L.) M. Bieb.; *Salsola sedoides* L.; *Schoberia linifolia* Nutt. ex Moq.; *Schoberia maritima* (L.) C.A. Mey.; *Suaeda australis* (R. Br.) Moq.; *Suaeda chenopodioides* Pall.; *Suaeda fernaldii* (Standl.) Standl.; *Suaeda indica* Moq.; *Suaeda nudiflora* Moq.; *Suaeda richii* Fernald

Common names: Alur (Indonesia), muoi bien (Vietnam); cha khraam (Thailand)

Habitat: Seaside, salty soils and swamps, mangroves, sea muds

Distribution: Central and Southwest Asia, Vietnam, Thailand, Indonesia, China, Siberia, Australia, Papua New Guinea

Botanical observation: This diffuse shrub grows up to about 1 tall. The stems are ribbed. The leaves are linear, simple, spiral, exstipulate, sessile, somewhat terete, with a very vague look of seaweed, 0.5–1.5 × 1–1.5 mm, and acute at apex. The inflorescences are axillary and minute. The perianth includes

5 light greenish and membranaceous tepals persistant in fruits. The androecium comprises 5 stamens. The 2 stigmas are brownish. The utricles are membranaceous and contain minute glossy seeds.

Medicinal uses: Hepatitis (India)

Weak broad-spectrum antibacterial fatty acids: Fatty acid fraction (containing mainly palmitic acid, lauric acid, linoleic acid) of shoots inhibited *Bacillus subtilis*, *Bacillus pumilus*, *Micrococcus luteus*, *Staphylococcus aureus*, *Pseudomonas aeruginosa*, *Klebsiella pneumonia*, *Escherichia coli*, and methicillin-resistant *Staphylococcus aureus* with the MIC values of 125/250, 250/500, 125/250, 125 /250, 500/1000, 500/1000, 500/1000, and 125/250 µg/mL, respectively (Chandrasekaran et al., 2008).

Moderate broad-spectrum antibacterial non-polar extract: Hexane extract of aerial parts inhibited the growth of *Bacillus subtilis* (ATCC 6051), *Staphylococcus haemolyticus* (MTCC 3383), *Staphylococcus aureus* (ATCC 25923), *Enterococcus faecalis* (ATCC 29212), *Citrobacter* sp., *Escherichia coli* (ATCC 25922), *Klebsiella pneumoniae* (ATCC 700603), *Klebsiella pneumoniae* KPC (ATCC BAA1705), *Pseudomonas* sp., and *Stenotrophomonas maltophila* (Nayak et al., 2018).

Moderate antifungal (yeast) non-polar extract: Hexane extract of aerial parts inhibited the growth of *Saccharomyces cerevisiae* with the MIC value of 280 µg/mL (Nayak et al., 2018).

Moderate anticandidal fatty acids: Fatty acid fraction (containing mainly palmitic acid, lauric acid, linoleic acid) of aerial parts inhibited the growth of *Candida albicans*, *Candida krusei*, *Candida tropicalis*, and *Candida parapsilosis* with the MIC values of 500 µg/mL, respectively (Chandrasekaran et al., 2008).

Broad-spectrum antibacterial halo developed by polar extract: Acetone extract of leaves (100 µL, 50 mg/mL/well) inhibited the growth of *Vibrio cholerae*, *Staphylococcus epidermidis*, *Bacillus subtilis*, and *Escherichia coli* with inhibition zone diameter of 11, 12, 14, and 14 mm, respectively (Patra et al., 2011).

Weak antiviral (enveloped monopartite double-stranded DNA) extract: Extract of leaves inhibited the secretion of Hepatitis B virus surface-antigen with the IC_{50} value of 325.9 µg/mL (Banerjee, et al., 2012).

Commentary: Methanol extract of roots of *Suaeda asparagoides* inhibited the Human immunodeficiency virus type-1 protease (Park, 2003).

REFERENCES

Banerjee, M.B., Ravikumar, S., Gnanadesigan, M., Rajakumar, B. and Anand, M., 2012. Antiviral, antioxidant and toxicological evaluation of mangrove associate from South East coast of India. *Asian Pacific Journal of Tropical Biomedicine*, *2*(3), pp. S1775–S1779.

Chandrasekaran, M., Kannathasan, K. and Venkatesalu, V., 2008. Antimicrobial activity of fatty acid methyl esters of some members of Chenopodiaceae. *Zeitschrift für Naturforschung C*, *63*(5–6), pp. 331–336.

Nayak, B., Roy, S., Roy, M., Mitra, A. and Karak, K., 2018. Phytochemical, antioxidant and antimicrobial screening of suaeda maritima l (dumort) against human pathogens and multiple drug resistant bacteria. *Indian Journal of Pharmaceutical Sciences*, *80*(1), pp. 26–35.

Park, J.C., 2003. Inhibitory effects of Korean plant resources on human immunodeficiency virus type 1 Protease activity. *Oriental Pharmacy and Experimental Medicine*, *3*(1), pp. 1–7.

Patra, J.K., Dhal, N.K. and Thatoi, H.N., 2011. *In vitro* bioactivity and phytochemical screening of *Suaeda maritima* (Dumort): a mangrove associate from Bhitarkanika, India. *Asian Pacific Journal of Tropical Medicine*, *4*(9), pp. 727–734.

8.7.8 FAMILY DROSERACEAE R. A SALISBURY (1808)

The family Droseraceae consists of 3 genera and about 100 species of spectacular carnivorous herbs. The leaves are simple, stipulate or not, basal, and spiral. The blade develops glandular hairs attracting, entraping, and digesting insects and are equipped with sensitive hairs that trigger insect sequestration. The inflorescence is a terminal cyme. The flowers are actinomorphic. The calyx and corolla consist of 4–8 sepals and petals, respectively. The androrecium comprises 4–20 stamens. The gynoecium is made of 3–5 carpels forming a 1–3 locular ovary with parietal of basal placentation, each locule containing 3 to numerous ovules. The fruits are dehiscent capsules.

8.7.8.1 *Drosera burmannii* Vahl

Synonym: *Drosera burmannii* DC.

Common names: Burmann's sundew; jin di luo (China); beda sudhari, mukhjali, suryasisir (India); punggu api (Indonesia); mok bo cay (Laos); chok bo wai (Thailand); carm dia la (Vietnam)

Habitat: Shady wet soil

Distribution: India, Bangladesh, Nepal, Cambodia, Laos, Vietnam, Thailand, China, Taiwan, Malaysia, Indonesia, the Philippines

Botanical observation: This is a minute herb. The stems are very much reduced. The leaves are simple, spiral, sessile, stipulate and form a round and flat rosette. The stipules are up to about 7 mm long and 3-fid. The blade is 6–10 mm× 5–7 mm, fleshy, spathulate, yellowish green, and covered with numerous slender glandular and somewhat reddish hairs. The cymes are basal and up to about 20 cm tall. The 5 sepals are about 3 mm long. The 5 petals are obovate, reddish to light pink, and up to about 4 mm long. Five stamens are present. The gynoecium is oblong and develops 5 slender and hand-like stigmas. The fruits are dehiscent capsules.

Medicinal uses: Cough (Vietnam); dysentery (India)

Commentary: The plant produces quercetin glycosides, ellagic acid (Braunberger et al., 2015) as well as plumbagin (see next) (Madhavan et al., 2008).

REFERENCES

Braunberger, C., Zehl, M., Conrad, J., Wawrosch, C., Strohbach, J., Beifuss, U. and Krenn, L., 2015. Flavonoids as chemotaxonomic markers in the genus Drosera. *Phytochemistry*, *118*, pp. 74–82.

Madhavan, V., Basnett, H., Kumar, A.C. and Yoganarasimhan, S.N., 2008. Fingerprinting of Plumbagin in lic> *Drosera burmannii* Vahl using high performance thin layer chromatography. *Indian Journal of Pharmaceutical Sciences*, *70*(6), p. 798.

8.7.8.2 *Drosera peltata* Thunb.

Synonyms: *Drosera lobbiana* Turcz.; *Drosera lunata* Buch.-Ham. ex DC.; *Drosera peltata* Willd.

Common names: Shield sundew; mao gao cai (China); Brahma suvarchala, mukhjali. (India); Pamga (Nepal); sanabugan (the Philippines)

Habitat: Grassy lands, roadsides, riverbanks

Distribution: India, Bangladesh, Nepal, Cambodia, Laos, Vietnam, Thailand, Malaysia, Indonesia, China, the Philippines, Taiwan

Botanical observation: This slender herb grows to about 30 cm long from a globose tuber. The stems are terete and smooth. The cauline leaves are simple, spiral, and exstipulate. The petiole is up to about 1.3 cm long. The blade is 2–3 × 4–5 mm, shield-like, and covered with numerous slender glandular hairs. The inflorescences are terminal cymes. The 5 to 7 sepals are about 4 mm long. The 5 petals are deltoid, whitish pink, and up to about 6 mm long. Five stamens are present. The gynoecium is oblong and develops 3 stigmas. The capsules are subglobose and about 4 mm long.

Medicinal uses: Bronchitis, cough, syphilis, tuberculosis (India); dysentery (China); toothache (the Philippines)

Strong broad-spectrum antibacterial mid-polar extract: Chloroform extract inhibited the growth of *Streptococcus mutans*, *Streptococcus sobrinus*, *Streptococcus rattus*, *Streptococcus cricetus*, *Streptococcus sanguinis*, *Streptococcus milleri*, *Streptococcus mitis*, *Streptococcus constellatus*,

Streptococcus oralis, *Streptococcus salivarus*, *Prevotella oris*, *Prevotella buccae*, and *Prevotella intermedia* with the MIC values of 31.2, 15.6, 125, 62.5, 125, 125, 250, 125, 125, 250, 125, 62.5, and 62.5 µg/mL, respectively (Didry et al., 1998).

Weak the antibacterial (Gram variable) naphthoquinone: Plumbagin inhibited the growth of *Gardnerella vaginalis* with a MIC/MBC value of 20/40 µM (Sobhani et al., 2018).

Strong antimycobacterial amphiphilic naphthoquinone: Plumbagin (Log D = 1.4 at pH 7.4; molecular mass = 188.1 g/mol) inhibited the growth of *Mycobacterium tuberculosis* (H37Rv) and multidrug-resistant *Mycobacterium tuberculosis* (IBIN 99) with the MIC values of 1.5 µg/mL (Uc-Cachón et al., 2014).

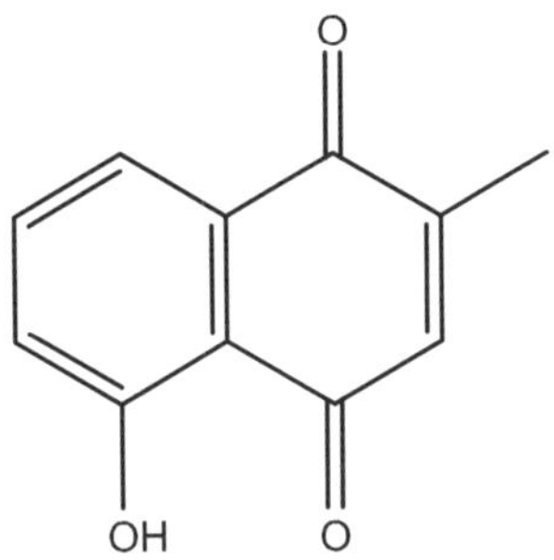

Plumbagin

Strong antimycobacterial hydrophilic naphthoquinone: Droserone (Log D = −1 at pH 7.4; molecular mass = 204.1 g/mol) inhibited the growth of multidrug-resistant *Mycobacterium tuberculosis* (IBIN 99) with the MIC value of 25 µg/mL (Uc-Cachón et al., 2014).

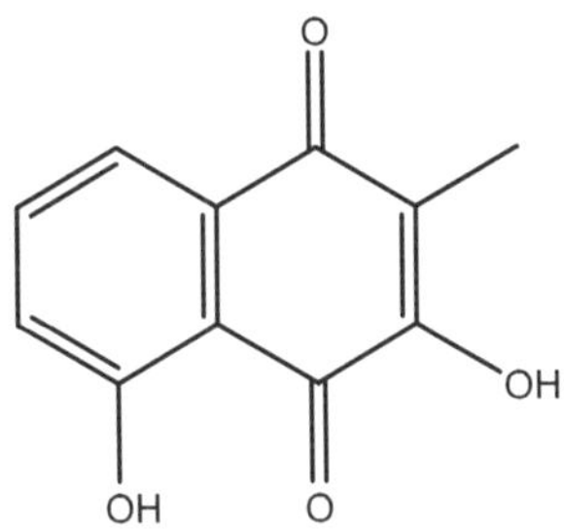

Droserone

Strong anticandidal amphiphilic naphthoquinone: Plumbagin inhibited the growth of *Candida albicans* with the MIC/MBC values of 2/2 µM (Sobhani et al., 2018).

Strong antiviral (enveloped monopartite linear single-stranded (−) RNA) hydrophilic naphthoquinone: Droserone at 10 µM inhibited by 50% the formation of syncitia in Vero cells infected with Measle virus (cytotoxicity > 60 µM) (Lieberherr et al., 2017). Droserone inhibited the synthesis of virus from infected cells with IC_{50} of 2 µM (Lieberherr et al., 2017).

Commentaries

i. The plant produces the naphthoquinones plumbagin (Shen et al., 2003), droserone, hydroxy-droserone and the flavonols quercetin, gossypetin, gossypin and isogossypitrin (Nair et al., 1990).

ii. One could have the opportunity to examine the antibacterial or even the antibiotic potentiator effects of droserone.
iii. *Drosera indica* L. is used in China (local name: *Hòu mí cǎo*) for ringworms.
iv. Consider that low molecular mass natural products with hydroxy or carbonyl groups are able to form hydrogen bonds with amino acids in the pocket of microbial enzymes whereas aromatic rings form van der Waals interactions and π-π interactions with the hydrophobic surface of the enzymatic pocket.

REFERENCES

Didry, N., Dubreuil, L., Trotin, F. and Pinkas, M., 1998. Antimicrobial activity of aerial parts of *Drosera peltata* Smith on oral bacteria. *Journal of Ethnopharmacology*, *60*(1), pp. 91–96.

Lieberherr, C., Zhang, G., Grafen, A., Singethan, K., Kendl, S., Vogt, V., Maier, J., Bringmann, G. and Schneider-Schaulies, J., 2017. The plant-derived naphthoquinone droserone inhibits *in vitro* measles virus infection. *Planta Medica*, *83*(3/4), pp. 232–238.

Nair, A.R., Shanmugasundaram, P. and Madhusudhanan, K.P., 1990. Naphthaquinones and flavonols from leaves of *Drosera peltata*. *Fitoterapia*, *61*(1), pp. 85–86.

Shen, Z., Dong, Z., Cheng, P., Li, L., Chen, Z. and Liu, J., 2003. Effects of plumbagin on platelet aggregation and platelet-neutrophil interactions. *Planta Medica*, *69*(7), pp. 605–609.

Sobhani, M., Abbas-Mohammadi, M., Ebrahimi, S.N. and Aliahmadi, A., 2018. Tracking leading anti-Candida compounds in plant samples; *Plumbago europaea*. *Iranian Journal of Microbiology*, *10*(3), p. 187.

Uc-Cachón, A.H., Borges-Argáez, R., Said-Fernández, S., Vargas-Villarreal, J., González-Salazar, F., Méndez-González, M., Cáceres-Farfán, M. and Molina-Salinas, G.M., 2014. Naphthoquinones isolated from *Diospyros anisandra* exhibit potent activity against pan-resistant first-line drugs *Mycobacterium tuberculosis* strains. *Pulmonary Pharmacology & Therapeutics*, *27*(1), pp. 114–120.

8.7.9 FAMILY MOLLUGINACEAE BARTLING (1825)

The family Molluginaceae consists of about 12 genera and 120 species of herbs or shrubs. The leaves are simple, alternate, or in rosettes, exstipulate, or stipulate. The inflorescences are mainly axillary cymes. The calyx comprises 5 sepals. The corolla is abscent or includes numerous petals. The androecium comprises 3–5 free stamens. The gynoecium is made of 2-many carpels fused into a pluricular ovary containing 1-many ovules attached to axil placentas. The capsules are dehiscent and contain tiny black and glossy seeds. Plants in this family produce triterpene saponins.

8.7.9.1 *Glinus oppositifolius* (L.) Aug. DC.

Synonyms: *Mollugo glinoides* A. Rich.; *Mollugo oppositifolia* L.; *Mollugo serrulata* Sond.; *Mollugo spergula* L.

Common names: Bitter leaf, Indian chickweed; dusera sag (Bangladesh); dang khöm (Cambodia); chang geng xing su cao (China); gima sask, kapithapatra, pampantra (India); rumput tapak burung (Indonesia); daang khôm (Laos); papait, sarsalida (the Philippines); phak khung (Thailand); au dắng dât (Vietnam)

Distribution: Tropical Asia

Habitat: Shady and moist spots

Botanical observation: This discrete creeping herb grows to 40 cm long. The stems are hairy when young and somewhat fleshy. The leaves are simple, sessile, opposite, or whorled, and exstipulate. The blade is spathulate to elliptic, dull green, 1–2.5 cm × 3–6 mm, irregularly and minutely serrate, attenuate at base, and obtuse at apex. The inflorescence is a cyme. The perianth is star-shaped, includes 5 tepals which are white, oblong, and about 5 mm long. The androecium presents 3–5 stamens. The gynoecium develops 3 styles. The capsules are deshiscent, minute, and shelter numerous tiny kidney-shaped seeds which are reniform.

Medicinal uses: Furuncles (China); leprosy (Myanmar); leucorrhea, wounds (Bangladesh); jaundice (Thailand)

Antibacterial broad-spectrum halo developed by polar extract: Methanol extract (500 μg/disc) inhibited the growth of *Escherichia coli*, *Pseudomonas aeruginosa*, *Salmonella typhi*, *Salmonella paratyphi*, *Staphylococcus aureus*, and *Sarcina lutea* with inhibition zone diameters of 6, 8, 6, 7, 6, and 6 mm, respectively (Ullah et al., 2013).

Weak broad-spectrum antibacterial (Gram-negative) mid-polar extract: Chloroform extract inhibited the growth of multidrug-resistant *Acinetobacter baumanii*, multidrug-resistant *Pseudomonas aeruginosa*, multidrug-resistant *Escherichia coli*, multidrug-resistant *Klebsiella pneumoniae*, *Escherichia coli* (ATCC 25922), *Klebsiella pneumoniae* (ATCC BAA1705), and *Pseudomonas aeruginosa* (ATCC 27853) with the MIC/MBC values of 2.5/10, 2.5/5, 5/10, 5/10, 1.2/5, 1.2/10, and 2/5 mg/mL, respectively (Martin-Puzon et al., 2015).

Anticandidal halo developed by mid-polar extract: Dichloromethane extract of whole plant (100 μg on TLC plate) inhibited the growth of *Candida albicans* (Diallo et al., 2001).

Strong antifungal (filamentous) oleanane-type triterpene saponins: Glinusopposide B inhibited the growth of *Microsporum gypseum* (CBS 118893) and *Trichophyton rubrum* (ATCC 4438) with the MIC_{50} values of 7.1 and 14.3 μg/mL, respectively (Zhang et al., 2019). Glinusopposide Q inhibited the growth of *Microsporum gypseum* (CBS 118893) and *Trichophyton rubrum* (ATCC 4438) with the MIC_{50} values of 6.7 and 13.4 μg/mL, respectively (Zhang et al., 2019). Glinusopposide T inhibited the growth of *Microsporum gypseum* (CBS 118893) and *Trichophyton rubrum* (ATCC 4438) with the MIC_{50} values of 6.8 and 11.9 μg/mL, respectively (Zhang et al., 2019).

Commentaries

i. Within the Caryophyllales are plants using pentacyclic triterpene saponins as phytoalexins (Dixon, 2001). Polar extracts from such plants are often broad-spectrum antibacterial and strong broad-spectrum antifungal and without much antiviral effects.
ii. *Glinus lotoides* L. is used in India (local name: *Siru seruppadai*) for the treatment of diarrhea.

REFERENCES

Dixon, R.A., 2001. Natural products and plant disease resistance. *Nature*, *411*(6839), pp. 843–847.

Diallo, D., Marston, A., Terreaux, C., Toure, Y., Smestad Paulsen, B. and Hostettmann, K., 2001. Screening of Malian medicinal plants for antifungal, larvicidal, molluscicidal, antioxidant and radical scavenging activities. *Phytotherapy Research*, *15*(5), pp. 401–406.

Martin-Puzon, J.J.R., Valle Jr, D.L. and Rivera, W.L., 2015. TLC profiles and antibacterial activity of *Glinus oppositifolius* L. Aug. DC. (Molluginaceae) leaf and stem extracts against bacterial pathogens. *Asian Pacific Journal of Tropical Disease*, *5*(7), pp. 569–574.

Ullah, M.O., Haque, M., Urmi, K.F., Zulfiker, A.H.M., Anita, E.S., Begum, M. and Hamid, K., 2013. Antibacterial activity and brine shrimp lethality bioassay of methanolic extracts of fourteen different edible vegetables from Bangladesh. *Asian Pacific Journal of Tropical Biomedicine*, *3*(1), pp. 1–7.

Zhang, D., Fu, Y., Yang, J., Li, X.N., San, M.M., Oo, T.N., Wang, Y. and Yang, X., 2019. Triterpenoids and their glycosides from *Glinus oppositifolius* with antifungal activities against *Microsporum gypseum* and *Trichophyton rubrum*. *Molecules*, *24*(12), p. 2206.

8.7.9.2 *Mollugo pentaphylla* L.

Synonyms: *Mollugo stricta* L.; *Mollugo trifolia* Schult.

Common names: Indian chickweed, mollugo; khet papra (Bangladesh); su mi cao (China); jalapapr, turapundu (India); malagoso, salsalida, salukot (Philippines); zakuro sou (Japan)

Habitat: Roadsides, wastelands, cultivated fields

Distribution: Pakistan, India, Sril Lanka, Bangladesh, Myanmar, Cambodia, Laos, Vietnam, Malaysia, Indonesia, Philippines; China, Korea, Japan, Taiwan

Botanical observation: This gracile and diffuse herb grows up to 30 cm long. The stems are slender and slightly angled. The leaves are simple and arranged in basal rosette or cauline verticils of 3–5. The blade is elliptic-lanceolate to oblanceolate, 1.5–3 cm × 3–7 mm, smooth, somewhat glossy, and acute at apex. The inflorescence is a cyme. The perianth includes 5 tepals which are about 2 mm long, elliptic, obtuse or rounded at apex, and white. Five stamens are present. The gynoecium is white and subglobose. The capsules are deshiscent, minute, membranaceous and contain tiny seeds which are reniform, warty, and dark brown.

Medicinal uses: Antiseptic (India); sores (Malaysia)

Antibacterial (Gram-negative) halo developed by polar extract: Methanol extract of aerial parts (5 μL of a 30 mg/mL solution/disc) inhibited the growth of *Escherichia coli* and *Pseudomonas aeruginosa* with inhibition zone diameters of 13 and 14 mm, respectively (Jijin & Biju, 2018).

Moderate antibacterial (Gram-positive) antibacterial polar extract: Ethanol extract inhibited the growth of *Propionibacterium acnes* with the MIC value of 250 μg/mL, and the growth of *Staphylococcus epidermidis* with the MIC value of 125 μg/mL (Kim et al., 2008).

Antifungal (filamentous) halo developed by hopane-type triterpene: Mollugogenol A (also known as (3β,6α)-hopa-15,17(21)-diene-3,6-diol ; Log D = 8.1 at pH 7.4; molecular mass = 440.7 g/mol) evoked an inhibition zone (TLC, 1.5 μg) against *Cladosporium cucumerinum* (Hamburger et al., 1989).

Commentaries

i. In India, *Mollugo cerviana* (L.) Ser. ex DC. (local name: *Grishma-sundara*) is used for fever, and *Mollugo spergula* L. (local name: *Parpata*) is used as antiseptic.
ii. Consider that oleanane-type triterpene saponins, such as glycyrrhizin (Log D = –2.2; molecular mass = 822.9 g/mol) inhibited the replication of the Severe acute respiratory syndrome-associated coronavirus *in vitro* (Hoever at al., 2005).

REFERENCES

Hamburger, M., Dudan, G., Nair, A.R., Jayaprakasam, R. and Hostettmann, K., 1989. An antifungal triterpenoid from *Mollugo pentaphylla*. *Phytochemistry*, *28*(6), pp. 1767–1768.

Hoever, G., Baltina, L., Michaelis, M., Kondratenko, R., Baltina, L., Tolstikov, G.A., Doerr, H.W. and Cinatl, J., 2005. Antiviral activity of glycyrrhizic acid derivatives against SARS–coronavirus. *Journal of Medicinal Chemistry*, *48*(4), pp. 1256–1259.

Jijin, R. and Biju, P., 2018. Antibacterial and anti-biofilm assay of methanolic extract of *Mollugo pentaphylla* Linn. *Trends in Biosciences*, *11*(7), pp. 1143–1150.

Kim, S.S., Kim, J.Y., Lee, N.H. and Hyun, C.G., 2008. Antibacterial and anti-inflammatory effects of Jeju medicinal plants against acne-inducing bacteria. *The Journal of General and Applied Microbiology*, *54*(2), pp. 101–106.

8.7.9.3 *Mollugo nudicaulis* L.

Synonym: *Paramollugo nudicaulis* (Lam.) Thulin

Common names: Daisy-leaved chickweed; wu jing su mi cao (China); parpadagam (India)

Habitat: Sandy shores, sandy soils

Distribution: Afghanistan, Pakistan; India, Ceylon, China, Pacific Islands

Botanical observation: This graceful slender herb grows to 20 cm long. The leaves are simple, sessile, and arranged in basal rosette. The blade is spathulate, 1–5 × 0.8–1.6 cm, dull darkish green, and obtuse at apex. The inflorescence is a cyme. The 5 tepals are pure white, minute, and elliptic. The androecium presents 3–5 stamens. The gynoecium is subglobose, 3-locular, and develops 3 styles. The capsules are deshiscent, minute membranaceous, ellipsoid and conceal minute, reniform, granulate, and black seeds.

Medicinal uses: Fever, cough, venereal diseases (India)

Antibacterial broad-spectrum halo developed by polar extract: Methanol extract (200 μg/mL/well) inhibited the growth of *Pseudomonas aeruginosa*, *Proteus sp.*, *Streptococcus sp.*, and *Enterobacter sp.* with inhibition zone diameters of 11, 10.2, 13.1, and 15.1 mm, respectively (Rameshkumar & Sivasudha, 2012).

Moderate antifungal (filamentous) polar extract: Ethanol extract inhibited the growth of *Trichophyton mentagrophytes*, *Trichophyton simii*, and *Microsporum gypseum* with the MIC/MFC values of 125/500, 250/1000, and 500/1000 μg/mL, respectively (Chellappandian et al., 2018).

Commentary: In the closely related family Gisekiaceae Nakai (1942), *Gisekia pharnaceoides* L. is used used in Iran for bronchitis and wounds.

REFERENCES

Chellappandian, M., Saravanan, M., Pandikumar, P., Harikrishnan, P., Thirugnanasambantham, K., Subramanian, S., Hairul-Islam, V.I. and Ignacimuthu, S., 2018. Traditionally practiced medicinal plant extracts inhibit the ergosterol biosynthesis of clinically isolated dermatophytic pathogens. *Journal de Mycologie Medicale*, *28*(1), pp. 143–149.

Rameshkumar, A. and Sivasudha, T., 2012. *In vitro* antioxidant and antibacterial activity of aqueous and methanolic extract of *Mollugo nudicaulis* Lam. leaves. *Asian Pacific Journal of Tropical Biomedicine*, *2*(2), pp. S895–S900.

8.7.10 FAMILY PHYTOLACCACEAE R. BROWN (1818)

The family Phytolacacceae includes 15 genera and 50 species of herbs, shrubs, or climbers. The leaves are simple, aternate, and exstipulate. The inflorescences are racemes or spikes. The flowers are minute. The perianth includes 4–5 tepals. The androecium comprises 4 free stamens. The gynoecium consists of 3–16 carpels free or fused forming a multilocular, ovary, each locule sheltering a single ovule. The fruits are fleshy drupes or berries which are often red and glossy. Plants in this family produce triterpene saponins.

8.7.10.1 *Phytolacca acinosa* Roxb.

Synonyms: *Phytolacca esculenta* Van Houtte; *Phytolacca latbenia* (Moq.) H. Walter; *Phytolacca pekinensis* Hance

Common name: Shang lu (China); jaringo (India)

Habitat: Slopes, forests margins, roadsides

Distribution: China, Bhutan, Myanmar, Sikkim, India, Vietnam, China, Japan, Korea

Botanical observation: This somewhat fleshy and gracious herb grows up to 0.5–1.5 m tall. The stems are terete, smooth, and glabrous. The leaves are simple, simple, exstipulate, and palatable. The petiole is up to about 3 cm long. The blade is lanceolate-elliptic, 10–30×4.5–15 cm, dull like green, cuneate at base, somewhat wavy, acute or acuminate at apex, and marked with about 10 pairs of secondary nerves. The racemes are fusiform, showy, many flowered, and up to about 20 cm long. The 5 tepals are white, pinkish, or yellowish green, elliptic, and up to about 4 mm long. The androecium includes 8–10 stamens, as long as the tepals and pinkinsh. Eight carpels are present. The berries are dark purplish, glossy, 8-lobed, about 7 mm in diameter, and contain minute kidney-shaped seeds.

Medicinal uses: Bronchitis, abscesses (China); cuts, diarrhea, wounds (India)

Commentary: The plant has not apparently been studied for antimicrobial activities. It produces saponins of which esculentoside T (He et al., 2011).

REFERENCE

He, J., Ma, J., Lai, D.W., Zhang, Y.M. and Sun, W.J., 2011. A new triterpenoid saponin from the roots of *Phytolacca acinosa*. *Natural Product Research*, *25*(18), pp. 1771–1775.

8.7.10.2 *Phytolacca americana* L.

Synonym: *Phytolacca decandra* L.

Common name: Pokeweed; chui xu shang lu (China); Amerikan şerbetçiboyası (Turkey)

Habitat: Slopes, forests margins, roadsides

Distribution: China, Taiwan

Botanical observation: This toxic herb grows to about 2 m tall. The stems are terete, green or pinkish, smooth, and glabrous. The leaves are simple, simple, and exstipulate. The petiole is up to about 4 cm long. The blade is lanceolate-elliptic, 9–18 × 4.5–10 cm, glossy, cuneate at base, acute at apex, and marked with 8–9 pairs of secondary nerves. The racemes are terminal and about 20 cm long. The 5 tepals are white, pinkish, or yellowish green, whitish-red, elliptic, and 4 mm long. Ten stamens are present. The gynoecium includes 10 carpels. The berries are dark purplish, glossy, 10-lobed, on pinkish stalks.

Medicinal use: Diarrhoea (Turkey)

Moderate antibacterial (Gram-positive) polar extract: Methanol extract of stems and leaves inhibited the growth of *Porphyromonas gingivalis* (ATCC 7BAA-1703) with an MIC value of 600 µg/mL (Patra et al., 2014).

In vivo antiviral (enveloped monopartite single-stranded (-) RNA) polar extract: Ethanol extract of leaves given orally at a daily at a dose of 1000 mg/Kg to mice infected with Rabies virus (CVS-11) for 7 days increased average survival time from 9 days (untreated) to 22.8 days (Admasu et al., 2014).

Commentary: The plant produces oleanane-type triterpene saponins (Wang et al., 2008). These may have antibacterial effects. Triterpene saponins consist of a non-polar triterpene moiety and a polar glycosidic moiety. The non-polar moiety binds to cholesterol in the phospholipid bilayer of mammalians cells to form pores through which escape the cytoplasmic content (Netala et al., 2015). Mixing saponins and erythrocytes evokes complete lysis of cells. Bacterial and fungal membrane do not contain cholesterol but hopanoids and ergosterol, respectively onto which the non-polar moiety of saponins binds to induce membrane insults (Mert-Türk et al., 2006). Do saponins alter viral envelops?

REFERENCES

Admasu, P., Deressa, A., Mengistu, A., Gebrewold, G. and Feyera, T., 2014. *In vivo* antirabies activity evaluation of hydroethanolic extract of roots and leaves of *Phytolacca dodecandra*. *Global Veterinaria*, *12*(1), pp. 12–18.

Mert-Türk, F., 2006. Saponins versus plant fungal pathogens. *Journal of Cell and Molecular Biology*, *5*, pp. 13–17.

Netala, V.R., Ghosh, S.B., Bobbu, P., Anitha, D. and Tartte, V., 2015. Triterpenoid saponins: a review on biosynthesis, applications and mechanism of their action. *International Journal of Pharmacy and Pharmaceutical Sciences*, *7*(1), pp. 24–28.

Patra, J.K., Kim, E.S., Oh, K., Kim, H.J., Kim, Y. and Baek, K.H., 2014. Antibacterial effect of crude extract and metabolites of *Phytolacca americana* on pathogens responsible for periodontal inflammatory diseases and dental caries. *BMC Complementary and Alternative Medicine*, *14*(1), p. 343.

Wang, L., Bai, L., Nagasawa, T., Hasegawa, T., Yang, X., Sakai, J.I., Bai, Y., Kataoka, T., Oka, S., Hirose, K. and Tomida, A., 2008. Bioactive triterpene saponins from the roots of *Phytolacca americana*. *Journal of Natural Products*, *71*(1), pp. 35–40.

8.7.11 FAMILY PLUMBAGINACEAE A. L DE JUSSIEU (1789)

The family Plumbaginaceae consists of about 30 genera and 900 species of herbs or shrubs. The leaves are simple, alternate, and exstipulate. The inflorescences are terminal or axillary panicles or spikes. The flowers are actinomorphic and showy. The calyx and corolla are 5-lobed. The androecium comprises 5 stamens. The gynoecium consists of 5 carpels forming a unilocular ovary containing 1 basal ovule and developing 5 stigmas. The fruits are capsules enclosed in persistant calyx. Members of this family produce antimicrobial naphthoquinones.

8.7.11.1 *Aegialitis rotundifolia* Roxb.

Common name: Club mangrove; banrua, kharsi (Bangladesh); bonruar, banoroa, banarua, dudumara, narikandam, satan, tara (India)

Habitat: Seashores, estuaries, swamps, sea muds

Distribution: India, Bangladesh, Andaman Islands

Botanical observation: This shrub grows up to 3 m tall. The leaves are simple, alternate, and exstipulate. The petiole is 5 cm long and sheathing at stem. The blade is broadly lanceolate to rotund, somewhat coriaceous, smooth, dull green, 3–9×2–7 cm, and marked with a conspicuous midrib. The flowers are arranged in terminal panicles. The calyx is tubular, about 1 cm long, 5-ribbed, and produces 5 lobes. The corolla is pure white, tubular, 5-lobed, the lobes are about 1 cm long, and recurved. The 5 stamens have somewhat reddish filaments and oblong anthers. The ovary is about 5 mm long, pentagonal with a capitate stigma. The capsules are characteristically falcate, yellowish, and up to 5 cm long

Medicinal use: Insect bites (Bangladesh)

Antibacterial (Gram-negative) polar extract: Methanol extract of leaves (500 μg/6 mm disc) inhibited the growth of *Pseudomonas fluorescens* and *Aeromonas hydrophila* (Choudhury et al., 2005).

Commentary: This interesting plant has apparently not been much studied for its antimicrobial properties and constituents.

REFERENCE

Choudhury, S., Sree, A., Mukherjee, S.C., Pattnaik, P. and Bapuji, M., 2005. *In vitro* antibacterial activity of extracts of selected marine algae and mangroves against fish pathogens. *Asian Fisheries Science*, *18*(3/4), p. 285.

8.7.11.2 *Limonium sinense* (Girard) Kuntze

Synonyms: *Statice fortunei* Lindl.; *Statice sinensis* Girard

Common names: Sea lavender; bu xue cao (China)

Habitat: Seashores, cultivated

Distribution: China, Taiwan, Vietnam, Korea, Japan

Botanical observation: This handsome herb grows up to 50 cm tall from somewhat reddish roots and has an aura of dryness. The leaves are simple, sessile, and basal. The blade is spathulate, 5–5×0.5–2.5 cm, and with conspicuous midrib. The corymbs are terminal, many flowered, beautiful, and colorful. The calyx is tubular, hairy, about 7 mm long, and produce 5 lobes. The corolla is whitish to yellowish, 5 lobed, infundibuliform, and about 1 cm long. The androecium includes 5 stamens adnate to corolla base. The gynoecium is obovoid and develops 5 linear stigmas. The capsules are ovoid.

Medicinal use: Bladder infection (Taiwan)

Strong antiviral (enveloped monopartite single-stranded (+) RNA) polar extract: Aqueous extract of roots inhibited the replication of the Hepatitis C virus in Huh-7.5 cells with the IC_{50} of 9.7 μg/mL with a selectivity index of 35.3 (Hsu et al., 2015). This extract was not virucidal and at 50 μg/mL and inhibited the entry of the virion in host cells (Hsu et al., 2015). This extract had also no effect on the postinfection stage of viral replication in Huh-7.5 cells (Hsu et al., 2015).

Strong antiviral (enveloped monopartite linear double standed DNA) polar extract: Ethanol extract inhibited the replication of the Herpes simplex virus type-1 in Vero cells with IC_{50} value of 25 μg/mL (Kuo et al., 2002).

Strong antiviral (enveloped monopartite single-stranded (+) RNA) hydrophilic phenolic acid: Gallic acid (Log D = −2.3 at pH 7.4; molecular mass= 170.1 g/mol) from the roots inhibited the replication of the Hepatitis C virus in Huh-7.5 cells with IC_{50} of 24.3 μg/mL and a selectivity index of 14.2 (Hsu et al., 2015). Gallic acid inactivated cell-free viral particles as well as blocked attachment of virus on Huh-7.5 cells (Hsu et al., 2015).

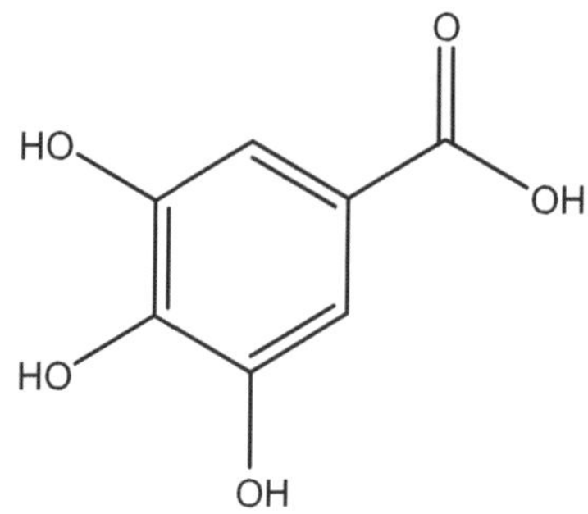

Gallic acid

Moderate antiviral (enveloped monopartite linear double-stranded DNA) condensed tannin: Galloyl-epigallocatechin([4β → 8, 2′ → 3″-O])-galloyl-epi-gallocatechin (also known as samarangenin B) inhibited the replication of Herpes simplex virus type-1 in Vero cells with an IC_{50} of 11.4 μM via inhibition of DNA synthesis and structure protein expression in Vero cells (Kuo et al., 2002).

Moderate antiviral (enveloped monopartite linear double-stranded DNA) flavanol: (−) Epigallocatechin 3-*O*-gallate from the roots inhibited Herpes simplex virus type-1 (KOS strain) with IC_{50} of 38.6 μM in Vero cells (Lin et al., 2000).

Commentaries

i. Methanol extract of roots of *Limonium tetragonium* Bullock at the concentration of 100 μg/mL inhibited RNA-dependent DNA polymerase and ribonuclease H activities of Human immunodeficiency virus type-1 reverse transcriptase by 90.7 and 37.1%, respectively (Min et al., 2001).
ii. Is gallic acid of value against coronaviruses?

REFERENCES

Hsu, W.C., Chang, S.P., Lin, L.C., Li, C.L., Richardson, C.D., Lin, C.C. and Lin, L.T., 2015. Limonium sinense and gallic acid suppress hepatitis C virus infection by blocking early viral entry. *Antiviral Research*, *118*, pp. 139–147.

Kuo, Y.C., Lin, L.C., Tsai, W.J., Chou, C.J., Kung, S.H. and Ho, Y.H., 2002. Samarangenin B from *Limonium sinense* suppresses herpes simplex virus type 1 replication in Vero cells by regulation of viral macromolecular synthesis. *Antimicrobial Agents and Chemotherapy*, *46*(9), pp. 2854–2864.

Lin, L.C., Kuo, Y.C. and Chou, C.J., 2000. Anti-herpes simplex virus type-1 flavonoids and a new flavanone from the root of *Limonium sinense*. *Planta Medica*, *66*(4), pp. 333–336.

Min, B.S., Kim, Y.H., Tomiyama, M., Nakamura, N., Miyashiro, H., Otake, T. and Hattori, M., 2001. Inhibitory effects of Korean plants on HIV-1 activities. *Phytotherapy Research*, *15*(6), pp. 481–486.

8.7.11.3 *Plumbago indica* L.

Synonyms: *Plumbago rosea* L.; *Plumbago rosea* var. *coccinea* (Lour.) Hook.; *Thela coccinea* Lour.

Common names: Rosy-flowered leadwort; aguni tita, rakt chirak (Bangladesh); zi hua dan (China); lal chitra, chitaka, rakta-chitraka (India); kant-choke-n (Myanmar); pampasapit (the Philippines); chettamun (Thailand)

Habitat: Wastelands

Distribution: India, Sri Lanka, Bangladesh, Laos, Cambodia, Vietnam, Thailand, Indonesia, The Philippines, China

Botanical observation: This graceful herb grows to about 1.5 tall. The stems are terete and striated. The leaves are simple, opposite, and exstipulate. The petiole is clasping at base. The blade is ovate, 7–9.5 × 3–4 cm, rounded or obtuse at base, marked with about 6 pairs of secondary nerves, and acute at apex. The inflorescences are terminal spikes. The calyx is tubular, up to 1 cm long, 5-Ribbed, covered with glands, and 5-lobed. The corolla is red to purplish, tubular, membranaceous, with a slender tube that grows to 2.5 cm long, and developing 5 obovate and mucronate lobes. The androecium includes 5 stamens with blue anthers. The ovary is ovoid, hairy, and develops 5 joined styles separating at filiform stigmas. The fruits are capsular and rarely seen.

Medicinal uses: Bronchitis, fever, dysentery, syphilis (India); diarrhea (Thailand); dysentery, toothache (Bangladesh); leprosy (India; Myanmar)

Moderate antibacterial (Gram-negative) antibacterial polar extract: Methanol extract of roots inhibited the growth of *Neisseria gonorrhea* with the MIC value of 253.5 µg/mL (Chomnawang et al., 2009).

Strong anticandidal amphiphilic naphthoquinone: Plumbagin (Log D = 1.4 at pH 7.4; molecular mass = 188.1 g/mol) inhibited the growth of *Candida albicans* (ATCC 10231) and *Candida albicans* (clinical strain) with the MIC values of 10.1 and 10.6 µg/mL, respectively (Hassan et al., 2016).

Moderate antiviral (monopartite linear dimeric single-stranded (+) RNA) non-polar extract: Hexane extract inhibited Human immunodeficiency virus type-1 reverse transcriptase with an IC_{50} value of 146.5 µg/mL (Silprasit et al., 2011).

Strong antiviral (enveloped monopartite single-stranded (+) RNA) amphiphilic naphthoquinone: Plumbagin inhibited the replication of the Hepatitis C virus with the IC_{50} value of 0.5 µg/mL and a selectivity index of 53.7 (Hassan et al., 2016).

Commentaries

i. The plant produces plumbagin, 3,3′-biplumbagin, and elliptinone, which are antibacterial (Kaewbumrung & Panichayupakaranant, 2012). Note that plumbagin inhibited AcrB-mediated substrate efflux in *Escherichia coli* (Ohene-Agyei et al., 2014). In cancer cells plumbagin is an inhibitor of the ATP-binding cassette transporter ABCG2 involved in efflux of drugs (Shukla and Ambudkar, 2007).
ii. Plumbagin is probably able to inhibit the replication of coronaviruses of which the Severe acute respiratory syndrome-associated coronavirus.

REFERENCES

Chomnawang, M.T., Trinapakul, C. and Gritsanapan, W., 2009. *In vitro* antigonococcal activity of *Coscinium fenestratum* stem extract. *Journal of Ethnopharmacology*, *122*(3), pp. 445–449.

Hassan, S.T., Berchová-Bímová, K. and Petráš, J., 2016. Plumbagin, a plant-derived compound, exhibits antifungal combinatory effect with amphotericin B against *Candida albicans* clinical isolates and anti-hepatitis C virus activity. *Phytotherapy Research*, *30*(9), pp. 1487–1492.

Kaewbumrung, S. and Panichayupakaranant, P., 2012. Isolation of three antibacterial naphthoquinones from *Plumbago indica* roots and development of a validated quantitative HPLC analytical method. *Natural Product Research*, *26*(21), pp. 2020–2023.

Ohene-Agyei, T., Mowla, R., Rahman, T. and Venter, H., 2014. Phytochemicals increase the antibacterial activity of antibiotics by acting on a drug efflux pump. *MicrobiologyOpen*, *3*(6), pp. 885–896.

Shukla, S. and Ambudkar, S., 2007. Plumbagin, a naturally occurring napthoquinone, is a specific inhibitor of the multidrug resistance-linked ABC drug transporter, ABCG2. *Cancer Research*, *67*(9 Supplement), p. 2287.

Silprasit, K., Seetaha, S., Pongsanarakul, P., Hannongbua, S. and Choowongkomon, K., 2011. Anti-HIV-1 reverse transcriptase activities of hexane extracts from some Asian medicinal plants. *Journal of Medicinal Plants Research*, *5*(17), pp. 4194–4201.

8.7.11.4 *Plumbago zeylanica* L.

Synonym: *Plumbago scandens* L.

Common names: Ceylon leadwort, aguni tita, aunitida, chitrak, lalchita, sadachita, chitamul (Bangladesh); bai hua dan (China); chitta chitra, chitraka, (India); daun encok (Indonesia); pokok celaka (Malaysia); kan-gyok-phyu (Myanmar); angdikit, talankan (the Philippines)

Habitat: Open lands, wastelands, cultivated

Distribution: Tropical Asia

Botanical observation: This beautiful herb grows to about 1.5 tall. The stems are woody at base and striated. The leaves are simple, opposite, and exstipulate. The petiole is clasping at base. The blade is ovate, 5–10×2–5cm, cuneate to obtuse at base, marked with about 4 pairs of secondary nerves, and acuminate and mucronate at apex. The inflorescences are terminal spikes. The calyx is tubular, up to 1.3cm long, 5-Ribbed, covered with glands, and 5-lobed. The corolla is tubular, membranaceous, with a slender tube that grows to 2.2cm long, white or with a unique type of light blue, and develop 5 obovate lobes. Five stamens with blue anthers are present. The ovary is 5-angular and develops 5-joined styles separating at filiform stigmas. The capsules are yellow-brown and oblong.

Medicinal uses: Abcesses, cold, cuts, fever, influenza (India); diarrhea, herpes, ringworm (Myanmar); leprosy (India; Myanmar)

Strong broad-spectrum antibacterial amphiphilic naphthoquinone: Plumbagin (Log D = 1.4 at pH 7.4; molecular mass = 188.1 g/mol) from the roots inhibited the growth of *Arthrobacter citrus, Bacillus cereus, Bacillus pumilus, Clostridium acetobutylicum, Staphylococcus aureus, Staphylococcus epidermidis, Escherichia coli, Klebsiella aerogenes, Pseudomonas aeruginosa,* and *Sarcina lutea* with the MIC values of 1.5, 3.1, 1.5, 1.5, 3.1, 0.7, 3.1, 1.5, 3.1, and 0.7 µg/mL, respectively (Jeyachandran et al., 2009). Plumbagin inhibited the growth of *Staphylococcus aureus* (Atcc 25923), *Streptococcus pneumoniae* (ATCC 27336), *Streptococcus pyogenes* (ATCC 12384), *Enterococcus faecalis* (ATCC 29212) with the MIC/MBC values of 16/32, 32/32, 16/32, and 32/32 µg/mL, respectively (Dey et al., 2014). Plumbagin inhibited the growth of *Escherichia coli* (ATCC 35218), *Klebsiella pneumoniae* (ATCC 27853), *Pseudomonas aeruginosa* (ATCC 27853), and *Proteus vulgaris* (ATCC 6896) with the MIC/MBC values of 16/16, 128/128, 128/256, and 32/64 µg/mL, respectively (Dey et al., 2014).

Strong broad spectrum antibacterial amphiphilic tetralones: Neoisoshinanolone (Log D = 1.9 at pH 7.4; molecular mass = 192.2 g/mol) inhibited the growth of *Bacillus cereus, Bacillus pumilus, Clostridium acetobutylicum, Staphylococcus aureus, Staphylococcus epidermidis, Escherichia coli, Klebsiella aerogenes, Pseudomonas aeruginosa,* and *Sarcina lutea* with the MIC values ranging from 50 to 100 µg/mL (Jetty et al., 2010). 1-Epineoisoshinanolone inhibited the growth of *Bacillus cereus, Bacillus pumilus, Clostridium acetobutylicum, Staphylococcus aureus, Staphylococcus epidermidis, Escherichia coli, Klebsiella aerogenes, Pseudomonas aeruginosa,* and *Sarcina lutea* with the MIC values of 12.5, 25, 25, 12.5, 25, 12.5, 25, 25, and 25 µg/mL, respectively (Jetty et al., 2010).

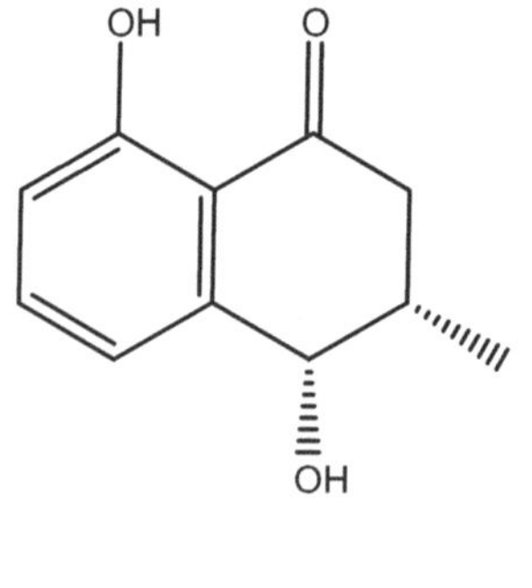

Neoisoshinanolone

Strong antimycobacterial amphiphilic naphthoquinone: Plumbagin inhibited the growth of *Mycobacterium tuberculosis* H37Rv with a MIC below 12.5 µg/mL (Mossa et al., 2004). Plumbagin inhibited the growth of *Mycobacterium tuberculosis* (H37Ra ATCC 25177) and multidrug-resistant *Mycobacterium tuberculosis* (clinical) with the MIC/MBC values of 8/16 and 0.2/1 5 µg/mL, respectively (Dey et al., 2014). Plumbagin inhibited the growth of *Mycobacterium tuberculosis* (H37Rv) and *Mycobacterium tuberculosis* (clinical isolate CIBIN 99, resistant to all first-line anti-tuberculosis drugs) with the MIC values of 1.5 and 1.5 µg/mL, respectively (Uc-Cachón et al., 2014).

Moderate antiviral (non-enveloped monopartite linear single-stranded (+) RNA) polar extract: Methanol extract of roots inhibited Coxsackie virus B3 with the IC_{50} of 82.1 µg/mL (Gebre-Mariam et al., 2006).

Commentaries

i. *Stenotrophomonas maltophilia* is an emerging Gram-negative opportunistic, multidrug resistant nosocomial bacteria responsible for hospital-acquired life threatening infections (Alqahtani, 2017).
ii. Neoisoshinanolone and l-epineoisoshinanolone are derivatives of plumbagin with only a single ketone group and these are less antibacterial than plumbagin implying that quinone moiety of the molecules combined with planarity are necessary for strong activity. Plumbagin is an example of secondary metabolite from plants with broad-spectrum antimicrobial effects *in vitro* and is probably a phytoanticipin. It is also a well-known cytotoxic molecule and is thus most probably the genetic material of bacteria, fungi, and viruses.

REFERENCES

Alqahtani, J.M., 2017. Emergence of *Stenotrophomonas maltophilia* nosocomial isolates in a Saudi children's hospital: risk factors and clinical characteristics. *Saudi Medical Journal*, *38*(5), p. 521.

Dey, D., Ray, R. and Hazra, B., 2014. Antitubercular and antibacterial activity of quinonoid natural products against multi-drug resistant clinical isolates. *Phytotherapy Research*, *28*(7), pp. 1014–1021.

Gebre-Mariam, T., Neubert, R., Schmidt, P.C., Wutzler, P. and Schmidtke, M., 2006. Antiviral activities of some Ethiopian medicinal plants used for the treatment of dermatological disorders. *Journal of Ethnopharmacology*, *104*(1–2), pp. 182–187.

Jetty, A., Subhakar, C., Rajagopal, D., Jetty, M., Subramanyam, M. and Marthanda Murthy, M., 2010. Antimicrobial activities of neo-and 1-epineo-isoshinanolones from *Plumbago zeylanica* roots. *Pharmaceutical Biology*, *48*(9), pp. 1007–1011.

Jeyachandran, R., Mahesh, A., Cindrella, L., Sudhakar, S. and Pazhanichamy, K., 2009. Antibacterial activity of plumbagin and root extracts of *Plumbago zeylanica* L. *Acta Biologica Cracoviensia Series Botanica*, *51*(1), pp. 17–22.

Mossa, J.S., El-Feraly, F.S. and Muhammad, I., 2004. Antimycobacterial constituents from *Juniperus procera*, *Ferula communis* and *Plumbago zeylanica* and their *in vitro* synergistic activity with isonicotinic acid hydrazide. *Phytotherapy Research: An International Journal Devoted to Pharmacological and Toxicological Evaluation of Natural Product Derivatives*, *18*(11), pp. 934–937.

Uc-Cachón, A.H., Borges-Argáez, R., Said-Fernández, S., Vargas-Villarreal, J., González-Salazar, F., Méndez-González, M., Cáceres-Farfán, M. and Molina-Salinas, G.M., 2014. Naphthoquinones isolated from *Diospyros anisandra* exhibit potent activity against pan-resistant first-line drugs *Mycobacterium tuberculosis* strains. *Pulmonary Pharmacology & Therapeutics*, *27*(1), pp. 114–120.

8.7.12 FAMILY POLYGONACEAE A.L DE JUSSIEU (1789)

The family Polygonaceae consists of about 55 genera and about 1100 species of herbs or shrubs. The stems are often hollowed, ribbed, zig-zag-shaped, juicy, and with an acidulate taste. The leaves are simple, alternate or spiral, and stipulate. The stipules are characteristically fused into a membranaceous tube around the stem known as the ocrea. The leaves are often fleshy. The inflorescences are terminal or axillary corymbs, panicles, heads, or spikes. The flowers are actinomorphic and small. The perianth comprises 3–6 tepals often persistant in fruits. The androecium is made of 4–9 stamens. The gynoecium consists of 2–3 carpels forming a unilocular ovary containing 1 ovule and developing a 2–4 styles. The fruits are trigonous achenes.

8.7.12.1 *Fagopyrum esculentum* Moench

Synonyms: *Fagopyrum emarginatum* (Roth) Meisn.; *Fagopyrum sagittatum* Gilib.; *Fagopyrum vulgare* T. Nees; *Fagopyrum zuogongense* Q.F. Chen; *Polygonum emarginatum* Roth; *Polygonum fagopyrum* L.

Common names: Buckwheat; qiao mai (China); kotu, mariwo (India); shari-mam (Myanmar); fapar (Nepal); brau (Pakistan)

Habitat: Cultivated

Distribution: Iran, Pakistan, India, Nepal, Sikkim, Bhutan, Myanmar, Korea, Mongolia, Russia

Botanical observation: This herb grows up to about 1 m tall. The stems are ribbed and reddish. The leaves are simple, spiral, and develop an ocrea which is about 5 mm long and membranaceous. The petiole is up to about 3.5 cm long. The blade is triangular, somewhat fleshy, glossy, 1.5–9.5 × 1–7.5 cm, with an anastomosis of sunken nerves above, cordate at base, and acuminate at apex. The inflorescences are axillary or terminal few flowered corymbs. The 5 petals are pinkish, elliptic, and 4 mm long. The androecium comprises 8 stamens. The gynoecium is trigonous and develops 3 styles. The achenes are edible, trigonous and up to 6 mm long.

Medicinal use: Diarrhea (India; Myanmar)

Moderate broad spectrum polar extract: Tannin fraction of seeds inhibited the growth of *Listeria monocytogenes*, *Staphylococcus aureus*, *Escherichia coli*, and *Salmonella typhimurium* with the MIC values of 62.5, 250, 500, and 125 μg/mL, respectively (Amarowicz et al., 2008).

Moderate broad-spectrum antibacterial essential oil: Essential oil of flowers inhibited the growth of *Agrobacterium tumefaciens* (ATCC 11158), *Escherichia coli* (ATCC 29425), *Pseudomonas lachrymans* (ATCC 11921), *Xanthomonas vesicatoria* (ATCC 11633), *Bacillus subtilis* (ATCC 11562), and *Staphylococcus aureus* (ATCC 6538) with the MIC values of 400, 600, 400, 400, 800, and 800 μg/mL, respectively (Amarowicz et al., 2008).

Strong broad-spectrum antifungal lipophilic naphthodianthrone piperidine alkaloid: Fagopyrin (Log D = 5.6 at pH 7.4; molecular mass= 670.7 g/mol) inhibited the growth of *Microsporum canis* (NRCC/F35892), *Trichophyton rubrum* (ATCC28188), *Exophiala dermatitidis* (ATCC28869), *Candida albicans* (ATCC14053) with the MIC/MFC values of 11.7/46.9, 46.9/46.9, 11.7/11.7, and 11.7/11.7 μg/mL, respectivel (Sytar et al., 2016).

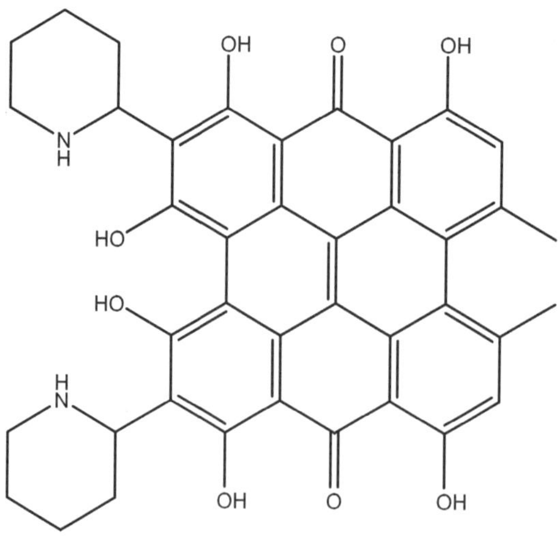

Fagopyrin

Strong broad-spectrum antibacterial flavonol glycoside: Rutin (Log D = –1.7 at pH 7.4; molecular mass=610.5 g/mol) inhibited the growth of *Escherichia coli* (ATCC 25922), *Escherichia coli* (clinical ESβL+), *Pseudomonas aeruginosa* (ATCC 10145), *Pseudomonas aeruginosa* (clinical isolate), *Proteus mirabilis* (ATCC 7002), *Proteus mirabilis* (clinical ESβL+), *Klebsiella pneumoniae* (RSKK 574), *Klebsiella pneumoniae* (clinical Extended-Spectrum β-lactamase), *Acinetobacter baumanii* (RSKK 02026), *Acinetobacter baumanii* (clinical isolate), *Staphylococcus aureus* (ATCC 25923), *Staphylococcus aureus* (clinical methicillin-resistant *Staphylococcus aureus*), *Enterococcus faecalis* (ATCC 29212), *Enterococcus faecalis* (clinical strain), and *Bacillus subtilis* (ATCC 6633) (Kreft et al., 1999).

Strong anticandidal flavonol glycoside: Rutin inhibited the growth of *Candida albicans* (ATCC 10231) and *Candida krusei* (ATCC 6258) with the MIC values of 16 and 32 μg/mL, respectively (Kreft et al., 1999).

Commentaries

i. The plant produces the flavonol glycoside rutin (Kreft et al., 2016) as well as fagopyrin (Sytar et al., 2016). One coud have the opportunity to examine the antibacterial and antiviral properties of fagopyrin which is structurally close to hypericin.
ii. *Fagopyrum dibotrys* (D. Don) H. Hara is used in India (local name: *Ban phapar*) for lung infections.

REFERENCES

Amarowicz, R., Dykes, G.A. and Pegg, R.B., 2008. Antibacterial activity of tannin constituents from *Phaseolus vulgaris*, *Fagoypyrum esculentum*, *Corylus avellana* and *Juglans nigra*. *Fitoterapia*, *79*(3), pp. 217–219.

Kreft, S., Knapp, M. and Kreft, I., 1999. Extraction of rutin from buckwheat (*Fagopyrum esculentum* Moench) seeds and determination by capillary electrophoresis. *Journal of Agricultural and Food Chemistry*, *47*(11), pp. 4649–4652.

Sytar, O., Švedienė, J., Ložienė, K., Paškevičius, A., Kosyan, A. and Taran, N., 2016. Antifungal properties of hypericin, hypericin tetrasulphonic acid and fagopyrin on pathogenic fungi and spoilage yeasts. *Pharmaceutical Biology*, *54*(12), pp. 3121–3125.

8.7.12.2 *Homalocladium platycladum* (F.J. Muell.) L.H. Bailey

Synonyms: *Coccoloba platyclada* (F. Muell.) F.J. Müll., *Muehlenbeckia platyclada* (F. Muell.) Meisn.; *Polygonum platycladum* F. Muell.

Common names: Ribbon bush, centipede plant, tapeworm-plant; jongli-bash (Bangladesh); fēi tiān wú (China); jangkang (Indonesia)

Habitat: Cultivated

Distribution: Native to Solomon Islands, Tropical Asia

Botanical observation: This odd-looking ornamental herb grows up to about 1 m tall. The plant has a positive aura of life. The stems are flat, tape-like, fleshy, striated, and articulated. The leaves are sessile, alternate, and simple. The ocreae are minute. The blade is lanceolate, cordate at base, and 1.5–6.5×1.5–3.5 cm. The inflorescences are axillary clusters. The perianth includes 5 broad lobes which are greenish and minute. Eight stamens are present. The ovary is elliptic and develops 3 stigmas. The nuts are embedded to the fleshy and purplish perianth.

Medicinal uses: Cough, sore throat (China); wounds (Bangladesh, Indonesia)

Broad-spectrum antibacterial halo developed by polar extract: Methanol extract inhibited the growth of *Bacillus subtilis* (ATCC 9466), *Staphylococcus aureus* (ATCC 25923), and *Salmonella typhi* (Nuria, 2010).

Moderate antibacterial (Gram-positive) naphthodianthrone: Hypericin inhibited the growth of *Staphylococcus aureus* (ATCC 29213) and *Staphylococcus epidermidis* (ATCC 12228) with the MIC values of 64 and 64 μg/mL, respectively (Feyzioğlu et al., 2013).

Strong broad-spectrum antifungal lipophilic naphthodianthrone: Hypericin (Log D = 7.1 at pH 7.4; molecular mass=504.4 g/mol) inhibited the growth of *Candida albicans*, *Exophiala dermatitidis* (ATCC 28869), *Microsporum canis* (NRCC/F35892), *Fusarium oxysporum* (NRCC/G-102), and *Trichophyton rubrum* (ATCC 28188) with the MIC/MFC values of 11.7/11.7, 0.1/0.7, 11.7/187.5, 46.9/187.5, and 2.9/46.9 μg/mL, respectively (Sytar et al., 2016).

Strong antiviral (enveloped segmented linear single-stranded (−) RNA) lipophilic naphthodianthrone: Hypericin (Log D = 7.1 at pH 7.4; molecular mass=504.4 g/mol) inhibited the replication of the Influenza virus A/Kitakyushu/159/93(H3N2) with the MIC of 2.1 ng/mL in Madin-Darby canine kidney cells (Yasuda et al., 2010).

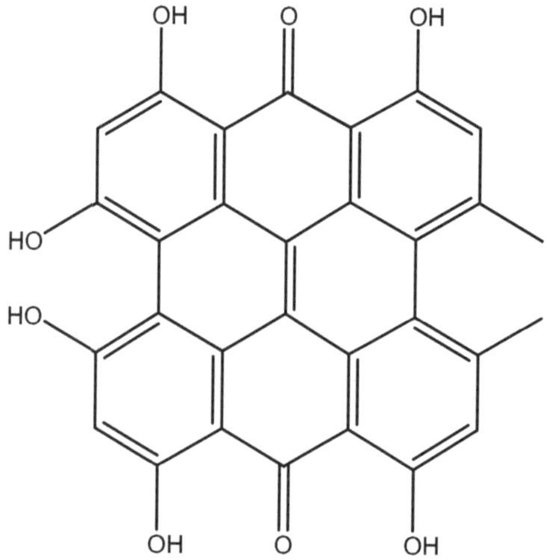

Hypericin

Commentaries

i. This plant does not seem to have been studied much for its potential antimicrobial value.
ii. The plant yields kaempferol and quercetin glycosides as well as (+)-catechin (Yen et al., 2019).
iii. Hypericin displayed broad-spectrum antiviral effects which were influenced by light exposure (Lopez-Bazzocchi et al., 1991). One could wonder if the presence of sunshine on plant tissues influences the activity of phytoalexins.

REFERENCES

Feyzioğlu, B., Demircili, M.E., Doğan, M. and Baykan, M., 2013. Antibacterial effect of hypericin. *African Journal of Microbiology Research*, *7*(11), pp.979–982.

Lopez-Bazzocchi, I., Hudson, J.B. and Towers, G.H.N., 1991. Antiviral activity of the photoactive plant pigment hypericin. *Photochemistry and Photobiology*, *54*(1), pp. 95–98.

Nuria, M.C., 2010. Antibacterial activities from Jangkang (*Homalocladium platycladum* (F. Muell) Bailey) leaves. *Mediagro*, *6*(2). 9–15

Sytar, O., Švedienė, J., Ložienė, K., Paškevičius, A., Kosyan, A. and Taran, N., 2016. Antifungal properties of hypericin, hypericin tetrasulphonic acid and fagopyrin on pathogenic fungi and spoilage yeasts. *Pharmaceutical Biology*, *54*(12), pp. 3121–3125.

Yasuda, T., Yamaki, M., Iimura, A., Shimotai, Y., Shimizu, K., Noshita, T. and Funayama, S., 2010. Anti-influenza virus principles from *Muehlenbeckia hastulata*. *Journal of Natural Medicines*, *64*(2), pp. 206–211.

Yen, C.T., Hsieh, P.W., Hwang, T.L., Lan, Y.H., Chang, F.R. and Wu, Y.C., 2009. Flavonol glycosides from *Muehlenbeckia platyclada* and their anti-inflammatory activity. *Chemical and Pharmaceutical Bulletin*, *57*(3), pp. 280–282.

8.7.12.3 *Persicaria barbata* (L.) H. Hara

Synonym: *Polygonum barbatum* L.

Common names: Water milkwort; bekh unjubach, pani morich (Bangladesh); kating he (Cambodia); mao liao (China); bhidiya, bish kantal, kunda mallier, niruganeru, yelang (India), jukut carang (Indonesia); pancis pancis (Malaysia); bish (Nepal); kanubsubang, saimbangan-tubig (the Philippines)

Habitat: Shady watery riverbanks, ponds, marshes

Distribution: Bhutan, China, India, Sri Lanka, Malaysia, Myanmar, Nepal, Thailand, Vietnam, Indonesia, the Philippines, Papua New Guinea

Botanical observation: This rhizomatous herb grows to about 60 cm tall. The stems are terete. The leaves are alternate and simple. The ocreae are tubular, up to about 2.5 cm long, and ciliate. The petiole is up to about 7 mm long and hairy. The blade is lanceolate, cuneate at base, ciliate at margin, acuminate at apex, and 6.5–15 × 1.5–4 cm. The spikes of racemes are terminal and up to about 10 cm long. The perianth includes 5 lobes which are greenish white and minute. The androecium comprises 5–8 stamens. The ovary is trigonous and develops 3 slender stigmas. The nuts are trigonous, black, minute, and glossy.

Medicinal uses: Boils, wounds, cuts (India); cholera (China); toothache (Pakistan); ulcers (Bangladesh; China)

Commentaries

i. The plant yields dihydrobenzofuran lignans (Farooq et al., 2019) that could be antimicrobial.
ii. In India, *Persicaria chinensis* L. (synonym: *Polygonum chinense* L.) (local names: *Anom yensil*; *Obei-o kati*) is used in to treat dysentery and wounds. In Myanmar, *Persicaria chinensis* L. is used to treat cuts, diarrhoea, and wounds (local name: *Vawm bam*) and in China (local name: *Huǒ shāo*) it is used for diarrhea, dysentery, and herpes. In Bangladesh, *Persicaria orientalis* (L.) Sach. is used for fever and *Persicaria limbata* (Meissn.) Hara. is used to wash ulcers.

REFERENCE

Farooq, U., Naz, S., Shams, A., Raza, Y., Ahmed, A., Rashid, U. and Sadiq, A., 2019. Isolation of dihydrobenzofuran derivatives from ethnomedicinal species Polygonum barbatum as anticancer compounds. *Biological Research*, *52*(1), p. 1.

8.7.12.4 *Persicaria hydropiper* (L.) Delarbre

Synonyms: *Polygonum hydropiper* L.; *Persicaria hydropiper* (L.) Opiz; *Persicaria hydropiper* (L.) Spach

Common names: Annual smartweed, marsh pepper knotweed; laliao (China); bishkatali; chotopanimorich; lalbiskatali; pani marich (Bangladesh); chok chine, paakur-muula (India); si tuba sawah (Indonesia); agagat, tuba (the Philippines); rumput tuboh (Malaysia); phak phai nam (Thailand); cây răm dại; nghể răm, nghể nước (Vietnam)

Habitat: Shady watery riverbanks, ponds, marshes

Distribution: Tropical Asia

Botanical observation: This erect herb grows to about 45 cm tall. The stems are smooth, glossy, glabrous, fleshy, and somewhat purplish. The leaves are simple, spiral, and stipulate. The ocreae are tubular and grow about 1.5 cm long and develop ciliae at apex. The petiole is about 5 mm long.

The blade is 1.5–8×0.4–2 cm, wavy, linear lanceolate, acuminate, and dotted with tiny glands. The spikes are 3–7 cm long. The perianth is tubular, white-pinkish, and develops 5 ovate lobes. The androecium includes 8 stamens attached to the periath tube. The ovary is trigonous and develops 3 styles with capitate stigmas. The nuts are minute and dark brownish.

Medicinal uses: Dysentery, wounds (Bangladesh); cold, cough, fever, fungal infection, jaundice (India); fever (Japan); boils, ulcers, wounds (Bangladesh; India)

Moderate broad-spectrum antibacterial drimane-type sesquiterpenes: The dialdehyde sesquiterpene polygodial inhibited the growth of *Bacillus subtilis* (ATCC 9372), *Bacillus thuringiensis* (ATCC 35646), *Brevibacterium ammoniagenes* (ATCC 6872), *Micrococcus luteus* (ATCC 4698), *Staphylococcus aureus* (ATCC 12598), *Staphylococcus aureus* (ATCC 25923), *Staphylococcus aureus* (ATCC 33591) (methicillin resistant), *Escherichia coli* (ATCC 29990), *Proteus vulgaris* (ATCC 13315), *Pseudomonas aeruginosa* (ATCC 14502), and *Salmonella choleraesuis* subsp. *choleraesuis* (ATCC 35640) (Kubo et al., 2005). Confertifolin from leaves inhibited the growth of *Staphylococcus epidermidis, Enterococcus faecalis, Pseudomonas aeruginosa*, and *Proteus vulgaris* with the MIC of 500, 31.2, 500 and 500 µg/mL, respectively (Duraipandiyan et al., 2010).

Strong broad-spectrum antifungal amphiphilic sesquiterpene: Polygodial (Log D = 3.8 at pH 7.4; molecular mass=234.3 g/mol)inhibited the growth of *Saccharomyces cerevisae* (IFO 0203), *Schizosaccharomyces pombe* (IFO 0342), *Hansenula anomala* (IFO 0136), *Candida utilis* (ATCC 42402), *Sclerotina libertiana* (SS), *Mucor mucedo* (IFO 7684), *Rhizopus chinensis* (IFO 4745), *Aspergillus niger* (ATCC 6275), and *Penicillium crustosum* with MIC values ranging from 0.7 to 25 µg/mL (Kubo & Taniguchi, 1988). Polygodial evoked fungal membrane insults resulting in cytoplasmic leakage in *Saccharomyces cerevisae* (IFO 0203) (Kubo & Taniguchi, 1988).

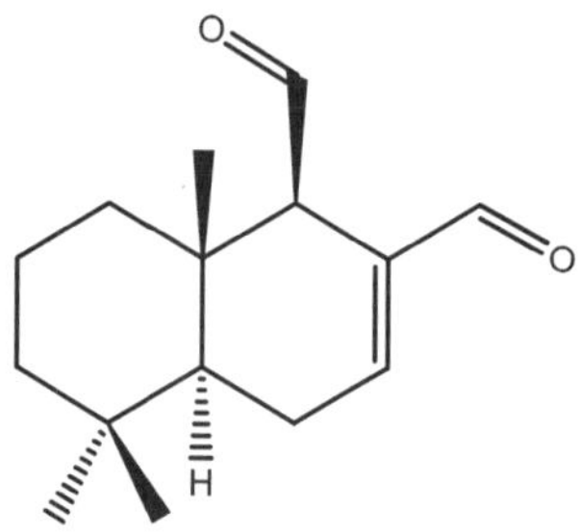

Polygodial

Antifungal potentiator sesquiterpene: Polygodial at 0.3 µg/mL (1/2 MIC) increased the sentitivity of *Saccharomyces cerevisae* (IFO 0203) to actinomycin D (Kubo & Taniguchi, 1988).

Commentary: The plant produces sesquiterpenes including polygodial (Asakawa et al., 2012).

REFERENCES

Asakawa, Y., Ludwiczuk, A., Harinantenaina, L., Toyota, M., Nishiki, M., Bardon, A. and Nii, K., 2012. Distribution of drimane sesquiterpenoids and tocopherols in liverworts, ferns and higher plants: Polygonaceae, Canellaceae and Winteraceae species. *Natural Product Communications, 7*(6), p. 1934578X1200700601.

Duraipandiyan, V., Indwar, F. and Ignacimuthu, S., 2010. Antimicrobial activity of confertifolin from *Polygonum hydropiper. Pharmaceutical Biology, 48*(2), pp. 187–190.

Kubo, I. and Taniguchi, M., 1988. Polygodial, an antifungal potentiator. *Journal of Natural Products, 51*(1), pp. 22–29.

Kubo, I., Fujita, K.I., Lee, S.H. and Ha, T.J., 2005. Antibacterial activity of polygodial. *Phytotherapy Research: An International Journal Devoted to Pharmacological and Toxicological Evaluation of Natural Product Derivatives, 19*(12), pp. 1013–1017.

8.7.12.5 *Persicaria perfoliata* H. Gross.

Synonyms: *Polygonum arifolium* var. perfoliatum L.; *Polygonum perfoliatum* (L.) L.

Common names: Devil's tail; cì lí tòu, gang ban gui (China); chepamullu kur, lilhar (India); bakhre aankhla (Nepal)

Distribution: China, Bangladesh, Bhutan, India, Indonesia, Japan, Korea, Malaysia, Nepal, New Guinea, Philippines, Russia, Sikkim, Thailand, Vietnam

Habitat: Watery and shady thickets, degraded forests

Botanical observation: This climber grows to about 2 m long. The stems are angled and prickly. The leaves are simple, spiral, edible (India), and stipulate. The ocreae are infundibuliform, green, and about 3 cm in diameter. The petiole is about 8 cm long. The blade is peltate, somewhat dull light green and fleshish, triangular, 3.5–6 × 5–8 cm, truncate at base and acute to round at apex. The spikes are terminal and up to about 3 cm long. The perianth is tubular, whitish, and develops 5 ovate lobes. The androecium includes 8 stamens attached to the periath tube. The ovary develops 3 styles. The achenes are embedded in a persistent perianth which is glossy, about 5 mm in diameter, bluish, and edible.

Medicinal uses: Cough, fever, tonsilitis, tracheitis (China); wounds (India).

Strong antibacterial (Gram-positive) polar extract: Methanol extract of stems inhibited the growth of *Staphylococcus aureus* (Newbould 305) with the MIC/MBC values of 7.8/ 31.2 µg/mL (Abdi & Dego, 2019).

In vivo antiviral (enveloped segmented linear single-stranded (−) RNA) flavonol glycoside: Quercetin-3-*O*-β-D-glucuronide (also known as miquelianin; Log D = – 4.2 at pH 7.4; molecular mass = 478.3 g/mol) given orally at a dose of 6 mg/Kg for 4 days protected mice against influenza A virus-induced lung edema by 27.9% (Fan et al., 2011).

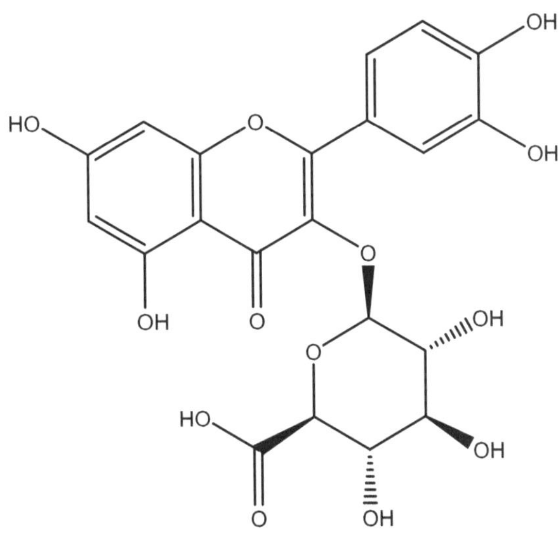

Quercetin-3-O-β-D-glucuronide

Viral enzyme inhibition by polar extract: Methanol extract of roots at 100 µg/mL inhibited Human immunodeficiency virus protease by 35% (Park et al., 2002).

REFERENCES

Abdi, R.D. and Dego, O.K., 2019. Antimicrobial activity of *Persicaria pensylvanica* extract against *Staphylococcus aureus*. *European Journal of Integrative Medicine*, *29*, p. 100921.

Fan, D., Zhou, X., Zhao, C., Chen, H., Zhao, Y. and Gong, X., 2011. Anti-inflammatory, antiviral and quantitative study of quercetin-3-*O*-β-D-glucuronide in *Polygonum perfoliatum* L. *Fitoterapia*, *82*(6), pp. 805–810.

Park, J.C., Hur, J.M., Park, J.G., Hatano, T., Yoshida, T., Miyashiro, H., Min, B.S. and Hattori, M., 2002. Inhibitory effects of Korean medicinal plants and camelliatannin H from Camellia japonica on human immunodeficiency virus type 1 protease. *Phytotherapy Research*, *16*(5), pp.422–426.

8.7.12.6 *Polygonum aviculare* L.

Synonyms: *Polygonum aequale* Lindm.; *Polygonum agreste* Sumner; *Polygonum aphyllum* Krock.; *Polygonum araraticum* Kom.; *Polygonum arenastrum* Boreau; *Polygonum berteroi* Phil.; *Polygonum heterophyllum* Lindm.; *Polygonum retinerve* Vorosch.; *Polygonum striatum* K. Koch

Common names: Prostrate knotweed, common knotgrass; bian xu (China); banalia, machuti, nisomali (India); atis shirin, pulpolak, sathing (Pakistan); kusdili (Turkey)

Habitat: Roadsides, wastelands

Distribution: Temperate Asia

Botanical observation: This discrete creeping herb grows to about 40 cm long. The stems are zigzag-shaped. The leaves are simple, edible (India), spiral, sessile, and stipulate. The ocrea is tubular, greenish-red, and membranaceous. The blade is lanceolate, spathulate, to elliptic, 1–4×0.5–1.5 cm, somewhat fleshy, dull light green, cuneate at base, and acute at apex. The inflorescences are axillary clusters of few minute flowers. The perianth is tubular, whitish-pink, and develops 5 elliptic lobes. Eight stamens are present. The ovary develops 3 styles. The achenes are trigonous, minute, and particularly affectionated by small birds.

Medicinal uses: Diarrhea, dysentery, wounds (Pakistan); fever (Korea); cough (Pakistan; Turkey)

Broad-spectrum antibacterial halo developed by mid-polar extract: Chloroform extract of stems (6 mm disc, 20 μL of a 10 mg/mL solution) inhibited the growth of *Escherichia coli*, *Proteus mirabilis*, *Pseudomonas aeruginosa*, *Salmonella typhii*, *Salmonella paratyphi*, *Shigella flexneri*, *Staphylococcus aureus*, *Bacillus subtilis*, and *Streptococcus pyogenes* (Salama & Marraiki, 2010).

Antifungal (filamentous) halo developed by a mid-polar extract: Chloroform extract of stems (6 mm disc, 20 μL of a 10 mg/mL solution) inhibited the growth of *Aspergillus flavus*, *Aspergillus fumigatus*, and *Aspergillus niger* with inhibition zone diameters of 17, 18, and 14 mm, respectively (Salama & Marraiki, 2010).

Moderate antibacterial (Gram-positive) polar extract: Methanol extract of aerial part inhibited the growth of *Staphylococcus aureus*, *Staphylococcus epidermidis*, *Staphylococcus saprophyticus*, with the MIC of 0.6, 0.6, and 1.2 mg/mL, respectively (Sassi et al., 2007).

Commentaries

i. The lignan glycoside aviculin and the flavonoid glycosides juglanin, avicularin, astragalin, and betmidin (Sassi et al., 2007), as well as quercetin-3-*O*-β-D-glucuronide are produced by this herb (Kim et al., 1994; Yang et al., 2014).

ii. Members of the genus *Polygonum* L. are often used for the treatment of microbial infections in Asia of which diarrhoea and dysentery (on account of tannins and anthraquinones). In Turkey, *Polygonum cognatum* Meisn. (local name: *Kus eppegi*) is used for abscesses and cough and *Polygonum lapathifolium* L. (local name: *Dereotu*) is used to treat fungal infections. In Iran, *Rheum turkestanicum* Janisch. (local name: *Eshghan*) is used to treat jaundice. In Pakistan, *Polygonum persicaria* L. is used for the treatment of diarrhoea and dysentery. In India, *Polygonum viviparum* L. (local name: *Maslun*) is used for the treatment of abscesses, bronchitis, diarrhoea, fever, sore throat, ulcers, and wounds, *Polygonum amplexicaule* D. Don (local name: *Masloon*) is used to treat cold and cough, *Polygonum runcinatum* Buch.-Ham. ex D. Don (local name: *Nobito*) is used as disinfectant, and *Polygonum paronychioides* C.A. Mey. is used for diarrhea and dysentery. In Bhutan, *Polygonum macrophyllum* D. Don (local name: *Spang-ram*) is used to treat diarrhoea and dysentery. In China, *Polygonum tinctorium* (local name: *Ta tsing*) is used to treat fever.

REFERENCES

Kim, H.J., Woo, E.R. and Park, H., 1994. A novel lignan and flavonoids from *Polygonum aviculare*. *Journal of Natural Products*, *57*(5), pp. 581–586.

Salama, H.M. and Marraiki, N., 2010. Antimicrobial activity and phytochemical analyses of *Polygonum aviculare* L. (Polygonaceae), naturally growing in Egypt. *Saudi Journal of Biological Sciences*, *17*(1), pp. 57–63.

Sassi, A.B., Harzallah-Skhiri, F. and Aouni, M., 2007. Investigation of some medicinal plants from Tunisia for antimicrobial activities. *Pharmaceutical Biology*, *45*(5), pp. 421–428.

Yang, H.H., Hwangbo, K., Zheng, M.S., Cho, J.H., Son, J.K., Kim, H.Y., Baek, S.H., Choi, H.C., Park, S.Y. and Kim, J.R., 2014. Quercetin-3-*O*-β-D-glucuronide isolated from *Polygonum aviculare* inhibits cellular senescence in human primary cells. *Archives of Pharmacal Research*, *37*(9), pp. 1219–1233.

8.7.12.7 *Polygonum bistorta* L.

Synonyms: *Bistorta lapidosa* Kitag.; *Bistorta major* Grey; *Bistorta officinalis* Raf.; *Persicaria bistorta* (L.) Samp.; *Polygonum lapidosum* (Kitag.) Kitag.

Common names: Great bistort, snake weed; quan shen (China); attalari (India); rishah-i-anjabar (Iran); bikh-anjubaz (Pakistan); cimen eveligi (Turkey)

Habitat: Grassy lands, meadows

Distribution: Iran, Pakistan, India, China, Japan, Mongolia

Botanical observation: This magnificent rhizomatous herb grows to about 40 cm tall and has somewhat a positive aura. The rhizome is stout and contorted. The stems are terete, smooth, and glabrous. The leaves are simple, spiral, basal, cauline, and stipulate. The ocreae is tubular, membranous, divided, and oblique at apex. The petiole is up to about 20 cm long. The blade is lanceolate to linear, 4.4 × 1.5–2.5 cm, decurrent at base, and acute-acuminate at apex. The spikes are terminal, cylindrical, and up to about 9 cm long. The perianth is tubular, pinkish, and develops 5 elliptic lobes. Eight stamens are present. The ovary develops 3 styles. The achenes are trigonous, minute, and glossy.

Medicinal uses: Antiseptic, cough, fever, jaundice, pneumonia (India); cough, sore throat (Turkey); diarrhoea, dysentery (Iran; Pakistan)

Broad antibacterial halo developed by polar extract: Methanol extract (50 mL per well of a 1 mg/5 mL solution) inhibited the growth of *Escherichia coli*, *Klebsiella pneumoniae*, *Pseudomonas aeruginosa*, *Proteus mirabilis*, *Salmonella typhi*, *Salmonella paratyphi* A, *Salmonella paratyphi* B, *Salmonella typhimurium*, *Shigella boydii*, *Shigella flexneri*, *Shigella sonnei*, *Staphylococcus aureus*, and *Streptococcus faecalis* (Satish et al., 2009).

Strong antiviral (enveloped monopartite linear single-stranded (+) RNA) lipophilic triterpenes: Friedelin (Log D = 10 at pH 7.4; molecular mass = 426.7 g/mol) and 3 β-friedelinol at the concentration of 5 μg/mL protected MRC-5 cells against the replication of the Human coronavirus (229E) by 132.4 and 109%, respectively (Chang et al., 2012).

Commentaries

i. The plant produces quercetin-3′-*O*-β-D-glucopyranoside as well as 3-methyl-gallic acid 4-*O*-β-D-(6′-*O*-3″-methyl-galloyl)-glucopyranoside (Liu et al., 2006) as well as 24(E)-ethylidenecycloartanone, 24(E)-ethylidenecycloartan-3alpha-ol, cycloartane-3,24-dione, 24-methylenecycloartanone, friedelin, 3β-friedelinol (Manoharan et al.,

2005), alnusenone and 3β-friedelinol (Duwiejua et al., 1999). Could *Polygonum bistorta* L. be of value against the Severe acute respiratory syndrome-associated coronavirus and other forthcoming pandemic coronaviruses? It is not unreasonable to think that one day teas of medicinal plants from Asia may effectively offer some protections against coronaviruses.

ii. In this volume, natural products able to inhibit coronaviruses *in vitro* are pentacyclic triterpenes, flavonols, coumarins, proanthocyanidins, and gallotannins.

REFERENCES

Chang, F.R., Yen, C.T., Ei-Shazly, M., Lin, W.H., Yen, M.H., Lin, K.H. and Wu, Y.C., 2012. Anti-human coronavirus (anti-HCoV) triterpenoids from the leaves of *Euphorbia neriifolia*. *Natural Product Communications*, *7*(11), p. 1934578X1200701103.

Duwiejua, M., Zeitlin, I.J., Grey, A.I. and Waterman, P.G., 1999. The anti-inflammatory compounds of *Polygonum bistorta*: Isolation and characterisation. *Planta Medica*, *65*(4), pp. 371–374.

Liu, X.Q., Hua, H.M., Liu, J., Chen, F.K. and Wu, L.J., 2006. A new tannin-related compound from the rhizome of *Polygonum bistorta* L. *Journal of Asian Natural Products Research*, *8*(4), pp. 299–302.

Manoharan, K.P., Benny, T.K.H. and Yang, D., 2005. Cycloartane type triterpenoids from the rhizomes of *Polygonum bistorta*. *Phytochemistry*, *66*(19), pp. 2304–2308.

Satish, S., Mohana, D.C., Raghavendra, M.P., Babu, S. and Raveesha, K.A., 2009. Antibacterial evaluation of some Iranian medicinal plants against some human pathogenic bacteria. *Asian Journal of Microbiology, Biotechnology and Environmental Sciences*, *11*(4), pp. 735–738

8.7.12.8 *Polygonum capitatum* Buch.-Ham. ex D. Don

Synonyms: *Cephalophilon capitatum* (Buch.-Ham. ex D. Don) Tzvelev; *Persicaria capitata* (Buch.-Ham. ex D. Don) H. Gross

Common names: Pinkhead smartweed; tou hua liao (China)

Habitat: Grassy and shady slopes, mountains

Distribution: India, Bhutan, Myanmar, Nepal, Sri Lanka, Thailand, Vietnam, China, Malaysia

Botanical observation: This pretty creeping herb grows to about 50 cm long. The stems are rooting at node, angled, and some what hairy. The leaves are simple, spiral, and stipulate. The ocrea is tubular to infundibuliform, membranaceous, up to 8 mm long, truncate and ciliate at apex. The petiole is up to 4 mm long. The blade is ovate-elliptic, 4.4 × 1.5–2.5 cm, hairy, cuneate at base, and acute at apex. The inflorescences are terminal heads which are up to about 1 cm in diameter. The perianth is tubular, pinkish, and develops 5 elliptic lobes. Eight stamens are present. The ovary develops 3 styles. The achenes are minute, trigonous, dark brown, glossy and embedded in persistant perianth.

Medicinal uses: Dysentery (China); fever, diarrhea (Pakistan); lung diseases (China); urinary tract infection (China; Pakistan)

Strong broad-spectrum antibacterial polar extract: Aqueous extract inhibited the growth of *Staphylococcus aureus*, *Escherichia coli*, *Neisseria gonorrhoeae*, and *Proteus mirabilis* with the MIC/MBC of 80/100, 80/100, 2/4, and 600/1300 µg/mL, respectively (Liao et al., 2011).

Strong antiviral (non-enveloped linear monopartite (+) RNA) amphiphilic flavanone: Taxifolin (LogD = 1.1 at pH 7.4; molecular mass 304.2 g/mol) at the concentration of 30 µg/mL inhibited the replication of the Hepatitis A virus (strain CF 53) by about 50% in PLC/PRF/5 cells. At 59 µg/mL, taxifolin decreased Log_{10} titer of the Hepatitis A virus by 1.5 Log_{10} (Biziagoset al., 1987).

Commentary: Gallic acid, epicatechin, catechin, rutin, quercetin-3-*O*-(2″-*O*-galloyl)-β-D-glucopyranoside and quercetin present in the plant are, at least in part, responsible for the antibacterial properties (Hu et al., 2016). The plant also produces taxifolin (Huang et al., 2015) which inhibits angiotensin converting enzyme (Arutyunyan et al., 2013) and acts against the (+) RNA virus Hepatitis A virus and as such could have potential value against coronaviruses.

REFERENCES

Arutyunyan, T.V., Korystova, A.F., Kublik, L.N., Levitman, M.K., Shaposhnikova, V.V. and Korystov, Y.N., 2013. Effects of taxifolin on the activity of angiotensin-converting enzyme and reactive oxygen and nitrogen species in the aorta of aging rats and rats treated with the nitric oxide synthase inhibitor and dexamethasone. *Age*, *35*(6), pp. 2089–2097.

Biziagos, E., Crance, J.M., Passagot, J. and Deloince, R., 1987. Effect of antiviral substances on hepatitis A virus replication in vitro. *Journal of medical virology*, *22*(1), pp. 57–66.

Hu, L., Zhang, J., Lin, L.C., Liao, S.G., Yan, Z., Xu, G.B. and Li, Y.J., 2016. A study on the antibacterial material basis of *Polygonum capitatum* by spectrum-effect relationship. *Zhong yao cai= Zhongyaocai= Journal of Chinese medicinal materials*, *39*(9), pp. 2037–2040.

Huang, G.H., Gao, Y., Wu, Z.J., Yang, Y., Huang, D.D., Chen, W.S. and Sun, L.N., 2015. Chemical constituents from *Polygonum capitatum* Buch-Ham. ex D. Don. *Biochemical Systematics and Ecology*, *59*, pp. 8–11.

Liao, S.G., Zhang, L.J., Sun, F., Zhang, J.J., Chen, A.Y., Lan, Y.Y., Li, Y.J., Wang, A.M., He, X., Xiong, Y. and Dong, L., 2011. Antibacterial and anti-inflammatory effects of extracts and fractions from *Polygonum capitatum*. *Journal of Ethnopharmacology*, *134*(3), pp. 1006–1009.

8.7.12.9 *Polygonum cuspidatum* Siebold & Zucc.

Synonyms: *Pleuropterus cuspidatus* (Siebold & Zucc.) H. Gross; *Pleuropterus cuspidatus* (Siebold & Zucc.) Moldenke; *Polygonum pictum* Siebold; *Polygonum reynoutria* Makino; *Polygonum zuccarinii* Small; *Reynoutria japonica* Houtt.

Common names: Japanese knotweed; hu zhang (China)

Habitat: Thickets, roadsides

Distribution: China, Korea, Japan

Botanical observation: This beautiful rhizomatous herb grows to about 1.5 m tall and has an aura of freshness. The stems are terete, reddish, and zig-zag-shaped. The leaves are simple, alternate, and stipulate. The ocrea is tubular, membranaceous, and truncate. The petiole is up to 2 cm long and somewhat reddish at base. The blade is ovate, 5–12×4–9.5 cm, cuneate, truncate, or rounded at base, and cuspidate at apex. The racemes are axillary and about 10 cm long. The perianth is tubular, whistish, and develops 5 elliptic lobes. Eight stamens are present. The ovary develops 3 styles. The achenes are minute, dark brown, glossy, and up to about 5 mm long.

Medicinal uses: Burns, bronchitis, jaundice (China)

Weak broad-spectrum antibacterial halo developed by polar extract: Methanol extract inhibited the growth of *Streptococcus mutans*, *Streptococcus sobrinus*, *Streptococcus pyogenes*, *Staphylococcus aureus*, *Actinomyces viscosus*, and *Porphyromonas gingivalis* with the MIC/MBC values of 2/4, 2>4, 4/>4, 2/>4, 0.5/2, and 1/4 mg/mL, respectively (Song et al., 2006). At sub-MIC levels, the extract inhibited virulence factor of *Streptococcus mutans* and *Streptococcus sobrinus* (Song et al., 2006).

Moderate broad-spectrum antibacterial mid-polar extract: Ethyl ether fraction of roots inhibited clinical drug resistant *Staphylococcus aureus*, *Acinetobacter baumannii*, and *Pseudomonas aeruginosa* with the MIC values of 0.2, 0.75, and 1.6 mg/mL, respectively (Su et al., 2015). This fraction was bactericidal and evoked at 2-fold MIC complete destruction of cell-shape of bacteria as well as atrophy and cytoplasmic agglutination suggesting activity on cell wall or cytoplasmic membrane (Su et al., 2015).

Antibacterial (Gram-positive) halo developed by anthraquinone: Emodin (50 μg/disc) inhibited the growth of methicillin-sensitive *Staphylococcus aureus* and methicillin-resistant *Staphylococcus aureus* with zone of inhibition of 19 and 30 mm, respectively (Joung et al., 2012).

Antibacterial (Gram-negative) halo developed by flavanol: Catechin (50 μg/disc) inhibited the growth of *Moraxella catarrhalis* with inhibition zone diameter of 12 mm (Pretorius et al., 2003).

Strong antibacterial (Gram-positive) amphiphilic anthraquinones: Emodin (Log D 1.2 at pH 7.4; molecular mass=270.3 g/mol) inhibited methicillin-sensitive and methicillin-resistant *Staphylococcus aureus* with the MIC of 25 µg/mL (Joung et al., 2012). Rhein inhibited methicillin-sensitive *Staphylococcus aureus* and methicillin-resistant *Staphylococcus aureus* with the MIC value of 15.6 µg/mL (Joung et al., 2012).

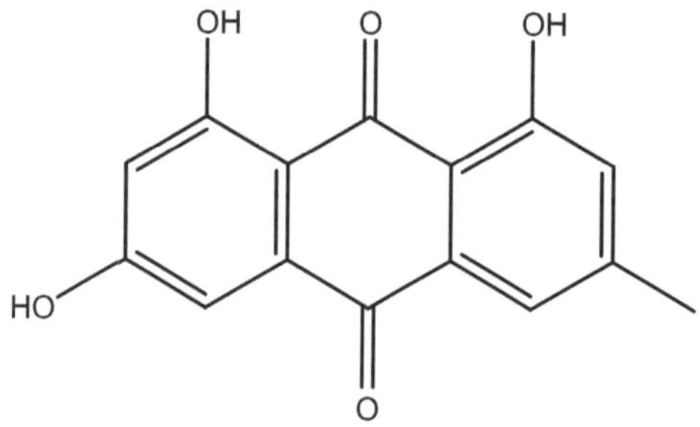

Emodin

Weak broad-spectrum antibacterial flavanol: Catechin inhibited the growth of *Staphylococcus aureus, Bacillus cereus, Staphylococcus epidermidis, Escherichia coli, Salmonella typhimurium, Shigella flexneri*, and *Pseudomonas aeruginosa* (Fu et al., 2016).

Antibiotic potentiator anthraquinone: With the combination on rhein, ampicillin inhibited methicillin-sensitive *Staphylococcus aureus* and methicillin-resistant *Staphylococcus aureus* (Joung et al., 2012). With the combination on rhein, oxacillin inhibited methicillin-sensitive *Staphylococcus aureus* and methicillin-resistant *Staphylococcus aureus* (Joung et al., 2012).

Strong antiviral (non-enveloped linear monopartite single-stranded (+) RNA) amphiphilic anthaquinone: Emodin inhibited the cytopathic effects of the Poliovirus type 3 in green monkey kidney cells with the IC_{50} value of 0.2 µg/mL and a selectivity index of 12.4 (Semple et al., 2001).

Strong antiviral (monopartite linear dimeric single-stranded (+) RNA) amphiphilic stilbene: (E)-Resveratrol (Log D=2.8) at pH 7.4; molecular mass=228.2 g/mol) inhibited Human immunodeficiency virus type-1-induced cytopathic effect in C8166 lymphocytes with the EC_{50} value of 4.3 µg/mL and a selectivity index of 8.1 (Lin et al., 2010).

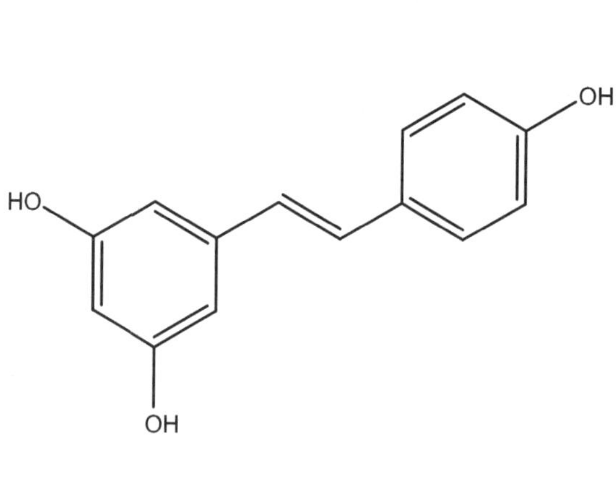

(*E*)-Resveratrol

Strong antiviral (monopartite linear dimeric single-stranded (+) RNA) benzofuran: 5,7-Dimethoxyphthalide inhibited the Human immunodeficiency virus type-1-induced cytopathic effect in C8166 lymphocytes with EC_{50} values of 9.9 2 μg/mL and a selectivity index above of 10 (Lin et al., 2010).

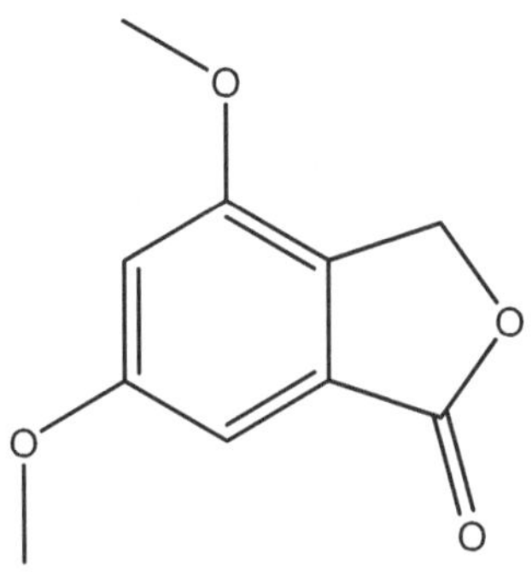

5,7-Dimethoxyphthalide

Strong antiviral (monopartite linear dimeric single-stranded (+) RNA) hydrophilic flavanol: Catechin (Log D=0.5 at pH 7.4; molecular mass=290.2 g/mol) inhibited the Human immunodeficiency virus type-1-induced cytopathic effect in C8166 lymphocytes with EC_{50} values of 14.4 μg/mL and the selectivity index of >13.8 (Lin et al., 2010).

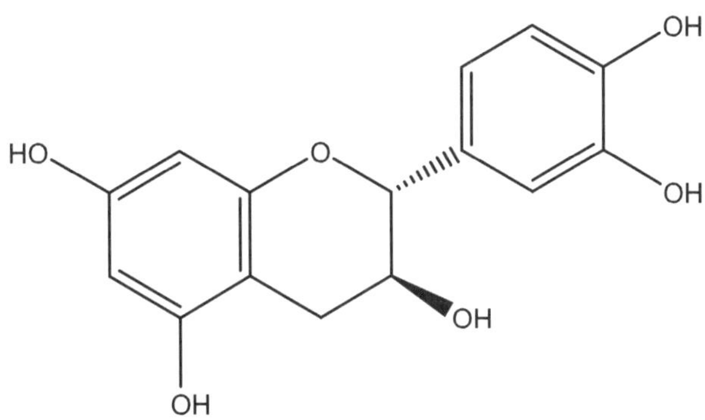

(+)-Catechin

Strong antiviral (monopartite linear dimeric single-stranded (+) RNA) anthraquinone glycoside: Emodin-8-*O*-β-D-glucopyranoside (Log D = −0.6 at pH 7.4; molecular mass=432.3 g/mol) inhibited the Human immunodeficiency virus type-1-induced cytopathic effect in C8166 lymphocytes with EC_{50} values of 11.2 μg/mL and selectivity index of above17 (Lin et al., 2010).

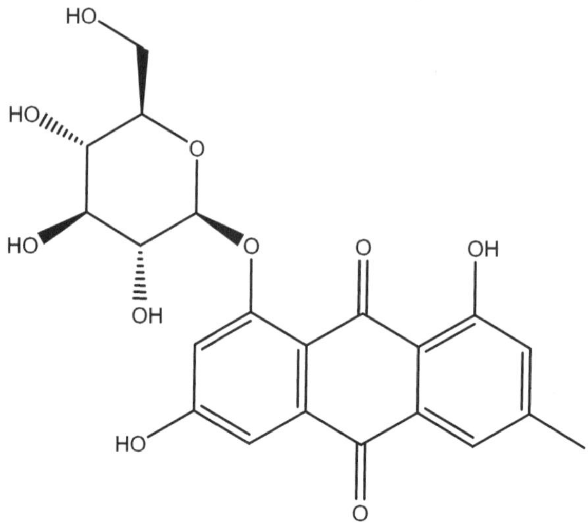

Emodin-8-*O*-*β*-D-glucopyranoside

Strong antiviral (enveloped segmented linear single-stranded (−) RNA) sesquiterpene: Reynoudiol from the roots inhibited the replication of the Influenza virus with an IC_{50} of 0.2 µM and a selectivity index of 172.4 (Nhiem et al., 2014).

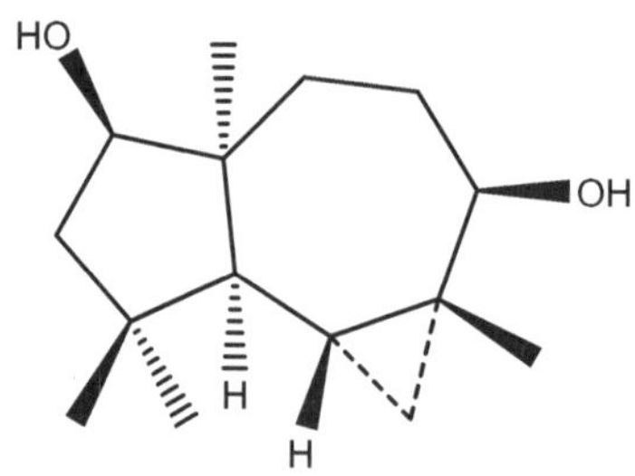

Reynoudiol

Moderate antiviral (non-enveloped monopartite linear single-stranded (+) RNA) anthraquinone: Emodin inhibited the replication of the Coxsackie B4 plaque reduction in Hep-2 cells with an EC_{50} of 12 µM and a selectivity index of 5 (Liu et al., 2013).

In vivo antiviral (non-enveloped monopartite linear single-stranded (+) RNA) anthraquinone: Emodin given orally to mice at a dose of 30 mg/Kg/day for 14 days increase lifetime as efficiently as Ribavirin at 10 mg/Kg/day against Coxsackie virus infection (Liu et al., 2013).

Viral enzyme inhibition by polar extract: Aqueous extract of root at the concentration of 250 µg/mL inhibited Human immunodeficiency virus type-1 protease by 51.7% (Xu et al., 1996).

Viral enzyme inhibition by anthraquinone: Emodin at the concentration of 50 µM inhibited Severe acute respiratory syndrome-associated coronavirus spike protein anchoring to angiotensin-converting enzyme by about 40% (Ho et al., 2007).

Commentaries

i. This plant is a typical example of medicinal plant producing a broad-array of antiviral principles of which emodin. One could wonder if this plant could be of value at phytomedication or herbal remedy for the prevention/management of viral pandemics of which COVID-19 and its mutants. Do the antiviral principles in this plant act synergistically *in vivo*?
ii. Emodin is planar and intercalate into microbial genetic material, it is thus mutagenic (Bösch et al., 1987). The plant produces 3,5-dihydroxybenzyl alcohol, the anthraquinone citreorosein, cis-resveratrol, trans-resveratrol, and trans-resveratrol-5-*O*-β-D-glucopyranoside which inhibited DNA topoisomerase II with the IC_{50} values of 0.5, 14, 15, 0.7, and 3 μM, respectively (Hwangboo et al., 2012).
iii. Severe acute respiratory syndrome-associated coronavirus spike protein is essential for the viral attachment to the host cell receptor angiotensin-converting enzyme and for mediating the fusion of viral and host membranes (Kirchdoerfer et al., 2016).

REFERENCES

Bösch, R., Friederich, U., Lutz, W.K., Brocker, E., Bachmann, M. and Schlatter, C., 1987. Investigations on DNA binding in rat liver and in Salmonella and on mutagenicity in the Ames test by emodin, a natural anthraquinone. *Mutation Research/Genetic Toxicology*, *188*(3), pp. 161–168.

Fu, L., Lu, W. and Zhou, X., 2016. Phenolic compounds and *in vitro* antibacterial and antioxidant activities of three tropic fruits: persimmon, guava, and sweetsop. *BioMed Research International*, *2016*. 1–9

Ho, T.Y., Wu, S.L., Chen, J.C., Li, C.C. and Hsiang, C.Y., 2007. Emodin blocks the SARS coronavirus spike protein and angiotensin-converting enzyme 2 interaction. *Antiviral Research*, *74*(2), pp. 92–101.

Hwangbo, K., Zheng, M.S., Kim, Y.J., Im, J.Y., Lee, C.S., Woo, M.H., Jahng, Y., Chang, H.W. and Son, J.K., 2012. Inhibition of DNA topoisomerases I and II of compounds from Reynoutria japonica. *Archives of pharmacal research*, *35*(9), pp.1583–1589.

Joung, D.K., Joung, H., Yang, D.W., Kwon, D.Y., Choi, J.G., Woo, S., Shin, D.Y., Kweon, O.H., Kweon, K.T. and Shin, D.W., 2012. Synergistic effect of rhein in combination with ampicillin or oxacillin against methicillin-resistant *Staphylococcus aureus*. *Experimental and Therapeutic Medicine*, *3*(4), pp. 608–612.

Kirchdoerfer, R.N., Cottrell, C.A., Wang, N., Pallesen, J., Yassine, H.M., Turner, H.L., Corbett, K.S., Graham, B.S., McLellan, J.S. and Ward, A.B., 2016. Pre-fusion structure of a human coronavirus spike protein. *Nature*, *531*(7592), pp. 118–121.

Lin, H.W., Sun, M.X., Wang, Y.H., Yang, L.M., Yang, Y.R., Huang, N., Xuan, L.J., Xu, Y.M., Bai, D.L., Zheng, Y.T. and Xiao, K., 2010. Anti-HIV activities of the compounds isolated from *Polygonum cuspidatum* and *Polygonum multiflorum*. *Planta Medica*, *76*(9), pp. 889–892.

Liu, Z., Wei, F., Chen, L.J., Xiong, H.R., Liu, Y.Y., Luo, F., Hou, W., Xiao, H. and Yang, Z.Q., 2013. *In vitro* and *in vivo* studies of the inhibitory effects of emodin isolated from *Polygonum cuspidatum* on Coxsakievirus B4. *Molecules*, *18*(10), pp. 11842–11858.

Nhiem, N.X., Van Kiem, P., Van Minh, C., Hoai, N.T., Duc, H.V., Tai, B.H., Quang, T.H., Le Anh, H.T., Yeo, S.G., Song, J.H. and Cheon, D.S., 2014. Anti-influenza sesquiterpene from the roots of *Reynoutria japonica*. *Natural Product Communications*, *9*(3), pp. 315–318.

Pretorius, J.C., Magama, S., Zietsman, P.C. and Jäger, A.K., 2003. Purification and identification of antibacterial compounds from *Euclea crispa* subsp. crispa (Ebenaceae) leaves. *South African Journal of Botany*, *69*(4), pp. 579–586.

Semple, S.J., Pyke, S.M., Reynolds, G.D. and Flower, R.L., 2001. *In vitro* antiviral activity of the anthraquinone chrysophanic acid against poliovirus. *Antiviral Research*, *49*(3), pp. 169–178.

Song, J.H., Kim, S.K., Chang, K.W., Han, S.K., Yi, H.K. and Jeon, J.G., 2006. *In vitro* inhibitory effects of *Polygonum cuspidatum* on bacterial viability and virulence factors of *Streptococcus mutans* and *Streptococcus sobrinus*. *Archives of Oral Biology*, *51*(12), pp. 1131–1140.

Su, P.W., Yang, C.H., Yang, J.F., Su, P.Y. and Chuang, L.Y., 2015. Antibacterial activities and antibacterial mechanism of *Polygonum cuspidatum* extracts against nosocomial drug-resistant pathogens. *Molecules*, *20*(6), pp. 11119–11130.

Xu, H.X., Wan, M., Loh, B.N., Kon, O.L., Chow, P.W. and Sim, K.Y., 1996. Screening of traditional medicines for their inhibitory activity against HIV-1 Protease. *Phytotherapy Research*, *10*(3), pp. 207–210.

8.7.12.10 *Polygonum dichotomum* Bl.

Synonyms: *Persicaria dichotoma* (Blume) Masam.; *Persicaria peduncularis* (Wall. ex Meisn.) Nemoto; *Polygonum pedunculare* Wall. ex Meisn.; *Polygonum tetragonum* Blume; *Tracaulon pedunculare* (Wall. ex Meisn.) Greene; *Tracaulon tetragonum* (Blume) Greene; *Truellum dichotomum* (Blume) Soják

Common names: Er qi liao (China); biskathali (India)

Habitat: Shady watery riverbanks, ponds, marshes

Distribution: India, Laos, Indonesia, Malaysia, Philippines, Thailand, Vietnam, China, Taiwan, Japan, Australia

Botanical observation: This magnificent herb grows to about 1 m in height. The stems are terete, somewhat reddish, gracile, and zig-zag-shaped. The leaves are simple, alternate, and stipulate. The ocrea is tubular, up to 2 cm long, membranaceous, rounded at base, somewhat wavy at margin, oblique at apex, and armed with few slender prickles at base. The petiole is up to 1 cm long. The blade is narrowly lanceolate, glossy, green to reddish, 5–10×1–3 mm, truncate at base, and acute at apex. The spikes are axillary or terminal. The 5 tepals are pinkish-white and about 3mm long. Five stamens are present. The ovary develops 2 styles and capitate stigmas. The achenes are minute.

Medicinal uses: Toothache (India)

Commentary: This plant has apparently not been studied for its potential antimicrobial properties.

8.7.12.11 *Polygonum equisetiforme* Sibth. & Sm.

Habitat: Deserts

Distribution: Turkey, Iran

Botanical observation: This diffuse shrub grows to about 1 m tall. The stems are terete. The leaves are simple, alternate, and stipulate. The ocrea is tubular, membranaceous, and up to about 1 cm long. The petiole is minute. The blade is lanceolate-linear to spathulate, about 2–4 cm×4–8 mm, and round at apex. The inflorescences are terminal spikes. The perianth is pinkish-white, darker in the middle, and develops 5 elliptic lobes. The androecium includes 5 stamens. The ovary develops 3 styles. The fruits are minute achenes.

Medicinal uses: Apparently none for microbial infections.

Weak antibacterial (Gram-positive) mid-polar extract: Chloroform fraction of aerial part inhibited the growth of *Bacillus cereus*, *Bacillus subtilis*, *Enterococcus faecalis*, *Staphylococcus aureus*, and *Staphylococcus epidermidis* with of 4, 4, 4, 2, and 2 mg/mL, respectively (Abdelah Bogdadi et al., 2007).

REFERENCE

Abdelah Bogdadi, H.A., Kokoska, L., Havlik, J., Kloucek, P., Rada, V. and Vorisek, K., 2007. *In vitro*. Antimicrobial activity of some Libyan medicinal plant extracts. *Pharmaceutical Biology*, *45*(5), pp. 386–391.

8.7.12.12 *Polygonum glabrum* Willd.

Synonyms: *Persicaria glabra* (Willd.) M. Gómez; *Polygonum densiflorum* Meisn.; *Polygonum glabrum* Cham. & Schltdl.; *Polygonum mexicanum* Small; *Polygonum portoricense* Bertero ex Endl.; *Polygonum portoricense* Bertero ex Small

Common names: Common marsh buckwheat; guang liao (China); bihagni, biskanta, sivapu kumbakodali (India)

Habitat: Shady watery riverbanks, ponds, marshes

Distribution: Afhghanistan, Pakistan, Bangladesh, India, Sri Lanka, Bhutan, Myanmar, the Philippines, Thailand, Vietnam; China, Taiwan, Australia, Pacific Islands

Botanical observation: This herb grows to about 1 m in height. The stems are terete and swollen at nodes. The leaves are simple, alternate, and stipulate. The ocrea is tubular, membranaceous, and up to about 3 cm long. The petiole is up to 1 cm long. The blade is narrowly lanceolate to oblong lanceolate, about 8–18 × 1.5–3 cm, cuneate at base, and acuminate at apex. The spikes are terminal and up to about 10 cm long. The 5 tepals are pinkish-white, fused at base, and 4 mm long. The androecium includes 6–8 stamens. The ovary develops 2 styles. The achenes are minute, glossy, tear-shaped and embedded in persistant perianth.

Medicinal uses: Cuts, jaundice, tonsilitis, wounds (India)

Strong antimycobacterial butetolide: 2-Methoxy-2-butenolide-3-cinnamate isolated from the aerial parts inhibited the growth of *Mycobacterium tuberculosis* (H37Ra) with the MIC of 1.4 µg/mL (Said et al., 2015).

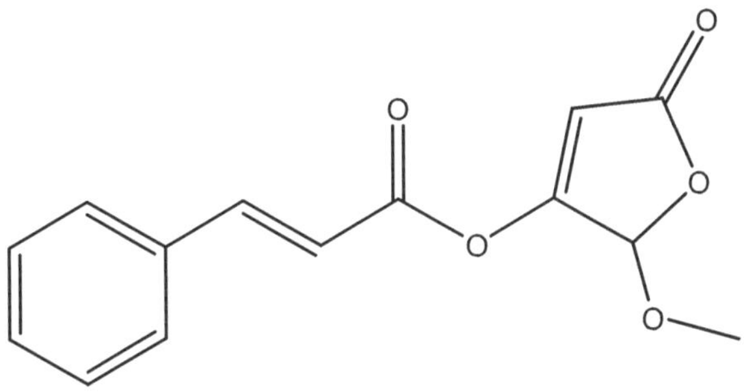

2-Methoxy-2-butenolide-3-cinnamate

Strong antimycobacterial amphiphilic stilbene: 3-Hydroxy-5-methoxystilbene (Log D = 3.8 at pH 7.4) molecular mass = 226.2 g/mol) inhibited *Mycobacterium tuberculosis* (H37Ra) with the MIC of 3.3 µg/mL (Said et al., 2015).

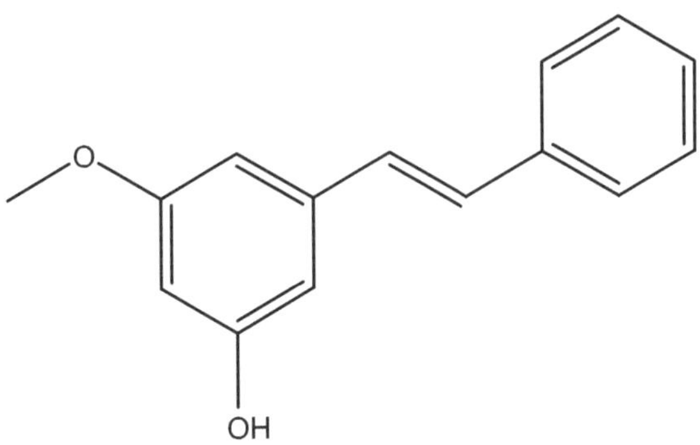

3-Hydroxy-5-methoxystilbene

Strong antimycobacterial amphiphilic flavanone: (–)-Pinocembrin (Log D = 2.9 at pH 7.4; molecular mass = 256.2 g/mol) inhibited *Mycobacterium tuberculosis* (H37Ra) with the MIC of 3.3 µg/mL (Said et al., 2015).

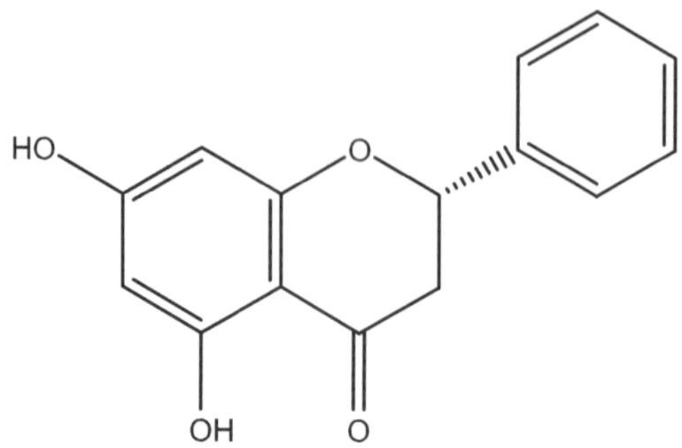

(-)-Pinocembrin

Strong antiviral (monopartite linear dimeric single-stranded (+) RNA) butetolide: 2-Methoxy-2-butenolide-3-cinnamate isolated from the aerial parts inhibited and Human immunodeficiency virus type-1 (VB59) with IC_{50} of 15.6 µg/mL (Said et al., 2015).

Strong antiviral (monopartite linear dimeric single-stranded (+) RNA) stilbene: 3-Hydroxy-5-methoxystilbene inhibited Human immunodeficiency virus type-1 (VB59) with the IC_{50} of 15.6 µg/mL (Said et al., 2015).

Strong antiviral amphiphilic flavanone: (–)-Pinocembrin inhibited Human immunodeficiency virus type-1 (VB59) with IC_{50} of 15.6 µg/mL (Said et al., 2015).

Commentary: The plant assembles drimane-type sesquiterpenes (Jacobsson & Muddathir, 1992) which are probably antimicrobial.

REFERENCES

Jacobsson, U. and Muddathir, A.K., 1992. Four biologically active sesquiterpenes of the drimane type isolated from *Polygonum glabrum. Phytochemistry*, *31*(12), pp. 4207–4211.

Said, M.S., Chinchansure, A.A., Nawale, L., Durge, A., Wadhwani, A., Kulkarni, S.S., Sarkar, D. and Joshi, S.P., 2015. A new butenolide cinnamate and other biological active chemical constituents from *Polygonum glabrum. Natural Product Research*, *29*(22), pp. 2080–2086.

8.7.12.13 *Polygonum orientale* L.

Synonyms: *Amblygonum orientale* (L.) Nakai ex T. Mori; *Lagunea cochinchinensis* Lour.; *Lagunea orientalis* (L.) Nakai; *Persicaria cochinchinensis* (Lour.) Kitag.; *Persicaria orientalis* (L.) Spach; *Persicaria orientalis* E. Vilm.; *Persicaria pilosa* (Roxb. ex Meisn.) Kitag.; *Polygonum amoenum* Blume; *Polygonum cochinchinense* (Lour.) Meisn.; *Polygonum cordobense* Lindau; *Polygonum pilosum* Roxb. ex Meisn.; *Polygonum torquatum* Bruijn

Common names: Oriental pepper, hatisur, pani morich, (Bangladesh); hong liao (China); vasak (India); oketade (Japan); pha kuang (Thailand); nghể bà (Vietnam)

Habitat: Wastelands, abandoned lots

Distribution: India, Sri Lanka, Bhutan, Bangladesh, Myanmar, Thailand, Vietnam

Indonesia, China, Japan, Korea, the Philippines,

Botanical observation: This pretty herb grows to about 1 m in height. The stems are terete and hairy. The leaves are simple, alternate, and stipulate. The ocrea is tubular, membranaceous, truncate, ciliate, with expanded margin, and up to about 2 cm long. The petiole is up to 10 cm long. The blade is ovate to elliptic, 10–20 × 5–12 cm, hairy, rounded at base, and acuminate at apex. The spikes are terminal, cylindrical, and a up to about 7 cm long. The perianth is pink, tubular, and develops 5 elliptic lobes which are about 4 mm long. Seven anthers are present. The ovary develops 2 styles. The achenes are minute, glossy, orbicular, and embedded in persistant perianth.

Medicinal uses: Wounds, fever (India)

Antiviral (enveloped segmented linear single-stranded (−) RNA) lignan glycoside: Arctiin (Log D=0.2 g/mol; molecular mass=534.5 g/mol) inhibited the replication of the Influenza A virus (A/NWS/33, H1N1) in Madin–Darby canine kidney cells with the IC_{50} value of 22 μM and a selectivity index of 13 (Hayashi et al., 2012).

In vivo (enveloped segmented linear single-stranded (−) RNA) lignan glycoside: Arctiin given to mice orally at the dose of 5 mg twice a day for 7 days decreased viral count in lung tissues (Hayashi et al., 2012).

Commentaries: The plant produces the lignans arctiin, lappaol B, and orientalin (Zheng et al., 1998), as well as the methoxylated flavonol exoticin, epicatechin-3-O-gallate, gallic acid, and 3,4-dihydroxy benzoic acid (Masum et al., 2019). Consider arctiin to be an active principle in *Arctium lappa* L. commonly known as burdock in Europe and used in China for the treatment of COVID-19 (Liu & Jiang, 2020). *Arctium lappa* L. was known of the Celts of the British Islands for the treatment of fever (Wagner et al., 2017) and used for the management of syphilis by French physicians up to the 19th century.

REFERENCES

Hayashi, K., Narutaki, K., Nagaoka, Y., Hayashi, T. and Uesato, S., 2010. Therapeutic effect of arctiin and arctigenin in immunocompetent and immunocompromised mice infected with influenza A virus. *Biological and Pharmaceutical Bulletin*, *33*(7), pp. 1199–1205.

Liu, J. and Jiang, L.Q., 2020. Research Progress of Chinese Medicine Used in Treating COVID-19. *Food Therapy and Health Care*, *2*(3), pp. 129–145.

Masum, M.N., Choodej, S., Yamauchi, K. and Mitsunaga, T., 2019. Isolation of phenylpropanoid sucrose esters from the roots of *Persicaria orientalis* and their potential as inhibitors of melanogenesis. *Medicinal Chemistry Research*, *28*(5), pp. 623–632.

Zheng, S., Wang, D., Meng, J. and Shen, X., 1998. Studies on the lignans of *Polygonum orientale*. *Acta Botanica Sinica*, *40*(5), pp. 466–469.

8.7.12.14 *Polygonum plebeium* R. Br.

Synonyms: *Polygonum changii* Kitag.; *Polygonum dryandri* Spreng.; *Polygonum parviflorum* Y.L. Chang & S.H. Li; *Polygonum roxburghii* Meisn.

Common names: Small knotweed; chikni-sag, chemti sak, munjara, pushitoa, rani phul (Bangladesh); tie ma bian (China); mouiara, sarpaakshi (India); hind rani, kheer wal, rani phal, (Pakistan)

Habitat: Sun exposed dry muds, riverbanks, around lakes

Distribution: Tropical Asia, Papua New Guinea, Australia, New Zealand

Botanical observation: This creeping herb grows to about 90 cm long. The leaves are simple, spiral, sessile, and stipulate. The stipules are united to form an ochrea which is white, minute, membranaceous and lacerated at apex. The blade is narrowly elliptic, edible (India), somewhat rosemary leaf-like, 5–15×2–4 mm, dull light green, narrowly cuneate at base, and obtuse or acute at apex. The inflorescences are axillary fascicles of few minute and graceful flowers. The perianth is tubular, white or light pink, and 5-lobed. The androecium comprises 5 stamens attached to the perianth tube. The ovary is minute and develops 3 styles with capitate stigmas. The achenes are dark brown, glossy, broadly ovoid, and trigonous.

Medicinal uses: Cholera, cough (Pakistan); diarrhoea, dysentery (India); pneumonia (Bangladesh, India, Pakistan)

8.7.12.15 *Polygonum viscosum* Buch.-Ham. ex D. Don

Synonyms: *Persicaria kuekenthalii* H. Lév.; *Persicaria viscosa* (Buch.-Ham. ex D. Don) H. Gross ex Nakai; *Polygonum kuekenthalii* H. Lév.

Common name: Bishkatali (Bangladesh); xiang liao (China)

Habitat: Riverbanks, roadsides, open grassy lands

Distribution: India, Nepal, Bangladesh, China, Korea, Taiwan, Japan

Botanical observation: This herb is odoriferous and grows to about 90 cm tall. The stems are terete and covered with hairs. The leaves are simple, spiral, subsessile, and stipulate. The stipules are united to form an ocrea which is tubular, up to about 1.2 cm long, hairy, and ciliate at apex. The blade is lanceolate, to ovate-elliptic, 5–15 × 2–4 cm, hairy, tapering at base, and acute at apex. The spikes are axillary and are up to about 4 cm long. The perianth is tubular, pinkish and 5-lobed, the lobes about 3 mm in lenght. The androecium comprises 8 stamens attached to the corolla tube. The ovary is minute and develops 3 styles with capitate stigmas. The achenes are minute, trigonous, black, glossy, and enveloped in the persistent perianth.

Medicinal uses: Apparently none for microbial infections

Strong broad-spectrum antibacterial amphiphilic cadidane-type sesquiterpene: Polygosumic acid (3-oxo-1-epi-sclerosporin) isolated from the aerial parts inhibited the growth of *Bacillus cereus* (NCTC 9689), *Citrobacter freundii* (NCTC 9750), *Escherichia coli* (NCIMB 8110), *Escherichia coli* (NCIMB 4174), *Klebsiella aerogenes* (NCTC 9528), *Pseudomonas aeruginosa* (NCTC 6750), *Staphylococcus aureus* (NCTC 10788), and methicillin-resistant *Staphylococcus aureus* (NCTC 11940) with the MIC values of 100, 250, 250, 50, 500, 500, 50, and 100 μg/mL, respectively (Datta et al., 2007).

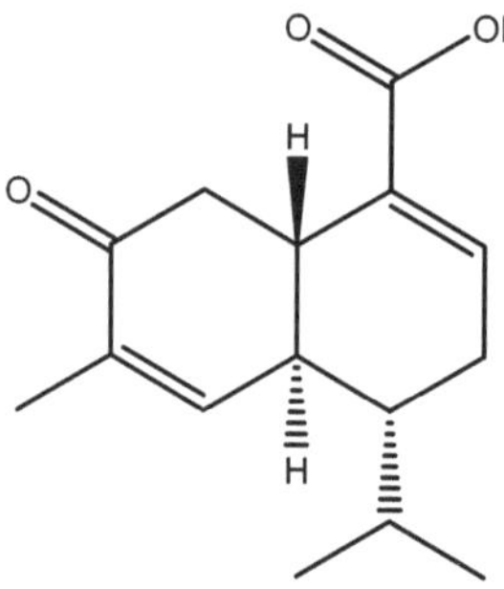

Polygosumic acid

Viral enzyme inhibition by sesquiterpene: Viscoazulone isolated from the aerial parts inhibited reverse transcriptase of the Human immunodeficiency virus type-1 with the IC_{50} values of 33.1 μg/mL (Datta et al., 2004).

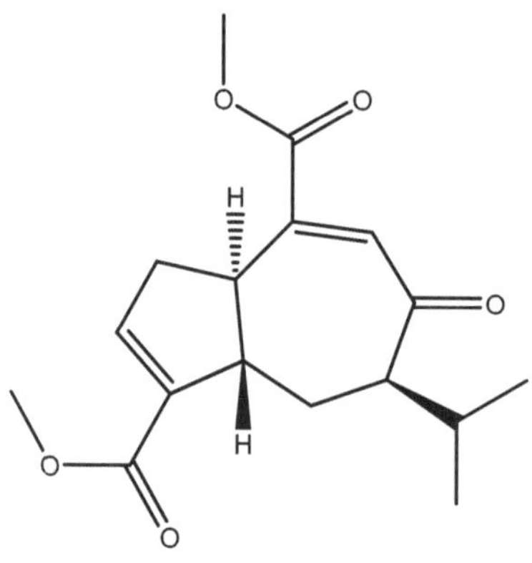

Viscoazulone

Viral enzyme inhibition by flavonol glycoside: Quercetin 3-*O*-(6″-feruloyl)-β-D-galactopyranoside isolated from the aerial parts inhibited reverse transcriptase of the Human immunodeficiency virus type-1 with IC_{50} values of 25.6 μg/mL (Datta et al., 2004).

Commentary: Consider the presence of conjugated quinone group favoring hydrogen bonds in polygosumic acid which is a typical example of broad-spectrum antibacterial sesquiterpene. Being planar, it may target bacterial DNA or enzymes involved in genetic material replication or protein synthesis. The presence of hydroxy, conjugated ketones, and aromatic moieties favor a good anchoring of natural products in the catalytic pockect of enzymes. Polygosumic acid is most probably antiviral and antifungal.

REFERENCE

Datta, B., Datta, S., Khan, T., Kundu, J., Rashid, M., Nahar, L. and Sarker, S., 2004. Anti-cholinergic, cytotoxic and anti-HIV-1 activities of sesquiterpenes and a flavonoid glycoside from the aerial parts of *Polygonum viscosum*. *Pharmaceutical Biology*, *42*(1), pp. 18–23.

8.7.12.16 *Rheum australe* D. Don

Synonym: *Rheum emodii* Wall. ex Meisn.

Common names: Indian rhubarb, Himalayan rhubarb; zang bian da huang (China); gandhini revatikka, revatchini (India); padamchaal (Nepal)

Distribution: Pakistan, India, Himalayas, Myanmar, Nepal, Sikkim, Bhutan, China

Habitat: Rocky or grassy mountain slopes

Botanical observation: This stout erect herb grows to about 2 m tall from a massive rhizome which is edible (Nepal). The stems are terete and striated. The leaves are simple, basal, spiral, and stipulate. The stipules are united to form a showy ocrea. The petiole is up to 50 cm long, edible, somewhat purplish, stout and striated. The blade is broadly ovate, 20–50 × 17.5–45 cm, hairy, with about 5–7 pairs of secondary nerves, somewhat coriaceous, dull dark green, wavy, cordate at base, incised, and obtuse to round at apex. The inflorescences are panicles of innumerable tiny flowers. The 6 tepals are about 3 mm in length, rounded and yellowish-white. The androecium comprises 9 stamens with slender filaments. The ovary is developing 3 styles with inflated stigmas. The achenes are trigonous, ellipsoid, up to about 1 cm long, somewhat purplish-red, and winged.

Medicinal use: Bronchitis (India)

Broad-spectrum antibacterial halo developed by oxanthrone ethers: Revandchinone-1 (30 μg/well) isolated from the rhizome inhibited the growth of *Bacillus subtilis* (MTCC 441), *Bacillus sphaericus* (MTCC 511), *Staphylococcus aureus* (MTCC 96), *Klebsiella aerogenes* (MTCC 39), and *Chromobacterium violaceum* (MTCC 2656) with inhibition zone diameter of 7, 8, 7, 7, and 7 mm, respectively (Babu et al., 2003). Revandchinone-4 (30 μg/well) isolated from the rhizome inhibited the growth of *Bacillus subtilis* (MTCC 441), *Bacillus sphaericus* (MTCC 511), *Staphylococcus aureus* (MTCC 96), *Klebsiella aerogenes* (MTCC 39), *Chromobacterium violaceum* (MTCC 2656), and *Pseudomonas aeruginosa* (MTCC 741) with inhibition zone diameter of 7, 7, 12, 7, 7, and 7 mm, respectively (Babu et al., 2003).

Broad-spectrum antibacterial halo developed by an anthraquinone: Revandchinone-3 (30 μg/well) isolated from the rhizome inhibited the growth of *Bacillus subtilis* (MTCC 441), *Bacillus sphaericus* (MTCC 51), *Staphylococcus aureus* (MTCC 96), *Klebsiella aerogenes* (MTCC 39), and *Chromobacterium violaceum* (MTCC 2656) (Babu et al., 2003).

Strong antiviral (enveloped monopartite linear single-stranded (+) RNA) amphiphilic anthraquinones: Chrysophanol (Log D = 2.5 at pH 7.4; molecular mass = 254.2 g/mol) and aloe-emodin inhibited Japanese encephalitis virus (T1P1) plaque reduction in BHK-21 with IC_{50} of 15.8 and 17.3 μg/mL, respectively (Chang et al., 2014). Chrysophanol and aloe-emodin were virucidal for Japanese encephalitis virus (T1P1) with IC_{50} of 0.7 and 0.4 μg/mL, respectively (Chang et al., 2014).

Strong antiviral (non-enveloped linar monopartite single-stranded (+) RNA) amphiphilic anthraquinone: Chrysophanic acid inhibited the replication of the Poliovirus types 2 and 3 in green monkey kidney cells with the EC_{50} values of 0.2 and 0.02 µg/mL, respectively (Semple et al., 2001).

Moderate antiviral (enveloped segmented linear single-stranded (−) RNA) stilbene: Resveratrol (Rokaya et al., 2012) inhibited the replication of the Influenza A virus (Oh7 strain) with ED_{50} of 45 µM (Kim et al., 2010).

Viral enzyme inhibition by anthraquinones: Chrysophanol inhibited Human immunodeficiency virus type-1 ribonuclease H and human immunodeficiency virus type-1 reverse transcriptase (RT)-associated DNA polymerase with IC_{50} values of 25 and 26.5 µM, respectively (Esposito et al., 2016). Aloe-emodin inhibited Human immunodeficiency virus type-1 ribonuclease H and Human immunodeficiency virus type-1 reverse transcriptase (RT)-associated DNA polymerase with IC_{50} values of 23 and 21 µM, respectively (Esposito et al., 2016). Aloe-emodin inhibited the Severe acute respiratory syndrome-associated coronavirus 3CL protease with the IC_{50} value of 35.7 µg/mL (Lin et al., 2005).

Commentaries

i. In Nepal the rhizome yields a dye, on probable account of anthraquinones which are orangish to red.
ii. Chrysophanol and aloe-emodin could be examined for their activity against the Severe acute respiratory syndrome-associated coronavirus.
iii. The petioles according to our peer the pharmacognosist Jonathan Peireira (1804–1853) afford the preparation of delicious tarts and puddings.
iv. One could endeaviour to examine the protective effect of this plant against COVID-19 and forthcoming viral pandemics.

REFERENCES

Babu, K.S., Srinivas, P.V., Praveen, B., Kishore, K.H., Murty, U.S. and Rao, J.M., 2003. Antimicrobial constituents from the rhizomes of *Rheum emodi*. *Phytochemistry*, *62*(2), pp. 203–207.

Chang, S.J., Huang, S.H., Lin, Y.J., Tsou, Y.Y. and Lin, C.W., 2014. Antiviral activity of Rheum palmatum methanol extract and chrysophanol against Japanese encephalitis virus. *Archives of Pharmacal Research*, *37*(9), pp. 1117–1123.

Esposito, F., Carli, I., Del Vecchio, C., Xu, L., Corona, A., Grandi, N., Piano, D., Maccioni, E., Distinto, S., Parolin, C. and Tramontano, E., 2016. Sennoside A, derived from the traditional chinese medicine plant *Rheum* L., is a new dual HIV-1 inhibitor effective on HIV-1 replication. *Phytomedicine*, *23*(12), pp. 1383–1391.

Kim, Y., Narayanan, S. and Chang, K.O., 2010. Inhibition of influenza virus replication by plant-derived isoquercetin. *Antiviral Research*, *88*(2), pp. 227–235.

Lin, C.W., Tsai, F.J., Tsai, C.H., Lai, C.C., Wan, L., Ho, T.Y., Hsieh, C.C. and Chao, P.D.L., 2005. Anti-SARS coronavirus 3C-like protease effects of Isatis indigotica root and plant-derived phenolic compounds. *Antiviral Research*, *68*(1), pp. 36–42.

Rokaya, M.B., Maršík, P. and Münzbergová, Z., 2012. Active constituents in *Rheum acuminatum* and *Rheum australe* (Polygonaceae) roots: a variation between cultivated and naturally growing plants. *Biochemical Systematics and Ecology*, *41*, pp. 83–90.

Semple, S.J., Pyke, S.M., Reynolds, G.D. and Flower, R.L., 2001. *In vitro* antiviral activity of the anthraquinone chrysophanic acid against poliovirus. *Antiviral Research*, *49*(3), pp. 169–178.

8.7.12.17 *Rheum officinale* Baills

Common names: Chinese rhubarb; yao yong da huang (China); usaare rewand (India); ravant çini (Turkey)

Habitat: forests, montains

Distribution: China

Botanical observation: This stout herb grows to about 2 m tall from a massive rhizome. The stems are terete and striated. The leaves are simple, basal, and cauline, spiral, subsessile, and stipulate. The stipules are united to form a massive ocrea which is up to about 15 cm long and hairy. The petiole is up to 50 cm long, stout, and striated. The blade is orbicular, 30–50 accross, hairy below, with about 5–7 pairs of secondary nerves, somewhat coriaceous, cordate at base, lobed, wavy, and acute at apex. The panicles present innumerable minute flowers. The 6 tepals are spathulate, 2.5 mm in length, and greenish. Nine stamens are present. The ovary is develops 3 styles with inflated stigmas. The achenes are trigonous, ellipsoid, up to about 1 cm long, and winged.

Medicinal use: Diarrhoea (India)

Broad-spectrum antibacterial hydrophilic anthraquinone: Rhein (Log D = – 0.5 at pH 7.4; molecular mass = 284.2 g/mol) from the roots inhibited the growth of *Bacteroides fragilis* and *Clostridium perfringens* with the MIC values of 1.5 and 50 μg/mL, respectively (Jong-Chol et al., 1987).

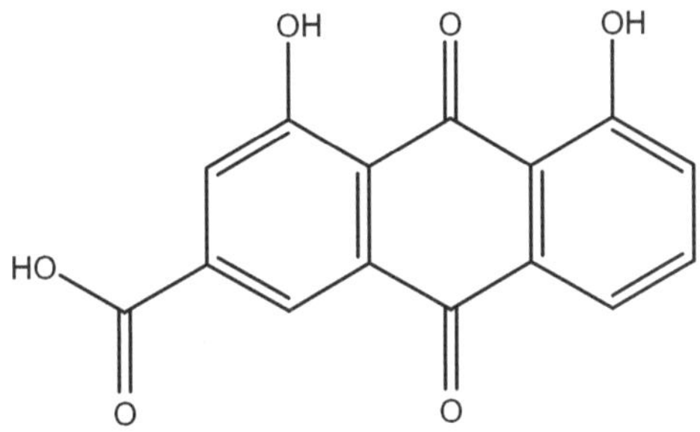

Rhein

Strong antiviral (enveloped monopartite linear double standed DNA) amphiphilic anthraquinone: Aloe-emodin (Log D = 1.2 at pH 7.4; molecular mass = 270.2 g/mol) inhibited the replication of the Herpes simplex virus type-1 (KOS), Herpes simplex virus type-2 (ATCC VR-734), Pseudorabies virus, Varicella-zoster virus (ATCC VR-916) with IC_{50} values of 1.6, 1.5, 5, and 6 μg/mL, respectively (Sydiscis et al., 1991)..

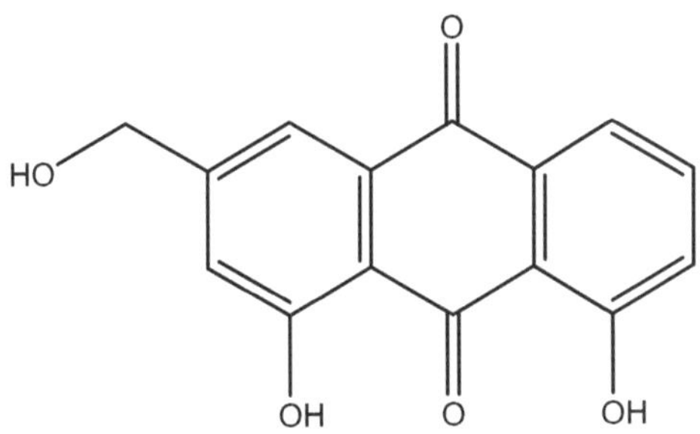

Aloe-emodin

Strong antiviral (enveloped segmented linear single-stranded (−) RNA) amphiphilic anthraquinone: Aloe-emodin inhibited the replication of the Influenza virus with IC_{50} value of 4.5 μg/mL (Sydiscis et al., 1991).

Viral enzyme inhibition by anthraquinone: Rhein inhibited Human immunodeficiency virus type-1 ribonuclease H with IC_{50} value of 60 μM (Esposito et al., 2016).

REFERENCES

Esposito, F., Carli, I., Del Vecchio, C., Xu, L., Corona, A., Grandi, N., Piano, D., Maccioni, E., Distinto, S., Parolin, C. and Tramontano, E., 2016. Sennoside A, derived from the traditional Chinese medicine plant *Rheum* L., is a new dual HIV-1 inhibitor effective on HIV-1 replication. *Phytomedicine*, *23*(12), pp. 1383–1391.

Jong-Chol, C., Tsukasa, M., Kazuo, A., Hiroaki, K., Haruki, Y. and Yasuo, O., 1987. Anti-bacteroides fragilis substance from rhubarb. *Journal of Ethnopharmacology*, *19*(3), pp. 279–283.

Sydiscis, R.J., Owen, D.G., Lohr, J.L., Rosler, K.H. and Blomster, R.N., 1991. Inactivation of enveloped viruses by anthraquinones extracted from plants. *Antimicrobial Agents and Chemotherapy*, *35*(12), pp. 2463–2466.

8.7.12.18 *Rheum palmatum* L.

Synonyms: *Rheum potaninii* Losinsk.; *Rheum qinlingense* Y.K. Yang, J.K. Wu & D.K. Zhang

Habitat: Grassy slopes, valleys

Common names: True rhubarb, palmated rhubarb; rasna (Bangladesh); jun mu zha, zhang ye da huang (China); rivand-i-chini (Iran)

Distribution: China

Botanical observation: This massive herb grows to about 2 m tall from a rhizome. The stems are terete and striated. The leaves are simple, spiral, basal, and stipulate. The stipules are united to form and muricate ocrea which grows up to about 15 cm long. The petiole is up to 60 cm long, stout, edible, and striated. The blade is orbicular, 40–60 cm accross, hairy below, with about 5 pairs of secondary nerves, somewhat coriaceous, wavy, dull green, cordate at base, pinnatisect, and acute at apex. The panicles are terminal and present numerous tiny flowers. The 6 tepals are spathulate, about 2.5 mm long, and purplish-red. Nine stamens are present. The ovary is develops 3 styles with inflated stigmas. The achenes are ellipsoid, up to about 1 cm long, and winged.

Medicinal uses: Jaundice, syphilis (Iran)

Strong antibacterial (Gram-positive) amphiphilic anthraquinones: Emodin (Log D = 1.7 at pH 7.4; molecular mass = 270.3 g/mol) inhibited the growth of *Staphylococcus aureus* (CMCC 26003) with a MIC value of 8 μg/mL (Yan et al., 2017). Aloe-emodin (Log D = 1.2 at pH 7.4; molecular mass = 270.2 g/mol) inhibited the growth of *Streptococcus mutans* (ATCC 25175) with the MIC values of 1.2 μg/mL (Zheng et al., 2011).

Strong antiviral amphiphilic anthraquinone: Emodin at the concentration of 50 μM protected by 94.1% Vero cells against the Severe acquired respiratory syndrome-associated corovirus infectivity and blocked the binding of spike protein to angiotensin-converting enzyme with the IC_{50} value of 200 μM (Ho et al., 2007).

Strong antiviral (enveloped monopartite linear double standed DNA) amphiphilic anthraquinone: Emodin halved the replication of Herpes simplex virus type-1 and Herpes simplex virus type-2 in cells at 50 μg/mL (Xiong et al., 2011).

In vivo (enveloped monopartite linear double standed DNA) antiviral anthraquinone: Emodin given orally at the dose of 6.7 g/Kg/day (3 times daily for 7 days beginning 24 hour post-virus exposure) increased the survival rate of mice intracerebrally infected by the Herpes simplex virus type-1 or type-2 to more than 35 days without signs of toxicity (Xiong et al., 2011). This effect was superior to Acyclovir at 0.1 g/Kg/day. Emodin decreased virus titer in various organs (Xiong et al., 2011).

Viral enzyme inhibition by dianthrone glycosides: Sennoside A inhibited the Human immunodeficiency virus type-1 ribonuclease H and Human immunodeficiency virus type-1 reverse transcriptase (RT)-associated DNA polymerase with the IC_{50} values of 1.9 and 5.3 μM, respectively (Esposito et al., 2016). The dianthrone glycoside sennoside B inhibited Human immunodeficiency virus type-1 ribonuclease H and human immunodeficiency virus type-1 RT-associated DNA polymerase with IC_{50} values of 2.1 and 2.3 μM, respectively (Esposito et al., 2016).

Viral enzyme inhibition by polar extract: Aqueous extract of rhizome at the concentration of 250 μg/mL inhibited the Human immunodeficiency virus type-1 protease by 65.5% (Xu et al., 1996).

Commentaries

i. The plant contains aloe-emodin, chrysophanol, emodin, physcion, and rhein (Rokaya et al., 2012).
ii. Several lines of evidence demonstrate that some endophytic fungi produce natural products similar to those produced by their host (Puri et al., 2005) and one could see here the assistance of symbiotic fungi to produce phytoalexins to protect its host. One example is rhein produced by endophytic *Fusarium solani* isolated from *Rheum palmatum* L. (You et al., 2013).
iii. A study reports the presence of Severe acute respiratory syndrome-associated coronavirus 3 chymotrypsin-like protease inhibitors from this plant (Luo et al., 2009) which is used for the treatment of COVID-19 in China (Wang et al., 2020). Consider that medicinal plants with activity against the Herpes simplex virus and the Human immunodeficiency virus are often active against Severe acute respiratory syndrome-associated coronavirus.

REFERENCES

Esposito, F., Carli, I., Del Vecchio, C., Xu, L., Corona, A., Grandi, N., Piano, D., Maccioni, E., Distinto, S., Parolin, C. and Tramontano, E., 2016. Sennoside A, derived from the traditional Chinese medicine plant *Rheum* L., is a new dual HIV-1 inhibitor effective on HIV-1 replication. *Phytomedicine*, *23*(12), pp. 1383–1391.

Ho, T.Y., Wu, S.L., Chen, J.C., Li, C.C. and Hsiang, C.Y., 2007. Emodin blocks the SARS coronavirus spike protein and angiotensin-converting enzyme 2 interaction. *Antiviral Research*, *74*(2), pp. 92–101.

Luo, W., Su, X., Gong, S., Qin, Y., Liu, W., Li, J., Yu, H. and Xu, Q., 2009. Anti-SARS coronavirus 3C-like protease effects of *Rheum palmatum* L. extracts. *Bioscience Trends*, *3*(4), pp. 124–126.

Puri, S.C., Verma, V., Amna, T., Qazi, G.N. and Spiteller, M., 2005. An endophytic fungus from *Nothapodytes foetida* that produces Camptothecin. *Journal of Natural Products*, *68*(12), pp. 1717–1719.

Wang, S.X., Wang, Y., Lu, Y.B., Li, J.Y., Song, Y.J., Nyamgerelt, M. and Wang, X.X., 2020. Diagnosis and treatment of novel coronavirus pneumonia based on the theory of traditional Chinese medicine. *Journal of Integrative Medicine*, *18*(4), pp. 275–283.

Xiong, H.R., Luo, J., Hou, W., Xiao, H. and Yang, Z.Q., 2011. The effect of emodin, an anthraquinone derivative extracted from the roots of *Rheum tanguticum*, against herpes simplex virus *in vitro* and *in vivo*. *Journal of Ethnopharmacology*, *133*(2), pp. 718–723.

Xu, H.X., Wan, M., Loh, B.N., Kon, O.L., Chow, P.W. and Sim, K.Y., 1996. Screening of traditional medicines for their inhibitory activity against HIV-1 Protease. *Phytotherapy Research*, *10*(3), pp. 207–210.

Yan, X., Gu, S., Shi, Y., Cui, X., Wen, S. and Ge, J., 2017. The effect of emodin on *Staphylococcus aureus* strains in planktonic form and biofilm formation *in vitro*. *Archives of Microbiology*, *199*(9), pp. 1267–1275.

You, X., Feng, S., Luo, S., Cong, D., Yu, Z., Yang, Z. and Zhang, J., 2013. Studies on a rhein-producing endophytic fungus isolated from *Rheum palmatum* L. *Fitoterapia*, *85*, pp. 161–168.

8.7.12.19 *Rheum ribes* L.

Common names: Warted-leaved rhubarb, Persian rhubarb; gul-i-livas, rivas, tukhm-i-livas (Iran); Rawash. (Pakistan); Isgin, uskun (Turkey)

Distribution: Turkey, Iran, Afghanistan, Pakistan

Habitat: Rocky slopes

Botanical observation: It is a herb that grows to about 1 m tall from a massive rhizome. The stems are terete and warty. The leaves are simple, spiral, subsessile, and stipulate. The stipules are united to form a massive ocrea. The petiole is up to 30 cm long, reddish, stout, and striated. The blade is orbicular, 30–50 ×25–40, warty below, somewhat coriaceous, wavy, cordate at base, and obtuse at apex. The inflorescences are panicles. The perianth presents 6 spathulate tepals, the lobes about 35 mm in length and purplish-red. Nine stamens are present. The ovary develops 3 styles with inflated stigmas. The achenes are oblong, up to about 1.5 cm long, reddish, and winged.

Medicinal uses: Gonorrhea, jaundice, urinary infection (Iran); ulcers, diarrhea (Turkey)

Moderate antibacterial (Gram-negative) polar extract: Hydroalcoholic extract inhibited *Acinetobacter baumanii* (ATCC 747) with the MIC/MBC values of 128/256 μg/mL (Bagheri et al., 2019).

Weak antibacterial (Gram-positive) anthraquinones: Aloe-emodin and emodin from the Rhizome inhibited *Staphylococcus aureus* with the MIC values of 250 and 63 μg/mL (Alaadin et al., 2007). Emodin inhibited the growth of *Streptococcus pneumoniae*, *Bacillus subtilis*, *Bacillus cereus*, *Enterobacter aerogene* with the MIC/MBC of 128/128, 128/256, 8/8, and 128/256 μg/mL (Dey et al., 2014).

REFERENCES

Alaadin, A.M., Al-Khateeb, E.H. and Jäger, A.K., 2007. Antibacterial activity of the Iraqi *Rheum ribes*. Root. *Pharmaceutical Biology*, *45*(9), pp. 688–690.

Bagheri, N., Safaei, N., Aleebrahim-Dehkordy, E., Khaledi, M., Madmoli, M. and Ansaripour, S., 2019. *In vitro* antibacterial activity of *Bunium persicum* and *Rheum ribes* on *Acinetobacter baumanii*. *International Journal of Ayurvedic Medicine*, *10*(1), pp. 47–51.

Dey, D., Ray, R. and Hazra, B., 2014. Antitubercular and antibacterial activity of quinonoid natural products against multi-drug resistant clinical isolates. *Phytotherapy Research*, *28*(7), pp. 1014–1021.

8.7.12.20 *Rumex acetosa* L.

Synonyms: *Acetosa fontano-paludosa* Holub; *Acetosa pratensis* Mill.; *Rumex fontanopaludosus* Kalela

Common names: Common sorrel; chukapalam (Bangladesh); suan mo (China); bardana, chutrika, khatta palak (India); Khatimal (Pakistan)

Habitat: Grassy roadsides, forests margins, mountain slopes

Distribution: Turkey, Iran, Kazakhstan, Kyrgyzstan, Pakistan, India, China, Mongolia, Korea, Japan

Botanical observation: This herb grows to about 1 m tall. The stems are terete, reddish, striated, and yield a sour taste. The leaves are simple, edible (India), spiral, edible, somewhat fleshy, and stipulate. The stipules are united to form a membranaceous and whitish ocrea. The petiole is up to 15 cm long, reddish, edible, stout, and striated. The blade is hastate, somewhat waffly (like most members in the genus *Rumex* L.), ovate to lanceolate, 2–3.5 × 5.5, wavy, and acute at apex. The inflorescences are panicles. The 6 tepals are persistant in fruits, the lobes about 3 mm in length and reddish. Six stamens are present. The ovary is ovoid and develops 3 styles. The achenes are dark brown, trigonous, smooth, and minute.

Medicinal uses: Bronchitis, diarrhea, fever (India); jaundice (Pakistan); skin infection (China)

Broad-spectrum antibacterial halo developed by mid-polar extract: Chloroform extract of roots (500 μg/ 6 mm disc) inhibited the growth of *Staphylococcus epidermidis* (ATCC1228), *Staphylococcus aureus* (ATCC 2921), methicillin-resistant *Staphylococcus aureus* (ATCC 43300), *Bacillus subtilis* (ATCC 6633), *Moraxella catarrhalis* (ATCC 43617), *Streptococcus pyogenes* (ATCC 19615), and *Streptococcus pneumoniae* (ATCC 49619) with inhibition zone diameter of 18.5, 16, 12.4, 12.2, 10.2, 8, and 8.5 mm, respectively (Orbán-Gyapai et al., 2017).

Strong antifungal (dimorphic) polar extract: Ethanol extract of aerial part inhibited the growth of *Paracoccidioides brasiliensis* (ATCC MYA826) with a MIC value of 60 μg/mL (Johann et al., 2010).

Moderate anticandidal polar extract: Ethanol extract inhibited the growth of *Candida albicans* (ATCC 2091), *Candida albicans* (ATCC 10231), and *Candida parapsilosis* (ATCC 22019) with the MIC values of 500 μg/mL (Johann et al., 2010).

Antibacterial (Gram-negative) halo developed by flavanol: Gallocatechin (50 μg/disc) inhibited the growth of *Moraxella catarrhalis* with inhibition zone diameter of 11 mm (Pretorius et al., 2003).

Strong antiviral (enveloped segmented linear single-stranded (−) RNA) amphiphilic flavanol: Epigallocatechin-3-*O*-gallate (Log D 1.6 at pH 7.4; molecular mass=458.3 g/mol) abrogated the replication of the Influenza A virus (H1N1) at the concentration of about 1 μM and inhibited the attachment of the virus to Madin-Darby canine cells (Derksen et al., 2014).

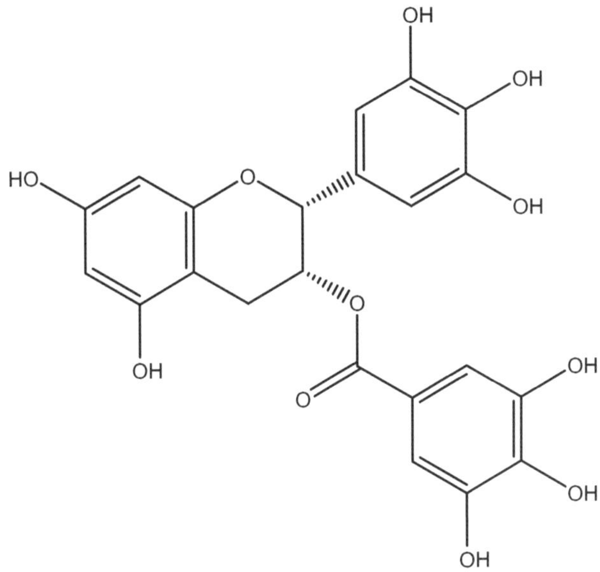

Epigallocatechin-3-*O*-gallate

Strong antiviral (enveloped segmented linear single-stranded (−) RNA) condensed tannins: Procyanidin B2-di-gallate, procyanidin B5, procyanidin C1, and procyanidin D1 inhibited the Influenza A virus (H1N1) with EC_{50} of ≤15, ≤163, 123, and ≤225 μM, respectively (Derksen et al., 2014). Procyanidin B2-di-gallate inhibited the attachment of the virus to Madin-Darby canine kidney cells (Derksen et al., 2014).

Strong antiviral (enveloped segmented linear single-stranded (−) RNA) hydrolyzable tannin: Pentagalloyl-glucose inhibited the Influenza A virus (H1N1) with EC_{50}≤22 μM (Derksen et al., 2014). Epigallocatechin-3-*O*-gallate and procyanidin B2-di-gallate inhibited the attachment of the virus to Madin-Darby canine kidney cells (Derksen et al., 2014).

Commentaries

i. The plant would be worth being investigated for anti-Human immunodeficiency virus properties as an ethanol of a member of the genus *Rumex* L. inhibited Human immunodeficiency virus type-1 (strain IIIB) with EC_{50} of 17.6 μg/mL and CC_{50} of 200 μg/mL in MT-4 cells (Cos et al., 2002), and this could be due to proanthocyanidins (Bicker et al., 2009).
ii. This plant is a typical example of phenolic-rich plant or "pan-antimicrobial plant". By opposition some plants elaborate a single or few compounds with either a strong antibacterial, or antifungal, or antiviral activity that could be called "narrow-spectrum antimicrobial plants".
iii. In Brittany (Morbihan, France) kids eat the stems in summer for the acidulate taste.
iv. This plant could be developed as phytomedication against Influenza. In this volume, natural products with strong activity against Influenza are mainly flavonoids or flavonoid glycosides.
v. The plant was known of our peer Dioscorides (See earlier).

REFERENCES

Bicker, J., Petereit, F. and Hensel, A., 2009. Proanthocyanidins and a phloroglucinol derivative from *Rumex acetosa* L. *Fitoterapia*, *80*(8), pp. 483–495.

Cos, P., Hermans, N., De Bruyne, T., Apers, S., Sindambiwe, J.B., Witvrouw, M., De Clercq, E., Berghe, D.V., Pieters, L. and Vlietinck, A.J., 2002. Antiviral activity of Rwandan medicinal plants against human immunodeficiency virus type-1 (HIV-1). *Phytomedicine*, *9*(1), pp. 62–68.

Derksen, A., Hensel, A., Hafezi, W., Herrmann, F., Schmidt, T.J., Ehrhardt, C., Ludwig, S. and Kühn, J., 2014. 3-*O*-galloylated procyanidins from *Rumex acetosa* L. inhibit the attachment of influenza A virus. *PloS One*, *9*(10), p.e110089.

Johann, S., Cisalpino, P.S., Watanabe, G.A., Cota, B.B., de Siqueira, E.P., Pizzolatti, M.G., Zani, C.L. and de Resende, M.A., 2010. Antifungal activity of extracts of some plants used in Brazilian traditional medicine against the pathogenic fungus *Paracoccidioides brasiliensis*. *Pharmaceutical Biology*, *48*(4), pp. 388–396.

Orbán-Gyapai, O., Liktor-Busa, E., Kúsz, N., Stefkó, D., Urbán, E., Hohmann, J. and Vasas, A., 2017. Antibacterial screening of Rumex species native to the Carpathian Basin and bioactivity-guided isolation of compounds from *Rumex aquaticus*. *Fitoterapia*, *118*, pp. 101–106.

Pretorius, J.C., Magama, S., Zietsman, P.C. and Jäger, A.K., 2003. Purification and identification of antibacterial compounds from *Euclea crispa* subsp. crispa (Ebenaceae) leaves. *South African Journal of Botany*, *69*(4), pp. 579–586.

8.7.12.21 *Rumex aquaticus* L.

Synonym: *Rumex protractus* Rech. f.

Common names: Water dock; shui sheng suan mo (China)

Habitat: Aquatic habitats, river, stream, ponds, lakes

Distribution: Kazakhstan, Kyrgyzstan, China, Korea, Japan

Botanical observation: This robust erect herb grows to about 2 m tall. The stems are terete, somewhat reddish, longitudinally striated. The leaves are simple, spiral, and stipulate. The stipules are united into a membranaceous and ephemeral ocrea. The petiole is up to 25 cm long. The blade is broadly lanceolate, somewhat fleshy, 10–30 × 4.5–15, cordate to round at base, wavy, and obtuse to acute at apex. The panicles are terminal or axillary, many flowered and about 30 cm long. The 6 tepals are persistant in fruits and about 6 mm in length. Six stamens are present. The ovary is ovoid and develops 3 styles. The achenes are trigonous, ellipsoid, about 4 mm long, and with reddish wings

Medicinal use: Skin infection (Kazakhstan, Kyrgyzstan)

Broad-spectrum antibacterial halo developed by non-polar extract: Hexane extract of aerials (500 µg/6 mm disc) inhibited the growth of *Staphylococcus epidermidis* (ATCC1228), *Staphylococcus aureus* (ATCC 2921), methicillin-resistant *Staphylococcus aureus* (ATCC 43300), *Bacillus subtilis* (ATCC 6633), *Moraxella catarrhalis* (ATCC 43617), and *Streptococcus pneumoniae* (ATCC 49619) with inhibition zone diameters of 12.3, 18.7, 9, 10.3, 12, and 10.5 mm, respectively (Orbán-Gyapai et al., 2017).

Weak broad-spectrum naphthalene: Nepodin (also known as musizin) isolated from the roots inhibited the growth of *Moraxella catarhalis* (ATCC 43617), *Staphylococcus epidermidis* (ATCC1228), *Staphylococcus aureus* (ATCC 2921), *Bacillus subtilis* (ATCC 6633), and methicillin-resistant *Staphylococcus aureus* with the MIC value of 57.8, 231.5, 231.5, 231.5, and 463 5 µM, respectively (Orbán-Gyapai et al., 2017).

Moderate broad-spectrum naphthoquinone: 2-Methoxystypandrone isolated from the roots inhibited the growth of *Moraxella catarhalis* (ATCC 43617), *Staphylococcus aureus* (ATCC 2921), *Bacillus subtilis* (ATCC 6633), and *Staphylococcus epidermidis* (ATCC 1228) with the MIC values of 48, 96, 96, and 192.3 µM (Orbán-Gyapai et al., 2017).

Moderate antiviral (enveloped monopartite linear single-stranded (−) RNA) anthraquinone: Emodin inhibited the replication of the Respiratory syncytial virus in Hep-2 cells with the EC_{50} value of 14.2 µM and a selectivity index of 5.3 (Liu et al., 2015).

Strong antiviral (non-enveloped monopartite linear single-stranded (+) RNA) anthraquinone: Emodin inhibited the replication of the Coxsackie B5 virus in Hep-2 cells with the EC_{50} value of 13 μM and a selectivity index of 5.5 (Liu et al., 2015).

REFERENCES

Liu, Z., Ma, N., Zhong, Y. and Yang, Z.Q., 2015. Antiviral effect of emodin from rheum palmatum against coxsakievirus b5 and human respiratory syncytial virus *in vitro*. *Journal of Huazhong University of Science and Technology [Medical Sciences]*, *35*(6), pp. 916–922.

Orbán-Gyapai, O., Liktor-Busa, E., Kúsz, N., Stefkó, D., Urbán, E., Hohmann, J. and Vasas, A., 2017. Antibacterial screening of Rumex species native to the Carpathian Basin and bioactivity-guided isolation of compounds from *Rumex aquaticus*. *Fitoterapia*, *118*, pp. 101–106.

8.7.12.22 *Rumex crispus* L.

Synonyms: *Lapathum crispum* (L.) Scop.; *Rumex hydrolapathum* Huds.; *Rumex magellanicus* Campd.; *Rumex patientia* Gay

Common names: Curled dock; bon palong (Bangladesh) zhou ye suan mo (China); ambavah, amrule, lal bibi, tandalak (India); naazdolat (Iran); sorijaeng (Korea); sag (Pakistan); evelik, kivircik labada, kuzu kulagi (Turkey)

Habitat: Rivers, streams, ponds, lakes, wastelands, grassy lands

Distribution: Turkey, Iran, Afghanistan, India, Kazakhstan, Kyrgyzstan, Myanmar, Thailand, China, Mongolia, Korea, Japan

Botanical observation: This herb grows to about 1.5 m tall. The stems are terete, somewhat reddish, and longitudinally striated. The leaves are simple, spiral, toxic, and stipulate. The stipules are united to form a membranaceous and ephemeral ocrea. The petiole is up to about 20 cm long. The blade is lanceolate, somewhat fleshy, 10–25 × 2–5, cuneate at base, curled at margin, and acute at apex. The inflorescences are terminal or axillary many flowered panicles. The 6 tepals are yellowish and about 5 mm in length. Six stamens are present. The ovary is ovoid and develops 3 styles. The achenes are dark brown, glossy, trigonous, about 2 mm long.

Medicinal uses: Cuts, wounds (India); skin diseases (Turkey, Pakistan); postpartum (Korea); skin infections (Bangladesh; Pakistan)

Broad-spectrum antibacterial halo developed by polar extracts: Acetone extract of roots (12.5 mg/mL, 0.1 mL in a 12 mm diameter well) inhibited the growth of *Staphylococcus aureus* (Cowan 1), *Micrococcus luteus* (LA 2971), *Mycobacterium smegmatis* (CCM 2067), *Bacillus subtilis* (IGM 22), *Bacillus subtilis* var. niger (ATCC 10), *Aeromonas hydrophila* (ATCC 7966), *Klebsiella pneumoniae* (FMC 5), and *Pseudomonas aeruginosa* (ATCC 27853) with the inhibition zone diameters of 23, 22, 21, 24, 24, 18, 20, and 16 mm, respectively (Ulukanli et al., 2005). Aqueous extract (500 μg/6 mm disc) inhibited the growth of *Staphylococcus epidermidis* (ATCC1228), *Staphylococcus aureus* (ATCC 2921), methicillin-resistant *Staphylococcus aureus* (ATCC 43300), *Bacillus subtilis* (ATCC 6633), and *Moraxella catarrhalis* (ATCC 43617) with inhibition zone diameters of 15.5, 12.4, 12, 10.2, and 10.6 mm, respectively (Orbán-Gyapai et al., 2017).

Antimycobacterial halo developed by polar extract: Acetone extract of roots (12.5 mg/mL, 0.1 mL in a 12 mm diameter well) inhibited the growth of *Mycobacterium smegmatis* (CCM 2067) with the inhibition zone diameter of 21 mm (Ulukanli et al., 2005).

Anticandidal halo developed by polar extract: Acetone extract of roots (12.5 mg/mL, 0.1 mL in a 12 mm diameter well) inhibited the growth of *Candida albicans* (ATCC 10231) with inhibition zone diameter of 17 mm (Ulukanli et al., 2005).

Commentaries

i. When the word "edible" is used in this series it does not mean that the plants are safe for food and likewise, "medicinal uses" does not claim any clinical activity but relates to the actual use in Asia and the Pacific.

ii. In the references consulted some authors write that *Rumex crispus* L. is used as vegetable whereas reports of fatal poisoning are available (Reig et al., 1990). In this context it is worth reminding our infaillible accreditation boards from the West of Europe coming to Asian countries and labelling pharmacognosy as an obscure topic (to serve the interest of the lobbies that finance them) that pharmacy students need robust teaching in the toxicity of medicinal plants (an mushrooms) to save lifes.

REFERENCES

Orbán-Gyapai, O., Liktor-Busa, E., Kúsz, N., Stefkó, D., Urbán, E., Hohmann, J. and Vasas, A., 2017. Antibacterial screening of Rumex species native to the Carpathian Basin and bioactivity-guided isolation of compounds from *Rumex aquaticus*. *Fitoterapia*, *118*, pp. 101–106.

Reig, R., Sanz, P., Blanche, C., Fontarnau, R., Dominguez, A. and Corbella, J., 1990. Fatal poisoning by *Rumex crispus* (curled dock): pathological findings and application of scanning electron microscopy. *Veterinary and Human Toxicology*, *32*(5), pp. 468–470.

Ulukanli, Z., Ulukanli, S., Ozbay, H., Ilcim, A. and Tuzcu, M., 2005. Antimicrobial activities of some plants from the Eastern Anatolia region of Turkey. *Pharmaceutical Biology*, *43*(4), pp. 334–339.

8.7.12.23 *Rumex dentatus* L.

Synonyms: *Rumex halacsyi* Rech. f.; *Rumex klotzschianus* Meisn.; *Rumex nipponicus* Franch. & Sav.

Common names: Toothed dock; chi guo suan mo (China); ambavah (India); chora chitra, sarkari palak, hola, abuj (Pakistan)

Habitat: Watery grassy slopes, valleys

Distribution: Afghanistan, Kazakhstan, Kyrgyzstan, Pakistan, India, Nepal, China, Taiwan

Botanical observation: This herb grows to about 70 cm tall. The stems are terete, somewhat reddish, and longitudinally striated. The leaves are simple, spiral, and stipulate. The stipules are united to form a membranaceous and ephemeral ochrea. The petiole is about 5 cm long. The blade is narrowly elliptic to oblong, 4–12 × 1.5–3 cm, subcordate, or truncate at base, constricted near the base, undulate at margin, and acute at apex. The inflorescences are many-flowered whorls at nodes (looking like verticils of Lamiaceae). The 6 tepals are yellowish, dentate, persistant in fruits, the lobes about 5 mm in length and toothed. Six stamens are present. The ovary is ovoid and develops 3 styles. The achenes are dark yellowish-brown, glossy, trigonous, and about 2.5 mm long.

Medicinal uses: Bronchitis, toothache (Pakistan); cuts, wounds (India)

Antibacterial (Gram-positive) halo developed by polar extract: Methanol extract (100 µg/disc) inhibited the growth of *Staphylococcus aureus* (ATCC 6538), *Bacillus bronchiseptica* (ATCC 6633), *Micrococcus luteus* (ATCC 10240) with inhibition zone diameters of 8, 7, and 9 mm, respectively (Khan et al., 2017).

Moderate antibacterial (Gram-negative) polar extract: Methanol extract inhibited the growth of *Shigella flexneri*, *Klebsiellae pneumonia*, and *Escherichia coli* with the MIC/MBC values of 225/425, 250/725, and 200/500 µL/mL, respectively (Humeera et al., 2013).

REFERENCES

Humeera, N., Kamili, A.N., Bandh, S.A., Lone, B.A. and Gousia, N., 2013. Antimicrobial and antioxidant activities of alcoholic extracts of *Rumex dentatus* L. *Microbial Pathogenesis*, *57*, pp. 17–20.

Khan, S., Ur-Rehman, T., Mirza, B., Ul-Haq, I. and Zia, M., 2017. Antioxidant, antimicrobial, cytotoxic and protein kinase inhibition activities of fifteen traditional medicinal plants from Pakistan. *Pharmaceutical Chemistry Journal*, *51*(5), pp. 391–398.

8.7.12.24 *Rumex japonicus* Houtt.

Synonyms: *Rumex cardiocarpus* Pamp.; *Rumex hadroocarpus* Rech. f.; *Rumex japonicus* Meisn.; *Rumex japonicus* Meisn.

Common names: Japanese dock; yang ti (China)

Habitat: Watery grassy slopes, field margins, valleys, riverbanks

Distribution: China, Mongolia, Korea, Japan, Taiwan

Botanical observation: This herb grows to about 1m tall. The stems are terete, somewhat reddish and longitudinally striated. The leaves are simple, spiral, and stipulate. The stipules are united to form a membranaceous and ephemeral ocrea. The petiole is up to about 15cm long and chanelled. The blade is lanceolate to oblong, 8–25×3–8cm, cordate or cuneate at base, dull green, marked with about 15 pairs of secondary nerves, wavy, and acute or obtuse at apex. The inflorescences are panicles. The 6 tepals are persistant in fruits, yellowish, and 6mm in length. Six stamens are present. The ovary is ovoid and develops 3 styles. The achenes are dark brown, glossy, trigonous, winged, and about 2.5mm long.

Medicinal use: Jaundice (China)

Broad-spectrum antibacterial halo developed by mid-polar extract: Ethylacetate extract (800 µg/disc) inhibited the growth of *Bacillus subtilis*, *Bacillus cereus*, and *Escherichia coli* with inhibition zone diameter of 15, 17, and 20mm, respectively (Elzaawely et al., 2005).

Strong antibacterial (Gram-positive) amphiphilic naphthoquinone: 2-Methoxystypandrone (Log D = 1.5 at pH 7.4; molecular mass=260.2 g/mol) from the roots inhibited the growth of *Staphylococcus aureus*, *Bacillus subtilis*, *Sarcina lutea*, and *Escherichia coli* with the MIC values of 40, <10, 40, and 1000ppm, respectively (Nishina et al., 1993).

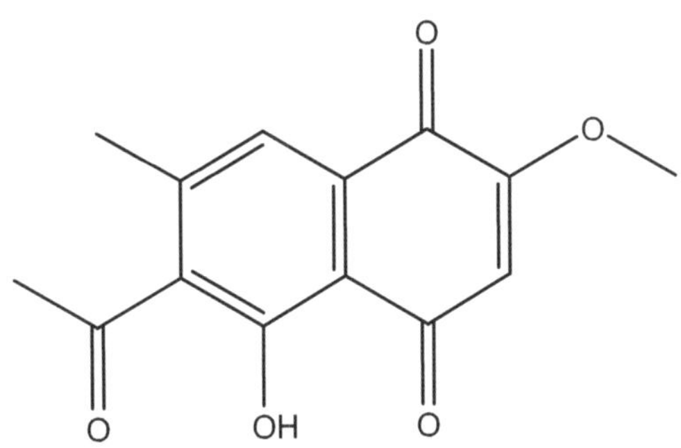

2-Methoxystypandrone

Moderate broad-spectrum antibacterial amphiphilic naphthalene: Torachrysone (also known as trachysone) inhibited the growth of *Staphylococcus aureus*, *Bacillus subitlis*, *Sarcina lutea*, *Escherichia coli*, and *Salmonella typhimurium* with the MIC values of 100, 100, 200, 100, and 100ppm respectively (Nishina et al., 1993).

Strong antifungal (yeast) amphiphilic naphthoquinone: 2-Methoxystypandrone inhibited the growth of *Saccharomyces cerevisae* and *Aspergillus niger* with the MIC value of 40 and 1000ppm (Nishina et al., 1993).

Strong broad-spectrum antifungal hydrophilic naphthalene: Torachrysone (Log D = 1 at pH 7.4; molecular mass=246.2 g/mol) inhibited the growth of *Saccharomyces cerevisae* and *Aspergillus niger* with the MIC values of 40 and 1000ppm (Nishina et al., 1993).

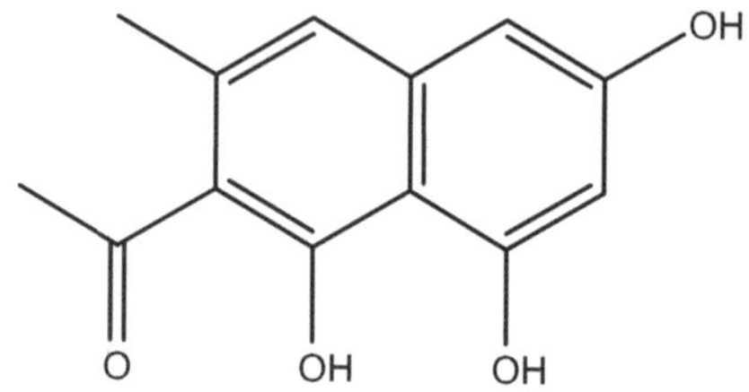

Trachrysone

Strong anticandidal hydrophilic naphthalene: Nepodin (also known as musizin; Log D = 0.9 at pH 7.4; molecular mass = 216.2 g/mol) isolated at 0.5 μg/mL abrogated the formation of biofilm by *Candida albicans* (DAY185) (Lee et al., 2005).

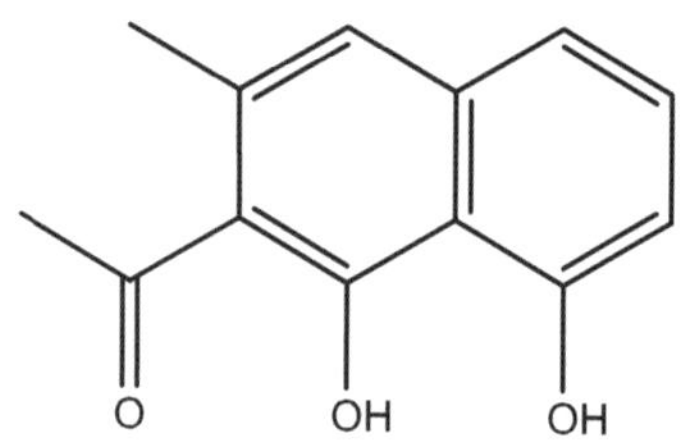

Nepodin

REFERENCES

Elzaawely, A.A., Xuan, T.D. and Tawata, S., 2005. Antioxidant and antibacterial activities of *Rumex japonicus* HOUTT. Aerial parts. *Biological and Pharmaceutical Bulletin, 28*(12), pp. 2225–2230.

Lee, J.H., Kim, Y.G., Khadke, S.K., Yamano, A., Watanabe, A. and Lee, J., 2019. Inhibition of biofilm formation by *Candida albicans* and *Polymicrobial Microorganisms* by Nepodin via hyphal-growth suppression. *ACS Infectious Diseases, 5*(7), pp. 1177–1187.

Nishina, A., Kubota, K. and Osawa, T., 1993. Antimicrobial components, trachrysone and 2-methoxystypandrone, in *Rumex japonicus* Houtt. *Journal of Agricultural and Food Chemistry, 41*(10), pp. 1772–1775.

8.7.12.25 *Rumex nepalensis* Spreng.

Synonyms: *Rumex bequaertii* De Wild.; *Rumex camptodon* Rech. f.; *Rumex quarrei* De Wild.

Common names: Nepal dock; pahari palang (Bangladesh); ni bo er suan mo (China); amlya, amlavetasa, sukkankeerai, jungli palak (India); ah dang, vawk ei an (Myanmar); ban haldi, hala hale, halhale sag, halya, halye (Nepal); shalkhay, hoola (Pakistan)

Habitat: Watery grassy slopes, field margins, valleys, riverbanks, ditches

Distribution: Turkey, Iran, Afghanistan, Pakistan, Tajikistan, Himalayas, India, Nepal, Sikkim, Bangladesh, Bhutan, Myanmar, Vietnam, China, Indonesia

Botanical observation: This herb grows to about 1.5 m tall from large roots. The stems are terete and longitudinally striated. The leaves are simple, spiral, and stipulate. The stipules are united to form a membranaceous and ephemeral ocrea. The petiole is about 10 cm long. The blade is obovate, lanceolate, or oblong, 9.5–15 × 4–8 cm, cordate at base, marked with about 6–10 pairs of secondary nerves, somewhat wavy, and obtuse at apex. The inflorescences are axillary fascicles. The 6 tepals are persistant in fruits, about 6 mm in length and toothed at margin. The androecium comprises 6 stamens. The ovary is ovoid and develops 3 styles. The achenes are dark brown, glossy, trigonous, and 3 mm long.

Medicinal uses: Cuts, diarrhea, jaundice (Myanmar); syphilis (India); hepatitis (Pakistan); wounds (India; Myanmar)

Broad-spectrum antibacterial halo developed by polar extracts: Methanol extract of roots at (1 mg/disc) inhibited the growth of *Bacillus subtilis*, *Staphylococcus aureus*, *Escherichia coli*, *Vibrio cholerae*, and *Shigella dysenteriae* with zones of inhibition of 10, 10.5, 10, 12.5, and 31.5 mm, respectively (Ghosh et al., 2003). Acetone extract of leaves (impregnated 6 mm disc with a 200 mg/mL soluttion) inhibited the growth of *Escherichia coli* (ATCC 25922), *Staphylococcus aureus* (ATCC 25923), and *Salmonella typhimurium* (ATCC 13311) and with the inhibition zone diameter of 10.3, 12, and 20.7 mm, respectively (Tura et al., 2003).

Strong antimycobacterial hydrophilic naphthalene: Torachrysone (Log D = 1 at pH 7.4; molecular mass = 246.2 g/mol) inhibited the growth of *Mycobacterium tuberculosis* with the MIC value of 6.1 μM (Liang et al., 2010).

Strong antimycobacterial naphthalene glycoside: Torachrysone-8-O-β-D-glucopyranoside inhibited the growth of *Mycobacterium tuberculosis* with the MIC valuesof 8.9 μM (Liang et al., 2010).

Strong antimycobacterial hydrophilic chromone: Aloesin (Log D −0.08 at pH 7.4; molecular mass = 394.3 g/mol) inhibited the growth of *Mycobacterium tuberculosis* with the MIC value of 2.8 μM (Liang et al., 2010).

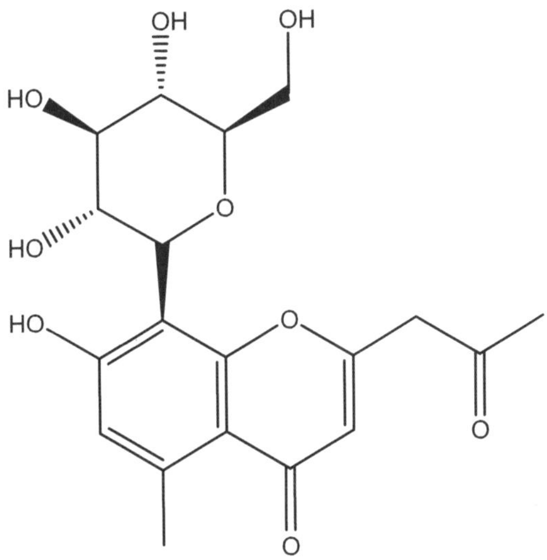

Aloesin

Strong antimycobacterial anthraquinone glycoside: Chrysophanol-8-*O*-β-D-(6′-O-acetyl) glucopyranoside inhibited the growth of *Mycobacterium tuberculosis* with the MIC value of 4.1 μM (Liang et al., 2010).

Commentary: The plant produces emodin, chrysophanol and physcion (Deng et al., 2016). Note that some members of the family Polygonaceae yield antimycobacterial phytosterols. 5α, 8α-Epidioxyergost-6,22-dien-3β-ol, 5α, 8α-epidioxystigmasta-6,22-dien-3β-ol, and stigmast-4-en-6β-ol-3-one which inhibited the growth of *Mycobacterium tuberculosis* with MIC values below 5 μg/mL (Woldemichael et al., 2003).

REFERENCES

Deng, L., Li, B., Wang, G., Zhang, J., Ge, J., Wang, H., Liao, Z. and Chen, M., 2016. Chemical constituents from roots of *Rumex nepalensis*. *Chinese Traditional and Herbal Drugs*, *47*(12), pp. 2095–2099.

Ghosh, L., Gayen, J.R., Sinha, S., Pal, S., Pal, M. and Saha, B.P., 2003. Antibacterial efficacy of *Rumex nepalensis* Spreng. roots. *Phytotherapy Research*, *17*(5), pp. 558–559.
Liang, H.X., Dai, H.Q., Fu, H.A., Dong, X.P., Adebayo, A.H., Zhang, L.X. and Cheng, Y.X., 2010. Bioactive compounds from Rumex plants. *Phytochemistry Letters*, *3*(4), pp. 181–184.
Tura, G.T., Eshete, W.B. and Tucho, G.T., 2017. Antibacterial efficacy of local plants and their contribution to public health in rural Ethiopia. *Antimicrobial Resistance & Infection Control*, *6*(1), p. 76.
Woldemichael, G.M., Franzblau, S.G., Zhang, F., Wang, Y. and Timmermann, B.N., 2003. Inhibitory effect of sterols from *Ruprechtia triflora* and diterpenes from *Calceolaria pinnifolia* on the growth of *Mycobacterium tuberculosis*. *Planta medica*, *69*(7), pp. 628–631.

8.7.12.26 *Rumex patientia* L.

Synonyms: *Lapathum hortense* Lam.; *Rumex callosus* (F. Schmidt ex Maxim.) Rech. f.; *Rumex interruptus* Rech. f.; *Rumex lonaczevskii* Klokov

Common names: Patience dock, monk's rhubarb; ba tian suan mo (China); shoma (India); at eveligi, evelik, labada (Turkey)

Habitat: Watery grassy slopes, field margins, valleys, ditches, riverbanks

Distribution: Turkey, Pakistan, Afghanistan, Kazakhstan, Kyrgyzstan, Tajikistan, India, China, Mongolia

Botanical observation: This herb grows to about 1.5 m tall from massive roots. The stems are terete and longitudinally striated. The leaves are simple, spiral, palatable, and stipulate. The stipules are united to form a membranaceous, up to 4 cm long, and ephemeral ocrea. The petiole is up to about 15 cm long. The blade is oblong lanceolate, 5–30 × 5–10 cm, round, cuneate or subcordate at base, with about 6–10 pairs of unconspicuous secondary nerves, somewhat wavy, and acute at apex. The inflorescences are terminal or axillary densely flowered elongated panicles. The 6 tepals are persistant in fruits, the lobes up to about 6 mm in length and toothed at margin. Six stamens are present. The ovary is ovoid and develops 3 styles. The achenes are dark brown, glossy, trigonous, about 3 mm long and with enlarged, cordate, somewhat reddish-green and reticulated persistant tepals.

Medicinal use: Wounds (Turkey)

Broad-spectrum antibacterial halo developed by polar extract: Aqueous fraction of flowers and fruits (500 µg/disc) inhibited the growth of *Staphylococcus epidermidis*, *Staphylococcus aureus*, methicillin-resistant *Staphylococcus aureus*, *Bacillus subtilis*, and *Moraxella catarrhalis* with inhibition zone diameter of 14.3, 12.2, 12.6, 11, and 10.6 mm, respectively (Orbán-Gyapai et al., 2017).

Moderate broad-spectrum antibacterial polar extract: Methanol extract of stems inhibited the growth of *Staphylococcus aureus* (ATCC 6538), *Proteus vulgaris* (NRRL B-123), *Salmonella typhimurium* (ATCC 13311), *Staphylococcus epidermidis* (ATCC 12228), *Bacillus subtilis* (NRRL B-4378), *Pseudomonas citronellosis* (NRRL B-2504), *Bacillus velezensis* (NRRL B-14580), *Gordonia rubripertincta* (NRRL B-3906), and *Escherichia coli* (ATCC 8739) with the MIC values of 500, 1000, 500, 1000, 1000, 500, 500, 1000, and 500 µg/mL, respectively (Ceylan et al., 2019).

Moderate broad-spectrum antifungal polar extract: Methanol extract of stems inhibited the growth of *Streptomyces griseolus* (NRRL B-1062), *Candida albicans* (ATCC 90028), *Candida glabrata* (ATCC 2001), and *Candida krusei* (ATCC 6258) with the MIC values of 500, 125, 125, and 125 µg/mL, respectively (Ceylan et al., 2019).

Commentary: The plant produces emodin-6-*O*-β-D-glucopyranoside, chrysophanol, physcion, emodin, chrysophanol-8-*O*-β-D-glucopyranoside, emodin-8-*O*-β-D-glucopyranoside, catechin, and orcinol (Demirezer et al., 2001a,b) as well as naphthalene glycosides of which nepodin-8-*O*-β-D-glucopyranoside and torachrysone-8-*O*-β-D-glucopyranoside (Demirezer et al., 2001a,b).

REFERENCES

Ceylan, S., Cetin, S., Camadan, Y., Saral, O., Ozsen, O. and Tutus, A., 2019. Antibacterial and antioxidant activities of traditional medicinal plants from the Erzurum region of Turkey. *Irish Journal of Medical Science (1971-)*, *188*, pp. 1303–1309.

Demirezer, L.Ö., Kuruüzüm-Uz, A., Bergere, I., Schiewe, H.J. and Zeeck, A., 2001a. The structures of antioxidant and cytotoxic agents from natural source: anthraquinones and tannins from roots of *Rumex patientia*. *Phytochemistry*, *58*(8), pp. 1213–1217.

Demirezer, L.Ö., Kuruüzüm-Uz, A., Bergere, I., Schiewe, H.J. and Zeeck, A., 2001b. Five naphthalene glycosides from the roots of *Rumex patientia*. *Phytochemistry*, *56*(4), pp. 399–402.

Orbán-Gyapai, O., Liktor-Busa, E., Kúsz, N., Stefkó, D., Urbán, E., Hohmann, J. and Vasas, A., 2017. Antibacterial screening of Rumex species native to the Carpathian Basin and bioactivity-guided isolation of compounds from *Rumex aquaticus*. *Fitoterapia*, *118*, pp. 101–106.

8.7.13 FAMILY PORTULACACEAE A. L. DE JUSSIEU (1789)

The family Portulacaceae consists of about 28 genera and 450 species of gracious and fleshy herbs or treelets. The leaves are simple, alternate or opposite, and exstipulate. The inflorescences are terminal or axillary cymes or panicles or solitary. The flowers are showy and variously colored. The calyx includes 2 sepals. The corolla includes 4–6 petals. The androrecium comprises 4 to many free stamens. The gynoecium consists of 2–8 carpels fused into an unilicoular ovary containing 2 or many ovules attached on a central placenta. The capsules are circumcissile capsules.

8.7.13.1 *Portulaca oleracea* L.

Synonyms: *Portulaca consanguinea* Schltdl.; *Portulaca intermedia* Link; *Portulaca marginata* Kunth; *Portulaca neglecta* Mack. & Bush; *Portulaca officinarum* Crantz; *Portulaca olitoria* Pall.; *Portulaca parvifolia* Haw.; *Portulaca pusilla* Kunth; *Portulaca retusa* Engelm.; *Portulaca sativa* Haw.; *Portulaca suffruticosa* Thwaites

Common names: Purslane; kulfi (Bangladesh); maông dâb phkaa (Cambodia); ma chi xian (China); badi loni (India); silobar pinggan (Indonesia); khorfeh (Iran); suberi-hiyu (Japan); rumput beremi (Malaysia); myet htauk (Myanmar); kulfa sag (Nepal); golasiman, ngalug (the Philippines); phakbia-yai (Thailand); rau sam (Vietnam)

Habitat: Disturbed lands, open fields, cultivates lands, drains, roadsides

Distribution: Subtropical and Tropical Asia

Botanical observation: This herb grows to about 25cm long and has a positive aura. The stems are fleshy and somewhat purplish. The leaves are simple, spiral or opposite, and exstipulate. The petiole is minute. The blade is fleshy, dull green, obovate, 1.5–8×3–25mm, attenuate at base, rounded at apex, and without visible secondary nerves. The inflorescences are axillary clusters of a few flowers. The 2 sepals are about 4mm long. The 5 petals are yellow, obovate, membranaceous, emarginate, yellow, and up to about 5mm long. The 7–12 stamens are up to about 1cm long. The ovary is minute and develops 5 stigmas. The capsules are ovoid, circumscissile, 8mm long, and contain numerous tiny glossy and black seeds.

Medicinal uses: Cough (Pakistan); cystitis, fever, jaundice, gonorrhoea (India); diarrhoea (Afghanistan; Pakistan) fever (Iran); intestinal and liver abscesses (Thailand); wounds (the Philippines)

Strong antifungal (filamentous) mid-polar extract: Ethylacetate extract at a contration of 100 µg/mL inhibited the growth of *Trichophyton mentagrophytes* (EV 861) and *Trichophyton mentagrophytes* (HIC 8185) by 64 and 70%, respectively (Oh et al., 2000).

Moderate broad-spectrum alkyl amine alkaloid glycosides: Portulacerebroside B inhibited *Escherichia coli*, *Staphylococcus aureus*, *Shigella flexneri*, and *Salmonella typhi* with the MIC/MBC values of 0.1/0.2, 0.1/0.5, 0.1/0.2, and 0.1/0.2mg/mL, respectively (Lei et al., 2015).

Portulacerebroside C inhibited the growth of *Escherichia coli*, *Staphylococcus aureus*, *Shigella flexneri*, and *Shigella typhi* with the MIC/MBC values of 0.1/0.2, 0.1/0.5, 0.1/0.2, and 0.1/0.2 mg/mL, respectively (Lei et al., 2015). Portulacerebroside D inhibited *Escherichia coli*, *Staphylococcus aureus*, *Shigella flexneri*, and *Salmonella typhi* with the MIC/MBC values of 0.1/0.2, 0.1/0.5, 0.1/0.2, and 0.1/0.2 mg/mL, respectively (Lei et al., 2015).

Weak broad-spectrum anti bacterial alkyl amine alkaloid: Portulaceramide A inhibited *Escherichia coli*, *Staphylococcus aureus*, *Shigella flexneri*, and *Salmonella typhi* with the MIC/MBC values of 0.3/0.5, 0.3/0.5, 0.1/0.2, and 0.3/0.5 mg/mL, respectively (Lei et al., 2015).

Antibiotic potentiator fatty acids: Linoleic acid and oleic acid at 32 and 64 µg/mL inhibited efflux pump in methicillin-resistant *Staphylococcus aureus* (Chan et al., 2015). Linoleic acid at 16 µg/mL reduced the MIC of erythromycin from 256 to 16 µg/mL, while oleic acid at 32 µg/mL could reduce the MIC of erythromycin from 256 to 32 µg/mL towards methicillin-resistant *Staphylococcus aureus* (Chan et al., 2015). Linoleic acid and oleic also evoked a synergistic effect with erythromycin against ATP-binding cassette efflux pump of methicillin-resistant *Staphylococcus aureus* (RN4220/pUL5054). (Fung et al., 2017).

Moderate antiviral (enveloped segmented linear single-stranded (−) RNA) polar extract: Aqueous extract inhibited the replication of the anti-influenza A/WSN/1933 (H1N1), A/California/07/2009 (H1N1), A/Perth/16/2009 (H3N2), and A/Brisbane/10/2007 (H3N2) with the EC_{50} value of 220.1, 121.6, 112.4, and 191.2 and selectivity indexes of 36.6, 66.3, 71.7, and 42.1 (Li et al., 2019).

Viral enzyme inhibition by polar extract: Aqueous extract of aerial parts at the concentration of 250 µg/mL inhibited the Human immunodeficiency virus type-1 protease (Xu et al., 1996). Methanol extract inhibited Hepatitis C virus NS3 serine protease (Noreen et al., 2015).

Commentaries

i. The plant produces unusual pyridinone alkaloids which should be studied for their possible antimicrobial effects. Consider that water is able to extract polysaccharides which are often antiviral.
ii. The plant was known of our peer the Persian physician Avicenna as *"Boghla-al-homgha"* and used for the treatment of uterine bleedings and is actually estrogenic (Mobli et al., 2015). Estrogenic substances tend to inhibit bacterial and fungal efflux pumps.

REFERENCES

Chan, B.C., Han, X.Q., Lui, S.L., Wong, C.W., Wang, T.B., Cheung, D.W., Cheng, S.W., Ip, M., Han, S.Q., Yang, X.S. and Jolivalt, C., 2015. Combating against methicillin-resistant *Staphylococcus aureus*–two fatty acids from Purslane (*Portulaca oleracea* L.) exhibit synergistic effects with erythromycin. *Journal of Pharmacy and Pharmacology*, *67*(1), pp. 107–116.

Fung, K.P., Han, Q.B., Ip, M., Yang, X.S., Lau, C.B.S. and Chan, B.C.L., 2017. Synergists from Portulaca oleracea with macrolides against methicillin-resistant *Staphylococcus aureus* and related mechanism. *Hong Kong Med J*, *23*(4 Supplement 5), pp. 38–42.

Lei, X., Li, J., Liu, B., Zhang, N. and Liu, H., 2015. Separation and identification of four new compounds with antibacterial activity from *Portulaca oleracea* L. *Molecules*, *20*(9), pp. 16375–16387.

Li, Y.H., Lai, C.Y., Su, M.C., Cheng, J.C. and Chang, Y.S., 2019. Antiviral activity of *Portulaca oleracea* L. against influenza A viruses. *Journal of Ethnopharmacology*, *241*, p. 112013.

Mobli, M., Qaraaty, M., Amin, G., Haririan, I., Hajimahmoodi, M. and Rahimi, R., 2015. Scientific evaluation of medicinal plants used for the treatment of abnormal uterine bleeding by Avicenna. *Archives of Gynecology and Obstetrics*, *292*(1), pp. 21–35.

Noreen, S., Hussain, I., Tariq, M.I., Ijaz, B., Iqbal, S., Ashfaq, U.A. and Husnain, T., 2015. *Portulaca oleracea* L. as a prospective candidate inhibitor of hepatitis C virus NS3 serine protease. *Viral Immunology*, *28*(5), pp. 282–289.

Oh, K.B., Chang, I.M., Hwang, K.J. and Mar, W., 2000. Detection of antifungal activity in Portulaca oleracea by a single-cell bioassay system. *Phytotherapy Research*, *14*(5), pp. 329–332.

Bibliography

Abe, R. and Ohtani, K., 2013. An ethnobotanical study of medicinal plants and traditional therapies on Batan Island, the Philippines. *Journal of Ethnopharmacology, 145*, pp. 554–565.

Abu Bakar, F.I., Abu Bakar, M.F., Abdullah, N., Endrini, S. and Rahmat, A., 2018. A review of Malaysian medicinal plants with potential anti-inflammatory activity. *Hindawi*, 13 pages, Article ID 8603602.

Acharya, K.P. and Acharya, M., 2010. Traditional knowledge on medicinal plants used for the treatment of livestock diseases in Sardikhola VDC, Kaski, Nepal. *Journal of Medicinal Plants Research, 4*(2), pp. 235–239.

Acharya, E. (Siwakoti) and Pokhrel, B., 2006. Ethno-medicinal plants used by Bantar of Bhaudaha, Morang, Nepal. *Our Nature, 4*, pp. 96–103.

Adams, M., Alther, W., Kessler, M., Kluge, M. and Hamburger, M., 2011. Malaria in the renaissance: Remedies from European herbals from the 16th and 17th century. *Journal of Ethnopharmacology, 133*, pp. 278–288.

Ahmad, S.S., 2007. Medicinal wild plants from Lahore-Islamabad Motorway (M-2). *Pakistan Journal of Botany*, 39(2), pp. 355–375.

Ahmad, K., Weckerle, C.S. and Nazir, A. Ethnobotanical investigation of wild vegetables used among local communities in northwest Pakistan. *Acta Societatis Botanicorum Poloniae, 88*(1), p. 3616.

Ahmed, N., Mahmood, A., Mahmood, A., Sadeghi, Z. and Farman, M., 2015. Ethnopharmacological importance of medicinal flora from the district of Vehari, Punjab province, Pakistan. *Journal of Ethnopharmacology, 168*, pp. 66–78.

Ahmed, M.J. and Murtaza, G., 2015. A study of medicinal plants used as ethnoveterinary: Harnessing potential phytotherapy in Bheri, District Muzaffarabad (Pakistan). *Journal of Ethnopharmacology, 159*, pp. 209–214.

Ainslie, W., 1813. *Materia Medica of Hindoostan, and Artisan's and Agriculturist's Nomenclature*. Madras: Government Press.

Al-Adhroey, A.H., Nor, Z.M., Al-Mekhlafi, H.M. and Mahmud, R., 2010. Ethnobotanical study on some Malaysian anti-malarial plants: A community based survey. *Journal of Ethnopharmacology, 132*, pp. 362–364.

Altundag, E. and Ozturk, M., 2011. Ethnomedicinal studies on the plant resources of east Anatolia, Turkey. *Procedia Social and Behavioral Sciences, 19*, pp. 756–777.

Amber, R., Adnan, M., Tariq, A. and Mussarat, S., A review on antiviral activity of the Himalayan medicinal plants traditionally used to treat bronchitis and related symptoms. *Journal of Pharmacy and Pharmacology*. doi: 10.1111/jphp.12669

Amini, M.H. and Hamdam, S.M., 2017. Medicinal plants used traditionally in Guldara District of Kabul, Afghanistan. *International Journal of Pharmacognosy and Chinese Medicine, 1*(3), 000118.

Amiri, M.S., Joharchi, M.R. and TaghavizadehYazdi, M.E., 2014. Ethno-medicinal plants used to cure jaundice by traditional healers of Mashhad, Iran. *Iranian Journal of Pharmaceutical Research, 13*(1), pp. 157–162.

Asiatic Society of Bengal, 1812. Asiatic researches or transactions of the Society instituted in Bengal, for inquiring into the history and antiquities, the arts, sciences, and literature, of Asia. The Bavarian State Library.

Atigur Rahman, M., Uddin, S.B. and Wilcok, C.C., 2007. Medicinal plants used by Chakma tribe in Hill tracts Districts of Bangladesh. *Indian Journal of Traditional Knowledge, 6*(3), pp. 508–517.

Ayyanar, M. and Ignacimuthu, S., 2011. Ethnobotanical survey of medicinal plants commonly used by Kani tribals in Tirunelveli hills of Western Ghats, India. *Journal of Ethnopharmacology, 134*, pp. 851–864.

Azhar, M.F., Aziz, A., Haider, M.S., Nawaz, M.F. and Zulfiqar, M.A., 2015. Exploring the ethnobotany of *Haloxylon recurvum* (KHAR) and *Haloxylon salicornicum* (LANA) in Cholistan desert, Pakistan. *Pakistan Journal of Agricultural Sciences, 52*(4), pp. 1085–1090.

Balfour, E., 1857. *The Cyclopaedia of India and of Eastern and Southern Asia*. Madras: Scottish Press.

Bentley, R., 1872. *Dr. Pereira's Elements of Materia Medica and Therapeutics: Abridged and Adapted for the Use of Medical and Pharmaceutical Practitioners and Students*. London: Longmans, Green, and Co.

Bhagya, B. and Sridhar, K.R., 2009. Ethnobiology of coastal sand dune legumes of Southwest coast of India. *Indian Journal of Traditional Knowledge, 8*(4), pp. 611–620.

Bhandary, M.J. and Chandrashekar, K.R., 2011. Herbal therapy for herpes in the ethno-medicine of Coastal Karnataka. *Indian Journal of Traditional Knowledge, 10*(3), pp. 528–532.

Bhatia, H., Sharma, Y.P., Manhas, R.K. and Kumar, K., 2014. Ethnomedicinal plants used by the villagers of district Udhampur, J&K, India. *Journal of Ethnopharmacology, 151*, pp. 1005–1018.

Bose, D., Roy, J.G., Mahapatra, S.D. (Sarkar), Datta, T., Mahapatra, S.D. and Biswas, H., 2015. Medicinal plants used by tribals in Jalpaiguri district, West Bengal, India. *Journal of Medicinal Plants Studies, 3*(3), pp. 15–21.

Bulut, G., Zeki Haznedaroğlu, M., Doğan, A., Koyu, H. and Tuzlacı, E., 2017. An ethnobotanical study of medicinal plants in Acipayam (Denizli-Turkey). *Journal of Herbal Medicine, 10*, pp. 64–81.

Chakraborty, M.K. and Bhattacharjee, A., 2006. Some common ethnomedicinal uses for various diseases in Purulia District, West Bengal. *Indian Journal of Traditional Knowledge, 5*(4), pp. 554–558.

Chang, C.-S., Kim, H. and Chang, K.S., 2014. *Provisional Checklist of Vascular Plants for the Korea Peninsula flora (KPF)*. 660 pp.

Chowdhury, M. and Mukherjee, R., 2010. Ethno-medicinal survey of Santal tribe of Malda District of West Bengal, India. *Journal of Economic and Taxonomic Botany, 34*(3).

Chowdhury, M.S.H., Koike, M., Muhammed, N., Halim, M.A., Saha, N. and Kobayashi H., 2009. Use of plants in healthcare: A traditional ethno-medicinal practice in southeastern rural areas of Bangladesh. *International Journal of Biodiversity Science and Management, United Kingdom, 5*(1), pp. 41–51.

Chowlu, K., Mahar, K.S. and Das, A.K., 2017. Ethnobotanical studies on orchids among the *Khamti* Community of Arunachal Pradesh, India. *Indian Journal of Natural Products and Resources, 8*(1), pp. 89–93.

Colonel Heber Drury, 1873. *The Useful Plants of India*. London: William H. Allen & Co.

Cox, P.A., 1993. Saving the ethnopharmacological heritage of Samoa. *Journal of Ethnopharmacology, 38*, pp. 181–188.

Das, D. and P. Ghosh., 2017. Some important medicinal plants used widely in Southwest Bengal, India. *International Journal of Engineering Science Invention, 6*(6), pp. 28–50.

Das, T., Mishra, S.B., Saha, D. and Agarwal, S., 2012. Ethnobotanical survey of medicinal plants used by ethnic and rural people in Eastern Sikkim Himalayan Region. *African Journal of Basic & Applied Sciences, 4*(1), pp. 16–20.

Das, A.K. and Tongbram, Y., 2014. Study on medicinal plants used by Meitei community of Bishnupur district, Manipur. *International Journal of Current Research, 6*(2), pp. 5211–5219.

Debala Devi, A., Ibeton Devi, O., Chand Singh, T. and Singh, E.J., 2014. A study of aromatic plant species especially in Thoubal District, Manipur, North East India. *International Journal of Scientific and Research Publications, 4*(6).

Defilipps, R.A. and Krupnick, G.A., 2018. The medicinal plants of Myanmar. *PhytoKeys, 102*, pp. 1–341.

Defilipps, R.A., Maina, S.L. and Pray, L.A., 1988. The palauan and yap medicinal plant studies of masayoshi okabe, 1941–1943. Atoll Research Bulletin. National Museum of Natural History Smithsonian Institution, Washington, D.C., October 1988.

Dey, A. and De, J.N., 2010. A survey of ethnomedicinal plants used by the tribals of Ajoydha Hill region, Purulia District, India. *American-Eurasian Journal of Sustainable Agriculture, 4*(3), pp. 280–290.

Dey, A. and De, J.N., 2012. Ethnobotanical survey of Purulia district, West Bengal, India for medicinal plants used against gastrointestinal disorders. *Journal of Ethnopharmacology, 143*, pp. 68–80.

Durmuşkahya, C. and Öztürk, M., 2013. Ethnobotanical survey of medicinal plants used for the treatment of diabetes in Manisa, Turkey. *Sains Malaysiana, 42*(10), pp. 1431–1438.

Dutta, S.K., Vanlalhmangaiha, Akoijam, R.S., Lungmuana, Boopathi, T. and Saha, S., 2018. Bioactivity and traditional uses of 26 underutilized ethno-medicinal fruit species of North-East Himalaya, India. *Journal of Food Measurement and Characterization, 12*, pp. 2503–2514.

Elliott, E., Chassagne, F., Aubouy, A., Deharo, E., Souvanasy, O., Sythamala, P., Sydara, K., Lamxay, V., Manithip, C., Torres, J.A. and Bourdy, G., 2020. Forest Fevers: Traditional treatment of malaria in the southern lowlands of Laos. *Journal of Ethnopharmacology, 247.*

Flora of China Editorial Committee, 2018. *Flora of China*. Missouri Botanical Garden and Harvard University, Herbaria, USA.

Fluchiger, F.A. and Hangury, D., 1874. *A History of the Principal Drugs of Vegetable Origin, Met with in Great Britain and British India*. London: Macmillan and Co.

Frederick Porter Smith, 1871. *Contributions towards the Materia Medica and Natural History of China*. Shanghai: American Presbyterian Mission Press.

Gagnepain, F., 1934. Iridacées, Amaryllidacées et Liliacées nouvelles d'Asie, *Bulletin de la Société Botanique de France, 81*(1), pp. 66–74. doi: 10.1080/00378941.1934.10833934

Gao, X-M., Ji, B-K., Li, Y-K., Ye, Y-Q., Jiang, Z-Y., Yang, H-Y., Du, G., Zhou, M., Pan, X-X., Liu, W-X. and Hu, Q-F., 2016. New biphenyls from *Garcinia multiflora*. *The Journal of the Brazilian Chemical Society, 27*(1), pp. 10–14.

Gautam, R., Saklani, A. and Jachak, S.M., 2007. Indian medicinal plants as a source of antimycobacterial agents. *Journal of Ethnopharmacology, 110*, pp. 200–234.

Ghosh, C., 2017. Ethnobotanical survey in the Bamangola Block of Malda District, West Bengal (India): II. Medicinal and Aromatic plants. *Pleione, 11*(2), pp. 249–267.

Giesen, W., Wulffraat, S., Zieren, M. and Scholten, L., 2007. *Mangrove Guidebook for Southeast Asia*. Dharmasarn Co., Ltd., Bangkok, Thailand.

Girach, R.D., Brahmam, M., Misra, M.K. and Ahmed, M., 1998. Indigenous phytotherapy for filariasis from Orissa. *Ancient Science of Life, 17*(3), pp. 224–227.

Goswami, D., Mukherjee, P.K. and Kar, A., 2016. Screening of ethnomedicinal plants of diverse culture for antiviral potentials. *Indian Journal of Traditional Knowledge, 15*(3), pp. 474–481.

Grosvenor, P.W., Gothard, P.K., McWilliam, N.C., Suprlono, A. and Gray, D.O., 1995. Medicinal plants from Riau Province, Sumatra, Indonesia. Part 1: Uses. *Journal of Ethnopharmacology, 45*, pp. 75–95.

Gruyal, G.A. del Roasario, R. and Palmes, N.D., 2014. Ethnomedicinal plants used by residents in Northern Surigao del Sur, Philippines. *Natural Products Chemistry & Research, 2*, p. 4.

Gunawardana, S.L.A. and Jayasuriya, W.J.A.B.N. 2019. Medicinally important herbal flowers in Sri Lanka. *Hindawi,* 18 pages. Article ID 2321961.

Hanbury, D., 1862. Notes on Chinese Materia Medica. In: John E. Taylor (ed.), *The Pharmaceutical Journal and Transactions.*

Hasan, Md. M., Hossain, Sk. A., Ali, Md. A. and Alamgir, A.N.M., 2014. Medicinal plant diversity in Chittagong, Bangladesh: A database of 100 medicinal plants. *Journal of Scientific and Innovative Research, 3*(5), pp. 500–514.

Hooper, D., 1937. *Useful Plants and Drugs of Iran and Iraq*. The Library of the University of Illinois.

Hope, J. 1970. *Lectures on the Material Medica: Containing the Natural History of Drugs, Their Virtues and Doses: Also Directions for the Study of the Materia Medica; and An Appendix on the Method of Prescribing*. London: Edward and Charles Dilly.

Ijaz, F., Iqbal, Z., Rahman, I.U., Alam, J., Khan, S.M., Shah, G.M., Khan, K. and Afzal, A., 2016. Investigation of traditional medicinal floral knowledge of Sarban Hills, Abbottabad, KP, Pakistan. *Journal of Ethnopharmacology, 179*, pp. 208–233.

Jahan, I., Rahman, M.A. and Hossain, M.A., 2019. Medicinal species of Fabaceae occurring in Bangladesh and their conservation status. *Journal of Medicinal Plants Studies, 7*(4), pp. 189–195.

Jain, S.K. and Srivastava, S., 2005. Traditional uses of some Indian plants among islanders of the Indian Ocean. *Indian Journal of Traditional Knowledge, 4*(4), pp. 345–357.

Jan, G., Khan, M.A. and Gul, F., 2008. Ethnomedicinal plants used against diarrhea and dysentery in Dir Kohistan Valley (NWFP), Pakistan. *Ethnobotanical Leaflets, 12*, pp. 620–637.

Jazani, A.M., Maleki, R.F., Kazemi, A.h., Matankolaei, L.g., Targhi, S.T., kordi, S., Rahimi-Esboei, B. and Azgomi, R.N.D., 2018. Intestinal Helminths from the viewpoint of traditional Persian medicine versus modern medicine. *African Journal of Traditional, Complementary and Alternative Medicines, 15*(2), pp. 58–67.

Ji, H., Long, C. and Pei, S., 2004. An ethnobotanical study of medicinal plants used by the Lisu people in Nujiang, Northwest Yunnan, China. *Economic Botany, 58*, pp. S253–264. doi: 10.1663/0013-000

Kadir, M.F., Sayeed, M.S.B., Setu, N.I., Mostaf, A. and Mia, M.M.K., 2014. Ethnopharmacological survey of medicinal plants used by traditional health practitioners in Thanchi, Bandarban Hill Tracts, Bangladesh. *Journal of Ethnopharmacology, 155*, pp. 495–508.

Kar, A. and Borthakur, S.K., 2008. Medicinal plants used against dysentery, diarrhoea and cholera by the tribes of erstwhile Kameng District of Arunachal Pradesh. *Natural Product Radiance, 7*(2), pp. 176–181.

Kar, T., Mandal, K.K., Reddy, C.S. and Biswal, A.K., 2013. Ethnomedicinal plants used to cure diarrhoea, dysentery and cholera by some tribes of Mayurbhanj District, Odisha, India. *Life Sciences Leaflets, 2*, pp. 18–28.

Karki, M., Hill, R., Xue, D., Alangui, W., Ichikawa, K. and Bridgewater, P., 2017. Knowing our Lands and Resources: Indigenous and Local Knowledge and Practices related to Biodiversity and Ecosystem Services in Asia. Published in 2017 by the United Nations Educational, Scientific and Cultural Organization, 7, place de Fontenoy, 75352 Paris 07 SP, France.

Kayani, S., Ahmad, M., Sultana, S., Shinwari, Z.K., Zafar, M., Yaseen, G., Hussain, M. and Bibi, T., 2015. Ethnobotany of medicinal plants among the communities of Alpineand Sub-alpine regions of Pakistan. *Journal of Ethnopharmacology, 164*, pp. 186–202.

Ketpanyapong, W. and Itharat, A., 2016. Antibacterial activity of Thai medicinal plant extracts against microorganism isolated from post-weaning diarrhea in piglets. *Journal of the Medical Association of Thailand, 99*(Suppl. 4), pp. S203–S210.

Khadka, D., Dhamala, M.K., Li, F., Aryal, P.C., Magar, P.R., Bhatta, S., Bhatta, S., Basnet, A., Shi, S. and Cui, D., 2005. The use of medicinal plant to prevent COVID-19 in Nepal. Research Square. doi:10.21203/rs.3.rs–88908/v1

Khan, S.U., Khan, R.U., Mehmood, S., Khan, A., Ullah, I. and Bokhari, T.Z. Medicinal plants used to cure diarrhea and dysentery by the local inhabitants of District Bannu, Khyber PakhtoonKhwa, Pakistan. *Advances in Pharmaceutical and Ethnomedicines, 1*(1), pp. 15–18.

Khan, S.W. and Khatoon, S., 2008. Ethnobotanical studies on some useful herbs of Haramosh and Bugrote valleys in Gilgit, Northern areas of Pakistan. *Pakistan Journal of Botany, 40*(1), pp. 43–58.

Khare, C.P. *Indian Medicinal Plants: An Illustrated Dictionary.* New York: Springer.

Khisha, T., Karim, R., Chowdhury, S.R. and Banoo, R., 2012. Ethnomedical studies of Chakma communities of Chittagong Hill tracts, Bangladesh. *Bangladesh Pharmaceutical Journal, 15*(1), pp. 59–67.

Kirtikar, K.R., Basu, B.D., An, I.C.S. *Indian Medicinal Plants.* Delhi: Jayyed Press.

Koch, M., Kehop, D.A., Kinminja, B., Sabak, M., Wavimbukie, G., Barrows, K.M., Matainaho, T.K., Barrows, L.R. and Rai, P.P., 2015. An ethnobotanical survey of medicinal plants used in the East Sepik province of Papua New Guinea. *Journal of Ethnobiology and Ethnomedicine, 11*, p. 79.

Korkmaz, M. and Karakuş, S., 2015. Traditional uses of medicinal plants of Üzümlü district, Erzincan, Turkey. *The Pakistan Journal of Botany, 47*(1), pp. 125–134.

Koteswara Rao, J., Suneetha, J., Seetharami Reddi, T.V.V. and Aniel Kumar, O., 2011. Ethnomedicine of the Gadabas, a primitive tribe of Visakhapatnam district, Andhra Pradesh. *International Multidisciplinary Research Journal, 1/2*, pp. 10–14.

Kültür, Ş., 2007. Medicinal plants used in Kırklareli Province (Turkey). *Journal of Ethnopharmacology, 111*, pp. 341–364.

Lemmens, R.H.M.J. and Bunyapraphatsara, N., 2003. *Plant Resources of South-East Asia: Medicinal and Poisonous Plants.* Leiden: Backhuys Publishers.

Lemmens, R.H.M.J., Soerianegara, I. and Wong, W.C., 1995. *Plant Resources of South-East Asia: Timber Trees: Minor Commercial Timbers.* Leiden: Backhuys Publishers.

Li, D-l. and Xing, F-w., 2016. Ethnobotanical study on medicinal plants used by local Hoklos people on Hainan Island, China. *Journal of Ethnopharmacology, 194*, pp. 358–368.

Li, D.-l., Zheng, X.-l., Duan, L., Deng, S.-w., Ye, W., Wang, A.-h. and Xinga, F.-w., 2017. Ethnobotanical survey of herbal tea plants from the traditional markets in Chaoshan, China. *Journal of Ethnopharmacology, 205*, pp. 195–206.

Liu, B., Zhang, X., Bussmaan, R.W., Hart, R.H., Li, P., Bai, Y. and Long, C., 2017. *Garcinia* in Southern China: Ethnobotany, management, and niche modeling. *Economic Botany, 70*(4), pp. 416–430.

Lokho, K. and Narasimhan, D., 2013. Ethnobotany of Mao-Naga Tribe of Manipur, India. *Pleione, 7*(2), pp. 314–324.

Lone, P.A. and Bhardwaj, A.K., 2013. Traditional herbal based disease treatment in some rural areas of Bandipora district of Jammu and Kashmir, India. *Asian Journal of Pharmaceutical and Clinical Research, 6*(4).

Long, C-l. and Li, R., 2004. Ethnobotanical studies on medicinal plants used by the Red-headed Yao People in Jinping, Yunnan Province, China. *Journal of Ethnopharmacology, 90*, pp. 389–395.

Mahbubur Rahman, A.H.M. and Akter, M., 2013. Taxonomy and medicinal uses of Euphorbiaceae (Spurge) family of Rajshahi, Bangladesh. *Research in Plant Sciences, 1*(3), pp. 74–80.

Mahmood, A., Mahmood, A., Shaheen, H., Qureshi, R.A., Sangi, Y. and Gilani, S.A., 2011. Ethno medicinal survey of plants from district Bhimber Azad Jammu and Kashmir, Pakistan. *Journal of Medicinal Plants Research, 5*(11), pp. 2348–2360.

Malla, B., 2015. Ethnobotanical study on medicinal plants in Parbat district of western Nepal. A Dissertation submitted for the Partial Fulfilment of the Requirement for the Degree of Doctor of Philosophy in Environmental Science.

Manilal, K.S. and Sabu, T., 1985. Cyclea Barbata Miers (Menispermaceae): A new record of a medicinal plant from south India. *Ancient Science of Life, IV*(4), pp. 229–231.

Merrill, E.D., 1903. *A Dictionary of the Plant Names of the Philippine Islands.* Manila: Bureau of Public Printing.

Mikaili, P., Shayegh, J. and Asghari, M.H., 2012. Review on the indigenous use and ethnopharmacology of hot and cold natures of phytomedicines in the Iranian traditional medicine. *Asian Pacific Journal of Tropical Biomedicine*, pp. S1189–S1193.

Minh, V.V., Yen, N.T.K. and Thoa, P.T.K., 2014. Medicinal plants used by the Hre community in the Ba to district of central Vietnam. *Journal of Medicinal Plants Studies, 2*(3), pp. 64–71.

Mishra, D.N., 2009. Medicinal plants for the treatment of fever (*Jvaracikitsā*) in the *Madhavacikitsā* tradition of India. *Indian Journal of Traditional Knowledge, 8*(3), pp. 352–361.

Mitra, S. and Mukherjee, S.Kr., 2010. Ethnomedicinal usages of some wild plants of North Bengal plain for gastro-intestinal problems. *Indian Journal of Traditional Knowledge, 9*(4), pp. 705–712.

Mohammad, N.S., Milow, P. and Ong, H.C., 2012. Traditional medicinal plants used by the Kensiu Tribe of Lubuk Ulu Legong, Kedah, Malaysia. *Studies on Ethno Medicine, 6*(3), pp. 149–153.

Morteza-Semnani, K., 2015. A Review on Chenopodium botrys L.: Traditional uses, chemical composition and biological activities. *Pharmaceutical and Biomedical Research, 1*(2), pp. 1–9.

Moungsrimuangdee, B., 2016. A survery of Riparian species in the Bodhivijjalaya college's forest, Srinakharinwirot University, Sa Kaeo. *Thai Journal of Forestry, 35*(3), 15–29.

Movaliya, V. and Zaveri, M., 2014. A review on the Pashanbheda plant "Aerva javanica". *International Journal of Pharmaceutical Sciences Review and Research, 25*(2), pp. 268–275. Article No. 51.

Nazan Çömlekçđoğlu and Sengül Karaman, 2008. Kahramanmaras Sehir Merkezindeki Aktar'larda Bulunan Tıbbi Bitkiler. *KSU Journal of Science and Engineering, 11*(1).

Neamsuvan, O. and Ruangrit, T., 2017. A survey of herbal weeds that are used to treat gastrointestinal disorders from southern Thailand: Krabi and Songkhla provinces. *Journal of Ethnopharmacology, 196*, pp. 84–93.

Neamsuvan, O., Sengnon, N., Seemaphrik, N., Chouychoo, M., Rungrat, R. and Bunrasri, S., 2015. A survey of medicinal plants around upper Songkhla lake, Thailand. *African Journal of Traditional, Complementary and Alternative Medicines, 12*(2), pp. 133–143.

Neamsuvan, O., Singdam, P., Yingcharoen, K. and Sengnon, N., 2012. A survey of medicinal plants in mangrove and beach forests from sating Phra Peninsula, Songkhla Province, Thailand. *Journal of Medicinal Plants Research, 6*(12), pp. 2421–2437.

Nguyen-Pouplin, J., Tran, H., Tran, H., Phan, T.A., Dolecek, C., Farrar, J., Tran, T.H., Caron, P., Bodo, B. and Grellier, P., 2007. Antimalarial and cytotoxic activities of ethnopharmacologically selected medicinal plants from South Vietnam. *Journal of Ethnopharmacology, 109*, pp. 417–427.

Noor Hassan Sajib, S. Uddin, B. and Islam, Md. M., 2014. Angiospermic plant diversity of Subarnachar Upazila in Noakhali, Bangladesh. *Journal of the Asiatic Society Bangladesh, Science, 40*(1), pp. 39–60.

O'shaughnessy, W.B., 1842. *Bengal Dispensatory*. Calcutta: W. Thacker and Co., St. Andrew's Library.

Ong, H.G., Ling, S.M., Win, T.T.M., Kang, D-H., Lee, J-H. and Kim, Y-D., 2018. Ethnomedicinal plants and traditional knowledge among three Chin indigenous groups in Natma Taung National Park (Myanmar). *Journal of Ethnopharmacology, 225*, pp. 136–158.

Panda, S.K., Das, R., Leyssen, P., Neyts, J. and Luyten, W., 2018. Assessing medicinal plants traditionally used in the Chirang Reserve Forest, Northeast India for antimicrobial activity. *Journal of Ethnopharmacology, 225*, pp. 220–233.

Pandit, P.K. Inventory of Ethno Veterinary Medicinal Plants of Jhargram Division, West Bengal, India.

Patra, J.K. and Thatoi, H.N., 2011. Metabolic diversity and bioactivity screening of mangrove plants: A review. *Acta Physiol Plant, 33*, pp. 1051–1061.

Pereira, J., 1854. *The Elements of Materia Medica and Therapeutics*. Philadelphia: Blanchard and Lea.

Pereira, J., 1855. *Materia Medica and Therapeutics*. London: Longman, Brown, Green, and Longmans.

Perry, L.M. and Metzger J., 1980. *Medicinal Plants of Asia and Southeast Asia*. MIT Press, USA.

Pomet, P., Histoire générale des drogues. A Paris : Chez Jean-Baptiste Loyson, & Augustin Pillon, sur le Pont au Change, à la Prudence. Et au Palais, chez Estienne Ducastin, dans la Gallerie des Prisonniers, au bon Pasteur 1664.

Pradhan, B.K. and Badola, H.K., 2008. Ethnomedicinal plant use by Lepcha tribe of Dzongu valley, bordering Khangchendzonga Biosphere Reserve, in North Sikkim, India. *Journal of Ethnobiology and Ethnomedicine, 4*, p. 22.

Prakasa Rao, J. and Satish, K.V., 2016. *Persicaria perfoliata* (L.) H. Gross (Polygonaceae): A species new to Eastern Ghats of India. *Tropical Plant Research, 3*(2), pp. 249–252.

Qasim, M., Gulzar, S. and Khan, M.A., 2011. Halophytes as medicinal plants. *Conference Paper*.

Qureshi, R. and Raza Bhatti, G., 2008. Ethnobotany of plants used by the Thari people of Nara Desert, Pakistan. *Fitoterapia, 79*, pp. 468–473.

Rahaman, C.H. and Karmakar, S., 2014. Ethnomedicine of Santal tribe living around Susunia hill of Bankura District, West Bengal, India: The quantitative approach. *Journal of Applied Pharmaceutical Science, 5*(02), pp. 127–136.

Rahman, M.A., 2010. Indigenous knowledge of herbal medicines in Bangladesh. 3. Treatment of skin diseases by tribal communities of the hill tracts districts. *Bangladesh Journal of Botany, 39*(2), pp. 169–177.

Rahmatullah, M., Azam, Md. N.K., Malek, I., Nasrin, D., Jamal, F., Rahman, Md. A., Khatun, Z., Jahan, S., Seraj, S. and Jahan, R., 2012. An ethnomedicinal survey among the Marakh Sect of the Garo Tribe of Mymensingh District, Bangladesh. *International Journal of PharmTech Research, 4*(1), pp. 141–149.

Rahmatullah, M., Haque, M.E., Kabir Mondol, M.R., Hasan, M., Aziz, T., Jahan, R. and Seraj, S., 2014. Medicinal formulations of the Kuch Tribe of Bangladesh. *The Journal of Alternative and Complementary Medicine, 20*(6), pp. 428–440.

Rahmatullah, M., Mollik, Md. A.H., Ahmed, Md. N., Bhuiyan, Md. Z.A., Hossain, Md. M., Azam, Md. N.K., Seraj, S., Chowdhury, M.H., Jamal, F., Ahsan, S. and Jahan, R., 2010. A survey of medicinal plants used by folk medicinal practitioners in two villages of Tangail district, Bangladesh. *American-Eurasian Journal of Sustainable Agriculture, 4*(3), pp. 357–362.

Rajendran, A., Ravikumar, K. and Henry, A.N., 2000. Plant genetic resources and knowledge of traditional medicine in Tamil Nadu. *Ancient Science of Life, XX*(1&2), pp. 25–28.

Rama Chandra Prasad, P., Sudhakar Reddy, C., Raza, S.H. and Dutt, C.B.S., 2008. Folklore medicinal plants of North Andaman Islands, India. *Fitoterapia, 79*, pp. 458–464.

Rao, P.K., Hasan, S.S., Bhellum, B.L. and Manhas, R.K., 2015. Ethnomedicinal plants of Kathua district, J&K, India. *Journal of Ethnopharmacology, 171*, pp. 12–27.

Ray, T., 2014. Customary use of mangrove tree as a folk medicine among the sundarban resource collectors. *International Journal of Research in Humanities, Arts and Literature (IMPACT: IJRHAL), 2*(4), pp. 43–48.

Reed, C.F., 1977. Economically Important Foreign Weeds. Potential Problems in the United States. United States Department of Agriculture.

Rehman, H., Begum, W., Anjum, F. and Tabasum, H., 2014. Rheum emodi (Rhubarb): A Fascinating Herb. *Journal of Pharmacognosy and Phytochemistry, 3*(2), pp. 89–94.

Romulo, A., Zuhud, E.A.M., Rondevaldova, J. and Kokoska, L., 2018. Screening of in vitro antimicrobial activity of plants used in traditional Indonesian medicine. *Pharmaceutical Biology, 56*(1), pp. 287–293.

Roy, S., 2015. An ethnobotanical study on the medicinal plants used by Rajbanshis of Coochbehar district, West Bengal, India. *Journal of Medicinal Plants Studies, 3*(5), pp. 46–49.

Safa, O., Soltanipoor, M.A., Rastegar, S., Kazemi, M., Dehkordi, K.N. and Ghannadi, A., 2012. An ethnobotanical survey on hormozgan province, Iran. *Avicenna Journal of Phytomedicine, 3*(1), Winter 2013, pp. 64–81.

Saha, M.R., De Sarker, D. and Sen, A., 2014. Ethnoveterinary practices among the tribal community of Malda district of West Bengal, India. *Indian Journal of Traditional Knowledge, 13*(2), pp. 359–367.

Saikia, A.P., Ryakala, V.K., Sharma, P., Goswami, P. and Bora, U., 2006. Ethnobotany of medicinal plants used by Assamese people for various skin ailments and cosmetics. *Journal of Ethnopharmacology, 106*, pp. 149–157.

Sandakan, S., 2005. Preferred check-list of Sabah trees.

Sankaranarayanan, S., Bama, P., Ramachandran, J., Kalaichelvan, P.T., Deccaraman, M., Vijayalakshimi, M., Dhamotharan, R., Dananjeyan, B. and Sathya Bama, S., 2010. Ethnobotanical study of medicinal plants used by traditional users in Villupuram district of Tamil Nadu, India. *Journal of Medicinal Plants Research, 4*(12), pp. 1089–1101.

Sapma, S., 2017. Indigenous plant. *ECHO Asia Conference*. October 2–6, 2017. Chiang Mai.

Savithramma, N., Yugandhar, P. and Linga Rao, M., 2014. Ethnobotanical studies on Japali Hanuman Theertham-A sacred grove of Tirumala hills, Andhra Pradesh, India. *Journal of Pharmaceutical Sciences and Research, 6*(2), pp. 83–88.

Shah, A., Ahmad, M., Marwat, S.K. and Zafar, M., 2013. Ethnobotanical study of medicinal plants of semi-tribal area of Makerwal & Gulla Khel (lying between Khyber Pakhtunkhwa and Punjab Provinces). *American Journal of Plant Sciences, 4*, pp. 98–116.

Shang, X., Tao, C., Miao, X., Wang, D., Tangmuke, Dawa, Wang, Y., Yang, Y. and Pan, H., 2012. Ethnoveterinary survey of medicinal plants in Ruoergai region, Sichuan province, China. *Journal of Ethnopharmacology, 142*, pp. 390–400.

Sharma, J., Gairola, S., Sharma, Y.P. and Gaur, R.D., 2014. Ethnomedicinal plants used to treat skin diseases by Tharu community of District Udham Singh Nagar, Uttarakhand, India. *Journal of Ethnopharmacology, 158*, pp. 140–206.

Simpson, D.A. and Inglis, C.A., 2001. Cyperaceae of economic, ethnobotanical and horticultural importance: A checklist. *Kew Bulletin, 56*(2), pp. 257–360.

Singh, J.P., Rathore, V.S. and Roy, M.M., 2015. Notes about Haloxylon salicornicum (Moq.) Bunge ex Boiss., a promising shrub for arid regions. *An International Journal, 62*, pp. 451–463.

Singh, B., Sultan, P., Hassan, Q.P., Gairola, S. and Bedi, Y.S. Ethnobotany, traditional knowledge, and diversity of wild edible plants and fungi: A case study in the Bandipora District of Kashmir Himalaya, India. *Journal of Herbs, Spices & Medicinal Plants.* DOI: 10.1080/10496475.2016.1193833

Song, M-J. and Kim, H., 2011. Ethnomedicinal application of plants in the western plain region of North Jeolla Province in Korea. *Journal of Ethnopharmacology, 137*, pp. 167–175.

Song, P., Sekhon, H.S., Lu, A., et al., 2007. M3 Muscarinic receptor antagonists inhibit small cell lung carcinoma growth and mitogen-activated protein kinase phosphorylation induced by acetylcholine secretion. *Cancer Research, 67*, pp. 3936–3944.

Srivastava, R.C., 2014. Family polygonaceae in India. *Indian Journal of Plant Sciences, 3*(2), pp. 112–150.

Sur, P.R., Sen, R., Halper, A.C. and Bandyopadhyay, S., 1987. Observation on the ethnobotany of malda-west Dinajpur Districts West Bengal-I. *Journal of Economic and Taxonomic Botany, 10*(2).

Tabata, M., Sezik, E., Yeşilada, E., et al. Traditional medicine in Turkey III. Folk medicine in East Anatolia, Van and Bitlis Provinces. *Pharmaceutical Biology.* doi: 10.3109/13880209409082966

Thapa, L.B., Dhakal, T.M., Chaudhary, R. and Thapa, H., 2013. Medicinal plants used by Raji ethnic tribe of Nepal in treatment of gastrointestinal disorders. *Our Nature, 11*(2), pp. 177–186.

Tolken, H.R. The species of Arthrocnemum and Salicornia (Chenopodiaceae) in Southern Africa. *Bothalia, 9*(2), pp. 255–307.

Tribedi, G.N., Mudgal, V. and Pal, D.C., 1993. Some less known ethnomedicinal uses of plants in sunderbans, India. *Bulletin of the Botanical Survey of India, 35*(1–4), pp. 6–10.

Tripathi, Y.C., Prabhu, V.V., Pal, R.S. and Mishra, R.N., 1996. Medicinal plants of Rajasthan in Indian system of medicine. *Ancient Science of Life, XV*, pp. 190–212.

Tumpa, S.I., Hossain, Md. I. and Ishika, T., 2014. Ethnomedicinal uses of herbs by indigenous medicine practitioners of Jhenaidah District, Bangladesh. *Journal of Pharmacognosy and Phytochemistry, 3*(2), pp. 23–33.

Uddin, S.N. and Hassan, Md. A., 2012. Angiosperm flora of Rampahar reserve forest under Rangamati district in Bangladesh. I. Liliopsida (Monocots). *Bangladesh Journal of Plant Taxonomy, 19*(1), pp. 37–44.

Uddin, M.Z., Hassan, Md. A. and Sultana, M., 2006. Ethnobotanical survey of medicinal plants in Phulbari Upazila of Dinajpur district, Bangladesh. *Bangladesh Journal of Plant Taxonomy, 13*(1), pp. 63–68.

Uddin, K., Mahbubur Rahman, A.H.M. and Rafiul Islam, A.K.M., 2014. Taxonomy and traditional medicine practices of Polygonaceae (Smartweed) Family at Rajshahi, Bangladesh. *International Journal of Advanced Research, 2*(11), pp. 459–469.

Ullah, M., Khan, M.U., Mahmood, A., Malik, R.N., Hussain, M., Wazir, S.M., Daud, M. and Shinwari, Z.K., 2013. An ethnobotanical survey of indigenous medicinal plants in Wana district south Waziristan agency, Pakistan. *Journal of Ethnopharmacology, 150*, pp. 918–924.

Ullah, S., Khan, M.R., Shah, N.A., Shah, S.A., Majid, M. and Farooq, M.A., 2010. Ethnomedicinal plant use value in the Lakki Marwat District of Pakistan. *Journal of Ethnopharmacology.*

Upadhyay, O.P., Kumar, K. and Tiwari, R.K., 2008. Ethnobotanical study of skin treatment uses of medicinal plants of Bihar. *Pharmaceutical Biology, 36*(3), pp. 167–172.

Vaidya, B.N. and Joshee, N., Report on Nepalese Orchid Species with Medicinal Properties. https://www.researchgate.net/publication/312888202.

Van Sam, H., Nanthavong, K. and Kessler, P.J.A., 2004. Trees of Laos and Vietnam: A field guide to 100 economically or ecologically important species. *Blumea, 49*, pp. 201–349.

Veldkamp, J.F. and Flipphi, R.C.H., 1987. A revision of Leptonychia (Sterculiaceae) in Southeast Asia. *BLUMEA, 32*, pp. 443–457.

Wagh, V.V. and Jain, A.K. Ethnopharmacological survey of plants used by the *Bhil* and *Bhilala* ethnic community in dermatological disorders in Western Madhya Pradesh, India. *Journal of Herbal Medicine.*

Wangchuk, P., Keller, P.A. Pyne, S.G., Taweechotipatr, M., Tonsomboon, A., Rattanajak, R. and Kamchonwongpaisan, S., 2011. Evaluation of an ethnopharmacologically selected Bhutanese medicinal plants for their major classes of phytochemicals and biological activities. *Journal of Ethnopharmacology, 137*, pp. 730–742.

Warrier, K.C.S. and Warrier, R.R., 2019. *Sacred Groves: Repositories of Medicinal Plants.* Coimbatore: Institute of Forest Genetics and Tree Breeding.

Watt, G., 1889. *A Dictionary of the Economic Products of India.* Calcutta: Government Printing.

Wazir, A.R., Shah, S.M. and Razzaq, A. Ethno botanical studies on plant resources of Razmak, North Waziristan, Pakistan.

WHO, 2009. *Medicinal Plants in Papua New Guinea*. Western Pacific Region:WHO.

Yaseen, G., Ahmad, M., Shinwari, S., Potter, D., Zafar, M., Zhang, G., Shinwari, Z.K. and Sultana, S., 2019. Medicinal plant diversity used for livelihood of Public health in deserts and arid regions of Sindh-Pakistan. *The Pakistan Journal of Botany, 51*(2), pp. 657–679.

Yazan, L.S. and Armania, N., 2014. *Dillenia* species: A review of the traditional uses, active constituents and pharmacological properties from pre-clinical studies. *Pharmaceutical Biology, 52*(7), pp. 890–897.

Zheng, X-l., Wei, J-h., Sun, W., Li, R-t., Liu, S-b. and Dai, H-f., 2013. Ethnobotanical study on medicinal plants around Limu Mountains of Hainan Island, China. *Journal of Ethnopharmacology, 148*, pp. 964–974.

Index

A

B

C

H

I

J

K

N

O

P

Q

R

S

T

U

V

W

X

Y

Z